THE EUKARYOTIC MICROBIAL CELL

Other Publications of the
Society for General Microbiology*
THE JOURNAL OF GENERAL MICROBIOLOGY
THE JOURNAL OF GENERAL VIROLOGY

SYMPOSIA

1 THE NATURE OF THE BACTERIAL SURFACE
2 THE NATURE OF VIRUS MULTIPLICATION
3 ADAPTATION IN MICRO-ORGANISMS
4 AUTOTROPHIC MICRO-ORGANISMS
5 MECHANISMS OF MICROBIAL PATHOGENICITY
6 BACTERIAL ANATOMY
7 MICROBIAL ECOLOGY
8 THE STRATEGY OF CHEMOTHERAPY
9 VIRUS GROWTH AND VARIATION
10 MICROBIAL GENETICS
11 MICROBIAL REACTION TO ENVIRONMENT
12 MICROBIAL CLASSIFICATION
13 SYMBIOTIC ASSOCIATIONS
14 MICROBIAL BEHAVIOUR, 'IN VIVO' AND 'IN VITRO'
15 FUNCTION AND STRUCTURE IN MICRO-ORGANISMS
16 BIOCHEMICAL STUDIES OF ANTIMICROBIAL DRUGS
17 AIRBORNE MICROBES
18 THE MOLECULAR BIOLOGY OF VIRUSES
19 MICROBIAL GROWTH
20 ORGANIZATION AND CONTROL IN PROKARYOTIC AND EUKARYOTIC CELLS
21 MICROBES AND BIOLOGICAL PRODUCTIVITY
22 MICROBIAL PATHOGENICITY IN MAN AND ANIMALS
23 MICROBIAL DIFFERENTIATION
24 EVOLUTION IN THE MICROBIAL WORLD
25 CONTROL PROCESSES IN VIRUS MULTIPLICATION
26 THE SURVIVAL OF VEGETATIVE MICROBES
27 MICROBIAL ENERGETICS
28 RELATIONS BETWEEN STRUCTURE AND FUNCTION IN THE PROKARYOTIC CELL
29 MICROBIAL TECHNOLOGY: CURRENT STATE, FUTURE PROSPECTS

* Published by the Cambridge University Press, except for the first Symposium, which was published by Blackwell's Scientific Publications Limited.

THE
EUKARYOTIC MICROBIAL
CELL

EDITED BY

G. W. GOODAY, D. LLOYD AND
A. P. J. TRINCI

THIRTIETH SYMPOSIUM OF THE
SOCIETY FOR GENERAL MICROBIOLOGY
HELD AT
THE UNIVERSITY OF CAMBRIDGE
MARCH 1980

Published for the Society for General Microbiology

CAMBRIDGE UNIVERSITY PRESS

CAMBRIDGE

LONDON   NEW YORK   NEW ROCHELLE

MELBOURNE   SYDNEY

Published by the Press Syndicate of the University of Cambridge
The Pitt Building, Trumpington Street, Cambridge CB2 1RP
32 East 57th Street, New York, NY 10022, USA
296 Beaconsfield Parade, Middle Park, Melbourne 3206, Australia

First published 1980

Printed in Great Britain by
Western Printing Services Ltd, Bristol

Library of Congress Cataloging in Publication Data

Society for General Microbiology.
The eukaryotic microbial cell.

(Society for General Microbiology symposium; 30)
Includes index.
1. Micro-organisms – Congresses. 2. Eukaryotic
cells – Congresses. I. Gooday, G. W., 1942–
II. Lloyd, David, 1940–   III. Trinci, A. P. J.
IV. Title. V. Series: Society for General
Microbiology. Symposium; 30.
QR1.S6233   no. 30 576'.08s [576]   79-20741
ISBN 0–521–22974–X

CONTRIBUTORS

BUCK, K. W., Department of Biochemistry, Imperial College of Science and Technology, London SW7 2AZ, UK

CARLILE, M. J., Department of Biochemistry, Imperial College of Science and Technology, London SW7 2AZ, UK

DODGE, J. C., Department of Botany, Royal Holloway College, Egham, Surrey, UK

EITLE, E., Biozentrum der Universität Basel, Klingelbergstrasse 70, 4056 Basel, Switzerland

GERISCH, G., Max-Planck-Institut für Biochemie, 8033 Martinsried bei München, West Germany

GOODAY, G. W., Department of Microbiology, Marischal College, University of Aberdeen, Aberdeen AB9 1AS, UK

GOODENOUGH, U. W., Department of Biology, Washington University, St Louis, Missouri 63130, USA

GOODWIN, B. C., School of Biological Sciences, University of Sussex, Falmer, Brighton BN1 9QG, UK

HOLWILL, M. E. J., Department of Physics, Queen Elizabeth College, Campden Hill Road, London W8 7AH, UK

KERRIDGE, D., Department of Biochemistry, University of Cambridge, Tennis Court Road, Cambridge CB2 1QW, UK

KORN, E. D., Laboratory of Cell Biology, National Heart, Lung and Blood Institute, National Institutes of Health, Bethesda, Maryland 20014, USA

LLOYD, D., Department of Microbiology, University College, Newport Road, Cardiff CF2 1TA, UK

MORRIS, N. R., Department of Pharmacology, College of Medicine and Dentistry of New Jersey, Rutgers Medical School, Piscataway, New Jersey 08854, USA

MÜLLER, M., The Rockefeller University, 1230 York Avenue, New York, NY 10021, USA

RAVEN, J. A., Department of Biological Sciences, The University, Dundee DD1 4HN, UK

ROSSIER, C., Friedrich Miescher Institut, PO Box 373, 4002 Basel, Switzerland

TRINCI, A. P. J., Department of Microbiology, Queen Elizabeth College, Campden Hill Road, London W1 7AH, UK

TURNER, G., Department of Bacteriology, University of Bristol, Bristol BS8 1TH, UK

VAN DRIEL, R., Max-Planck-Institut für Biochemie, 8033 Martinsreid bei München, West Germany

VICKERMAN, K., Department of Zoology, The University, Glasgow G12 8QQ, UK

CONTENTS

EDITORS' PREFACE

This Symposium aims at reflecting the excitement of current advances in the cell biology of the eukaryotic microbes – the algae, protozoa, filamentous fungi and yeasts. As shown by previous Symposia in this series, these organisms have always been of great interest to microbiologists, as they play central roles in many important activities, such as primary productivity, pathogenesis, and nutrient recycling.

In recent years, however, the tremendous range of structure and function shown within this group of microbes has attracted the attention of a wider audience of biologists, biochemists and biophysicists, who have enthusiastically employed these organisms as models for similar or identical activities occurring in 'higher organisms'. The validity and success of their roles as models do not concern us here, but as microbiologists we can welcome the extra attention that these eukaryotic protists have consequently received, and we can especially welcome those 'higher biologists' who clearly have become captivated by the lives of their microbial models.

This Symposium forms a natural partner of that for 1978, 'Relations between Structure and Function in the Prokaryotic Cell'. Taken together, these two symposia show us the great wealth of knowledge provided by the microbes.

We hope that this volume will point to the future, showing the reader systems which promise rich dividends. We are aware, however, of the many other organisms that have yet to be investigated, and we encourage any uncommitted young microbiologists, or scientists looking for new horizons, to observe the algae, fungi and protozoa; they will very soon become captivated and find fruitful new topics on which to work.

Finally, we thank the authors for their timely and valuable contributions, and the many people in the Society for General Microbiology and Cambridge University Press who have made this Symposium possible.

G. W. Gooday

D. Lloyd

A. P. J. Trinci

FROM PROKARYOTE TO EUKARYOTE: GAINS AND LOSSES

MICHAEL J. CARLILE

Department of Biochemistry, Imperial College of Science and Technology, London SW7 2AZ, UK

INTRODUCTION

Ten years ago, the Society held a symposium on 'Organisation and Control in Prokaryotic and Eukaryotic Cells'. In the opening contribution Stanier (1970) stressed the profound differences between the two cell types and emphasised those structural features lacking in prokaryotes but common to all eukaryotes. Subsequent research has confirmed the validity of Stanier's views. It is unnecessary here to provide a detailed description of such features; instead I shall consider how they have opened evolutionary possibilities, improved performance and led to eukaryotic dominance in some activities and habitats. However, a major reorganisation, such as occurred in the origin of the eukaryotic cell must involve losses as well as gains, and the ways in which eukaryotes are less effective than prokaryotes will also be considered.

PROKARYOTE CAPABILITIES AND LIMITATIONS

Before considering the capabilities and limitations of eukaryotes it seems worthwhile to examine briefly those of prokaryotes. Perhaps the most striking conclusion is that prokaryotes can maintain a biosphere without any assistance from eukaryotes. They can utilise solar energy, either under anaerobic conditions (purple and green photosynthetic bacteria) or by oxygenic photosynthesis (cyanobacteria = blue-green algae). Carbon dioxide can be reduced to organic compounds and a wide range of organic compounds can be oxidised to carbon dioxide. Molecular nitrogen can be fixed and can be returned again to the atmosphere (denitrification). Ammonia can be oxidised to nitrate and nitrate reduced to ammonia. Inorganic sulphur compounds can be both oxidised and reduced. It is clear that the metabolic activities of prokaryotes do not result in any essential element being converted irrevocably to a particular oxidation state – there are other prokaryotes that can carry out further reactions. The metabolic capabilities of present day prokaryotes are hence sufficient to maintain all the natural cycles of essential elements. It is certain that at one time, prior to the origin of eukaryotes, that they did so, first in an anaerobic and later in a predominantly aerobic environment. They could probably do so again, if eukaryotes were destroyed.

The metabolic capabilities of prokaryotes are very great. Energy can be obtained not only by photosynthesis but by the oxidation of hydrogen, reduced inorganic compounds and methane. Almost any naturally occurring and a great many new synthetic organic compounds can be degraded. Glucose can be metabolised by a variety of routes, and under anaerobic conditions a range of end products can result. The biosynthetic capabilities of prokaryotes are also extensive. There are few areas of bio-organic chemistry, synthetic or degradative, that have not been explored by prokaryotes.

Many prokaryotes have very high growth rates (generation times of 20 min or less), and can thus rapidly exploit newly available resources. On the other hand, bacterial endospores may persist for decades in a dormant state awaiting favourable conditions for growth. Moreover, some bacteria can adjust their metabolism from a state appropriate for 'famine' to that suitable for participation in a 'feast' with great rapidity (Koch, 1971). The metabolic versatility and genetic plasticity of bacteria permit them to respond effectively to new opportunities and hazards, such as novel substrates or antimicrobial agents.

It would seem that at the biochemical level the capabilities of prokaryotes are those of life itself; limitations appear only when morphological features – size and structural complexity – are considered. Prokaryotic cells are small compared with those of eukaryotes, volumes of one or a few cubic micrometres being typical. Those prokaryotes with larger cells are often very difficult to cultivate, through lacking adaptive flexibility and ability to accommodate to minor environmental changes (Stanier, Adelberg & Ingraham, 1977) suggesting that an upper size limit is being approached. A limited capacity for cellular interaction would appear to restrict the capacity for constructing true colonies with functional differentiation or multicellular organisms with co-ordinated cellular activities. The limited achievements of existing prokaryotes (especially actinomycetes, cyanobacteria and myxobacteria) in this direction have been discussed elsewhere (Carlile, 1979). However, any prokaryotes which specialised in activities that eukaryotes now perform outstandingly well probably became extinct after the origin of eukaryotes. Conclusions about prokaryote limitations are hence necessarily tentative.

THE ORIGIN OF EUKARYOTES

Speculation on the origin of eukaryotes is now a flourishing activity – see, for example, Margulis (1970) and Cavalier-Smith (1975) for the lucid presentation of differing viewpoints. Curiosity about origins in general and family trees in particular is an irrepressible human feature, but is it scientific and is it useful? Sneath (1974) contemplates the 'wrecks of broken theories' but concludes that by accumulating molecular detail and applying methods developed from numerical taxonomy, gradual progress towards a scientific phylogeny is possible. Stanier (1970), on the other hand, regards evolutionary speculation as an activity peripheral to science which is harmless provided that the microbiologist does not become an addict.

Phylogeny can be fully scientific in a group of organisms with a good fossil record. Under these circumstances an hypothesis can be demolished by a new discovery; philosophers stress that for a viewpoint to be scientific it must be capable of disproof. But the prospect of a scientific phylogeny of micro-organisms based on the fossil record is remote. The most hopeful alternative approach would seem to be the numerical treatment of the molecular features of existing forms as discussed by Sneath (1974); such a treatment yields an objective assessment of similarity together with a conclusion as to the most probable route by which such similarity arose. Such data have been assembled for the evolution of a few widespread molecules such as cytochrome. It remains uncertain, however, as to the extent to which such painstaking studies will illuminate major evolutionary steps. A new development, perhaps in a hitherto obscure group of organisms, might lead to a succession of rapid changes, not fully conforming to the usual tempo of molecular evolution. Meanwhile biologists will continue to speculate on the basis of non-quantitative data.

Such speculation may well be of value: evolution has undoubtedly occurred, natural selection takes place, and on these premises it is reasonable for a biologist to speculate on what might have happened. A fascinating phylogenetic theory, such as that of the symbiotic origin of the chloroplast and the mitochondrion, stimulates worthwhile investigations and serves a mnemonic role, making memorable a range of hitherto unconnected facts. However, evolutionary speculation, as Stanier (1970) concludes, is perhaps best regarded as a metascience, not itself strictly scientific but deeply influencing the activities of scientists.

I will not attempt here to give a full account of how the various eukaryote organelles may have evolved. Instead some of the steps which could have led to the origin of eukaryotes will be discussed, with an emphasis on how some developments could open the way for others.

The ancestor of the eukaryotes

Eukaryotes are usually regarded as having arisen from prokaryote ancestors c. 1–2 × 10⁹ years ago. Some authors (e.g. Darnell, 1978), noting certain biochemical features of the present day prokaryotes and eukaryotes, have seen major difficulties in deriving the latter from the former and propose instead that both are derived from some earlier form of life. While it is reasonable to accept that the ancestor of the eukaryotes may have differed greatly from any present day prokaryote, nevertheless it would have resembled the prokaryotes in lacking the elaborate nuclear and cytoplasmic organisation of eukaryotes. It hence seems better to accept the conventional view that eukaryotes arose from prokaryotes rather than to designate as a new major group organisms whose features must remain hypothetical.

The prokaryotic occurrence of essential eukaryotic constituents

There are some lipids and proteins which are of striking importance in eukaryotic cellular organisation but are often regarded as absent from prokaryotes. They, or similar substances, do however occur in at least some present day prokaryotes. They may hence have been present in the ancestor of the eukaryotes and available to assume new roles as evolution occurred.

Sterols and polyunsaturated fatty acids are normal components of most eukaryotes, being an essential part of the flexible and fusible eukaryotic cell membranes. They are often regarded as being absent or at very low concentrations in prokaryotes but there are exceptions, especially in organisms with extensive membrane invaginations. *Methylococcus capsulatus*, for example, synthesises sterols in amounts similar to those present in eukaryote cells (Bird *et al.*, 1971) and cyano-bacteria contain both sterols and polyunsaturated fatty acids (Ragan & Chapman, 1978). Mycoplasmas have large amounts of sterols in their membranes, but obtain these from their hosts. There is at least one group of eukaryotes which neither synthesises nor requires sterols for vegetative development. The Pythiaceae (Pythium and Phytophthora) cannot synthesise sterols and their vegetative hyphae can grow indefinitely in the absence of sterols (Elliot, 1977). Sterols are however needed in this group for asexual and sexual sporulation.

Microfilaments composed of actin and microtubules composed of tubulin have a vital role in cell structure (the cytoskeleton), motility and nuclear division in eukaryotes. Cavalier-Smith (1978a) observed that a filament is optimal for resisting stretch, and a tube for resisting bending or compression. He proposed that these structures originally had complementary cytoskeletal functions, and that a role in motility in association with myosin (microfilaments) and dynein (microtubules) came later. Various types of fibrils have been observed in the proto-plasmic cylinder of spirochaetes (Holt, 1978) and proteins with some actin-like features (see however Rodwell, Rodwell & Archer, 1979) have been demonstrated in mycoplasmas (Niemark, 1977; Searcy, Stein & Green, 1978). In *Escherichia coli*, Minkoff & Damadian (1976) found an actin-like protein which they suggest is able, through contraction, to regulate cell volume, water content and ion uptake. Both spirochaetes (Holt, 1978) and mycoplasmas (Razin, 1978) are able to undergo flexing movements. Microtubules have been observed in some large spiro-chaetes and a gliding bacterium (Margulis, To & Chase, 1978); there is some evidence that these are composed of a tubulin-like protein.

Histones, an essential component of the nucleosome, have been demonstrated in many eukaryotic micro-organisms (Horgen & Silver, 1978; Morris; this volume). Histone-like proteins have recently been demonstrated in several prokaryotes (see Seavey et al., 1978 and refer-ences in Horgen & Silver, 1978).

The evolution of cell envelope and cytoplasmic organisation

Peptidoglycans do not occur in the walls of eukaryotic cells. Hence it is reasonable to postulate as ancestral to the eukaryotes a prokaryote-like organism that lacked such cell walls. Such organisms may once have been common. Cell walls might well be of little value in large volumes of water (in which desiccation or exposure to osmotic fluctuations is unlikely) in a world in which predators did not yet exist. Further, since cell wall synthesis is expensive in energy, carbon and nitrogen, it could well be advantageous not to possess them. Organisms lacking cell walls may have arisen from bacteria with walls; mutants lacking peptidogly-cans (L-forms) are widespread in bacteria. On the other hand it is possible that both eukaryotes and bacteria with walls arose from wall-less organisms, and it has been suggested that mycoplasmas are the modern descendants of such organisms (Morowitz & Wallace, 1973). Whether or not this is true, the increasing volume of information available about mycoplasmas (Razin, 1978) is of interest in relation to the problems faced by wall-less prokaryotes.

A wall-less prokaryote will be vulnerable to rapid changes in environ-mental osmotic pressure, being liable, as are bacterial protoplasts, to shrinkage if the medium becomes hypertonic and bursting through sudden influx of water if it becomes hypotonic. Most modern myco-plasmas escape these problems by living in a uniform environment inside plants and animals, an option not open to the ancestors of eukaryotes. Large bodies of water may remain osmotically uniform for long periods, but smaller volumes may fluctuate rapidly in osmotic pressure. Gradual adaptation to changed ambient osmotic pressure could be achieved by taking up or excreting salts – osmotic pressure adjustment by excretion of Cl^+ and K^+ has been demonstrated in an alga (Nuccitelli & Jaffe, 1976) – or by synthesising or breaking down soluble organic compounds. Some volume change whilst adaptation is occurring could be tolerated provided that the plasma membrane is flexible. Large cells are likely to be more tolerant to an osmotic change since they will have a smaller surface to volume ratio than small cells and, given the same membrane permeability to water, a lower rate of influx per unit volume. If cells, in adapting to changed osmotic pressure, excreted salts into membrane invaginations and vesicles, then water would follow by osmosis. Given flexible and fusible membranes, a swelling water-filled vesicle on coming into contact with the cell surface might fuse with it and discharge its contents. Some such origin for water expulsion vesicles (contractile vacuoles) is a possibility. Thus the development in wall-less cells of a capability for coping with environmental osmotic changes may have been a factor in the evolution of a tough, flexible and fusible plasma membrane, exocytosis, a more dilute and hence less viscous cytoplasm, and larger size.

The origin of the eukaryote nucleus

The key event in the origin of eukaryotes is, by definition, the origin of the eukaryon, the true nucleus. Copious information is now appearing on both the molecular biology (Horgen & Silver, 1978; Morris, this volume) and ultrastructure (Kubai, 1978; Dodge & Vickerman, this volume) of the nuclei of eukaryotic micro-organisms, and this will not be repeated here. It is however, worth indicating how the origin of the eukaryote nucleus could have been yet another fruitful response to the problems of a wall-less prokaryote.

Nucleoid segregation following chromosome replication in prokary-otes involves the separation of chromosome attachment sites on the plasma membrane. Cavalier-Smith (1975) suggests that with the evolu-tion of a more fluid plasma membrane the process would become inefficient and envisages the development of a microtubule mechanism for pushing the two chromosome attachment sites apart. Once such a system was established, endocytosis of the attachment sites could occur with the invaginated membrane becoming the nuclear envelope. Cell fusion, feasible with the plastic cell envelope of an amoeboid organism, could lead to a fully diploid condition, and nuclear division in the absence of chromosome replication a return to the haploid condition. Cavalier-Smith (1974) has also discussed how endonuclease action could convert the circular prokaryotic chromosome into several linear chromosomes and how other features of eukaryotic nuclear organisation could have originated (Cavalier-Smith, 1975, 1978a).

The origin of phagocytosis, predation and endosymbiosis

Discussion has so far centred on coping with the disadvantages of a wall-less condition, rather than any advantage which might lead to a primitive eukaryote succeeding in competition with prokaryotes and leaving diverse descendants. The key advantage as appreciated by Stanier (1970), must surely have been the possibility of endocytosis and its exploitation in phagocytosis, predation and the origin of endosymbiosis.

Many bacteria produce extracellular enzymes which enable macromolecules to be broken down into small molecules that can be assimilated. Attack on much larger objects is also possible, as with the bacteriolytic and cellulolytic myxobacteria. The more intimate the contact between 'predator' and 'prey' (whether organic detritus or other bacteria) the more efficient and economical will be the utilisation of the predator's extracellular enzymes. A wall-less organism with a flexible membrane could be at an advantage, and it can be envisaged that intimate contact could evolve into partial invagination and finally into phagocytosis. The advantages of this form of nutrition would clearly be immense: the capture, in an environment often depleted of nutrients by small rapidly growing organisms, of all the required types of nutrient in a concentrated form. It is this capacity for phagocytosis that is most likely to have accounted for the initial success of eukaryotes in a prokaryotic world. Some organisms captured by a predator, however, might resist digestion and survive. These might be excreted or persist in the cell to become, if their metabolism harmed the host, the first intracellular parasites, and if it benefited the hosts, the first mutualistic endosymbionts, or perhaps each in turn. This can happen swiftly; Jeon & Jeon (1976) found that an accidental and harmful bacterial infection of a strain of Amoeba proteus had become an endosymbiont essential for the survival of the amoeba within five years – less than 1000 generations.

The Symbiotic Theory of the origin of eukaryotic organelles

Mitochondria are of almost universal occurrence in eukaryotes, and show a remarkable uniformity in the components of their electron-transfer paths, in contrast to the diversity shown in aerobic prokaryotes. This suggests that they were acquired very early in the evolution of eukaryotes, and being already highly effective, subsequently were little modified. Whatley, John & Whatley (1979) point out that the mito-chondrial electron transfer path closely resembles that of *Paracoccus denitrificans*. They also draw attention to the amoeba *Pelomyxa palustris* which has many primitive features and which lacks mitochondria: perhaps they were never acquired or possibly they have been lost. *P. palustris* does, however, contain two types of endosymbiotic bacteria, one of which in its physical features and relationship with the membrane systems of its host suggests how mitochondria could have evolved from endosymbiotic bacteria. The view that mitochondria evolved from endosymbiotic aerobic bacteria (Margulis, 1970; Whatley *et al.*, 1979) very early in eukaryote evolution is now a popular one, although sceptics remain (Raff & Mahler, 1975; Cavalier-Smith, 1975). Such a primitive eukaryote would probably already have been able to tolerate oxygen, and perhaps to obtain energy by oxidation, through the possession of oxidative enzymes such as occur in peroxisomes and other eukaryote microbodies (Müller, 1975) and in some mycoplasmas (Searcy *et al.*, 1978).

A good case can be made for the origin of the chloroplasts of eukary-otes from endosymbionts having oxygenic photosynthesis. The various major groups of photosynthetic eukaryotes, however, differ from each other in the structure and pigmentation of their chloroplasts, hence it seems likely that each group independently acquired chloroplasts by the capture of a different type of photosynthetic organism. Stanier (1974) has considered in detail the resemblance between cyanobacteria and the chloroplasts of red algae (Rhodophyta); the origin of the chloro-plasts of other groups has been discussed by Margulis (1970) and Whatley *et al.* (1979).

A symbiotic origin has also been postulated for other eukaryotic organelles including the hydrogen-evolving organelle (hydrogenosome) (Müller, this volume) of anaerobic ciliates (Whatley *et al.*, 1979) and eukaryotic flagella (Margulis, Chase & To, 1979) – but see Cavalier-Smith (1978*b*) for criticism of this suggestion. Whatever the validity of each specific proposal, the eukaryotic cell has undoubtedly exercised frequently the capacity to acquire, benefit from and become wholly dependent upon endosymbionts (e.g. Buchner, 1965).

SIZE, GROWTH RATES AND NATURAL SELECTION

Size, protoplasmic streaming and growth rate

The size of a spherical prokaryote is limited by problems of uptake, diffusion and excretion. Diffusion, efficient for a very small cell, becomes less so with increasing linear dimensions, and as size increases, surface to volume ratio decreases, bringing about problems of transport across the cell membrane (Koch, 1971). The surface to volume ratio can be increased by departing from the spherical form, but this will increase the maximum internal distance introducing problems of metabolic coordination.

Eukaryotic cells have a less viscous cytoplasm than prokaryotes and protoplasmic streaming is widespread (e.g. Allen, R. D. & Allen, N. S., 1978; Allen, N. S. & Allen, R. D., 1978). One of the factors determining streaming velocity will be the size of the channel in which it occurs; other factors being equal, the rate of flow in a cylindrical tube varies as the square of the radius of the tube (Hagen–Poiseuille Law). Hence flows in small cells may well be imperceptible but in large cells, where resistance to flow is less, it can be spectacular, reaching 1 mm s^{-1} in myxomycete plasmodia. It is of course in large cells that rapid protoplasmic streaming is needed to supplement diffusion, and the large sizes attainable by eukaryotic cells as compared with prokaryotes may be attributable to protoplasmic streaming. Nevertheless, protoplasmic streaming cannot wholly overcome the consequences of greater linear distances and lower surface to volume ratios in large cells, and hence eukaryotes tend to have lower metabolic rates per unit mass than do prokaryotes. Schmidt-Nielsen plotted the metabolic rate and body mass of a wide range of unicellular and other organisms and found in a double logarithmic plot a slope of 0.75 showing that metabolic rates do not increase proportionately to size (see Wilkie, 1977). The decrease in metabolic rate per unit mass in large organisms must set an upper limit on their growth rates.

The way in which size affects reproductive rates was illustrated for a range of micro-organisms, animals and plants by Bonner (1974) who showed in a double logarithmic plot an approximately linear relationship between body length and generation time. Under ideal conditions many prokaryotes can double their population several times per hour, and generation times of under ten minutes occur, e.g. for *Benekea natriegens* (Eagon 1962). The shortest generation times reported for eukaryotes are about one hour: Griffin, Timberlake & Cheney (1974) found 57 minutes for the water mould *Achlya bisexualis*, Trinci (1972)

66 minutes for the yeast-like fungus *Geotrichum candidum* and Fulton (1977) 1.7 hours for the amoebo-flagellate *Naegleria gruberi*. The most rapidly growing phototrophs have rather longer generation times (see van Baalen, 1974). A reasonable summary would be that under ideal conditions fast-growing prokaryotes can divide several times per hour, fast-growing eukaryotes several times per day.

Eukaryotes can, however, achieve far higher localised growth rates than prokaryotes. A hypha, for example, can increase in length far more rapidly than can a bacterial cell; the hyphae of *Neurospora crassa* can grow at 100 μm min^{-1} and the sporangiophores of *Phycomyces blakesleeanus* at 60 μm min^{-1}. This is because materials are transported by protoplasmic streaming to the hyphal apex where extension occurs from a growth zone which may extend for several millimetres behind the apex (Trinci, 1978*a*). An extreme form of such polarised growth can be seen in fungus colonies under conditions of very low nutrient concentration: the colony margin continues to advance but hyphae at the centre of the colony are emptied of protoplasm. The polarised growth of eukaryotic micro-organisms has some of the expected attributes of growth (biosynthesis, especially of new wall material) but also some of the attributes of motility, i.e. some protoplasm, and in extreme cases, all viable material, is moved from its site of synthesis. The effect has something in common with the movement of an amoeba or a myxomycete plasmodium: in one instance empty hyphal walls are left behind, in the other slime from the glycocalyx, and in both protoplasmic streaming is involved.

r-selection and K-selection

The concepts of *r*-selection and *K*-selection were introduced by Mac Arthur & Wilson (1967) in relation to island biogeography and have since been found fruitful in other areas of animal ecology (e.g. Pianka, 1970; Southwood, 1977) but with a few exceptions (e.g. Cavalier-Smith, 1978*b*) have received little attention from microbiologists. *r*-selection (*r* represents the intrinsic rate of increase) will be experienced in its most intense form by new arrivals in an uncolonised environment; the most successful will be those that multiply fastest. *K*-selection (*K* represents the carrying capacity of the environment) operates in crowded conditions where resources are scarce, and attributes other than a high growth rate are important; for example, the efficient use of resources, an ability to use resources unavailable to other organisms and effectiveness in attack and defense with respect to competitors. Any species will be exposed to both *r*- and *K*-selection, and depending on the intensity of each, will be located at some point on an r–K-selection spectrum.

The supreme r-strategists will clearly be found among the prokaryotes with their capacity for high growth rates. Structural considerations, whether the organs required in a mammal or the organelles needed in a yeast cell, will limit the miniaturisation and hence the growth rate of eukaryotes in response to r-selection. Eukaryotes, on the other hand, can respond to K-selection by taking advantage of all the opportunities, direct and indirect, offered by large size. A discussion of the exploitation of these opportunities occupies much of the remainder of this chapter.

MOTILITY AND BEHAVIOUR

Swimming

Many bacteria are motile by means of one or more flagella, and can reach speeds of about 50 μm s^{-1} (Smith, 1978). Typically, bacteria swim in an approximately straight line (a 'run') for a few seconds, change direction abruptly (a 'tumble') and then resume smooth swimming in another direction, i.e. they perform a random walk. Some are capable of reversing their swimming direction. Most prokaryotes are small enough to show Brownian movement and hence precise control of swimming direction is not possible. Eukaryotes being larger are little affected by thermal agitation, hence many are able to swim smoothly or to change direction by a required amount with respect to their former path.

Swimming in small eukaryotic micro-organisms is brought about by the action of one or a few flagella and in larger ones by the co-ordinated action of numerous cilia (see Holwill, this volume). Eukaryotic flagella are very different from bacterial flagella in their structure and mode of action; their possible evolutionary origin has been discussed by Cavalier-Smith (1978b). Ciliated micro-organisms often reach speeds of 1 mm s^{-1} (Sleigh & Blake, 1977) whilst flagellate micro-organisms attain speeds that are rather lower but considerably greater than those of bacteria. Ciliates can determine their direction of movement through reversal of ciliary beating or through differential activity of cilia on opposite sides of the body. Most flagellates rotate about their long axis as they move, and asymmetric action by a flagellum at an appropriate point in the roll, repeated as often as necessary, can bring about a turn of the required magnitude; such a mechanism occurs in *Euglena gracilis* (Diehn, 1973).

Movements on surfaces

The myxobacteria and many cyanobacteria are capable of gliding at speeds of about 3 μm s^{-1} if in contact with a surface (Smith, 1978). The mechanism of gliding and other wriggling, flexing and twitching movements observed in various bacteria is unknown.

Gliding movements occur in some eukaryotic micro-organisms, such as the diatoms. The widespread and characteristic mode of eukaryotic motility on surfaces, however, is amoeboid movement. Some students of amoeboid movement have centred their interest on the movement of protoplasm within the cell (Allen, R. D. & Allen, N. S., 1978), and some on the way in which the cell makes and breaks contact with the substratum (Rees, Lloyd & Thom, 1977), but by relating the two approaches a convincing picture of the mechanism of amoeboid movement is emerging from fibroblasts (e.g. Abercrombie, Dunn & Heath, 1977). The manifestations of amoeboid movement among animal cells and micro-organisms are, however, very varied and differences in detail, along with fundamental similarities, are to be expected. The striking feature about amoeboid movement as compared with the other forms of locomotion discussed is the precision with which directional movement may be controlled by advancing or withdrawing pseudopodia at appropriate points on the body surface. One successful life-style is that of the amoebo-flagellate; such organisms (e.g. *Naegleria gruberi*) can crawl as amoebae on a damp surface and acquire flagella and swim if flooded (Fulton, 1977).

Taxes and tropisms

Random motility is of some value to an organism; it disperses a dense population more rapidly than will Brownian movement, following which some individuals may by chance reach a favourable environment. However, if an organism possesses a sensory system, motility or growth may be guided so that it reaches favourable sites and avoids harmful conditions. The advantages of such an arrangement are clearly great, and a wide range of micro-organisms have been shown to respond to a variety of environmental clues by appropriate movement (taxis) or oriented growth (tropism). For a broad survey of such responses and their significance see Carlile (1975).

Eukaryotic micro-organisms respond by taxis or tropism to a wide range of physical stimuli, including light, gravity, contact or pressure, heat and weak electric currents. Oriented responses by prokaryotes to such stimuli are usually less obvious and have received relatively little attention. Prokaryotes with gas vacuoles are able to adjust their position by a buoyancy mechanism (Walsby, 1978) – this cannot, however, be considered a sensory response. The only unquestionable and well-studied taxis to a physical agent in prokaryotes is phototaxis and this has been most intensively studied in flagellate photosynthetic bacteria (Hildebrand, 1978). In some species a reversal of swimming direction occurs on encountering a lower light intensity, in others an increased rate of turning, both of which tend to return the organism to higher intensities. Phototaxis also occurs in gliding prokaryotes, both cyanobacteria (e.g. Waterbury & Stanier, 1978) and myxobacteria (Dworkin, 1973). These deserve further study, since precise orientation and movement towards the light source can occur. It is likely that the slow movement and large size of individuals permit simultaneous comparisons of light intensity at different points on the cell. In eukaryotes both phototaxis (Lenci & Colombetti, 1978) and phototropism (Cerdá-Olmedo, 1977; Foster, 1977) are widespread and cells are commonly large enough for light absorption across the cell to be considerable, enabling the direction of the stimulus to be determined, and precisely oriented growth or movement to occur.

The chemotaxis of a few species of flagellate bacteria, especially *Escherichia coli* and *Salmonella typhimurium*, has been intensively studied (e.g. Hazelbauer & Parkinson, 1977) but work on chemotaxis in gliding bacteria remains sparse. Chemotaxis has been demonstrated in a very wide range of swimming and crawling eukaryotic micro-organisms (Bonner, 1977; Levandowsky & Hauser, 1978). A chemotactic response requires that concentrations of the attractant or repellent be measured through interaction with chemoreceptors, and comparisons made either between different points on the surface of the organism (spatial sensing) or at different times (temporal sensing); the latter process yields information on the spatial distribution of a stimulant if the organism is moving in a uniform direction. Flagellate bacteria utilise temporal sensing in chemotaxis, and a 'run' is longer if it is in the 'right' direction, i.e. if an increasing concentration of an attractant or a diminished concentration of a repellent is experienced. Thus they will approach the source of an attractant, or escape from a repellent by means of a biassed random walk. It is clear that fast-moving bacteria could not use a spatial sensing mechanism. Berg & Purcell (1977) show that although it would be just possible for a cell the size of *E. coli* to discriminate between concentrations at the front and rear of the cell, sensing takes a finite time, of the order of a second. During this time an *E. coli* cell (length *c.* 2 μm) moving at *c.* 30 μm s^{-1} will have travelled 15 body lengths which will bring about far greater changes in the concentration of a stimulant than any difference detectable between the ends of the cell. Taxis based on the comparison of the intensities of stimulation at different points on the body surface is hence unlikely to occur in swimming prokaryotes; it should, however, be sought in large slow-moving gliding bacteria.

Eukaryotes tend to swim faster than prokaryotes in terms of μm s^{-1}, but cover fewer body-lengths per second. Hence there is a greater possibility, especially in the larger species, of making front-to-rear or even side-to-side comparisons of stimulus intensity. Eukaryote swimming, unlike that of prokaryotes, is little affected by Brownian movement, so eukaryotic micro-organisms can persist in a straight path for much longer than prokaryotes, or undertake a controlled turn rather than a 'tumble'. This provides an opportunity for spatial information about a stimulus to be translated into a direct approach to the source of the stimulus. More tracking experiments on eukaryotic micro-organisms are needed, however, to assess the relative importance of biased random walks and more direct approaches in swimming eukaryotic micro-organisms.

Studies on the chemotaxis of amoeboid micro-organisms have centred on the cellular slime mould *Dictyostelium discoideum* (Bonner, 1977; Newell, 1977; Darmon & Brachet, 1978). Here cells can advance directly towards an attractant. These amoebae move slowly (0.2 μm s^{-1}) in relation to their length (*c.* 10 μm) and Berg & Purcell (1977) conclude that simultaneous spatial comparisons of stimulus intensities would be effective. They do not however exclude the possibility of temporal comparisons by crawling or advancing a pseudopodium.

The bacterial attractants that have so far been identified are common nutrients or related substances which lead the bacteria to food sources. Chemotactic or chemotropic responses to nutrients are common also in eukaryotic micro-organisms, many of which, however, also respond to highly specific substances released by other members of the same species. Such behaviour is a form of cellular interaction and is discussed below.

Cellular interactions and cell–cell recognition

When cells are in close proximity interactions must inevitably occur; one cell will influence another through reducing the concentration of nutrients in its immediate environment and by producing diffusible metabolites. In addition to such fortuitous interactions there are more meaningful interactions in which a substance produced by one cell acts as a signal to which a second cell responds – an act of cell–cell recognition (Curtis, 1978). Such substances may be present at a cell surface and only produce an effect when two cells come into contact, or they may diffuse and act on cells at a distance. Cellular interactions between cells of the same species occur in mating, in morphogenesis, and in coordinating activities in a colony or multicellular individual. They will occur between species in the establishment of a mutualistic or parasitic relationship.

The capacity of prokaryotes for specific cellular interactions seems limited. High specificity interactions involving cell surfaces and appendages are involved in mating (Achtmann & Skurray, 1977) and recipient strains of *Streptococcus faecalis* have been shown to produce a factor which causes donor strains to become adherent (Dunny, Brown & Clewell, 1978). Cellular interaction must also be involved in pattern formation in cyanobacterial filaments, as for example in the spacing of heterocysts (Willcox, Mitchison & Smith, 1975), although the metabolites involved may not necessarily be of high specificity. Probably the most sophisticated cellular interactions that occur in prokaryotes are in cell aggregation and fruiting body formation in myxobacteria, a subject that deserves more detailed study.

In eukaryotic micro-organisms, gametic approach is often brought about through chemotactic or chemotropic agents that are of high specificity and active at very low concentrations (Gooday, 1978; Kochert, 1978). For example, the sex attractant in the water mould *Allomyces*, sirenin, is produced by the female gametangia and gametes and attracts only male gametes and not other motile stages in the *Allomyces* life cycle; it is active at concentrations as low as 10^{-10}M and none of the closely related compounds that have been synthesised are active. Sexual interaction in eukaryotes can also involve the exchange of morphogenetic signals; in the water mould *Achlya bisexualis* the female produces the steroid antheridiol, which induces sexual differentiation in the male, which then produces another steroid, oogoniol, which induces differentiation in the female. Antheridiol also acts as an attractant, causing chemotropism of the male hyphae towards the female. In addition to signals that act as sex attractants or induce sexual differentiation, there are specific substances which produce agglutination and cell surface interactions between sexually complementary cells, as in yeasts (Crandell, Egel & Mackay, 1976; Manney & Meade, 1977) and in algae (Goodenough, this volume).

Sophisticated cellular interactions also occur in non-sexual morphogenesis, as in fruiting in the cellular slime moulds (Bonner, 1977;

Newell, 1977; Darmon & Brachet, 1978; Gerisch, this volume). Cellular aggregation is brought about by attractants termed acrasins; some acrasins appear to be peptides but that in the larger species of *Dictyostelium* is cyclic AMP. In *Dictyostelium discoideum* it has been shown that cyclic AMP is also involved in the co-ordination of subsequent morphogenetic events. Cell surface interactions are also of importance in differentiation, and if two *Dictyostelium* species are brought together through responding to the same attractant, cell sorting occurs (Garrod *et al.*, 1978) and separate fruiting bodies are produced. Interactions between vegetative cells also have an essential role in colonies of the higher fungi in which tropism and anastomosis produces a three-dimensional hyphal network (Carlile & Gooday, 1978). This topic is discussed further (p. 18).

Circadian Rhythms

Many eukaryotic organisms, both unicellular and multicellular, show a daily rhythmicity in their behaviour (Brady, 1979). In some instances these rhythms are exogenous, being produced by daily fluctuations in light intensity or temperature. Often, however, they are endogenous and will persist with a periodicity close to 24 h (i.e. circadian) if the organism is transferred from a daily alternation of light and darkness to constant conditions of darkness or dim light; continuous bright light, however, will suppress the rhythm. Many types of activities show rhythmicity. The dinoflagellate *Gonyaulax polyedra* has daily rhythms of photosynthetic activity (mid-day peaks), luminescent glow (peak near dawn), flashing in response to disturbance (peak early in night) and undergoes cell division at dawn. Daily rhythms of sporulation are shown by the *Oedogonium cardiacum* and of phototactic sensitivity by *Euglena gracilis*. Overt rhythms appear to be controlled by an underlying oscillator or clock which persists in its natural period of near 24 h even if the rhythm is not expressed. The clock is temperature-independent in the sense that it has a temperature coefficient of close to unity but temperature-sensitive in that it can be reset by a temperature change as well as by changes in light intensity. Progress towards elucidating the biochemical basis of the 'clock' is slow, but physiological work suggests a basis in oscillations in the permeability to ions of the plasma membrane or perhaps other membrane systems (Njus, Sulzman & Hastings, 1974; Njus *et al.*, 1976). H. G. Schweiger & M. Schweiger (1977) have suggested a coupled translation–membrane model of the clock in which crucial features are cytoplasmic protein synthesis with a temperature coefficient greater than unity, and protein incorporation into membranes with a coefficient of less than unity to account for temperature compensation.

There have been occasional claims for the occurrence of circadian rhythms in prokaryotes, including *Escherichia coli* and *Klebsiella* sp., but the effects reported are not striking and are regarded with scepticism (see Hastings & Schweiger, 1976, p. 57). Indeed, circadian rhythmicity in prokaryotes with such short generation times would seem surprising on both evolutionary and mechanistic grounds. Perhaps the prokaryotic cellular organisation is inadequate for sustaining circadian rhythms; if such rhythms do occur in prokaryotes they are most likely to be found in slow-growing or long-lived forms with either a light requirement or complex fruiting bodies, i.e. the cyanobacteria and the myxobacteria.

Circadian biological clocks are very widespread in eukaryotic organisms but appear not to be universal. They are unlikely to occur in unicellular micro-organisms with short generation times such as yeasts but can occur in colonial organisms which have doubling times of much less than a day, e.g. in *Neurospora crassa* which shows a daily zonation in its growth pattern. The evolutionary origin of circadian biological clocks has been discussed in Hastings & Schweiger (1976, pp. 52–55). It is pointed out that 'stable steady state (non-oscillating) operation is vanishingly rare in interactive systems of any complexity' and that 'cell-types incapable of a true steady state operation might be the rule rather than the exception' and it is suggested that 'the business of natural selection might be more to eliminate maladaptive oscillations, stabilize the periods (e.g. with respect to temperature) of those remaining, and couple them (e.g. by photoreceptor) to a synchronizing environment than to engineer a clock in cells that would otherwise not oscillate'. On this view one might expect the evolution of a biological clock was almost inevitable in organisms with an active life of more than a day, and may have occurred more than once, with differences in biochemical mechanisms. Equally, its loss when fast-growing unicellular organisms evolve from colonial forms (as may have happened repeatedly in the evolution of the various yeasts from filamentous fungi) is to be expected.

COLONIES, MULTICELLULAR ORGANISMS AND SYMBIOSIS

Populations and colonies

The ability of cells to interact permits the evolution of colonies and multicellular organisms. The colonial condition can be regarded as being intermediate between that of a population of unicellular

individuals and a multicellular organism (Carlile, 1979). In a population
the arrangements of cells is random, cellular interaction fortuitous, and
competition between cells total. In a colony there is a fairly well-defined
distribution of cells, significant cell–cell interaction and co-operation,
but also some competition. In a multicellular organism form is well-
defined and the suppression of cellular autonomy and competition
between cells is virtually complete. A spectrum of forms occurs between
what is indisputably a population of unicellular individuals through the
colonial condition to undoubtedly multicellular organisms.

On the criteria indicated above the colonial condition is not common
in prokaryotes – a typical bacterial 'colony' on agar media is essentially
a heap of cells. More acceptable colonies are found, however, in actinomycetes, cyanobacteria and myxobacteria. Among actinomycetes the
Streptomyces colony has a well-defined structure, with differentiation
into both hyphae spreading upon and penetrating the substratum and
aerial mycelium bearing spores. Cyanobacterial filaments may show
differentiation into several types (Rippka *et al.*, 1979). In myxobacteria
the highly co-ordinated movements of swarms and development of
fruiting bodies indicates a degree of co-ordination approaching that of a
multicellular organism and it is possible that further studies will show
that the myxobacteria possess the most sophisticated sensory systems
in prokaryotes.

Eukaryote micro-organisms, exploiting their capacity for various
forms of cell interaction, have evolved larger and more complex
colonies than have the prokaryotes, and such colonies often approach a
multicellular organism in complexity. Multinucleate hyphae occur in
some algae but are most fully exploited in the colonies of filamentous
fungi (Trinci, 1978*b*; Carlile, 1979). The hyphae branch in such a way
as to drain the substrate of nutrients most effectively, and the spacing of
hyphae in a growing colony seems to be controlled by negative autotropism. In the higher fungi (basidiomycetes, ascomycetes and related
fungi imperfect) fusion between vegetative hyphae (hyphal anastomosis), involving positive autotropism (Carlile & Gooday, 1978) is common. This will convert the branching pattern of the colony to a three-
dimensional network, which permits protoplasm and nutrients to be
brought from a wide area of mycelium to a few points for the construction of fruiting bodies. Such fruiting bodies can be of large size, as in
mushrooms and puff-balls, with a structure precise enough to qualify
them as multicellular organisms. Colonies constructed from hyphae
are an effective device; the fungi are a highly successful group as judged
by numbers of species, adaptation to diverse habitats, and biomass.

A less widespread device is the plasmodium, which reaches its most spectacular development as part of the myxomycete life cycle. It arises from the amoeboid phase of the life cycle and increases in size through growth and nuclear division unaccompanied by cell division. Fusion between genetically similar plasmodia further increases size until a plasmodium may contain millions of nuclei and cover many square centimetres. Such plasmodia can engulf and digest objects far larger than those that can be attacked by the amoeboid phase and can migrate rapidly over long distances in search of suitable sites for sporulation.

As indicated above, fusion between vegetative cells occurs as a normal morphogenetic event in both higher fungi and myxomycetes. It is, however, limited to genetically similar cells by somatic incompatibility (Carlile & Gooday, 1978), which may either prevent cell fusion or destroy one genotype soon after fusion. Complete genetic identity is not required for cell fusion, only identity at the various somatic incompatibility loci. Hence in these multi-nucleate organisms there is the possibility of heterokaryosis. Heterokaryons seem quite common among fresh isolates of fungi that can exist in nature either as parasites or saprophytes (Carlile, 1979) and perhaps a heterokaryotic colony can adapt to changed conditions by changes in the ratio between nuclear types – the concept of a balanced hererokaryon.

The cellular slime moulds have received extensive discussion elsewhere (Bonner, 1977; Darmon & Brachet, 1978; Newell, 1977; Gerisch, this volume). It is, however, worth noting that the developmental sequence of vegetative amoebae, aggregation and grex illustrates a sequence of population, colony and multicellular organism in a single life cycle, and that these organisms provide an admirable illustration of the role of intercellular signals and surface interactions in the coordination of the activities of colonies and multicellular organisms. The cells in the cellular slime moulds remain separate and do not fuse, an arrangement that clearly has the greatest evolutionary potentialities, as shown by the development of higher animals and plants. A wide range of colony types occurs also in algae and protozoa, as discussed, for example, by Curds (1979).

Mutualistic symbiosis

The term symbiosis was originally used to indicate an intimate association of two organisms, which could be beneficial to both (mutualism) or to one only (parasitism). Subsequently symbiosis came to mean solely mutualistic associations, but recently some important discussions of the topic have led to reversion to the earlier usage (e.g. Jennings &

Lee, 1975; Stanier *et al.*, 1977). This has merit, since microscopic examination can establish that two organisms are intimately associated but extensive research may be needed to determine whether one or both organisms benefit from the association. Such research has, however, established the occurrence of numerous examples of mutualistic symbiosis (Table 1), some of great ecological importance.

Table 1. *Examples of mutualistic symbiosis, with role of smaller partner indicated*

Organisms	Ectosymbiosis	Endosymbiosis
Both partners prokaryotes	'Chlorochromatium' (photosynthesis)	Not known

One partner prokaryote, one eukaryote	Spirochaetes on termite flagellates (motility)
	    Lichens with cyanobacterial components (photosynthesis, nitrogen fixation)
	    Luminous bacteria of flashlight fish (signalling)
	    Rumen bacteria of herbivores (cellulose digestion)
		Cyanelles of protozoa (photosynthesis)
		    Rhizobium in legume and Frankia in non-legume root nodules (nitrogen fixation)
		    Kappa in *Paramecium aurelia* (destruction of competing *P. aurelia*)

Both partners eukaryotes	Lichens with green alga component (photosynthesis)	Green and brown algae of protozoa, corals, flatworms, molluscs (photosynthesis)
	Ectomycorrhizal fungi of higher plants (mineral uptake)	Endomycorrhizal fungi of higher plants (mineral uptake)

Beneficial interactions between prokaryotes are common, but intimate physical associations of two species rare, or at least little studied. One example is the 'chlorochromatium' association, in which a large colourless rod-shaped bacterium is coated with regularly arranged cells of the photosynthetic green bacterium *Chlorobium*. Symbiosis between two eukaryotes or between a eukaryote and a prokaryote are, however, both common and important, whether or not one accepts that chloroplasts and mitochondria originated as prokaryotic symbionts within early eukaryotes (page 8). Frequently, one partner (the host) is much larger than the other (the symbiont); under these circumstances part at least of the host's contribution is providing a site in or on which the smaller partner can thrive. Symbionts may be either ectosymbionts (on the surface of the host, in the host's body cavities, or between host cells) or endosymbionts (within the host cells). Either the larger or smaller partner may contribute photosynthates to the association. Nitrogen fixation can only be carried out by a prokaryotic partner but an animal can provide a photosynthetic endosymbiont with excretory products that are used as nitrogen sources. The enzymic capabilities of rumen bacteria enable herbivores to harvest grass effectively. The hyphal systems of mycorrhizal fungi permit plants to extract mineral nutrients from soil efficiently. Lichens can survive in environments too arduous for either fungi or algae. The capabilities of symbiotic associations are amazingly varied (Buchner, 1965; Henry, 1966, 1967; Jennings & Lee, 1975; Cooke, 1977; Stanier *et al.*, 1977) and seem able to overcome almost any metabolic limitations of eukaryotes.

NUTRITIONAL AND METABOLIC CAPABILITIES

Much of the basic work on energy sources and metabolic pathways was worked out on a few eukaryotic systems, such as yeast, vertebrate muscle, and *Chlorella*. Major departures from the systems occur mainly in prokaryotes – the basic biochemistry of eukaryotes is far less diverse especially with respect to energy sources and metabolism in anaerobic conditions.

Energy sources

Eukaryotes are photolithotrophs (energy from oxygenic photosynthesis with water as the ultimate reductant) or chemorganotrophs (energy from saprophytic, parasitic, mutualistic or predatory activity). These activities occur in prokaryotes, but they also carry out photolithotrophy using reduced sulphur compounds instead of water as the reductant; they carry out photoorganotrophy (photosynthesis with organic hydrogen donors) and chemolithotrophy (energy obtained by the oxidation of reduced inorganic compounds, such as hydrogen sulphide and sulphur, ammonia and nitrite, hydrogen and ferrous salts, and even hydrogen itself) (Table 2). Hydrogen is oxidised by a wide range of bacteria. Methylotrophs, obtaining energy by the oxidation of methane and other C-1 compounds include both prokaryotes and eukaryotes.

The respiratory electron-transport systems of prokaryotes are diverse, with vitamin K or derivatives sometimes taking the place of ubiquinone, and cytochromes showing a variety of absorption spectra. This contrasts with the uniformity of the cytochromes and other components of the mitochondrial electron-transfer paths in eukaryotes. Many bacterial electron-transport systems can moreover use terminal hydrogen acceptors other than oxygen, for example sulphate (*Desulfovibrio*), nitrate (denitrifying bacteria) or even carbon dioxide (methanogens).

This permits oxidative metabolism under anaerobic conditions, which is absent from eukaryotes.

Table 2. *The basic types of energy metabolism*

Reductant	Energy source	
	Light (phototrophs)	Chemical (chemotrophs)
Inorganic (lithotrophs)	Photosynthetic eukaryotes and cyanobacteria (H_2O); photosynthetic sulphur bacteria (mainly H_2S)	Prokaryotes only; e.g. non-photosynthetic sulphur oxidising bacteria, nitrifying bacteria, iron bacteria and hydrogen bacteria
Organic (organotrophs)	Prokaryotes only; photosynthetic bacteria utilising reduced organic compounds (e.g. isopropanol)	Many eukaryotes and prokaryotes (saprophytes, symbionts and predators)

Data on metabolic efficiency, which can be assessed in a variety of ways, remains sparse for eukaryotes, although some values are given by Stouthamer (1977) and Harrison (1978). Growing on methanol, *Pseudomonas extorquens* yields 0.40 g cells and the yeast *Hansenula anomola* 0.54 g cells per gramme of substrate. Most data concerns growth under anaerobic conditions with glucose as a substrate, and here also the yields obtained with yeasts and bacteria do not differ greatly. The maintenance coefficients – the energy supply needed to keep eukaryotic cells alive and at constant mass – are lower for the yeast *Saccharomyces cerevisiae* than for bacteria under comparable conditions (see Stouthamer, 1977, Table 7; Harrison, 1978; Table 7.1). This may be due to the relatively large yeast cells having a lower surface-to-volume ratio than bacteria, and consequently being able to 'keep the environment at bay' with less expenditure of energy; or it may be a more fundamental consequence of eukaryote cellular organisation.

Nutrient sources and metabolic paths

Eukaryotic capabilities for attacking a wide range of carbon sources are almost as great as those of prokaryotes. Various fungi for example are able to attack such abundant but refractory carbon sources as cellulose, chitin and lignin. Eukaryotes lack, however, one important method of glucose metabolism, the Entner–Doudoroff pathway, and the fermentative dissimilation of pyruvate, which occurs in a wide variety of ways in prokaryotes, is limited to alcohol or lactate production.

Eukaryotes are also capable of using a wide range of organic and inorganic nitrogen sources, with the striking omission of nitrogen itself. Why do all eukaryotes lack the ability to fix nitrogen? Were *nif* genes never present or lost in the line from which eukaryotes evolved? Possibly, but then why have *nif* genes not been evolved or acquired in 2×10^9 years of evolution? A possible answer is that the eukaryote cell is too aerobic for nitrogenase to function. The most likely explanation, however, is that eukaryotes have evolved effective alternatives to nitrogen fixation and predators, phagotrophs and parasites can acquire all the organic nitrogen they need, and root systems, mycorrhiza and fungal hyphae are effective in scavenging combined nitrogen from a wide area. Under such circumstances the energetically expensive fixation of gaseous nitrogen would not be an advantage. Such eukaryotes as have entered environments seriously devoid of combined nitrogen have acquired ectosymbiotic or endosymbiotic prokaryotes with nitrogen-fixing ability. Endosymbiosis would seem to be a typically eukaryotic way of acquiring an organelle which contains a *nif* gene and nitrogenase in a compartment which can be protected from excessive penetration by oxygen – a swifter solution than *de novo* evolution of nitrogenase and analogous with the postulated acquisition of chloroplast and mitochondrion. Indeed if the evolution of the obligately symbiotic nitrogen-fixing actinomycete *Frankia* had proceeded a little further we might be debating the symbiotic or endogenous origin of the 'nitrogenosome' of non-leguminous root nodules.

GENETIC ORGANISATION AND THE CONTROL OF METABOLISM

The study of eukaryotic DNA and its transcription and translation is one of the most active areas of molecular biology, and is revealing many striking contrasts, summarised in Table 3, between eukaryotic and prokaryotic organisation. Much of the work has been carried out with vertebrates rather than micro-organisms. Although rapid progress is being made, the general biological significance of many of the findings is still unclear, hence the brevity of the treatment below. Two topics, however, that of the amount of DNA in eukaryotic cells, and the recent discovery of introns, deserve comment.

A eukaryotic cellular organisation can be sustained by an amount of haploid DNA no greater than that present in some prokaryotes. The nucleus of a haploid cell of *Saccharomyces cerevisiae* contains only three to five times as much DNA as does a cell of *Escherichia coli*, an amount similar to the highest values recorded in prokaryotes (Table 4).

Table 3. *Features of genetic organisation and control of metabolism in eukaryotes*

DNA organisation	More than one chromosome. Several replicons per chromosome. DNA associated with proteins as chromatin. Chromatin organised into nucleosomes. DNA in excess of that needed to code for proteins, tRNA and rRNA. No operons – genes concerned with a metabolic path not clustered. Non-coding sequences (introns) within genes. Repetitive sequences common. Numerous copies (e.g. *c.* 100) of the DNA coding for rRNA – repetitions may be tandem (the frog *Xenopus laevis*) or inverted (the myxomycete *Physarum polycephalum* and the protozoan *Tetrahymena pyriformis*)
Transcription	Separate RNA polymerases for messenger, transfer and ribosomal RNA. Primary transcript often much longer than the final functional mRNA, but codes for one polypeptide only i.e. is monocistronic
Post-transcriptional modification of primary RNA transcript	Capping of 5′ end of mRNA with methylated structure Addition of *c.* 100–200 adenine nucleotides to 3′ end of mRNA. Excision of sequences specified by introns and splicing to give tRNA, rRNA and functional mRNA
Translation	Ribosomes larger and of more complex structure than in prokaryotes. Species differ considerably in ribosome structure. Initiation of translation limited to a single site on the mRNA

Almost all the above features are discussed in a two-volume symposium on chromatin (Cold Spring Harbor Symposium, 1978), but usually with reference to vertebrate material. For micro-organisms see Horgen & Silver (1978), Fincham, Day & Radford (1979) and contributions in this volume by Morris (pp. 41–76) and by Dodge & Vickerman (pp. 77–102), and for recent accounts of ribosomes and translation see Cox (1977) and Kozak (1978).

Yet whereas prokaryotes have a single chromosome, the replication of which starts at a single point, haploid *S. cerevisiae* has 18 chromosomes, with several replicons per chromosome. *S. cerevisiae* is one of the fastest growing eukaryotes, and it is probable that to facilitate rapid growth non-coding DNA has been reduced to a minimum. *S.cerevisiae* lacks extensive repetitive sequences, as do some other fast growing eukaryotic micro-organisms, such as *Neurospora crassa* and *Aspergillus nidulans* which have haploid DNA contents (often referred to as DNA C-values) only about twice that of *S. cerevisiae*.

The DNA C-values among eukaryotes, however, vary greatly, in contrast to the relatively limited range in prokaryotes (Table 4; Cavalier-Smith, 1978*a*). The organisms with high C-values contain far more DNA than is needed to code for the estimated 4000 to 30 000 proteins of a eukaryotic cell. Moreover, the 5000-fold range of C-values

Table 4. *Haploid DNA contents (C-values) in various groups of organisms*

Organisms	C-values		
	Lowest (fg)	Highest (fg)	Ratio Highest/Lowest
Prokaryotes	1.7	17	10
Eukaryotes	5	350 000	70 000
Fungi	5	190	38
Algae	40	200 000	5 000
Protozoa	60	350 000	5 800
Angiosperms	1 000	89 000	89
Mammals	3 000	5 800	1.9

Data mainly from Cavalier-Smith (1978a), expressed in femtograms (10^{-15}g), or, more approximately, millions of base pairs. On the basis of 650 daltons per base pair, a femtogram represents about 920 000 base pairs, which could code for 1000 proteins containing on average 307 amino acid residues. *Escherichia coli* contains about 4 fg DNA. The highest prokaryote value is calculated from Herdman *et al.* (1979). The lowest fungus value is for a yeast; values cited for the intensively studied yeast *Saccharomyces cerevisiae* range from 13–22 fg, calculated from data in Fincham, Day & Radford (1979). About 15% of the yeast DNA is mitochondrial.

in unicellular algae, which do not vary greatly in developmental complexity, is not explicable in terms of requirements for differing numbers of protein-coding genes or differing complexity of transcriptional control. Cavalier-Smith (1978a) finds that small fast-growing cells have low C-values and large slow-growing cells high C-values. He suggests that the evolution of large cells necessitates concomitant evolution of large nuclei which require large amounts of DNA with an essentially skeletal function (nucleoskeletal DNA). Conversely, selection for rapid growth rates will require small cells with small nuclei and a reduction in the amount of DNA. Hence the DNA content of eukaryotic cells may be determined mainly by *K*-selection for large size or *r*-selection for high growth rates.

A recent development of great evolutionary interest has been the discovery that there are sequences within eukaryotic genes which are not expressed in the final product – protein, tRNA or rRNA. These unexpressed sequences have been termed *introns* (intragenic regions) and the expressed sequences *exons* (expressed regions) by Gilbert (1978). Introns are copied into the primary RNA transcript but are then excised and the RNA spliced to give messenger, transfer or ribosomal RNA. Introns varying in length from 10 to 10 000 bases have been found and Gilbert suggests that there may be five to ten times as much DNA in introns as in exons – an estimate that may well be correct for animal cells but not for micro-organisms with low DNA C-values. It is becoming clear that exons correspond to structural or functional domains within a protein (Blake, 1979), i.e. that introns do not occur within a sequence the integrity of which is essential for a protein function. Doolittle (1978) and Darnell (1978) point out that it is very difficult to envisage how introns could arise in an organism previously lacking them, and suggest that 'genes in pieces' was the normal situation in the common ancestor of prokaryotes and eukaryotes. Doolittle (1978) suggests that in these primitive organisms DNA replication, transcription and translation would have been of low fidelity and that redundancy, in the form of reiterated exons, along with introns and gene splicing, would have been needed to produce adequate amounts of functional protein. In the course of subsequent evolution prokaryotes would have eliminated introns and repetitive sequences, and eukaryotes would have retained and exploited them. Gilbert (1978) has discussed how splicing errors can generate new proteins assembled from the functional parts of old ones without sacrificing the original gene, and how minor mutations within an intron could 'fix' a beneficial new arrangement. We can hence envisage that in prokaryotes r-selection gradually eliminated introns, repetitive sequences and other redundant DNA, whereas in eukaryotes in which large size was crucial redundant DNA was increased and exploited rather than eliminated. Fast growing eukaryotes in which repetitive sequences are few and there is little redundant DNA are already known as indicated above. Are there prokaryotes in which redundant DNA persists? Mitochondria (slow growing prokaryotes in a non-competitive environment?) have been shown to contain introns (see Lloyd & Turner, this volume). It is possible that detailed studies on large and relatively slow growing prokaryotes may reveal that introns, repetitive sequences or other forms of redundant DNA do occur in some prokaryotes.

MECHANISMS FOR GENETIC CHANGE

The ultimate basis for genetic change is mutation. Too low a mutation rate would limit the capacity of a species to evolve. Too high a rate might handicap almost all the individuals in a species with harmful mutations. Drake (1974) has therefore suggested that organisms evolve an optimal mutation rate, which provides adequate novelty without being excessively disruptive. He found that among some well studied micro-organisms the mutation rate per genome per generation was remarkably uniform at about 0.001. Thus *Neurospora crassa*, which contains approximately ten times as much DNA as *Escherichia coli*, has a mutation rate per base pair replication of about one-tenth that of *E. coli*. This inverse relationship holds also for viral genomes with about one-hundredth the DNA content of *E. coli*. A departure from the relationship may well occur in organisms in which a large proportion of the DNA is non-coding; higher mutation rates per genome per generation would seem likely. Hence it is probable that both prokaryotes and eukaryotes with low DNA C-values have similar mutation rates per genome per generation and that any wide departure from that rate is not tolerable. Eukaryotes, however, usually have longer generation times than prokaryotes and, as a consequence of their size, smaller populations. Hence in a given time the number of mutations in a eukaryote population is likely to be less than in a prokaryote population able to occupy the same environment. Eukaryotes, especially large and slow growing ones, therefore need to conserve and utilise mutations far more effectively than do prokaryotes. The devices for doing so, heterokaryosis, sexual and parasexual recombination and diploidy, all depend on the ability of eukaryotic cells to fuse with each other. The genetic and biochemical control of both somatic and sexual cell fusion in myxo-mycetes and fungi has been discussed by Carlile & Gooday (1978); for sexual fusion in algae see Goodenough (this volume).

The sexual process dominates much of eukaryote biology. The basic feature is that two haploid cells fuse, their nuclei fuse to give a diploid nucleus, and meiosis follows (Dodge & Vickerman, this volume). During meiosis homologous chromosomes pair; crossing over with reciprocal exchange of segments of the chromosomes occurs, and two successive nuclear divisions accompanied by only one set of chromo-some divisions restores the haploid condition, yielding four daughter cells, which, if the two parental cells had substantial genetic differences, will differ genetically from each other and from the parent cells. The genetic material of the two parents is thus conserved and recombined in various ways. This contrasts with prokaryote sexuality, in which a segment of DNA contributed by a donor cell replaces the corresponding segment on the recipient's chromosome. Moreover, whereas in pro-karyotes sexual recombination is a relatively infrequent event, in many eukaryotes sexuality is an integral part of the life cycle.

The sexual process will only bring about recombination and vari-ability if the parent cells differ genetically. Hence outbreeding mechan-isms have evolved to reduce the chance that genetically similar cells will undergo sexual fusion (Esser, 1974). They are generally based on homo-genic incompatibility – fusion cannot be successfully completed unless the cells differ at certain loci. These mechanisms are often very elaborate in the fungi, especially the basidiomycetes (Casselton, 1978; Fincham, Day & Radford, 1979; Schwalb & Miles, 1978) in which a species may have hundreds of mating types. The sexual process in fungi leads to the production of spores adapted either for dormancy or dispersal, and often to elaborate structures to facilitate dispersal. Hence sexuality has here acquired a secondary function, the initiation of morphogenetic steps leading to a crucial phase in the life cycle, spore production. The accomplishment of this phase will be more certain if it is not dependent on an encounter between two mating types. This is the probable explanation for the widespread occurrence in fungi of self-fertility, which wholly defeats the genetic purpose of sexuality. Presumably in highly adapted organisms in a uniform environment variability has temporarily ceased to be required for the survival of a species, and a breakdown in outbreeding mechanisms has been encouraged rather than opposed by natural selection.

Mating in many eukaryotic micro-organisms is between cells that are indistinguishable in size and form (isogamy), but often mating cells do differ in size (anisogamy or heterogamy). The circumstances under which anisogamy would evolve have been discussed by Maynard Smith (1978). Genera often contain both isogamous and anisogamous species, so it is likely that anisogamy has evolved on many occasions, culminating in a clear-cut structural and functional differentiation into static female gametes, with abundant food reserves to initiate development, and small mobile male gametes, that can be mass-produced and have the role of transporting their genome to the female. (In non-mobile organisms that show sexual differentiation, such as many fungi, a slender male hypha grows towards a large female cell.) Sexual differentiation into male and female is basically a tactical device and not an outbreeding mechanism; it occurs in many self-fertile organisms, and many self-sterile organisms produce both male and female structures. Confusion between mating type and differentiation into male and female has often occurred, an error perhaps excusable since in the evolution of animals nature has achieved the same confusion, namely two mating types only, with one male and one female. In micro-organisms it is important to avoid confusion. The various possible combinations of mating system and sexual differentiation have been tabulated by Carlile & Gooday (1978).

In many micro-organisms nuclear fusion is promptly followed by meiosis, so the vegetative phase is exclusively haploid (Fig. 1a). In others, however, the zygote may undergo cell division so that there is in the life cycle an alternation of haploid and diploid vegetative generations (Fig. 1*b*), often distinguishable only by nuclear and cell size or, if conditions are appropriate, by the events that lead to the next phase in the life cycle. In yet other micro-organisms, the cells produced by meiosis are capable only of mating so the vegetative phase is exclusively diploid (Fig. 1*c*). It seems certain that the evolution of diploidy has occurred

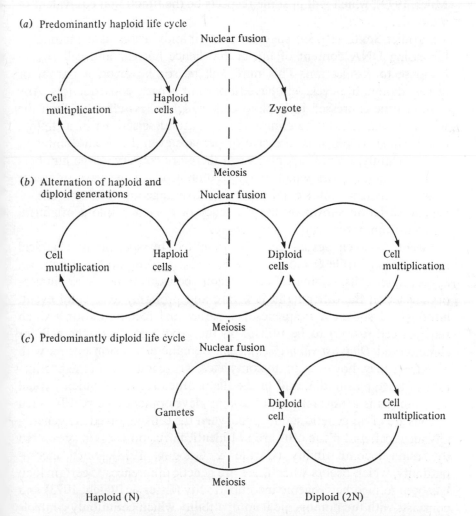

Fig. 1. Evolution of diploidy. (*a*) Haploid cells, sometimes specialised gametes, fuse to give a diploid cell (zygote). This, sometimes after a delay, undergoes meiosis to give haploid cells which multiply. Yeast example: *Schizosaccharomyces octosporus*. (*b*) As (*a*), but the diploid cell is capable of prolonged vegetative multiplication before some by meosis yield haploid cells. Yeast example: *Saccharomycopsis lipolytica*. (*c*) The haploid cells are incapable of, or do not normally show, vegetative multiplication. Yeast example: *Saccharomycodes ludwigii*.

many times among eukaryotic micro-organisms. All three types of life cycle are found among ascomycete yeasts, although the vegetative phase of filamentous ascomycetes is haploid. Diploidy also occurs in several other groups of fungi. A curious alternative to a lengthy diploid phase occurs in basidiomycetes; a prolonged dikaryosis between cell fusion and nuclear fusion (Casselton, 1978; Fincham et al., 1979; Schwalb & Miles, 1979), which will in some respects be the functional equivalent of a diploid.

Cavalier-Smith (1978b) suggests that diploidy arises as a method of increasing DNA content of nuclei and hence nuclear and cell size in response to K-selection. This may well be so; however, a partial or largely diploid life-cycle will have important genetic consequences. Any mutant gene expressed in haploid cells which is less effective than the wild-type will rapidly be eliminated by natural selection; in a diploid cell it will survive if it is recessive or perhaps even if it is co-dominant. Predominantly diploid organisms can therefore conserve large numbers of such mutant genes which may, recombined in new ways or under changed environmental conditions, be advantageous. The amount of genetic variation shown by the progeny of a single diploid organism isolated from nature can be impressive.

Fusion between sexual cells is promoted by specific biochemical mechanisms (Carlile & Gooday, 1978) and is a common event between appropriate cells. Somatic cell fusion between unicellular micro-organisms on the other hand is a rare and probably accidental event, although of sufficient frequency for parasexual recombination (which requires cell fusion) to be utilised in the genetic study of the cellular slime mould *Dictyostelium discoideum*. Somatic cell fusion occurs with high frequency, however, in myxomycetes, where it assists in the production of large plasmodia, and in the filamentous fungi, in which hyphal anastomosis is a normal part of colony development (see p. 18). Such fusions are of no genetic significance when the cells involved are genetically identical, and if the cells are not identical, fusion is often prevented by fusion incompatibility (Carlile & Gooday, 1978). Such incompatibility, which occurs when there are genetic differences at certain loci, has been termed heterogenic incompatibility (Esser & Blaich, 1973) and contrasts with the homogenic incompatibility which commonly controls sexual fusion. The role of somatic fusion incompatibility may be to prevent the spread of infectious cytoplasmic factors including viruses.

The immediate consequence of somatic fusion between genetically unlike cells is the establishment of a heterokaryon, which, like diploidy, can permit the survival of recessive genes that are not immediately advantageous. Heterokaryons may also permit the co-existence of two nuclear types appropriate for two different modes of life, such as saprophytic and parasitic (p. 19; Carlile, 1979). Heterokaryosis is also a necessary preliminary for the parasexual cycle – fusion of unlike haploid nuclei, mitotic crossing over, and haploidisation. The frequency of each of these events is rather low, the figures cited for *Aspergillus nidulans* per mitosis being about 1 in 10^6–10^7, 1 in 500 and 1 in 10^3, respectively (Fincham *et al.*, 1979). There are, however, ways of increasing the frequency of these events and they may be commoner in species which, unlike *A. nidulans*, lack a sexual cycle.

Heterogenic incompatibility preventing somatic cell fusion has been discussed above. It can also occur after cell fusion, other loci being involved (Carlile & Gooday, 1972). In filamentous fungi hyphal anastomosis between incompatible strains may cause the death of cells immediately adjacent to the anastomosis, other parts of the colonies involved being unaffected. In the ascomycete *Podospora anserina* some of the genes that can thus terminate somatic cell fusion are constructively involved in sexual morphogenesis. In *Neurospora crassa* the mating type alleles *A* and *a*, which must be brought together for sexual fusion to succeed, cause an incompatibility reaction if brought together at any other time in the life cycle. A hint as to the significance of this curious association between somatic incompatibility and sexual compatibility comes from work on diploid plasmodia of the myxomycete *Physarum polycephalum*. Here fusion between a strain (the 'killer') carrying a dominant incompatibility gene (e.g. *let A*; *let* stands for lethal) and another (the 'sensitive') which is homozygous for the corresponding recessive allele (*let a*) results in the destruction of nuclei from the sensitive strain (Lane & Carlile, 1979). The interaction between the alleles, however, causes very high frequency somatic nuclear fusion, involving both nuclear types. Since the dominant and recessive incompatibility alleles can co-exist in the same nucleus without causing the incompatibility reaction, it is possible that this may result in the survival of the chromosomes of any 'sensitive' nucleus which fuses with one from the 'killer' strain. The net effect of the reaction may therefore be to destroy all 'sensitive' nuclei that have failed to fuse. Since there are several *let* loci in *P. polycephalum* it is possible to produce and fuse two strains (e.g. with genotypes *let A let b* and *let a let B*) each of which is a 'killer' with respect to the other strain's nuclei (Carlile, 1976). Attempts to recover viable material from the area of interaction of two such strains have not yet been carried out, but it would seem likely that only nuclei that had resulted from somatic nuclear fusion between the two strains would survive. Post-fusion incompatibility may have arisen as a means of limiting somatic cell fusion by localised protoplasmic destruction following fusion, or of destroying defective nuclei that can arise and persist in multinucleate organisms. However, as indicated above, it can facilitate somatic nuclear fusion and destroy nuclei that have failed to fuse. It is possible to envisage that such a process could bring about the evolution of high frequency sexual recombination from low frequency parasexual recombination; this would account for some of the somatic incompatibility genes in *N. crassa* and *P. anserina* having a positive role in sexual morphogenesis.

ECOSYSTEMS, HABITATS AND LIFE-STYLES

Present-day aerobic ecosystems are dominated by eukaryotes, either microbial or multicellular, as will be demonstrated by a brief consideration of food chains. The primary production of organic matter is overwhelmingly through oxygenic photosynthesis, mainly by higher plants on land and eukaryotic algae, especially plankton, in aquatic environments; the contribution by cyanobacteria is usually small. The primary producers are grazed by eukaryotes, mainly higher animals (vertebrates and insects) on land and mainly protozoa in aquatic habitats; the carnivores that consume the herbivores are larger eukaryotes. Primary producers and members of the grazing food chain, even if they do not succumb to attack by larger organisms, ultimately die, and while they are alive excrete waste products, providing materials for the support of the detritus food chain. It is as the first members of the detritus food chain that bacteria are strikingly active, their high metabolic rates and fast growth being advantageous in exploiting food supplies that may be copious but which are erratic in their appearance. Land plants, however, have tough tissues which need to be broken up mechanically (e.g. by insects such as termites) or penetrated and partially lysed by fungal hyphae before significant bacterial attack can occur. Hence in some soils, such as in woodlands (Lynch & Poole, 1979), the fungal biomass is greater than that of bacteria, even though, because of their high metabolic activity, the energy flow through bacteria probably remains greater. Bacteria are preyed upon by protozoa which in turn are consumed by larger eukaryotes – the detritus food chain. The effectiveness of protozoan predation can be spectacular; an individual of the ciliate Tetrahymena can consume 150 bacteria per hour, clearing bacteria from about 20 000 times its own volume (Fenchel & Jørgensen, 1977).

Anaerobic environments are dominated by prokaryotes. Primary production of organic matter in such environments is very small compared with that in aerobic environments, and results from the activities of prokaryotes: anoxygenic photosynthesis and anaerobic chemolithotrophic processes. However, a great deal of organic matter, produced under aerobic conditions, enters anaerobic environments, whether the bottom of a lake or the gut of an animal. The anaerobic decomposition of this organic matter can support large bacteria populations, which are grazed by various anaerobic eukaryotes, especially ciliate protozoa.

Various other transient or extreme habitats are dominated by prokaryotes. The combination of photosynthetic activity and nitrogen fixation may enable cyanobacteria to be the first colonists in new habitats such as was provided on the tropical volcanic island of Krakatoa, although they were not conspicuous during recolonisation of the Icelandic volcanic island Surtsey (Fridriksson, 1975). Chemolithotrophic activity also permits the exploitation of resources not available to eukaryotes. Some bacteria can grow at temperatures approaching boiling point and hence can thrive in hot springs without competition from eukaryotes. The upper temperature for eukaryotic activity, perhaps as a consequence of membrane structure, is about 60 °C (Tansey & Brock, 1978). Eukaryotes as well as prokaryotes have been able to evolve tolerance of most other environmental extremes such as cold, high pressure, extreme pH values, high concentrations of salt or other solutes and desiccation (Kushner, 1978); yeasts for example are very well adapted for coping with extreme water stress (Brown, 1978). It is likely, however, that prokaryotes will be more successful than eukaryotes in exploiting an unusual environment of limited volume or uncommon substrates that may be transient. Due to the small size of bacteria such an environment or substrate can generate rapidly a population probably 1000 times as large as would occur with eukaryotes. Hence there are far more individuals to permit a few to evade chance extinction and reach suitable new sites.

In the course of evolution eukaryotes have replaced prokaryotes as primary producers and have come to dominate aerobic ecosystems in terms of biomass, number of species and morphological diversity. Their evolution has, however, also provided opportunities for prokaryotes as symbionts (both parasitic and mutualistic) and as decomposers of the detritus produced by eukaryotes. In so doing, eukaryotes will have permitted further prokaryote evolution. Streptomyces, for example, with hyphae simulating in miniature those of fungi, is admirably adapted for attacking small fragments of higher plant material. Even *Escherichia coli*, for many people the standard prokaryote, will have evolved to take advantage of a rather late development in evolution, the gut of warm-blooded animals.

CONCLUSIONS

Early in evolution prokaryotes developed a great biochemical diversity, including anoxygenic and oxygenic photosynthesis, chemolithotrophy, a range of aerobic and anaerobic pathways and nitrogen fixation. Selection for high growth rates (r-selection) has probably been of major importance among prokaryotes, resulting in miniaturisation and the elimination of redundant features. The metabolic versatility of prokaryotes permits them to sustain a biosphere without assistance from eukaryotes, and to the present day some steps in the cycling of nitrogen and sulphur are carried out solely by prokaryotes, which also retain a quantitatively crucial role in decomposing organic materials.

Eukaryotes are likely to have evolved from a wall-less prokaryote. Initial success was probably due to a capacity for feeding on prokaryotes by phagocytosis, and their subsequent diversity through a structural organisation which permitted increased size and the exploitation of possibilities not open to very small organisms or to organisms with a rigid cell wall. Among the attributes acquired by eukaryotes were swifter movement, a more sophisticated sensory system including directional perception, increased capabilities for cellular interaction, the possibility of cell fusion, of endosymbiosis, circadian rhythms, true diploidy and reciprocal genetic recombination, and a capacity for forming complex colonies and multicellular organisms. Eukaryote evolution has been dominated by selection for the advantages conferred by large-size (K-selection) and redundant features have often been exploited rather than eliminated. Contrary to the situation in prokaryotes, biochemical innovation seems limited compared with structural, organisational and behavioural innovation. Such striking prokaryotic activities as nitrogen fixation and chemolithotrophy are lacking; they may well have been absent in the line from which eukaryotes evolved, but if so, why have they not been acquired subsequently? Possibly eukaryote cell organisation is incompatible with such activities, but it seems more likely that the eukaryote ability to take advantage of prokaryotic activity through predation, symbiosis or habitat selection rendered such acquisitions unnecessary. Eukaryotes cannot compete with the fast growing prokaryotes in the rapid colonisation of transient resources but their attributes permit their success in more stable environments, leading to domination of the biosphere as regards both biomass and number of species.

I wish to thank Dr K. W. Buck, Dr T. Cavalier-Smith and Dr A. P. J. Trinci for helpful discussion of the manuscript.

REFERENCES

ABERCROMBIE, M., DUNN, G. A. & HEATH, J. P. (1977). The shape and movement of fibroblasts in culture. In *Cell and Tissue Interactions*, ed. J. W. Lash and M. M. Burger, pp. 57–70. New York: Raven Press.

ACHTMANN, M. & SKURRAY, R. (1977). A redefinition of the mating phenomenon in bacteria. In *Microbial Interactions*, ed. J. L. Reissig, pp. 233–79. London: Chapman & Hall.

ALLEN, N. S. & ALLEN, R. D. (1978). Cytoplasmic streaming in green plants. *Annual Review of Biophysics and Bioengineering*, 7, 497–526.

ALLEN, R. D. & ALLEN, N. S. (1978). Cytoplasmic streaming in amoeboid movement. *Annual Review of Biophysics and Bioengineering*, 7, 469–95.

BERG, H. C. & PURCELL, E. M. (1977). Physics of chemoreception. *Biophysical Journal*, 20, 193–219.

BIRD, C. W., LYNCH, J. M., PIRT, F. J., REID, W. W., BROOKS, C. J. W. & MIDDLEDITCH, B. S. (1971). Steroids and squalene in *Methylococcus capsulatus* grown on methane. *Nature, London*, 230, 473–4.

BLAKE, C. C. F. (1979). Exons encode protein functional units. *Nature, London*, 277, 598.

BONNER, J. T. (1974). *On Development: the Biology of Form*. Cambridge, Massachusetts: Harvard University Press.

BONNER, J. T. (1977). Some aspects of chemotaxis using the cellular slime moulds as an example. *Mycologia*, 69, 443–59.

BRADY, J. B. (1979). *Biological Clocks*. London: Edward Arnold.

BROWN, A. D. (1978). Compatible solutes and extreme water stress in eukaryotic micro-organisms. *Advances in Microbial Physiology*, 17, 181–242.

BUCHNER, P. (1965). *Endosymbiosis of Animals with Plant Micro-organisms*. New York: Wiley.

CARLILE, M. J. (1975). Taxes and tropisms: diversity, biological significance and evolution. In *Primitive Sensory and Communication Systems: the Taxes and Tropisms of Micro-organisms and Cells*, ed. M. J. Carlile, pp. 1–28. London: Academic Press.

CARLILE, M. J. (1976). The genetic basis of the incompatibility reaction following plasmodial fusion between different strains of the myxomycete *Physarum polycephalum*. *Journal of General Microbiology*, 71, 581–90.

CARLILE, M. J. (1979). Bacterial, fungal and slime mould colonies. In *Biology and Systematics of Colonial Organisms*, ed. G. Larwood & B. R. Rosen, pp. 3–27. London: Academic Press.

CARLILE, M. J. & GOODAY, G. W. (1978). Cell fusion in myxomycetes and fungi. In *Membrane Fusion*, ed. G. Poste and G. L. Nicholson, pp. 219–65. New York: Elsevier/North-Holland Biomedical Press.

CASSELTON, L. A. (1978). Dikaryon formation in the higher Basidiomycetes. In *The Filamentous Fungi*, vol. 3, *Developmental Mycology*, ed. J. E. Smith & D. R. Berry, pp. 275–97. London: Edward Arnold.

CAVALIER-SMITH, T. (1974). Palindromic base sequences and replication of eukaryote chromosome ends. *Nature, London*, **250**, 467–70.

CAVALIER-SMITH, T. (1975). The origin of nuclei and of eukaryotic cells. *Nature, London*, **256**, 463–8.

CAVALIER-SMITH, T. (1978*a*). Nuclear volume control by nucleoskeletal DNA, selection for cell volume and cell growth rate, and the solution of the DNA C-value paradox. *Journal of Cell Science*, **34**, 247–78.

CAVALIER-SMITH, T. (1978*b*). The evolutionary origin and phylogeny of micro-tubules, mitotic spindles and eukaryote flagella. *Biosystems*, **10**, 93–114.

CERDÁ-OLMEDO, E. (1977). Behavioural genetics of Phycomyces. *Annual Review of Microbiology*, **31**, 535–47.

COLD SPRING HARBOR SYMPOSIUM (1978). Chromatin. *Symposia on Quantitative Biology*, **42**, pts. 1 & 2. New York State: Cold Spring Harbor Laboratory.

COOKE, R. (1977). *The Biology of Symbiotic Fungi*. London: Wiley.

COX, R. A. (1977). Structure and function of prokaryotic and eukaryotic ribosomes. *Progress in Biophysics and Molecular Biology*, **32**, 193–231.

CRANDALL, M., EGEL, R. & MACKAY, W. (1976). Physiology of mating in three yeasts. *Advances in Microbial Physiology*, **15**, 307–98.

CURDS, C. R. (1979). Group phenomena in the phylum Protozoa. In *Biology and Systematics of Colonial Organisms*, ed. G. Larwood & B. R. Rosen, pp. 29–37. London: Academic Press.

CURTIS, A. S. G. (ed.) (1978). *Cell–Cell Recognition. Symposium of the Society for Experimental Biology*, **32**. London: Cambridge University Press.

DARMON, M. & BRACHET, P. (1978). Chemotaxis and differentiation during the aggregation of *Dictyostelium discoideum* amoebae. In *Taxis and Behaviour*, ed. G. L. Hazelbauer, pp. 101–39. London: Chapman & Hall.

DARNELL, J. E. (1978). Implications of RNA–RNA splicing in evolution of eukaryotic cells. *Science*, **202**, 1257–60.

DIEHN, B. (1973). Phototaxis and sensory transduction in Euglena. *Science*, **181**, 1009–15.

DOOLITTLE, W. F. (1978). Genes in pieces: were they ever together? *Nature, London*, **272**, 581–2.

DRAKE, J. W. (1974). The role of mutation in microbial evolution. In *Evolution in the Microbial World. Symposium of the Society for General Microbiology* **24**, ed. M. J. Carlile & J. J. Skehel, pp. 41–58. London: Cambridge University Press.

DUNNY, G. M., BROWN, B. L. & CLEWELL, D. B. (1978). Induced cell aggregation and mating in *Streptococcus faecalis*: evidence for a bacterial sex pheromone. *Proceedings of the National Academy of Sciences, USA*, **75**, 3479–83.

DWORKIN, M. (1973). Cell–cell interactions in the myxobacteria. In *Microbial Differentiation. Symposium of the Society for General Microbiology* **23**, ed. J. M. Ashworth & J. E. Smith, pp. 125–42. London: Cambridge University Press.

EAGON, R. G. (1962). *Pseudomonas natriegens*, a marine bacterium with a generation time of less than 10 minutes. *Journal of Bacteriology*, **83**, 736–7.

ELLIOT, C. G. (1977). Sterols in fungi: their functions in growth and reproduction. *Advances in Microbial Physiology*, **15**, 121–73.

ESSER, K. (1974). Breeding systems and evolution. In *Evolution in the Microbial World. Symposium of the Society for General Microbiology* 24, ed. M. J. Carlile & J. J. Skehel, pp. 87–104. London: Cambridge University Press.

ESSER, K. & BLAICH, R. (1973). Heterogenic incompatibility in plants and animals. *Advances in Genetics*, 17, 107–52.

FENCHEL, T. M. & JØRGENSEN, B. B. (1977). Detritus food chains of aquatic ecosystems: the role of bacteria. *Advances in Ecological Research*, 1, 1–58.

FINCHAM, J. R. S., DAY, P. R. & RADFORD, A. (1979). *Fungal Genetics*, 4th edn. Oxford: Blackwell Scientific Publications.

FOSTER, K. W. (1977). Phototropism of coprophilous zygomycetes. *Annual Review of Biophysics and Bioengineering*, 6, 419–43.

FRIDRIKSSON, S. (1975). *Surtsey: Evolution of Life on a Volcanic Island*. London: Butterworth.

FULTON, C. (1977). Cell differentiation in *Naegleria gruberi*. *Annual Review of Microbiology*, 31, 597–629.

GARROD, D. R., SWAN, A. P., NICOL, A. & FORMAN, D. (1978). Cellular recognition in slime and mould development. In *Cell–Cell Recognition. Symposia of the Society for Experimental Biology* 32, ed. A. S. G. Curtis, pp. 173–202. London: Cambridge University Press.

GILBERT, W. (1978). Why genes in pieces? *Nature, London*, 271, 501.

GOODAY, G. W. (1978). Microbial hormones. In *Companion to Microbiology*, ed. A. T. Bull & P. M. Meadow, pp. 207–20. London: Longman.

GRIFFIN, D. H., TIMBERLAKE, W. E. & CHENEY, J. C. (1974). Regulation of macromolecular synthesis, colony development, and specific growth rate of *Achlya bisexualis* during balanced growth. *Journal of General Microbiology*, 80, 381–8.

HARRISON, D. E. F. (1978). Efficiency of microbial growth. In *Companion to Microbiology*, ed. A. T. Bull & P. M. Meadow, pp. 159–79. London: Longman.

HASTINGS, J. W. & SCHWEIGER, H. G. ed. (1976). *The Molecular Basis of Circadian Rhythms*. Berlin: Dahlem Konferenzen.

HAZELBAUER, G. L. & PARKINSON, J. S. (1977). Bacterial chemotaxis. In *Microbial Interactions*, ed. J. L. Reissig, pp. 61–98. London: Chapman & Hall.

HENRY, S. M. (ed.) (1966, 1967). *Symbiosis*, vols. 1 & 2. New York: Academic Press.

HERDMAN, M., JANVIER, M., RIPPKA, R. & STANIER, R. Y. (1979). Genome size of cyanobacteria. *Journal of General Microbiology*, 111, 73–85.

HILDEBRAND, E. (1978). Bacterial phototaxis. In *Taxis and Behaviour*, ed. G. L. Hazelbauer, pp. 35–73. London: Chapman & Hall.

HOLT, S. C. (1978). Anatomy and chemistry of spirochaetes. *Microbiological Reviews*, 42, 114–60.

HORGEN, P. A. & SILVER, J. C. (1978). Chromatin in eukaryotic microbes. *Annual Review of Microbiology*, 52, 249–84.

JENNINGS, D. H. & LEE, D. L. (ed.) (1975). *Symbiosis. Symposium of the Society for Experimental Biology*, 29. London: Cambridge University Press.

JEON, K. W. & JEON, M. S. (1976). Endosymbiosis in amoebae: recently established endosymbionts have become required cytoplasmic components. *Journal of Cellular Physiology*, 89, 337–44.

KOCH, A. L. (1971). The adaptive responses of *Escherichia coli* to a feast and famine existence. *Advances in Microbial Physiology*, 6, 147–217.

KOCHERT, G. (1978). Sexual pheromones in algae and fungi. *Annual Review of Plant Physiology*, **29**, 461–8.

KOZAK, M. (1978). How do eukaryotic ribosomes select initiator regions in messenger RNA? *Cell*, **15**, 1109–23.

KUBAI, D. F. (1978). Mitosis and fungal phylogeny. In *Nuclear Division in the Fungi*, ed. I. Brent Heath, pp. 177–229. New York: Academic Press.

KUSHNER, D. J. (ed.) (1978). *Microbial Life in Extreme Environments*. London: Academic Press.

LANE, E. B. & CARLILE, M. J. (1979). Post-fusion somatic incompatibility in plasmodia of *Physarum polycephalum*. *Journal of Cell Science*, **35**, 339–54.

LENCI, F. & COLOMBETTI, G. (1978). Photobehavior of micro-organisms: a biophysical approach. *Annual Review of Biophysics and Bioengineering*, **7**, 341–61.

LEVANDOWSKY, M. & HAUSER, D. C. R. (1978). Chemosensory responses of swimming algae and protozoa. *International Review of Cytology*, **53**, 145–210.

LYNCH, J. M. & POOLE, N. J., ed. (1979). *Microbial Ecology: a Conceptual Approach*. Oxford: Blackwell Scientific Publications.

MAC ARTHUR, R. H. & WILSON, E. O. (1967). *The Theory of Island Biogeography*. Princeton, New Jersey: Princeton University Press.

MANNEY, T. R. & MEADE, J. H. (1977). Cell–cell interactions during mating in *Saccharomyces cerevisiae*. In *Microbial Interactions*, ed. J. L. Reissig, pp. 281–321. London: Chapman & Hall.

MARGULIS, L. (1970). *Origin of Eukaryotic Cells*. New Haven, Connecticut: Yale University Press.

MARGULIS, L., CHASE, D. & TO, L. P. (1979). Possible evolutionary significance of spirochaetes. *Proceedings of the Royal Society of London, Series B*, **204**, 189–98.

MARGULIS, L., TO, L. & CHASE, D. (1978). Microtubules in prokaryotes. *Science*, **200**, 1118–24.

MAYNARD SMITH, J. (1978). *The Evolution of Sex*. Cambridge: Cambridge University Press.

MINKOFF, L. & DAMADIAN, R. (1976). Actin-like protein from *Escherichia coli*: concept of cytotonus as the missing link between cell metabolism and the biological ion-exchange resin. *Journal of Bacteriology*, **125**, 353–65.

MOROWITZ, H. J. & WALLACE, D. C. (1973). Genome size and life cycle of the mycoplasma. *Annals of the New York Academy of Sciences*, **225**, 62–73.

MÜLLER, M. (1975). Biochemistry of protozoan microbodies: peroxisomes, α-glycerophosphate oxidase bodies, hydrogenosomes. *Annual Review of Microbiology*, **29**, 467–83.

NEIMARK, H. C. (1977). Extraction of an actin-like protein from the prokaryote *Mycoplasma pneumoniae*. *Proceedings of the National Academy of Sciences, USA*, **74**, 4041–5.

NEWELL, P. C. (1977). Aggregation and cell surface receptors in cellular slime molds. In *Microbial Interactions*, ed. J. L. Reissig, pp. 3–49. London: Chapman & Hall.

NJUS, D., GOOCH, V. D., MERGENHAGEN, D., SULZMAN, F. & HASTINGS, J. W. (1976). Membranes and molecules in circadian systems. *Federation Proceedings*, **35**, 2353–7.

NJUS, D., SULZMAN, F. M. & HASTINGS, J. W. (1974). Membrane model for the circadian clock. *Nature, London*, **248**, 116–20.

NUCCITELLI, R. & JAFFE, L. F. (1976). Current pulses involving chloride and potassium relieve excess pressure in Pelvetia embryos. *Planta*, **131**, 315–20.

PIANKA, E. R. (1970). On r- and K-selection. *American Naturalist*, **104**, 592–7.

RAFF, R. A. & MAHLER, H. R. (1975). The symbiont that never was; an enquiry into the evolutionary origin of the mitochondrion. In *Symbiosis. Symposia of the Society for Experimental Biology*, **29**, ed. D. H. Jennings & D. L. Lee, pp. 41–92. London: Cambridge University Press.

RAGAN, M. A. & CHAPMAN, D. J. (1978). *A Biochemical Phylogeny of Protists*. New York: Academic Press.

RAZIN, S. (1978). The mycoplasmas. *Microbiological Reviews*, **42**, 414–70.

REES, D. A., LLOYD, C. W. & THOM, D. (1977). Control of grip and stick in cell adhesion through lateral relationships of membrane glycoproteins. *Nature, London*, **267**, 124–8.

RIPPKA, R., DERUELLES, J., WATERBURY, J. B., HERDMAN, M. & STANIER, R. Y. (1979). Generic assignments, strain histories and properties of pure cultures of cyanobacteria. *Journal of General Microbiology*, **111**, 1–61.

RODWELL, A. W., RODWELL, E. S. & ARCHER, D. B. (1979). Mycoplasmas lack a protein which closely resembles α-actinin. *FEMS Microbiology Letters*, **5**, 235–8.

SCHWALB, M. N. & MILES, P. G., ed. (1978). *Genetics and Morphogenesis in the Basidiomycetes*. New York: Academic Press.

SCHWEIGER, H. G. & SCHWEIGER, M. (1977). Circadian rhythms in unicellular organisms: an endeavor to explain the molecular mechanism. *International Review of Cytology*, **51**, 315–42.

SEARCY, D. G., STEIN, D. B. & GREEN, G. R. (1978). Phylogenetic affinities between eukaryotic cells and a chemophilic mycoplasma. *Biosystems*, **10**, 19–28.

SLEIGH, M. A. & BLAKE, J. R. (1977). Methods of ciliary propulsion and their size limitations. In *Scale Effects in Animal Locomotion*, ed. T. J. Pedley, pp. 243–56. London: Academic Press.

SMITH, D. G. (1978). Bacterial motility and taxis. In *Companion to Microbiology*, ed. A.T. Bull & P. M. Meadow, pp. 321–41. London: Longman.

SNEATH, P. H. A. (1974). Phylogeny of micro-organisms. In *Evolution in the Microbial World. Symposium of the Society for General Microbiology*, **24**, ed. M. J. Carlile & J. J. Skehel, pp. 1–39. London: Cambridge University Press.

SOUTHWOOD, T. R. E. (1977). Habitat, the templet for ecological strategies? *Journal of Animal Ecology*, **46**, 337–65.

STANIER, R. Y. (1970). Some aspects of the biology of cells and their possible evolutionary significance. In *Organisation and Control in Prokaryotic and Eukaryotic Cells. Symposium of the Society for General Microbiology*, **20**, ed. H. P. Charles & B. C. J. G. Knight, pp. 1–38. Cambridge University Press.

STANIER, R. Y. (1974). The origins of photosynthesis in eukaryotes. In *Evolution in the Microbial World. Symposium of the Society for General Microbiology*, **24**, ed. M. J. Carlile & J. J. Skehel, pp. 229–40. London: Cambridge University Press.

STANIER, R. Y., ADELBERG, E. A. & INGRAHAM, J. L. (1977). *General Microbiology*, 3rd edn. London: Macmillan.

STOUTHAMER, A. H. (1977). Energetic aspects of the growth of micro-organisms. In *Microbial Energetics, Symposium of the Society for General Microbiology*, **27**, ed. B. A. Haddock & W. A. Hamilton, pp. 285–315. London: Cambridge University Press.

TANSEY, M. R. & BROCK, T. D. (1978). Microbial life at high temperatures: ecological aspects. In *Microbial Life in Extreme Environments*, ed. D. J. Kushner, pp. 159–216. London: Academic Press.

TRINCI, A. P. J. (1972). Culture turbidity as a measure of mould growth. *Transactions of the British Mycological Society*, **58**, 467–73.

TRINCI, A. P. J. (1978a). Wall and hyphal growth. *Science Progress, Oxford*, **65**, 75–99.

TRINCI, A. P. J. (1978b). The duplication cycle and vegetative development in moulds. In *The Filamentous Fungi*, vol. 3, *Developmental Mycology*, ed. J. E. Smith & D. R. Berry, pp. 132–63. London: Edward Arnold.

VAN BAALEN, C. (1974). Growth, photosynthetic and respiratory rates of the microalgae. In *Handbook of Micro-biology*, vol. 4, ed. A. I. Laskin & H. A. Lechavalier, pp. 21–8. Cleveland, Ohio: CRC Press.

WALSBY, A. E. (1978). The gas vacuoles of aquatic prokaryotes. In *Relations between Structure and Function in the Prokaryotic Cell. Symposium of the Society for General Microbiology*, **28**, ed. R. Y. Stanier, H. J. Rogers & B. J. Ward, pp. 327–57. London: Cambridge University Press.

WATERBURY, J. B. & STANIER, R. Y. (1978). Patterns of growth and development in pleurocapsalean cyanobacteria. *Microbiological Reviews*, **42**, 2–44.

WHATLEY, J. M., JOHN, P. & WHATLEY, F. R. (1979). From extracellular to intracellular: the establishment of mitochondria and chloroplasts. *Proceedings of the Royal Society of London, Series B*, **204**, 165–87.

WILKIE, D. R. (1977). Metabolism and body size. In *Scale Effects in Animal Locomotion*, ed. T. J. Pedley, pp. 23–36. London: Academic Press.

WILLCOX, M., MITCHISON, G. J. & SMITH, R. (1975). Spatial control of differentiation in the blue-green alga Anabaena. In *Microbiology* 1975, ed. D. Schlessinger, pp. 453–63. Washington: American Society for Microbiology.

CHROMOSOME STRUCTURE AND THE MOLECULAR BIOLOGY OF MITOSIS IN EUKARYOTIC MICRO-ORGANISMS

N. RONALD MORRIS

Department of Pharmacology,
College of Medicine and Dentistry of New Jersey,
Rutgers Medical School,
PO Box 101,
Piscataway, NJ 08854, USA

INTRODUCTION

Mitosis is one of the most important but least well understood of biological processes. The morphological features of mitosis, first described about 100 years ago (Flemming, 1878, 1880), are well-known to all biologists. Yet the biochemistry of mitosis is for the most part as obscure as the morphology is familiar. For example, almost nothing certain is known about the biochemical events involved in chromosomal condensation and relaxation, or assembly and disassembly of the mitotic spindle and movement of the chromosomes to the poles. Equally unclear are the processes by which mitosis is triggered and by which it is synchronised with DNA synthesis and cytoplasmic division. Another mystery is the topological mechanism which ensures that following each nuclear division, one nucleus, possibly a specific nucleus, is partitioned to each daughter cell. These mechanisms are as vital to the heredity of cells and organisms as the orderly replication of DNA itself but they are far less well understood.

Mitotic mutants of microbial eukaryotes

Recently several laboratories, including my own, have initiated a new approach to the study of mitosis by isolating mutants of microbial eukaryotes that are temperature-sensitive with respect to mitosis and/or cell division, and mutants with altered sensitivities to antimitotic drugs. The first, largest and most extensively studied set of mutants was isolated in *Saccharomyces cerevisiae* by Hartwell and his colleagues (Hartwell, Culotti & Reid, 1970; Hartwell *et al.*, 1973; Hartwell *et al.*, 1974; Hartwell, 1978; Pringle, 1978; Simchen, 1978; Moir, Stewart & Botstein, 1978). Interesting mutants have also been isolated of the fission yeast, *Schizosaccharomyces pombe* (Lederberg & Stetten, 1970; Nurse, Thuriaux & Nasmyth, 1976; Thuriaux, Nurse & Carter, 1978), in the filamentous fungus *Aspergillus nidulans* (Orr & Rosenberger, 1976*a*, *b*; Morris, 1976*a*, *b*; van Tuyl, 1977; Sheir-Neiss, Lai & Morris, 1978; Morris, Lai & Oakley, 1979), in the flagellate *Chlamydomonas reinhardii* (Warr & Durber, 1971; Warr & Gibbons, 1974; Warr, Flanagan & Quinn, 1978; Howell & Naliboff, 1973; Flavin & Slaughter, 1974; Sato, 1976), in the ciliates *Paramecium tetraaurelia* (Peterson & Berger, 1976) and *Tetrahymena pyriformis* (Frankel *et al.*, 1976; Frankel, Jenkins & DeBault, 1976; Bruns & Sanford, 1978), and in the slime mould *Physarum polycephalum* (Gingold *et al.*, 1976; Wheals, Grant & Jockusch, 1976). Most of the work that has so far been done with these mutants has concerned morphological aspects of cells blocked in mitosis, the timing of gene function with respect to the cell cycle and the interrelationship between various gene functions (see reviews by Hartwell, 1978; Pringle, 1978; Simchen, 1978). Surprisingly little use has been made of these mutants to study the biochemistry of mitosis, but this situation has now started to change. The analysis by two-dimensional gel electrophoresis of mutants of *A. nidulans* resistant to antimicrotubule drugs has provided a key to the biochemical genetics of tubulin (Sheir-Neiss *et al.*, 1978; Morris *et al.*, 1979). Moreover, the methods developed for analysing these tubulin mutants would appear to have general applicability to the study of the biochemistry of mitosis and cell division in other types of mutants and in mutants of other organisms as well. These methods are potentially so powerful that in principle it should now be possible to identify almost any protein necessary for mitosis or cell division for which there exists an appropriate temperature-sensitive mutant. In this chapter, I will review recent work on mitotic and antimicrotubule-drug-resistant mutants of *A. nidulans* and other microbial eukaryotes that can be manipulated genetically, with special emphasis upon these new methods.

Mitosis in microbial eukaryotes as a model for mitosis in man

My own work on mitosis in *Aspergillus nidulans* has been strongly motivated by a desire to understand mitosis in higher eukaryotes, especially in man; and I imagine that this may be true of other investigators studying mitosis in simple eukaryotes. We have elected, however, to study mitosis in microbial eukaryotes because it is difficult to isolate mitotic mutants in higher eukaryotes and because the genetic systems of most higher eukaryotes are technically less amenable to genetic analysis than those of microbial eukaryotes. Thus the rationale for using microbial eukaryotes to study the molecular biolgy of mitosis is founded in some measure upon the expectation that the biochemistry of mitosis in simple organisms, e.g. *A. nidulans*, will be similar to the biochemistry of mitosis in higher eukaryotes, e.g. man. What are the chances that this will turn out to be correct and on what evidence do we base our hopes? First of all, we know that despite considerable diversity of detail (see Dodge & Vickerman, this volume pp. 77–105; Kubai, 1975; Fuller, 1976; Heath, 1978), the fundamental morphological features of mitosis are similar in most eukaryotes. For example, in *A. nidulans*, as in man, a microtubular mitotic spindle forms and the chromosomes condense at mitosis. The chromosomes are attached to the microtubules of the spindle by a special structure, the kinetochore; and the chromosomes move to the poles at anaphase (B. Oakley & N. R. Morris, unpublished). Additional encouragement for the belief that mitosis in *A. nidulans* and other microbial eukaryotes is representative of mitosis in higher organisms comes directly from studies of the biochemistry of their nuclei.

The molecular biology of the nucleus

Evidence that the molecular biology of the nucleus in *Aspergillus nidulans* and other microbial eukaryotes resembles that of higher eukaryotes has been forthcoming only recently. Before 1975, the bulk of the available evidence was against the existence of a normal complement of histones in fungi; there had been no studies of chromatin structure in a lower eukaryote and tubulin had not been well characterised in any of the important, genetically manipulable, non-flagellated, microbial eukaryotes. In 1975, our laboratory initiated a study of the histones, chromatin structure and tubulin of *A. nidulans*, the results of which clearly demonstrated that *A. nidulans* was biochemically as well as morphologically eukaryotic. This work has been reviewed in detail elsewhere (Morris *et al.*, 1977) and will only be summarised here. For a more complete review of chromatin in eukaryotic microbes see Horgan & Silver (1978).

HISTONES AND CHROMATIN STRUCTURE

The first level of chromatin organisation in higher eukaryotes is the nucleosome (Kornberg, 1977; Chambon, 1977); 140 base pairs of DNA wrapped about an octomeric protein core containing two copies each of the four core histones, H2A, H2B, H3, and H4 (Kornberg, 1977). Nucleosomes are connected by DNA linkers, usually about 60 base pairs long, whose configuration in relation to the core is not known, and histone H1 is associated with the linker region (Varshavsky, Bakayev & Georgiev, 1976; Noll & Kornberg, 1977; Sommer, 1978). The amount of DNA in the core plus the linker is known as the nucleosome repeat length.

During the past several years, it has been demonstrated that the chromatin of many microbial eukaryotes is organised like the chromatin of higher eukaryotes (Horgan & Silver, 1978). Microbial eukaryotes have been shown to have essentially the same histones as the higher eukaryotes and the organisation of their chromatin has been shown to be nucleosomal. Although the early studies of the nuclear proteins of microbial eukaryotes failed to demonstrate histone-like proteins or identified an incomplete set of histones, we now know that this was because histones are easily degraded during isolation by cellular proteases.

The first demonstration of a complete set of histones in a microbial eukaryote was in *Aspergillus nidulans* by Felden, Sanders & Morris (1976). The *A. nidulans* histones were isolated from nuclei which had been purified from frozen, vegetative mycelia under conditions designed to minimise the activity of both proteases and nucleases (Gealt, Sheir-Neiss & Morris, 1975). The DNA from these nuclei was shown to be of high molecular weight ($> 20 \times 10^6$) and the nuclear morphology was similar to that of nuclei in the fixed mycelium. A set of small, acid-soluble, basic proteins from these isolated nuclei were extracted and were identified as histones by their elution patterns from columns of BioGel P-10, P-60, and Sephadex G-100, by their mobilities in several electrophoretic systems and by amino acid analysis (Felden *et al.*, 1976). Almost immediately thereafter histones were identified in *Neurospora crassa* by Goff (1976) and in *Saccharomyces cerevisiae* by Thomas & Furber (1976) and Moll & Wintersberger (1976). Histones have since been described in a variety of fungi and other eukaryote microbes (Horgan & Silver, 1978). In most cases the histones of microbial eukaryotes resemble those of higher eukaryotes and in general, the four core histones, H2A, H2B, H3 and H4 are similar to the core histones of higher eukaryotes. However, the fungal histones are generally less basic than the histones of higher eukaryotes and exhibit some differences in amino acid composition (Felden, *et al.*, 1976; Goff, 1976). Histone H4 from *Tetrahymena thermophila* has been partially sequenced and differs from that of calf thymus at 15 out of 66 sequenced amino acid locations (Glover & Gorovsky, 1979). Compared to the almost total conservation of amino acid sequences of histone H4 between calf thymus and pea seedling, the difference between *T. thermophila* and calf thymus is large. As might be expected from this large sequence difference, histone H4 from *T. thermophila* is not capable of replacing that from calf thymus in nucleosome reconstitution experiments (Gorovosky *et al.*, 1977).

The fifth histone, H1, has been difficult to identify with surety in lower eukaryotes because it is polymorphic; it is much less highly conserved evolutionarily than the core histones and it is very susceptible to proteolytic degradation. Nevertheless histone H1-like proteins have been demonstrated in some lower eukaryotes (Charlesworth & Parish, 1975; Felden *et al.*, 1976; Sommer, 1978; Tessier *et. al.*, 1978) Whether these proteins are functionally similar to the histones H1 of higher eukaryotes must depend upon the development of a clearer general understanding of their function in chromatin.

The first biochemical studies of chromatin structure in a microbial eukaryote were in *Saccharomyces cerevisiae* by Lohr & Van Holde (1975), who demonstrated that yeast chromatin had a nucleosomal organisation with a nucleosome repeat that was shorter than the 200 base pair repeat previously reported for mammalian chromatin (see also Thomas & Furber, 1976; Lohr, Kovacic & Van Holde, 1977; Nelson, Beltz & Rill, 1977). *Aspergillus nidulans* chromatin was also found to have a nucleosomal organisation with a short nucleosome repeat of 154 base pairs (Morris, 1976c) and *Neurospora crassa* chromatin gave a third short nucleosome repeat of 170 base pairs (Noll, 1976). In *A. nidulans*, *N. crassa* and *S. cerevisiae*, the nucleosome core was measured as 140 base pairs of DNA. Consequently the short nucleosome repeat lengths found in these organisms were ascribed to differences in the length of the linker between DNA cores. The primary structure of chromatin has now been studied in many other microbial eukaryotes, including *Physarum polycephalum* (Stadler & Braun, 1978) and *Dictyostelium discoideum* (Parish, Stalder & Schmidlin, 1977; Bakke, Wu & Bonner, 1978; see review by Horgan & Silver, 1978) and the pattern of a short nucleosome repeat combined with a normal 140 base pair nucleosome core appears to be common, but not universal. For example, the ciliates *Tetrahymena pyriformis* and *Stylonichia mytilis* are single-celled organisms with two nuclei in each cell, a transcriptionally inactive sexual micronucleus and a transcriptionally active, vegetative macronucleus. In both organisms the active macronucleus has a long nucleosome repeat of about 215 base pairs, whereas the inactive micronucleus has a much shorter repeat of about 200 base pairs (Lipps & Morris, 1976; Gorovsky *et al.*, 1977). The macronuclear nucleosome repeat length has been reported to change as a function of growth rate in both

T. pyriformis and *Paramecium aurelia* (Prince, Cummings & Seale 1977) suggesting that the repeat length may be related to rate of transcription. How these different nucleosome repeats are established and whether they are related to the different transcriptional activities of the micro- and macronuclei provide fascinating investigative opportunities.

One consequence of finding short nucleosome repeats in fungi was a re-examination of the possibility that there was variation in the nucleo- some repeat length in higher eukaryotes. It was subsequently found that nucleosome repeat lengths did vary in higher eukaryotes and that variation could be found between different cell types and tissues (Morris, 1976*d*; Compton, Bellard & Chambon, 1976; Spadafora *et al.*, 1976) and even between different types of nuclei in the same tissue (Thomas & Thompson, 1977; Ermini & Kuenzle, 1978), leading to the suggestion that nucleosomes may function to regulate gene transcription (Morris, 1976*d*; Kornberg, 1975). Investigation of the nucleosomal organisation of specific genes as a function of gene activity is feasible in microbial eukaryotes like *Saccharomyces cerevisiae* and *Aspergillus nidulans* which have genomes of low complexity and such studies should be useful in determining whether nucleosomes are important in the regulation of transcription.

The higher and lower eukaryotes are similar with respect to the nucleosomal organisation of chromatin, but are they also similar with respect to those features which are responsible for the transition from the diffuse interphase to the highly organised metaphase structure of chromosomes? Since we do not yet know what these features are this question cannot be very clearly answered at the present time, but it may be useful to consider briefly what little is known about the higher order structure of mitotic chromosomes in higher eukaryotes.

Laemmli and his coworkers have recently demonstrated that Hela chromosomes depleted of histones by treatment with dextran sulphate and heparin maintain a highly-folded configuration which retains chromosome morphology (Adolph, Cheng & Laemmli, 1977). In the electron microscope these histone-depleted chromosomes exhibit a core or 'scaffolding' from which emerge and re-enter loops or domains of DNA 30 to 90 kilobases long (Paulson & Laemmli, 1977). The DNA loops can be digested with micrococcal nuclease, and a scaffolding which is recognisably chromosomal in structure persists (Adolph *et al.*, 1977*b*). The chromosomal scaffolding contains relatively few proteins, but these few proteins are probably responsible for the maintenance of chromo- somal structure.

The chromosome scaffolding, by organising the DNA into loops, determines the basic shape of the chromosome. The DNA loops are normally compacted by the core histones into 10 nm nucleosome fibres which in turn are further organised, probably by histone H1, into higher order 20–30nm structures, possibly solenoids (Finch & Klug, 1976) or superbeads (Renz, Nihls & Hozier, 1977). Thus there are at least three levels of compaction of DNA in metaphase chromosomes i.e. the nucleosomes, the 20–30nm fibres and the scaffolding.

Little is known about the biochemical processes which lead to chromosome condensation at metaphase or chromosome relaxation at interphase. Presumably the bases of the DNA loops come together during mitosis to organise the scaffolding and the higher order histone-dependent compaction of the loop into nucleosomes and fibres also contributes to the condensed appearance of the metaphase chromosomes. A change in the nucleosomal repeat is apparently not involved in the interphase to metaphase transition of chromatin, since interphase chromatin and mitotic chromatin have been shown to have the same nucleosomal repeat length (Compton *et al.*, 1976; Vogt & Braun, 1976). However, phosphorylation of histone H1 has been shown to increase greatly at mitosis in *Physarum polycephalum* and mammalian cells, and it has been suggested that H1 phosphorylation may play a role in chromosome condensation (Bradbury *et al.*, 1973; Bradbury, 1975; Gurley *et al.*, 1978). Mitotic mutants of eukaryotic microbes should be useful in testing this hypothesis.

Although the microbial eukaryotes with their minute chromosomes would seem to be ideal organisms for investigating chromosome structure, few such studies have been reported. In *Saccharomyces cerevisiae*, folded chromosomes containing all five histones have been isolated (Piñon & Salts, 1977). The DNA of these chromosomes, like that of higher eukaryotes, has been shown to be superhelical and to be organised into domains. Folded chromosomes have been characterised from the G0, G1 and G2 stages of the cell cycle mutants; and these differ in their sedimentation velocities and presumably in their structures (Piñon, 1978). The biochemical and ultrastructural details of the yeast folded chromosome and how the structure relates to the structure of chromosomes from higher eukaryotes are not known.

TUBULIN

The best known of the proteins important to mitosis is tubulin. Tubulin from mammalian sources has been studied extensively (Snyder & McIntosh, 1976). It is a dimeric protein with a molecular weight of approximately 110 000 which is believed to be a heterodimer of two similar 55 000 mol. wt subunits, α- and β-tubulin. Tubulin can be polymerised to microtubules *in vitro* in the presence of GTP and Mg^{2+} and can be purified from tissues that are rich in tubulin, e.g. mammalian or avian brain, by a procedure which utilises several cycles of microtubule polymerisation and depolymerisation to separate tubulin and microtubule-associated proteins from other cellular proteins. The tubulin concentration of cell extracts must be high for in-vitro polymerisation of microtubules to occur. Because the non-flagellated microbial eukaryotes contain very low concentrations of tubulin in comparison to mammalian brain, it has not been possible to polymerise tubulin directly from these organisms. However, tubulins from widely diverse species are able to recognise each other in the polymerisation reaction and it has been possible to identify and characterise tubulin in eukaryotic microbes by copolymerising radioactively-labelled microbial proteins with tubulin from mammalian brain. The microbial tubulins are enriched by copolymerisation and in this way cytoplasmic tubulin has been identified and characterised in several eukaryotic microbes including *Aspergillus nidulans* (Sheir-Neiss *et al.*, 1976, 1978; Morris, *et al.*, 1979), *Saccharomyces cerevisiae* (Water & Kleinsmith, 1976; Baum, Thorner & Honig, 1978) and *Dictyostelium discoideum* (Cappuccinelli, Marinotti & Hames, 1978). Flagellar tubulins have been characterised from *Chlamydomonas reinhardii* (Piperno & Luck, 1976, 1977) and from *Naegleria gruberi* (Kowit & Fulton, 1974) and regulation of flagellar microtubule protein synthesis has been studied in these organisms (Weeks & Collis, 1976, 1979, Weeks, Collis & Gealt, 1977, Lefebvre *et al.*, 1978).

The most thoroughly characterised of the microbial eukaryotic cytoplasmic tubulins are those of *Aspergillus nidulans* (Sheir-Neiss *et al.*, 1976, 1978; Morris *et al.*, 1979). Two-dimensional gel electrophoresis of radioactively labelled, *A. nidulans* proteins purified by copolymerisation with porcine brain tubulin showed at least five radioactive proteins in the tubulin region of the gel, which were greatly enriched by the copolymerisation procedure (Sheir-Neiss *et al.*, 1978). Three radioactive spots (α1, α2, and α3) were seen to be at or near the position of porcine brain α-tubulin and two other radioactive spots (β1 and β2) were seen at the region of porcine β-tubulin. The *A. nidulans* protein spots in the tubulin regions were re-identified on two-dimensional gels of non-copolymerised total *A. nidulans* proteins. In these experiments the protein spots in the tubulin region were better resolved than in the copolymerisation experiments and it could be seen that the α1-tubulin spot had an acidic satellite spot and the β1-tubulin had both acidic and basic satellite spots. Thus *A. nidulans* appears to have eight electrophoretically separable tubulins, four α-tubulins and four β-tubulins. The major *A. nidulans* α- and β-tubulins (α1 and β1 and β2), when codigested with brain α- and β-tubulins by *Staphylococcus aureus* protease, gave essentially identical peptide maps (Sheir-Neiss *et al.*, 1978) indicating that the distribution of acidic residues is conserved in *A. nidulans* and brain tubulins.

MITOTIC AND CELL DIVISION MUTANTS

The use of mutants of simple organisms to investigate the biochemistry of cellular processes is a well-known and generally productive technique. The virtue of this approach is that it is both systematic and comprehensive. In principle, it should be possible to find mutants altered or defective in every significant enzyme or other macromolecule involved in each of the biochemical events necessary for mitosis and cell division. Because mitosis is a vital cellular function, the mitotic mutants of *Aspergillus nidulans*, of *Saccharomyces cerevisiae* and of other microbial eukaryotes have necessarily been isolated as conditional lethals. Most are temperature-sensitive (ts) mutants, but cold-sensitive (cs) mutants have also been isolated in *S. cerevisiae* (Moir *et al.*, 1978) and *A. nidulans* (C. F. Roberts, personal communication).

Temperature-sensitive mitotic mutants have generally been isolated by direct visual identification. For example, the ts mitotic mutants of *Aspergillus nidulans* were identified by screening under the light microscope about 1000 unselected, conditional lethal mutant strains blocked at the restrictive temperature (42 °C). Each of the ts mutants was fixed, stained with aceto-orcein to demonstrate nuclear chromatin, and examined for abnormalities of nuclear structure, nuclear division, nuclear distribution or septation (Morris, 1976*a*). In *A. nidulans* 45 mitotic mutants were found and assigned to 38 complementation groups (Table 1). Nine of these in six complementation groups were blocked during mitosis. These mutants, which were assigned the gene symbol *bim* for *blocked in mitosis*, could be divided on morphological grounds into three subclasses blocked in early, medial or late phases of nuclear division (see p. 51). In addition to the nine ts *bim* mutants isolated in my laboratory, a tenth was isolated by Orr & Rosenberger (1976*a*) and a cs *bim* mutant has been isolated by C. F. Roberts (personal communication). Several *ts ben*A (β-tubulin) mutants also exhibit a *bim*-like phenotype at the restrictive temperature (Morris *et al.*, 1979; B. Oakley

& N. R. Morris, unpublished; see p. 59). Twenty-six mutants in 23 complementation groups, which were unable to enter mitosis at the restrictive temperature, were assigned the gene symbol *nim* for *never in mitosis*. Five mutants in four complementation groups failed to septate at 42 °C and were given the gene symbol *sep* for *sep*tationless; and five mutants in five complementation groups exhibited an abnormal distribution of nuclei along the mycelium and were given the gene symbol

Table 1. *Comparison of* Aspergillus nidulans *mitotic mutants with* Saccharomyces cerevisiae *cell division cycle mutants*

Phenotype at restrictive temperature	A. nidulans[a]	S. cerevisiae
Blocked in interphase	nimA-W	cdc 25, 28, 33, 35, tra3 (start) 1?, 22?
Blocked in early nuclear division	bimC	cdc 4
Blocked in medial nuclear division	bimA	cdc 2[b], 6[b], 7[b], 8[b], 9[b], 13 16, 17, 20, 21[b], 23, 40[b](?)
Blocked in late nuclear division stage I	bimB (III), bimF (II)	cdc 5, 14, 15
Blocked in late nuclear division stage II	bimD (IV)	
Slow nuclear division	bimE (VI)	
Altered α-tubulin	tubA (VIII)	
Altered β-tubulin	benA (VIII)	
Unable to form septa	sepA-D	cdc 3, 10, 11, 12, 24
Blocked in nuclear movement	nudA-E	

[a]Roman numerals in parentheses designate linkage groups. [b]These strains have primary defects in DNA synthesis at restrictive temperature.

nud for abnormal *nu*clear *d*istribution. All of these mutants were recessive for temperature sensitivity when tested in diploids and most were in different complementation groups, indicating that the *A. nidulans* mutant collection is not yet complete. There was no evidence for genetic clustering of mutations with similar phenotypes and mutations were identified in all eight linkage groups.

The cell division cycle (cdc) mutants of *Saccharomyces cerevisiae* comprise another important group of ts mitotic mutants (see reviews by Hartwell, 1978; Pringle, 1978; Simchen, 1978). In *S. cerevisiae*, 148 cdc mutants have been assigned to 32 different complementation groups (Table 1). More than 70 mutants in 16 different complementation groups exhibited a *bim*-like block in nuclear division at the restrictive temperature. These have been subclassified by ultrastructural studies into early, medial and late nuclear division mutants (Byers & Goetsch, 1974). Seventeen mutants in seven complementation groups exhibited an apparent *nim*-like block, the most interesting ones being blocked at a point early in the cycle designated as 'start'. Twenty-seven mutants in five complementation groups were blocked in cytokinesis and were unable to complete cell division. Comparison of the *S. cerevisiae* cdc collection with the *Aspergillus nidulans* collection (Table 1) shows several interesting differences. The most noticeable difference is that in the yeast collection there are many fewer genes which cause a *nim*-like block and many more genes which cause a *bim*-like block. There are also no nuclear distribution (*nud*-like) mutants in the yeast collection. These differences may be due to the different methods used to identify mutants in yeast and *A. nidulans*, since the yeast mutants were mostly identified by their termination phenotypes (final blocked morphologies) whereas the *A. nidulans* mutants were identified by a variety of criteria, some independent of termination phenotype (Morris, 1976a). Cell division cycle mutants with phenotypes similar to most of those discussed above have also been characterised in *Schizosaccharomyces pombe* (Lederberg & Stetten, 1970; Nurse *et al.*, 1976), *Chlamydomonas reinhardii* (Warr & Durber, 1971; Howell & Naliboff, 1973; Sato, 1976), *Paramecium tetraaurelia* (Peterson & Berger, 1976), *Tetrahymena pyriformis* (Frankel *et al.*, 1976a, b) and *Physarum polycephalum* (Gingold *et al.*, 1976; Wheals *et al.*, 1976).

Two mutations of *Schizosaccharomyces pombe* are unique and merit special attention. These mutations, *wee* 1 and *wee* 2 regulate the timing of mitosis in relation to cell size (Nurse, 1975; Thuriaux *et al.*, 1978), and cells carrying *wee* 1 or *wee* 2 are phenotypically smaller than the wild-type. The *wee* 1 gene product appears to be a negative control element, since loss of the *wee* 1 function in temperature-sensitive or nonsense mutations induces premature mitosis at an unusually small cell mass. Thus *wee* 1 in the wild type must function to delay mitosis. Mutations at the *wee* 2 locus also cause premature mitosis, but in contrast to *wee* 1, *wee* 2 appears to be a positive control element required for initiation of mitosis. The evidence for this is that *wee* 2 is allelic to a ts mitotic mutant, *cdc*-2, which at restrictive temperature fails to enter mitosis. Thus the *wee*-2/*cdc*-2 gene product must be required for mitosis. These mutants of *S. pombe* promise to be important for analysing the relationship between growth rate, control of cell size and the mitotic trigger.

Ultrastructure of normal and blocked mitosis

Ultrastructural analysis of blocked mitosis in mitotic mutants can help to define the normal functions of the products of mutant genes. The ultrastructure of mitosis has been studied in normal and mutant strains of both *Aspergillus nidulans* (Robinow & Caten, 1969; B. Oakley &

N. R. Morris, unpublished) and *Saccharomyces cerevisiae* (Robinow & Marak, 1966; Moens & Rapport, 1971; Peterson & Ris, 1976; Byers & Goetsch, 1974, 1975) and is generally similar in both organisms (Table 1). In this paper, I will mostly discuss the ultrastructure of mitosis in *A. nidulans*. To solve the technical problem of finding normal mitotic figures (only 3–4% of nuclei are in mitosis in *A. nidulans* under ordinary growth conditions) a mutant, *nim*A 5 (ts333) which is blocked just before mitosis was used. This mutant, if held at restrictive temperature for one generation time to accumulate premitotic nuclei, gives a nearly synchronous burst of mitotic figures about 5 min after a shift-down from restrictive to permissive temperature, allowing mitotic nuclei to be easily located in material sectioned for the electron microscope. Presumably the mitotic figures which appear after release of the block are normal. This is supported by ultrastructural studies which show that the synchronised mitotic nuclei are morphologically very similar to unsynchronised mitotic nuclei of *A. nidulans* and also to mitotic nuclei of *S. cerevisiae* and other closely related fungi (Kubai, 1975; Fuller, 1976; Heath, 1978). As in other ascomycetes, the nuclear membrane of *A. nidulans* remains intact during mitosis. The mitotic spindle when fully formed consists of about 50 microtubules terminating at either end in a flattened electron-dense, disc-like spindle pole body. There is a central spindle consisting of interpolar microtubules and more peripherally a group of kinetochore microtubules, eight to each pole body. Since *A. nidulans* has eight linkage groups, this suggests that there is one kinetochore microtubule for each chromosome. A numerical equivalence between kinetochore microtubules and chromosomes has also been demonstrated in *S. cerevisiae* (Peterson & Ris, 1976). In fixed and stained mycelia of *A. nidulans*, the chromosomes appear condensed at mitosis under the light microscope, but not under the electron microscope, even when using conditions of fixation and staining that are optimal for demonstrating chromosome condensation in other fungi. In *S. cerevisiae* condensed chromosomes have not been visualised by either light or electron microscopy (Robinow & Marak, 1966). The conspicuous failure to visualise condensed chromosomes by electron microscopy in these organisms raises the possibility that some feature of chromosome organisation responsible for the usual condensed appearance of chromosomes in higher eukaryotes may be different in the chromosomes of *A. nidulans* and *S. cerevisiae* (see p. 46).

The sequence of morphological events during mitosis in *Aspergillus nidulans* is as follows. The spindle pole bodies must replicate about one-half way through the cell cycle, since in one-half of randomly selected wild-type nuclei the pole body was single. After replication, the spindle pole bodies remain joined by a bridge until the beginning of mitosis when they separate and become decorated with short microtubules which interact to form a short nuclear spindle. The spindle then lengthens, the pole bodies move apart, and at approximately the same time the kinetochore microtubules shorten, pulling the chromosomes to the poles. Finally the elongated nucleus splits into two daughter nuclei. The spindle then disappears leaving each daughter nucleus with a single pole body.

Because they are blocked during mitosis the *bim* mutants of *Aspergillus nidulans* and the *bim*-like mutants of *Saccharomyces cerevisiae* are the most likely to give information about the biochemical mechanics of mitosis. The *bim* gene of *A. nidulans* which functions earliest in mitosis is *bim*C, (*bim*C4, *ts*244). At the restrictive temperature the spindle pole bodies of *bim*C4 separate and become decorated with a few short microtubules; however, the microtubules appear not to interact and the

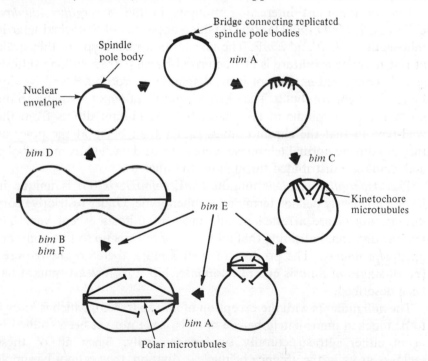

Fig. 1. Terminal phenotypes of *bim*A–F and *nim*A mutants of *Aspergillus nidulans* at restrictive temperature as determined from light and electron microscopy. The steps at which various mutations block the nuclear division cycle are indicated. The terminal phenotypes displayed behind the *bim* C block (short non-interacting microtubules) and the *bim* D block (an extremely elongated nucleus without kinetochore microtubules) have not been observed in wild-type nuclei undergoing mitosis.

short nuclear spindle fails to form (Fig. 1). *Bim*C4 resembles the *cdc*4 mutation of *S. cerevisiae* in its blocked morphology (Byers & Goetsch, 1974, 1975). The *A. nidulans* gene next in order of function after *bim*C is *bim*A (*bim*A1, *ts*69) which at restrictive temperature generates a short, well-formed nuclear spindle morphologically similar to the wild-type spindle of early mitosis, but fails to progress to the long nuclear spindle stage. *Bim*A is morphologically similar to the medial nuclear division mutants of *S. cerevisiae*, many of which have a primary block in DNA synthesis (see Table 1) (Byers & Goetsch, 1974). Whether or not *bim*A is blocked in DNA synthesis is not yet known. Two mutants, *bim*B (*bim*B2, *ts*136) and *bim* F (*bim* F8, *ts*967) are blocked somewhat later in nuclear division. In these mutants elongation of the nuclear spindle occurs at restrictive temperature, but the kinetochore microtubules fail to shorten and the extremely long spindle seen in the wild-type at late telophase is not made (Fig. 1). *Bim*B and *bim*F are classified as late nuclear division mutants.

One of the most interesting mutants in the *Aspergillus nidulans* collection is *bim*D (*bim*D5, *ts*537) which appears to be blocked later in mitosis than *bim*B and *bim*F. The termination phenotype of this strain at restrictive temperature is characterised by an extremely long mitotic spindle composed entirely of central spindle microtubules and lacking kinetochore microtubules. The blocked spindle of *bim*D is similar to the normal mitotic spindle at telophase, but the mutant differs from the wild-type in that the chromosomes, rather than being at the poles as they are during normal telophase, appear to be detached from the poles and randomly distributed throughout the nucleus.

The *Aspergillus nidulans* mutant *bim*E (*bim*E7, *ts*706) is unique in having no clearly defined termination phenotype. Under restrictive conditions, nuclei are arrested at all stages of mitosis except very late nuclear division, indicating that all stages of mitosis up to this point are greatly prolonged. The product of *bim*E7 must, therefore, be required for all stages of mitosis except telophase. No similar yeast mutant has been described.

The *nim* mutants with the exception of *nim*A5 (*ts*333), which is known to be blocked immediately before mitosis, have not yet been studied in detail either ultrastructurally or biochemically. Since all of these mutants grow in the absence of nuclear division, they cannot be simple metabolic mutants or RNA- or protein-synthesis defective mutants, but must be defective in biochemical reactions that are specific to or prerequisite to mitosis.

The *sep* mutants of *Aspergillus nidulans* have not been studied except for *sep*A2 (Trinci & Morris, 1979). *Sep*A2 fails to septate at restrictive temperature. However, when a blocked culture of *sep*A2 is downshifted to permissive temperature, septation of the 'old' mycelium occurs. The resulting septa are formed with a normal distribution and at the correct time in the duplication cycle. Thus, although septation *per se* does not occur, septation sites are produced and distributed normally under restrictive conditions in this mutant (Trinci & Morris, 1979). Whether a signal for septation is not given or not properly received under these conditions or whether the biochemical machinery for building a septum is itself temperature sensitive is not known. However, one might expect to find, among the *sep* mutants, examples of all of these possible mechanisms as well as mutants which prevent replication or distribution of the septation site. The *sep* mutants of *A. nidulans* are formally analogous to the cytokinesis mutants of *Saccharomyces cerevisiae* and other microbial eukaryotes. An ultrastructural abnormality related to defective cytokinesis has recently been described in yeast. In wild-type *S. cerevisiae* a ring of membrane-associated 10-nm filaments is found in the wild type encircling the base of the bud (Byers & Goetsch, 1976*a*). That this ring is important to cytokinesis is indicated by the fact that in four different cytokinesis mutants (*cdc* 3, 10, 11, and 12) the ring does not form at restrictive temperature (Byers and Goetsch, 1976*b*).

The *nud* mutants of *Aspergillus nidulans* are to my knowledge unique among eukaryotic cell cycle mutants. Although nuclear migration has been studied in other microbial eukaryotes, only in *A. nidulans* have mutants affecting this process been isolated. The existence of a special mechanism for the distribution of nuclei along the *A. nidulans* mycelium was first postulated by Rosenberger & Kessel (1968), based on autoradiographic studies of DNA distribution among nuclei during spore germination. The *nud* mutants directly confirm the existence of a special mechanism for nuclear movement in *A. nidulans*. When conidiospores (which are uninucleate) are germinated at restrictive temperature, the spore nucleus undergoes division. In the wild type, the daughter nuclei migrate into the germ tube, but in *nud* mutants nuclear migration does not occur. The block is reversible, since when the germlings are returned to permissive temperature the nuclei redistribute to give an essentially normal distribution pattern along the mycelium.

Nothing is yet known about the biochemical defects in any of the *nud* mutants. There is, however, evidence which demonstrates that functioning microtubules are required for nuclear distribution in *Aspergillus nidulans* (B. Oakley & N. R. Morris, unpublished). The antimicrotubule drug benomyl, which is known to bind the β-subunit of tubulin (see p. 60), blocks not only nuclear division but also nuclear migration in *A. nidulans*. This represents a specific effect of benomyl on tubulin rather than on some other process, since benomyl resistant (*ben*A) mutants, resistant because of a mutation in β-tubulin, exhibit normal nuclear distribution (and nuclear division) in the presence of benomyl. We also know that *ts ben*A mutations block both nuclear division and nuclear migration. Thus we can confidently state that β-tubulin is involved in both nuclear migration and nuclear division in *A. nidulans*. Antimicrotubule drugs have been shown to inhibit nuclear migration in other microbial eukaryotes (Aronson, 1971; Kiermayer & Heppler, 1970; Kiermayer, 1972; Kiermayer & Fedtke, 1977); therefore, the involvement of microtubules in nuclear migration may be a general phenomenon in these organisms. The existence in *A. nidulans bim* and *nud* mutations, affecting respectively, nuclear division and nuclear distribution, demonstrates that the microtubules involved in nuclear division and migration are controlled independently of each other.

As a guide to future biochemical investigations of the *Aspergillus nidulans bim* genes, it is useful to ask what types of biochemical lesions might cause the observed phenotypes of these mutations. The earliest acting of the *bim* mutations, *bim*C, allows spindle pole body separation at restrictive temperature but *bim*C makes only a few short microtubules and fails to generate a spindle. The terminal phenotype of this strain suggests that the *bim*C gene product may be required for microtubule polymerisation. The protein requirements for polymerisation *in vitro* (of mammalian brain microtubules) are the α, β-dimer of tubulin and one or more microtubule-associated proteins (MAPs) (Snyder & McIntosh, 1976). If the *bim*C mutant produced either a mutant tubulin subunit, an improperly modified tubulin subunit (see Cleveland, Kirschner & Cowan, 1978) or a defective MAP, the observed phenotype might result. It is, however, unlikely that *bim*C affects the primary structure of tubulin since *bim*C maps on linkage group II whereas *tub*A and *ben*A, which are the major structural genes for α- and β-tubulins map on linkage group VIII (see below). The next mutant, *bim*A, is able to form a short mitotic spindle but at restrictive temperature is unable to undergo normal spindle elongation. Although the mechanism of spindle elongation is not known at the biochemical level in any organism, recent studies of the highly organised mitotic spindles of diatoms have demonstrated that antipolar sliding of interdigitated microtubules plays a role in spindle elongation (McDonald *et al.*, 1977; Pickett-Heaps & Tippit, 1978). It has been suggested that this is caused by a force generating interaction between adjacent microtubules similar to that responsible for the movement of flagellar axonemes in which flagellar movement results from ATP-dependent cyclic movement of dynein cross-bridges between adjacent microtubules (Gibbons & Fronk, 1979; Gibbons & Gibbons, 1979; see Holwill, this volume). There has been some support for this model by recent experiments demonstrating dynein immunofluorescence in the mitotic spindle, mitotic inhibition by antidynein antibody, and mitotic inhibition by vanadate which inhibits dynein ATPase (Sakai *et al.*, 1976; McIntosh, Cande & Snyder, 1975; Cande & Wolniak 1978; Mohri *et al.*, 1976). However, convincing evidence for the involvement of dynein in mitosis has not yet been presented. If a dynein–tubulin system or some other system which relies on a microtubule associated protein for force generation is involved in mitosis, a short-spindle termination phenotype such as that of *bim*A could represent a defect in such a system. For example, *bim*A might code for a defective dynein-like protein or for a tubulin with a defective dynein binding site. However, since neither *tub*A nor *ben*A is located on chromosome I where *bim*A maps, the latter possibility is unlikely.

*Bim*B and *bim*F are clearly not defective with regard to the initial steps of spindle elongation, but appear to be blocked later in their ability to bring the chromosomes to the poles of the spindle. There is abundant evidence that chromosomal spindle fibres (kinetochore fibres) are responsible for pulling chromosomes to the poles in many eukaryotes (Nicklas, 1971) and a similar mechanism may be operative here, since the kinetochore microtubules of *Aspergillus nidulans* shorten at anaphase (B. Oakley & N. R. Morris, unpublished). Two speculative hypotheses proposed to explain chromosome movement to the poles are: (i) that an actomyosin-dependent force generating system moves the chromosomes poleward (Sanger, 1975; Forer, 1976; Cande, Lazarides & McIntosh, 1977); and (ii) that selective depolymerisation of kinetochore microtubules generates the poleward force (Inoué, 1976). Thus abnormal actin or myosin-like proteins, abnormalities in tubulin or abnormal MAPs might be looked for in *bim*B and *bim*F. Once again, because *bim*B and *bim*F map on chromosomes III and II respectively, whereas the known genes for α- and β-tubulin map on VIII (see below) it is not likely that the defects in these mutants involve the primary structure of tubulin.

The termination phenotype of *bim*D, which lacks kinetochore microtubules, suggests that *bim*D may be a gene for some essential component of the kinetochore such that at the restrictive temperature kinetochore microtubules fail to be nucleated. Alternatively, *bim*D could code defectively for a structural component found in the kineto-chore microtubules, but not in the microtubules of the central spindle. A model for this situation exists in *Polytomella agilis* in which flagellar and cytoplasmic microtubules have been shown to have different α-tubulins (McKeithan & Rosenbaum, 1978). Sea urchin mitotic and ciliary tubulins also exhibit tubulin microheterogeneity (Bibring *et al.*, 1976). Since *bim*D maps on chromosome IV and the genes for α- and β-tubulins map on VIII (see below) the defective component, however, is probably not tubulin.

Identification of gene products using two-dimensional gel electrophoresis

Perhaps the most difficult problem faced by investigators studying the genetics of mitosis and/or the cell cycle is how to get beyond genetic and morphological description to biochemical experimentation. How can the products of the genes involved in mitosis be identified since there are no available assays for the functions of these genes? The approach we have taken in our laboratory is as direct as possible. We have used the two-dimensional gel electrophoresis systems of O'Farrell and colleagues (O'Farrell, 1975; O'Farrell, Goodman & O'Farrell, 1977) to look for the mutant gene products.

Most temperature-sensitive mutations are missense mutations in which the consequence of an abnormal amino acid substitution is instability of the functional gene product at restrictive temperature. A wide range of missense mutations should be detectable by two-dimensional gel electrophoresis. Amino acid substitutions that alter amino acid charge will be detected as changes in isoelectric point; those that alter protein configuration will cause changes in electrophoretic mobility, and substitutions that increase the susceptibility of proteins to proteolytic degradation will either alter molecular weight, and therefore mobility, or cause disappearance of protein spots from the gel. The usefulness of two-dimensional gels is, however, limited by the following considerations. We assume that the mutant gene products are proteins, that they are present in sufficiently large amounts to be easily detectable as spots on gels and that they are not obscured by other proteins at the same location on the gels. The original O'Farrell gel system displays only acidic proteins, therefore, a different gel system is required for detection of basic proteins (O'Farrell *et al.*, 1977) and no single system provides a general screen for all proteins. There are also some biological problems. The genetic code dictates that only about one-third of amino acid substitutions can be expected to result in a change in amino acid charge, therefore, many, if not most, nonsense mutations will not be detected after electrophoresis. Another problem is how to determine whether an altered protein seen on a gel represents the gene product of the mutation in question or reflects an alteration in post translational modification. Finally there is the logistical problem of finding a single altered mutant protein among the 1000 or more proteins separated by the two-dimensional gels. In order to determine whether, despite these potential difficulties, two-dimensional gels could be used to identify and characterise mutant gene products in *Aspergillus nidulans*, we have studied a series of *ben*A (benomyl and/or thiabendazole resistant) mutants. These mutants were chosen for study because benomyl binding experiments (see below) indicated that *ben*A was probably a gene for tubulin; therefore the chances of finding a mutant gene product by two-dimensional gel electrophoresis seemed good. This approach which was suggested by my colleague, Geraldine Sheir-Neiss, has proven to be highly productive.

Tubulin mutants

Benomyl and thiabendazole belong to a family of benzimidazole derivatives which block mitosis by inhibiting microtubule polymerisation. (Davidse, 1977; Davidse & Flach, 1978; Hoebke, van Nijen & de Brabander, 1976). A series of mutants of *Aspergillus nidulans* resistant to benomyl and/or thiabendazole has been isolated and characterised genetically by van Tuyl (1977). Most of these mutants mapped to the *ben*A locus in linkage group VIII; however, two other genes, *ben*B and *ben*C, were identified in other linkage groups. The first indication that some of these mutants might have abnormal tubulins came from studies of the binding of radioactively-labelled methyl benzimidazole carbamate (MBC), the active metabolite of benomyl, to the proteins of *ben*A mutants (Davidse & Flach, 1977). In cell-free extracts of wild-type *A. nidulans*, MBC bound to a protein with a molecular weight of 110 000; presumably the tubulin dimer. In extracts prepared from *ben*A mutants the binding affinity for MBC was inversely correlated with benomyl-sensitivity. The binding affinity for MBC was low in a benomyl resistant mutant *ben*A15 (BEN-13, R); whereas in a benomyl supersensitive mutant, *ben*A16 (BEN-14, 186), the binding affinity was high. These findings led Davidse & Flach (1977) to conclude that *ben*A was probably a gene for tubulin, but they noted that *ben*A could equally well be a gene involved in the post-transcriptional modification of tubulin.

The *ben*A mutants of *Aspergillus nidulans* provided an ideal test of the feasibility of using two-dimensional gel electrophoresis to identify mutant gene products. Since we had previously characterised the *A. nidulans* tubulins electrophoretically (see p. 48), we could undertake a directed search for abnormal tubulins among the *ben*A mutants. We first examined the two *ben*A mutants studied by Davidse & Flach (1977). The benomyl supersensitive mutant *ben*A16 (BEN-14, 186) was normal with respect to the electrophoretic pattern of its tubulins. However, the benomyl-resistant mutant *ben*A15 (BEN-13, R) had an electrophoretically abnormal (1⁻) β-tubulin. Twenty-four additional, independently isolated *ben*A mutants were subsequently examined by two dimensional gel electrophoresis of which 17 had electrophoretic

Table 2. *Characteristics of the benA mutants of* Aspergillus nidulans

Allele	Benomyl[a]	Thiabendazole[b]	FPA[c]	Tubulin	Other
*ben*A+	S	S	S	WT	
*ben*A3	RR	S	S	WT	
*ben*A4	R	R	S	1⁻	
*ben*A5	RR	R	S	WT	
*ben*A6	R	R	S	1⁻	
*ben*A7	R	R	S	β2	
*ben*A8	R	R	S	1⁻	
*ben*A9	R	R	S	1⁻	CS
*ben*A10	R	R	S	1⁻	
*ben*A11	R	R	R	WT	TS
*ben*A12	R	R	S	1⁻	
*ben*A13	R	R	S	1⁺	
*ben*A14	R	R	S	twin β1	
*ben*A15	RR	R	S	1⁻	
*ben*A16	SS	R	S	WT	
*ben*A17	R	R	R	WT	TS, CS
*ben*A18	RR	S	S	WT	
*ben*A19	SS	R	S	WT	
*ben*A20	R	R	S	1⁺	
*ben*A21	RR	R	R	WT	TS, CS
*ben*A22	R	R	S	2⁺	
*ben*A23	R	R	S	1⁻	
*ben*A24	RR	R	S	1⁻	
*ben*A25	R	R	S	1⁻	
*ben*A26	R	R	S	1⁻	
*ben*A27	R	R	S	1⁻	
*ben*A28	R	R	S	1⁻	

[a]Wild-type *A. nidulans* is 50% inhibited with respect to colony diameter (ID_{50}) by 0.4 μg ml⁻¹ of benomyl. R signifies benomyl resistance with an ID_{50} of 10 μg ml⁻¹; RR signifies an ID_{50} greater than 10 μg ml⁻¹; SS signifies supersensitivities with ID_{50} of 0.1 μg ml⁻¹.
[b]Wild-type *A. nidulans* is 50% inhibited by 8 μg ml⁻¹ of thiabendazole. R signifies thiabendazole resistances with ID_{50} of 60–125 μg ml⁻¹.
[c]Wild-type *A. nidulans* is 50% inhibited by 90 μg ml⁻¹ of *p*-fluorophenylalanine (FPA). R, signifies FPA resistances with $ID_{50} > 200$ μg ml⁻¹.
S, wild-type sensitivity.
Modified from Sheir-Neiss *et al.* (1978). *Cell*, **15**, 639–47. © The MIT Press.

abnormalities in β-tubulin and none had any abnormality in α-tubulin Plate 1*A*, Table 2) (Sheir-Neiss *et al.*, 1978). Because of these results we concluded (i) that the benomyl binding site must be located, at least in part, on the β-subunit of tubulin, and (ii) that since benomyl-like drugs compete for the same tubulin binding site as colchicine in organisms in which both drugs bind to tubulin (Hoebke *et al.*, 1976), the colchicine-binding site in such organisms would also be found on the β-tubulin subunit. The finding of an electrophoretically altered β-tubulin in a strain of CHO cells resistant to colchicine is in agreement with this prediction (Cabral & Gottesmann, 1978).

The electrophoretic abnormalities in the β-tubulins of the *ben*A mutants are for the most part manifested as changes in isoelectric point equivalent to plus or minus one or two unit charges (Plate 1*A*, Table 2). In the absence of any additional data these changes could represent either amino acid substitutions in tubulin or charge-altering post translational modifications of tubulin. To determine whether *ben*A was a structural gene for β-tubulin or, alternatively, whether it coded for or affected the activity of a tubulin-modifying enzyme, a heterozygous *ben*A22 (2$^+$ tubulin)/wild-type diploid was constructed and subjected to two dimensional gel electrophoresis. Since genes on both chromosomes are generally expressed in *Aspergillus nidulans* diploids, the wild-type and 2$^+$ tubulin spots should be co-expressed and both would be seen on the gel if *ben*A were a structural gene, but if *ben*A coded for a diffusible, tubulin-modifying enzyme all β-tubulins would be subject to the same modifications and only one set of β-tubulin spots would be seen. The result of this experiment (Plate 1*B*) was that both wild-type and mutant (2$^+$) β-tubulins appeared in the two dimensional gels from the diploid. Thus the altered β-tubulin is not the result of a post-translational modification and *ben*A (as defined by the 2$^+$ *ben*A22 mutation) must therefore be a structural gene for β-tubulin in *A. nidulans*.

Revertants of temperature-sensitive ben*A mutations*

A number of revertants of *ben*A, including both back-mutations and indirect suppressor mutations, have been isolated and characterised in our laboratory. There were two reasons for wanting to examine *ben*A revertants; (i) to determine whether back-mutations could be used to identify mutations in *ben*A which were 'electrophoretically silent', i.e. in which the amino acid substitution had no effect on charge; (ii) to determine whether extragenic suppressor mutations could be used to identify proteins which were functionally interactive with β-tubulin (see Hartman & Roth, 1973; Jarvik & Botstein, 1975).

'Electrophoretically silent' mutations pose a significant obstacle to the identification of mutant gene products by two-dimensional gel electrophoresis. As noted above, the genetic code specifies that only about one-third of random mutations will cause charge-altering amino acid substitutions. Thus for any single mutation the chances are poor that the product of the mutant gene will be electrophoretically altered and this theoretical consideration severely limits the expected productivity of the technique. It should be possible, however, to increase the chances for successfully identifying unknown gene products on two-dimensional gels by looking not at single mutants but at a mutant strain plus a number of related strains carrying back mutations in the same gene. Even if the original mutation were 'electrophoretically silent', approximately one-third of the back mutations should have charge alterations detectable by electrophoresis. Finding the same protein electrophoretically altered in two or more related strains carrying back mutations in the same gene would greatly strengthen the identification of such a protein as the gene product of the original mutation. We have used the *ben*A system to test the effectiveness of this theoretically plausible strategy.

Three of the *ben*A mutants *ben*A11, 17, and 21 which have electrophoretically normal tubulins were found to be temperature-sensitive (ts⁻) for growth (Morris *et al.*, 1979) and incidentally, also to be moderately resistant to *p*-fluorophenylalanine (FPA) (Table 2; Morris & Oakley, 1979). From these strains a set of ts⁺ revertants was selected and tested genetically by out-crossing to wild-type (ts⁺) strains to determine whether they represented back-mutations or carried extragenic suppressor mutations (Morris *et al.*, 1979). The appearance of ts⁻ segregants among the progeny of such a cross was presumptive evidence for an extragenic suppressor mutation which segregated independently. The absence of ts⁻ segregants indicated either a back mutation or a very closely linked suppressor mutation. Four of the revertants were determined to be back-mutations and of the four, two exhibited electrophoretically altered *β*-tubulins (Table 3, Plate 2*A*). This experiment demonstrates that two-dimensional gel analysis of back-mutations can be used to identify the gene products of electrophoretically silent mutations. Since revertants of ts⁻ mitotic mutants from ts⁻ to ts⁺ can be selected with equivalent ease, there is every reason to expect that the same strategy will be generally useful in identifying the gene products of *ts⁻ bim*, *nim*, *sep* and *nud* mutations which affect mitosis and cell division.

The strategy of using suppressor mutations to identify genes for proteins which interact with β-tubulin has also been successful in its initial trial. Although four of the ts+ revertants of the ts− benA mutants had back-mutations in β-tubulin, fourteen of the ts+ revertants of benA11, 17, and 21 were found to carry indirect suppressor mutations at loci other than benA (Table 3) (Morris et al., 1979). Thirteen of these strains carrying suppressors of benA had electrophoretically normal

Table 3. *Revertants of ts benA mutants of* Aspergillus nidulans

Revertant	Strain	Benomyl resistance[a]	Temperature sensitivity[b]	Mutation[c]	Tubulin[d]	
					α-	β-
benA11		R	−		wt	wt
	R1	R	+	Su	wt	wt
	R7	R	+	Su	1+	wt
	R8	R	+	Su	wt	wt
benA17		R	−		wt	wt
	R3	R	+	Su	wt	wt
	R6	S	CS	Su	wt	wt
	R7	R	+	Su	wt	wt
	R8	R	+	Su	wt	wt
	R9	R	+	Su	wt	wt
	R10	S	+	bm	wt	1−
	R11	R	+	bm	wt	wt
	R12	R	+	Su	wt	wt
	R13	S	+	bm	wt	wt
benA21		R	−		wt	wt
	R7	R	+	bm	wt	2−
	R8	R	+	Su	wt	wt
	R9	R	+	Su	wt	wt
	R10	R	+	Su	wt	wt
	R11	S	CS	Su	wt	wt
	R12	R	+	Su	wt	wt

[a]R, resistant; S, sensitive. [b]+, wild-type; CS, cold-sensitive. [c]Su, suppressor; bm, back mutation. [d]wt, wild-type. Modified from Morris et al. (1979). *Cell*, **16**, 437–42. © The MIT Press.

tubulins, but one strain, benA11R7, was found to have an abnormal (1+) α1-tubulin (Plate 2B). The 1+ α-tubulin co-segregated perfectly with the suppressor mutation when outcrossed to a wild-type strain and was co-expressed with wild-type α1-tubulin in a heterozygous (1+/wild-type) diploid strain (Plate 2c). From these two experiments, it can be concluded that the suppressor mutation in benA11R7 identifies a structural gene for α1-tubulin in *A. nidulans*. This R7 suppressor mutation has since been designated tubA1 (for α-tubulin). *Tub* A has been mapped to chromosome VIII, but is not closely linked to benA. The identification of a mutation in a structural gene for α-tubulin among the extragenic suppressors of a temperature-sensitive β-tubulin mutation is most easily understood as a case of compensatory suppression in which a mutation affecting one protein is suppressed by a mutation which causes a compensatory alteration in a second protein with which it interacts physically (Hartman & Roth, 1973; Jarvik & Botstein, 1975). Since α- and β-tubulin presumably interact to form dimers and probably interact at additional sites during microtubule polymerisation, it is not unexpected that certain mutations in β-tubulin can be suppressed by mutations in α-tubulin. We believe firstly that the ts *ben*A mutation in *ben*A11 alters a site on β-tubulin such that at restrictive temperature one of the interactions between α- and β-tubulin is disrupted, and secondly, that the *tub*A mutation introduces a compensatory change in the structure of α-tubulin which restores the defective interaction. Only *tub*A1 among the 14 suppressors of *ben*A that we have examined by electrophoresis has an abnormal α1-tubulin. Whether the other suppressors of *ben*A represent 'electrophoretically silent' mutations in α1-tubulin or mutations in other proteins is currently being tested.

One interesting property of *tub*A1 is that it increases the sensitivity of strains carrying this mutation to griseofulvin, an antimitotic fungicide (Gull & Trinci, 1973; 1974*a*, *b*) whose mechanism of action is not well understood. For examples of apparently contradictory data concerning the mechanism by which griseofulvin inhibits microtubule polymerisation compare Roobol, Gull & Pogson (1977) with Wehland, Herzog & Weber (1977). Interestingly, the increased griseofulvin sensitivity of strains carrying *tub*A1 is suppressed by *ben*A11. These data indicate that griseofulvin inhibits growth by an effect on the interaction between α- and β-tubulin. Since *tub*A increases sensitivity to griseofulvin, it has been possible to select revertants to griseofulvin resistance from *tub*A1, among which we expect to find both back-mutations in *tub*A and suppressor mutations in other genes. Analysis of the suppressors of *tub*A1 should be useful in determining the molecular mechanism of action of griseofulvin.

That antimitotic drugs could be used to generate mutants with altered mitotic proteins has been recognised for some time (see Warr, 1974). On the basis of this paradigm, Lederberg & Stetten (1970) isolated colcemid-resistant mutants of *Schizosaccharomyces pombe*; Warr's and other groups have isolated colchicine-resistant, colchicine-sensitive, colchicine-dependent and vinblastine-resistant mutants of *Chlamydomonas reinhardii* (Adams & Warr, 1971; Flavin & Slaughter, 1974; Warr *et al.*, 1978), and Sato (1976) has isolated an interesting colchicine-resistant mutant of *C. reinhardii* which is also temperature-sensitive for cell division. In addition to these mutants antimitotic drug-resistant mutants have also been found in other microbial eukaryotes including *Neurospora crassa* (Borck & Braymer, 1974) and *Ustilago maydis* (van Tuyl, 1977), usually in connection with studies of resistance to antimitotic agricultural fungicides, e.g. benomyl and thiabendazole. The colcemid resistant mutants of *S. pombe* have been shown not to differ from the parental, wild-type strain with respect to either permeability or detoxification of colcemid (Lederberg, Gourse & Sackett, 1977). Similarly, colchicine uptake has been shown to be normal in most of the resistant mutants of *C. reinhardii* (Warr *et al.*, 1978). These results suggest that many of the mutants of *S. pombe* and *C. reinhardi* may have mutations affecting tubulin and that the biochemical characterisation of the microtubule proteins of these strains should be informative.

DISCUSSION

The emphasis of this chapter has been on the use of genetics as a tool to study the biochemistry of mitosis in eukaryotic micro-organisms, primarily *Aspergillus nidulans*. Purely conceptual genetic studies have been reviewed elsewhere (Johnston, 1978; Hartwell, 1978; Pringle, 1978; Simchen, 1978). Here, I have tried to present a convincing case that *A. nidulans* and other microbial eukaryotes are valid model systems in which to study mitosis; I have described the ts mitotic mutants and the tubulin mutants of *A. nidulans* in detail, and I have described a general methodology for finding the products of genes involved in mitosis. Although no gene product of any of the ts mitotic mutants has yet been identified, the potential usefulness of this approach has been confirmed by our success in identifying the *ben*A gene product as the β-subunit of tubulin and by our finding of a new gene, *tub*A, which codes for the α-subunit of tubulin. I should like to conclude this chapter by speculating about the future role of microbial eukaryotes in elucidating mitosis, and on some of the implications and potential uses of the *A. nidulans* tubulin mutants.

We can confidently predict that mutations in structural genes for α- and β-tubulins similar to the *tub*A and *ben*A mutations of *Aspergillus nidulans* will be identified in other microbial eukaryotes. The colchicine-resistant mutants of *Chlamydomonas reinhardii* (Adams & Warr, 1972; Warr & Gibbons, 1974; Warr *et al.*, 1978; Sato, 1976), the colcemid resistant mutants of *Schizosaccharomyces pombe* (Lederberg & Stetten, 1970), and the benomyl- and thiabendazole-resistant mutants of *Neurospora crassa* (Borck & Braymer, 1974), and *Ustilago maydis* (van Tuyl, 1977) all provide excellent opportunities for finding mutations in structural genes for tubulins. We can also predict with some certainty that mutants resistant to other important antimitotic drugs, e.g. vinblastine and griseofulvin, will be isolated in one or more of these organisms and that the tubulins and microtubule associated proteins (MAPs) of these mutants will be characterised. In this way, a collection of tubulin (and MAP) mutants will be developed in which the mutations define and characterise sites at which various antimitotic drugs are bound.

The next useful step will be to examine the chemistry of these drug binding sites in detail. For example, it is important to know where the benomyl binding site is on β-tubulin, what amino acid sequences are involved, and which amino acid substitutions increase and which decrease the binding affinity of tubulin for benomyl. This ought not to be a technically difficult problem to solve. The amino acids involved in benomyl binding should be locatable by a combination of genetic fine structure mapping, peptide mapping and DNA sequencing. Comparison of the amino acid sequences of the benomyl binding site in different benA mutants should then lead to a molecular map of the benomyl binding site and an understanding of how various amino acid changes and other alterations affect the binding of benomyl and its congeners. The same methodology should be applicable to the analysis of binding sites for other drugs if appropriate drug-resistant mutants become available.

One important consequence of discovering that tubulin mutants are easily produced in Aspergillus nidulans and that many different mutant tubulins apparently function quite normally in the living organism is that these findings contradict the previously held conception of tubulin as a protein which is very highly conserved during evolution. If tubulin mutations can be generated almost at will in the laboratory, then it is fair to conclude that tubulin need not be so well conserved during evolution as was once believed (Luduena & Woodward, 1975). The fact that fungal tubulins bind colchicine weakly if at all and that different benzimidazole compounds exhibit selective inhibition of tubulins from different sources supports this conclusion (Haber et al., 1972; Davidse, 1975; Davidse & Flach, 1977; Davidse, 1977). Because tubulin has been thought to be a conserved molecule it has not been seriously considered as a target for pharmacological attack, yet tubulin-containing eukaryotic micro-organisms, are a major cause of disease in plants, animals and man: malaria, leishmaniasis and trypanosomiasis are all caused by microtubule-containing parasites. Our finding that the benomyl binding site is highly mutable in Aspergillus nidulans suggests that tubulin may indeed be a very useful target for pharmacological attack and that a systematic, detailed study of the chemical anatomy of antimicrotubule drug-binding sites in microbial mutants and in different organisms may be useful in designing new and more selective anti-microtubule drugs for use against eukaryotic pathogens of crops, livestock and man. Since benomyl interferes with tubulin polymerisation, and since some of the *ben*A mutants appear to be temperature-sensitive for microtubule polymerisation, we may also expect to gain important information about the molecular biology of tubulin function by studying those sites.

What can we expect to find out about the ts mitotic and cell division mutants of *Aspergillus nidulans, Saccharomyces cerevisiae, Schizosaccharomyces pombe, Chlamydomonas reinhardii, Paramecium tetraaurelia*, and *Tetrahymena pyriformis*? In this chapter I have stressed one specific methodological approach, the use of genetics and two-dimensional gel electrophoresis to identify the protein gene products of the ts mitotic and cell division mutants and of the benomyl resistant *ben*A mutants and revertants. The theoretical advantage of this approach is that it is both systematic and general. In principle we should be able to identify many, if not all, of the protein gene products involved in mitosis and cell division and the same methodology can be used to look for the gene product of any genetically well characterised ts or non-sense mutation. The weakness of this approach is that it is designed to identify gene products but not necessarily gene functions. However, once a gene product is known, it can be located by cell fractionation or immunocytochemistry. It can be purified and characterised, compared to known proteins (e.g. actin or dynein), assayed for potentially interesting enzymatic activities and so on. Identification of the proteins involved in mitosis is a first step toward understanding how they function.

The molecular biology of mitosis and cell division is still mostly *terra incognita*. The biochemistry will undoubtedly be complicated, and it will probably take a very long time before we understand fully the molecular biology of spindle assembly, chromosome condensation, anaphase movement, cell division (septation), the controls over these processes and the mechanisms by which they are integrated with DNA and protein synthesis. The era of intense morphological investigation of mitosis has lasted 100 years. Will the molecular era be any shorter?

This work was supported by U.S.P.H.S. grant GM 23060.

REFERENCES

ADAMS, M. & WARR, J. R. (1972). Colchicine-resistant mutants of *Chlamydomonas reinhardii*. *Experimental Cell Research*, **71**, 473–5.

ADOLPH, K. W., CHENG, S. M. & LAEMMLI, U. K. (1977a). Role of nonhistone proteins in metaphase chromosome structure. *Cell*, **12**, 805–16.

ADOLPH, K. W., CHENG, S. M., PAULSON, J. R. & LAEMMLI, U. K. (1977b). Isolation of a protein scaffold from mitotic HeLa cell chromosomes. *Proceedings of the National Academy of Sciences, USA*, **74**, 4937–41.

ARONSON, J. F. (1971). Demonstration of a colcemid-sensitive attractive force between the nucleus and a center. *Journal of Cell Biology*, **51**, 579–83.

BAKKE, A. C., WU, J. R. & BONNER, J. (1978). Chromatin structure in the cellular slime mold *Dictyostelium discoideum*. *Proceedings of the National Academy of Sciences, USA*, **75**, 705–9.

BAUM, P., THORNER, J. & HONIG, L. (1978). Identification of tubulin from the yeast *Saccharomyces cerevisiae*. *Proceedings of the National Academy of Sciences, USA*, **75**, 4962–66.

BIBRING, T., BAXANDALL, J., DENSLOW, S. & WALKER, B. (1976). Heterogeneity of the alpha-subunit of tubulin and the variability of tubulin within a single organism. *Journal of Cell Biology*, **69**, 301–12.

BORCK, K. & BRAYMER, H. D. (1974). The genetic analysis of resistance to benomyl in *Neurospora crassa*. *Journal of General Microbiology*, **85**, 51–6.

BRADBURY, E. M. (1975). Histones, chromatin structure, and control of cell division. *Current Topics in Developmental Biology*, **9**, 1–13.

BRADBURY, E. M., INGLIS, R. J., MATHEWS, H. R. & SARNER, N. (1973). Phosphorylation of very-lysine-rich histone in *Physarum polycephalum*: correlation with chromosome condensation. *European Journal of Biochemistry*, **33**, 131–9.

BRUNS, P. J. & SANFORD, Y. M. (1978). Mass isolation and fertility testing of temperature-sensitive mutants in *Tetrahymena*. *Proceedings of the National Academy of Sciences, USA*, **75**, 3355–8.

BYERS, B. & GOETSCH, L. (1974). Duplication of spindle plaques and integration of the yeast cell cycle. *Cold Spring Harbor Symposia in Quantitative Biology*, **38**, 123–37.

BYERS, B. & GOETSCH, L. (1975). Behavior of spindles and spindle plaques in the cell cycle and conjugation of *Saccharomyces cerevisiae*. *Journal of Bacteriology*, **124**, 511–23.

BYERS, B. & GOETSCH, L. (1976a). A highly ordered ring of membrane-associated filaments in budding yeast. *Journal of Cell Biology*, **69**, 717–21.

BYERS, B. & GOETSCH, L. (1976b). Loss of the filamentous ring in cytokinesis defective mutants of the budding yeast. *Journal of Cell Biology*, **70**, 35a.

CABRAL, F. & GOTTESMANN, M. M. (1978). An alteration in β-tubulin of CHO cells associated with resistance to colcemid or griseofulvin. *Journal of Cell Biology*, **79**, 307a.

CANDE, W. Z., LAZARIDES, E. & MCINTOSH, J. R. (1977). A comparison of the distribution of actin and tubulin in the mammalian mitotic spindle as seen by indirect immuno-fluorescence. *Journal of Cell Biology*, **72**, 532–67.

CANDE, W. Z. & WOLNIAK, S. M. (1978). Chromosome movement in lysed mitotic cells is inhibited by vanadate. *Journal of Cell Biology*, **79**, 573–80.

CAPPUCCINELLI, P., MARINOTTI, G. M. & HAMES, B. D. (1978). Identification of cytoplasmic tubulin in *Dicteostelium discoideum. FEBS Letters*, **91**, 153–7.

CHAMBON, P. (1977). Summary: The molecular biology of the eukariotic genome is coming of age. *Cold Spring Harbor Symposia on Quantitative Biology*, **42**, 1209–34.

CHARLESWORTH, M. C. & PARISH, R. W. (1975). Further studies on basic nucleo-proteins from the cellular slime mold *Dictyostelium discoideum. European Journal Biochemistry*, **75**, 241–50.

CLEVELAND, D. W., KIRSCHNER, M. W. & COWAN, N. J. (1978). Isolation of separate mRNAs for α- and β-tubulin and characterization of the corresponding *in vitro* translation products. *Cell*, **15**, 1021–31.

COMPTON, J. L., BELLARD, M. & CHAMBON, P. (1976). Biochemical evidence of variability in the DNA repeat length in the chromatin of higher eukariotes. *Proceedings of the National Academy of Sciences, USA*, **73**, 4382–6.

COMPTON, J. L., HANCOCK, R., OUDET, P. & CHAMBON, P. (1976). Biochemical and electron microscopic evidence that the subunit structure of chinese-hamster-ovary interphase chromatin is conserved in mitotic chromosomes. *European Journal of Biochemistry*, **70**, 555–68.

DAVIDSE, L. C. (1975). Antimitotic activity of methyl benzimidazol-2-yl carbamate (MBC) in fungi and its binding to cellular protein. In *Microtubules and Micro-tubule Inhibitors*, ed. M. Bergers & M. de Brabander, pp. 483–95. Amsterdam: North Holland.

DAVIDSE, L. C. (1977). Mode of action, selectivity and mutagenicity of benzimida-zole compounds. *Netherlands Journal of Plant Pathology* (suppl. 1), 135–44.

DAVIDSE, L. C. & FLACH, W. (1977). Differential binding of methyl benzimidazole-2-yl carbamate to fungal tubulin as a mechanism of resistance to this antimitotic agent in mutant strains of *Aspergillus nidulans. Journal of Cell Biology*, **72**, 174–93.

DAVIDSE, L. C. & FLACH, W. (1978). Interaction of thiabendazole with fungal tubulin. *Biochimica et Biophysica Acta*, **543**, 82–90.

ERMINI, M. & KUENZLE, C. C. (1978). The chromatin repeat length of cortical neurons shortens during early post natal development. *FEBS Letters*, **90**, 167–72.

FELDEN, R. A., SANDERS, M. M. & MORRIS, N. R. (1976). Presence of histones in *Aspergillus nidulans. Journal of Cell Biology*, **68**, 430–9.

FINCH, J. T. & KLUG, A. (1976). Solenoidal model for superstructure in chromatin. *Proceedings of the National Academy of Sciences, USA*, **73**, 1897–1901.

FLAVIN, M. & SLAUGHTER, C. (1974). Microbial assembly and function in *Chlamy-domonas* inhibition of growth and flagellar regeneration by antitubulins and other drugs and isolation of resistant mutants. *Journal of Bacteriology*, **118**, 59–69.

FLEMMING, W. (1878, 1880). Beitrage zur Kenntniss der Zelle und ihrer hebenser-scheinungen, Theil I Archiv. Eur. Mikroskopische Anatomie, **16**, 302–436; **18**, 151–259. Reprinted in English in *Journal of Cell Biology*, **25** (suppl. 1), 1–69.

FORER, A. (1976). Actin filaments and birefringent spindle fibers during chromo-some movements. In *Cell Motility, Cold Spring Harbor Conference on Cell Proliferation 3*, pp. 1273–94. New York: Cold Spring Harbor Laboratory.

FRANKEL, J., JENKINS, L. M. & DEBAULT, L. E. (1976a). Causal relations among cell cycle processes in *Tetrahymena pyriformis*: an analysis employing temperature-sensitive mutants. *Journal of Cell Biology*, **71**, 242–60.

FRANKEL, J., JENKINS, L. M., DOERDER, F. P. & NELSEN, E. M. (1976b). Mutations affecting cell division in *Tetrahymena pyriformis*. I. Genetic analysis. *Genetics*, **83**, 489–506.

FULLER, M. S. (1976). Mitosis in fungi. *International Review of Cytology*, **45**, 113–55.

GEALT, M. A., SHEIR-NEISS, G. & MORRIS, N. R. (1975). The isolation of nuclei from the filamentous fungus *Aspergillus nidulans*. *Biochemical and Biophysical Research Communications*, **69**, 285–90.

GIBBONS, I. R. & FRONK, E. (1979). A latent adenosine triphosphatase form of dynein I from sea urchin sperm flagella. *Journal of Cell Biology*, **254**, 187–96.

GIBBONS, B. H. & GIBBONS, I. R. (1979). Relationship between the latent adenosine triphosphatase state of dynein I and its ability to recombine functionally with KCl-extracted sea urchin sperm flagella. *Journal of Biological Chemistry*, **254**, 197–201.

GINGOLD, E. C., GRANT, W. D., WHEALS, A. E. & WREN, M. (1976). Temperature sensitive mutants of the slime mold *Physarum polycephalum* II. Mutants of the plasmodial phase. *Molecular and General Genetics*, **149**, 115–19.

GLOVER, C. & GOROVSKY, M. A. (1979). Amino-acid sequence of *Tetrahymena* histone H4 differs from that of higher eukaryotes. *Proceedings of the National Academy of Sciences, USA*, **76**, 585–9.

GOFF, B. (1976). The histones of *Neurospora crassa*. *Journal of Biological Chemistry*, **251**, 4131–8.

GOROVSKY, M. A., GLOVER, C., JOHMANN, C. A., KEEVART, J. B., MATHIS, D. J. & SAMUELSON, M. (1977). Histones and chromatin structure in *Tetrahymena* macro- and micronuclei. *Cold Spring Harbor Symposia on Quantitative Biology*, **42**, 493–503.

GULL, K. & TRINCI, A. P. J. (1973). Griseofulvin inhibits fungal mitosis. *Nature, London*, **244**, 292–4.

GULL, K. & TRINCI, A. P. J. (1974a). Effects of griseofulvin on the mitotic cycle of the fungus *Basidiobolus ranarum*. *Archive für Microbiologie*, **95**, 57–65.

GULL, K. & TRINCI, A. P. J. (1974b). Ultrastructural effects of griseofulvin on the myxomycete *Physarum polycephalum*. *Protoplasma*, **81**, 37–48.

GURLEY, L. R., D'ANNA, J. A., BARHAM, S. S., DEAVEN, L. L. & TOBEY, R. A. (1978). Histone phosphorylation and chromatin structure during mitosis in chinese hamster cells. *European Journal of Biochemistry*, **35**, 1–15.

HABER, J. E., PELOQUIN, J. G., HALVORSON, H. O. & BORISY, G. G. (1972). Colcemid inhibition of cell growth and the characterization of a colcemid-binding activity in *Saccharomyces cerevisiae*. *Journal of Cell Biology*, **55**, 355–67.

HARTMAN, P. E. & ROTH, J. R. (1973). Mechanisms of suppression. *Advances in Genetics*, **17**, 1–105.

HARTWELL, L. H. (1974). *Saccharomyces cerevisiae* cell cycle. *Bacteriological Reviews*, **38**, 164–98.

HARTWELL, L. H. (1978). Cell division from a genetic perspective. *Journal of Cell Biology*, **77**, 627–37.

HARTWELL, L. H., CULOTTI, J. & REID, B. (1970). Genetic control of the cell division cycle in yeast. I. Detection of mutants. *Proceedings of the National Academy of Sciences, USA*, **66**, 352–9.

HARTWELL, L. H., MORTIMER, R. K., CULOTTI, J. & CULOTTI, M. (1973). Genetic control of the cell division cycle in yeast. V. Genetic analysis of *cdc* mutants. *Genetics*, **74**, 267–86.

HARTWELL, L. H., CULOTTI, J., PRINGLE J. R. & REID, B. J. (1974). Genetic control of the cell division cycle in yeast. *Science*, **183**, 46–51.

HEATH, I. B. (1978). Experimental studies of mitosis in the fungi. *In Nuclear Division in the Fungi*, ed. I. B. Heath, pp. 89–176. New York: Academic Press.

HOEBEKE, J., VAN NIJEN, G. & DE BRABANDER, M. (1976). Interactions of oncodazole (R 17934), a new antitumoral drug, with rat brain tubulin. *Biochemical and Biophysical Research Communications*, **69**, 319–24.

HORGAN, P. A. & SILVER, J. C. (1978). Chromatin in eukaryotic microbes. *Annual Review of Microbiology*, **32**, 249–84.

HOWELL, S. H. & NALIBOFF, J. A. (1973). Conditional mutants in *Chlamydomonas reinhardi* blocked in the vegetative cell cycle. I. An analysis of cell cycle block points. *Journal of Cell Biology*, **57**, 760–72.

INOUÉ, S. (1976). Chromosome movement by reversible assembly of microtubules. In *Cell Motility, Cold Spring Harbor Conference on Cell Proliferation*, 3, pp. 1317–28. New York: Cold Spring Harbor Laboratory.

JARVIK, J. & BOTTSTEIN, D. (1975). Conditional lethal mutations that suppress genetic defects in morphogenesis by altering structural proteins. *Proceedings of the National Academy of Sciences, USA*, **72**, 2738–42.

JOHNSTON, L. H. (1978). *S. cerevisiae* mutant *cdc* 9 is defective in DNA ligase. 9th International conference on yeast genetics and molecular biology, 139a.

KIERMAYER, O. & HEPLER, P. K. (1970). Hemmung der Kernmigration bei Jochalgen (Micrasteriase) durch isopropyl-*N*-phenyl carbamat. *Naturwissenschaften*, **57**, 252–7.

KIERMAYER, O. (1972). Beeinflussung der postmitotischen Kernmigration von *Micrasterias denticulata* Breb. durch das Herbizid Trifluralin. *Protoplasma*, **75**, 421–6.

KIERMAYER, O. & FEDTKE, C. (1977). Strong antimicrotubule action of amiprophos-methyl (APM) in *Micrasterias. Protoplasma*, **92**, 163–6.

KORNBERG, R. D. (1975). In *The Eukaryotic Chromosome*, ed. W. J. Peccode & R. D. Brock, Canberra: ANU Press.

KORNBERG, R. D. (1977). Structure of Chromatin. *Annual Review of Biochemistry*, **46**, 931–54.

KOWIT, J. D. & FULTON, C. (1974). Purification and properties of flagellar outer doublet tubulin from *Naegleria gruberi* and a radioimmune assay for tubulin. *Journal of Biological Chemistry*, **249**, 3638–46.

KUBAI, D. F. (1975). The evolution of the mitotic spindle. *International Review of Cytology*, **43**, 167–227.

LEDERBERG, S. & STETTEN, G. (1970). Colcemid sensitivity of fission yeast and the isolation of colcomid-resistant mutants. *Science*, **168**, 485–7.

LEDERBERG, D., GOURSE, R. L. & SACKETT, D. L. (1977). Colcemid sensitivity of fission yeast: permeability and detoxification properties of resistant mutants. *Journal of Bacteriology*, **129**, 198–201.

LEFEBVRE, P. A., NORDSTROM, S. A., MOULDER, J. C. & ROSENBAUM, J. (1978). Flagellar elongation and shortening in *Chlamydomonas*. IV. Effects of flagellar detachment, regeneration and resorption on the induction of flagellar protein synthesis. *Journal of Cell Biology*, **78**, 8–27.

LIPPS, H. J. & MORRIS, N. R. (1977). Chromatin structure in the nuclei of the ciliate

Stylonichia mytillis. Biochemical and Biophysical Research Communications, **74**, 230–4.

LOHR, D. & VAN HOLDE, K. E. (1975). Yeast chromatin subunit structure. *Science*, **188**, 165–6.

LOHR, D., CORDON, J., TATCHELL, K., KOVACIC, R. T. & VAN HOLDE, K. E. (1977). Comparative subunit structure of HeLa yeast and chicken erythrocyte chromatin. *Proceedings of the National Academy of Sciences, USA*, **74**, 79–83.

LOHR, D., KOVACIC, R. T. & VAN HOLDE, K. E. (1977). Quantitative analysis of the digestion of yeast chromatin by Staphyloccal nuclease. *Biochemistry*, **16**, 463–71.

LUDUENA, R. F. & WOODWARD, D. O. (1975). α- and β-tubulin; separation and partial sequence analysis. *Annals of the New York Academy of Sciences*, **253**, 272–83.

McDONALD, K., PICKETT-HEAPS, J., McINTOSH, J. R. & TIPPIT, D. (1977). On the mechanism of anaphase spindle elongation in *Diatoma vulgare. Journal of Cell Biology*, **73**, 377–88.

McINTOSH, J. R., CANDE, W. Z. & SNYDER, J. A. (1975). Structure and physiology of the mammalian mitotic spindle. In *Molecules and Cell Movements*, ed. S. Inoué & R. E. Stephens, pp. 31–76. New York: Raven Press.

McKEITHAN. T. W. & ROSENBAUM, J. L. (1978). Tubulin heterogeneity in *Polytomella agilis. Journal of Cell Biology*, **79**, 297a.

MOENS, P. B. & RAPPORT, E. (1971). Spindles, spindle plaques and meiosis in the yeast *Saccharomyces cerevisiae* (Hansen). *Journal of Cell Biology*, **50**, 344–61.

MOHRI, H., MOHRI, T., MABUCHI, I., YAZAKI, I., SAKAI, H. & OGAWA, K. (1976). Localization of dynein in sea urchin eggs during cleavage. *Development, Growth and Differentiation*, **18**, 392–7.

MOIR, D., STEWART, S. & BOTTSTEIN, D. (1978). Isolation of cold-sensitive cell division mutants of *S. cerevisiae*. 9th International Conference on Yeast Genetics and Molecular Biology, p. 69.

MOLL, R. & WINTERSBERGER, E. (1976). Synthesis of yeast histones in the cell cycle. *Proceedings of the National Academy of Sciences, USA*, **73**, 1863–7.

MORRIS, N. R. (1976a). Mitotic mutants of *Aspergillus nidulans. Genetical Research*, **26**, 237–54.

MORRIS, N. R. (1976b). A temperature-sensitive mutant of *Aspergillus nidulans* reversibly blocked in nuclear division. *Experimental Cell Research*, **98**, 204–10.

MORRIS, N. R. (1976c). Nucleosome structure in *Aspergillus nidulans. Cell*, **8**, 357–64.

MORRIS, N. R. (1976d). A comparison of the structure of chicken erythrocyte and chicken liver chromatin. *Cell*, **9**, 627–32.

MORRIS, N. R., FELDEN, R. A., GEALT, M. A., NARDI, R. V., SHEIR-NEISS, G. & SANDERS, M. M. (1977). The *Aspergillus* nucleus: Histones, Chromatin, and Tubulin. In *Genetics and Physiology of Aspergillus*, ed. J. E. Smith & J. A. Pateman, pp. 267–79. New York, London & San Francisco: Academic Press.

MORRIS, N. R., LAI, M. H. & OAKLEY, C. E. (1979). Identification of a gene for α-tubulin in *Aspergillus nidulans. Cell*, **16**, 437–42.

MORRIS, N. R. & OAKLEY, C. E. (1979). Evidence that *p*-fluorophenylalanine has a direct effect on tubulin in *Aspergillus nidulans. Journal of General Microbiology*, in press.

NELSON, D. A., BELTZ, W. R. & RILL, R. L. (1977). Chromatin subunits from baker's yeast: Isolation and partial purification. *Proceedings of the National Academy of Sciences, USA,* **74,** 1343–7.

NICKLAS, R. B. (1971). Mitosis. In *Advances in Cell Biology,* eds. D. M. Presott, L. Goldstein & E. McConkey, vol. 2, pp. 225–97. New York: Appleton-Century-Crofts.

NOLL, M. (1976). Differences and similarities in chromatin structure of *Neurospora crassa* and higher eukariotes. *Cell,* **8,** 349–56.

NOLL, M. & KORNBERG, R. D. (1977). Action of micrococal nuclease on chromatin and the location of histone H1. *Journal of Molecular Biology,* **109,** 393–404.

NURSE, P. (1975). Genetic control of cell size at cell division in yeast. *Nature, London,* **256,** 547–55.

NURSE, P., THURIAUX, P. & NASMYTH, K. (1976). Genetic control of the cell division cycle in the fission yeast *Schizosaccharomyces pombe. Molecular and General Genetics,* **146,** 167–78.

O'FARRELL, P. H. (1975). High resolution two-dimensional electrophoresis of proteins. *Journal of Biological Chemistry,* **250,** 4007–21.

O'FARRELL, P. Z., GOODMAN, H. M. & O'FARRELL, P. H. (1977). High resolution two-dimensional electrophoresis of basic as well as acidic proteins. *Cell,* **12,** 1133–42.

ORR, E. & ROSENBERGER, R. F. (1976*a*). Initial characterization of *Aspergillus nidulans* mutants blocked in the nuclear replication cycle. *Journal of Bacteriology,* **126,** 895–902.

ORR, E. & ROSENBERGER, R. F. (1976*b*). Determination of the executive points of mutations in the nuclear replication cycle of *Aspergillus nidulans. Journal of Bacteriology,* **126,** 903–6.

PAULSON, J. R. & LAEMMLI, U. K. (1977). The structure of histone-depleted metaphase chromosomes. *Cell,* **12,** 817–28.

PARISH, R. W., STALDER, J. & SCHMIDLIN, S. (1977). Biochemical evidence for a DNA repeat length in the chromatin of *Dictyostelium discoideum. FEBS Letters,* **84,** 63–6.

PETERSON, E. L. & BERGER, J. D. (1976). Mutational blockage of DNA synthesis in *Paramecium tetraaurelia. Canadian Journal of Zoology,* **54,** 2089–97.

PETERSON, J. B. & RIS, H. (1976). Electron microscopic study of the spindle and chromosome movement in the yeast *Saccharomyces cerevisiae. Journal of Cell Science,* **22,** 219–42.

PICKETT-HEAPS, J. D. & TIPPIT, D. H. (1978). The diatom spindle in perspective. *Cell,* **14,** 455–68.

PIPERNO, G. & LUCK, D. (1976). Phosphorylation of axonemal proteins in *Chlamydomonas reinhardtii. Journal of Biological Chemistry,* **251,** 2161–7.

PIPERNO, G. & LUCK, D. (1977). Microtubular proteins of *Chlamydomonas reinhardtii:* an immunochemical study based on the use of an antibody specific for the β-tubulin subunit. *Journal of Biological Chemistry,* **252,** 383–91.

PIÑON, R. & SALTS, Y. (1977). Isolation of folded chromosomes from the yeast *Saccharomyces cerevisiae. Proceedings of the National Academy of Sciences, USA,* **74,** 2850–4.

PIÑON, R. (1978). Folded chromosomes in *Saccharomyces cerevisiae:* a probe into nuclear events during mitosis and meiosis. 9th International Conference on yeast genetics and molecular biology, p. 107 abstr.

PRINCE, D. J., CUMMINGS, D. J. & SEALE, R. L. (1977). Analysis of chromatin repeat units in logarithmically and stationary growing cells of *Paramecium aurelia* and *Tetrahymena pyriformis*. *Biochemical and Biophysical Research Communications*, **79**, 190–7.

PRINGLE, J. R. (1978). The use of conditional lethal cell cycle mutants for temporal and functional sequence mapping of cell cycle events. *Journal of Cell Physiology*, **95**, 393–400.

RENZ, M., NEHLS, P. & HOZIER, J. (1977). Histone H1 involvement in the structure of the chromatin fiber. *Cold Spring Harbor Symposia on Quantitative Biology*, **42**, 245–52.

ROBINOW, C. F. & CATEN, C. E. (1969). Mitosis in *Aspergillus nidulans*. *Journal of Cell Science*, **5**, 403–37.

ROBINOW, C. F. & MARAK, J. (1966). A fiber apparatus in the nucleus of the yeast cell. *Journal of Cell Biology*, **29**, 129–51.

ROOBOL, A., GULL, K. & POGSON, C. I. (1977). Evidence that griseofulvin binds to a microtubule associated protein. *FEBS Letters*, **75**, 149–53.

ROSENBERGER, R. F. & KESSEL, M. (1968). Non-random sister chromatid segregation and nuclear migration in hyphae of *Aspergillus nidulans*. *Journal of Bacteriology*, **96**, 1208–13.

SAKAI, H., MABUCHI, I., SHIMODA, S., KURIYAMA, R., OGAWA, K. & MOHRI, H. (1976). Induction of chromosome motion in the glycerol-isolated mitotic apparatus: nucleotide specificity and effects of anti-dynein and myosin sera on the motion. *Development, Growth and Differentiation*, **18**, 211–19.

SANGER, J. W. (1975). Presence of actin during chromosomal movement. *Proceedings of the National Academy of Sciences, USA*, **72**, 2451–5.

SATO, C. (1976). A conditional cell division mutant of *Chlamydomonas reinhardii* having an increased level of colchicine resistance. *Experimental Cell Research*, **101**, 251–9.

SHEIR-NEISS, G., NARDI, R. V., GEALT, M. A. & MORRIS, N. R. (1976). Tubulin-like proteins from *Aspergillus nidulans*. *Biochemical and Biophysical Research Communications*, **96**, 285–90.

SHEIR-NEISS, G., LAI, M. H. & MORRIS, N. R. (1978). Identification of a gene for β-tubulin in *Aspergillus nidulans*. *Cell*, **15**, 638–47.

SIMCHEN, G. (1978). Cell Cycle Mutants, *Annual Review of Genetics*, **12**, 161–91.

SNYDER, J. A. & MCINTOSH, J. R. (1976). Biochemistry and physiology of microtubules. *Annual Review of Biochemistry*, **45**, 699–720.

SOMMER, A. (1978). Yeast chromatin: a search for histone H1. *Molecular and General Genetics*, **161**, 323–31.

SPADAFORA, C., BELLARD, M., COMPTON, J. L. & CHAMBON, P. (1976). The DNA repeat lengths in chromatins from sea urchin sperm and gastrula cells are markedly different. *FEBS Letters*, **69**, 281–5.

STADLER, J. & BRAUN, R. (1978). Chromatin structure of *Physarum polycephalum* plasmodia and amoebas. *FEBS Letters*, **90**, 223–7.

TESSIER, A., ROLAND, B., ANDERSON, W. A. & PALLOTTA, D. (1978). The isolation, identification, and characterization of H1, H2B, H3, and H4-like histones from the baker's yeast. *Saccharomyces cerevisiae*. *Journal of Cell Biology*, **79**, 106a.

THOMAS, J. O. & FURBER, V. (1976). Yeast chromatin structure. *FEBS Letters*, **66**, 274–80.

THOMAS, J. O. & THOMPSON, R. J. (1977). Variation in chromatin structure in two cell types from the same tissue: A short repeat length in cortical neurons. *Cell*, **10**, 633–40.

THURIAUX, P., NURSE, P. & CARTER, B. (1978). Mutants altered in the control co-ordinating cell division with cell growth in the fission yeast *Schizosaccharomyces pombe*. *Molecular and General Genetics*, **161**, 215–20.

TRINCI, A. P. J. & MORRIS, N. R. (1979). Morphology and growth of a temperature sensitive mutant of *Aspergillus nidulans* which forms aseptate mycelia at non-permissive temperatures. *Journal of General Microbiology*, **114**, 53–9.

VAN TUYL, J. M. (1977). Genetics of fungal resistance to systemic fungicides. Thesis. Department of Phytopathology, Department of Genetics, Agricultural University, Wageningen, The Netherlands.

VARSHAVSKY, A. J., BAKAYEV, V. V. & GEORGIEV, G. P. (1976). Heterogeneity of chromatin subunits *in vitro* and location of histone H1. *Nucleic Acid Research*, **3**, 477–92.

VOGT, V. M. & BRAUN, R. (1976). Repeated structure of chromatin in metaphase nuclei of *Physarum*. *FEBS Letters*, **64**, 190–2.

WARR, J. R. (1974). Genetic approaches to the study of microtubule structure and function. *Subcellular Biochemistry*, **3**, 149–54.

WARR, J. R. & DURBER, S. (1971). Studies on the expression of a mutant with abnormal cell division in *Chlamydomonas reinhardii*. *Experimental Cell Research*, **64**, 463–9.

WARR, J. R. & GIBBONS, D. (1974). Further studies on colchicine-resistant mutants of *Chlamydomonas reinhardii*. *Experimental Cell Research*, **85**, 117–22.

WARR, J. R., FLANAGAN, D. & QUINN, D. (1978). Mutants of *Chlamydomonas reinhardii* with altered sensitivity to antimicrotubular agents. *Experimental Cell Research*, **11**, 37–46.

WATER, R. D. & KLEINSMITH, L. J. (1976). Identification of α- and β-tubulin in yeast. *Biochemical and Biophysical Research Communications*, **70**, 704–8.

WEEKS, D. P. & COLLIS, P. S. (1976). Induction of microtubule protein synthesis in *Chlamydomonas reinhardii* during flagellar regeneration. *Cell*, **9**, 15–27.

WEEKS, D. P. & COLLIS, P. S. (1979). Induction and synthesis of tubulin during the cell cycle and life cycle of *Chlamydomonas reinhardii*. *Developmental Biology*, **69**, 400–7.

WEEKS, D. P., COLLIS, P. S. & GEALT, M. A. (1977). Control of induction of tubulin synthesis in *Chlamydomonas reinhardii*. *Nature, London*, **268**, 667–8.

WEHLAND, J., HERZOG, W. & WEBER, K. (1977). Interaction of griseofulvin with microtubules, microtubule protein and tubulin. *Journal of Molecular Biology*, **111**, 329–42.

WHEALS, A. E., GRANT, W. D. & JOCKUSCH, B. M. (1976). Temperature-sensitive mutants of the slime mold *Physarum polycephalum*. I. Mutants of the ameobal phase. *Molecular and General Genetics*, **149**, 110–14.

EXPLANATION OF PLATES

Plate 1*A*. Tubulin region of two-dimensional gels of electrophoretic variants of β1/β2 tubulin in *ben*A mutants of *Aspergillus nidulans*: (*a*), wild-type (B3); (*b*) *ben*A27 (1⁻); (*c*) *ben*A13 (1⁺); (*d*) *ben* A22 (2⁻); (*e*) *ben* A7 (β2 missing); (*f*) *ben* A14 (split β1). Insert also shows *ben*A14 – long electrofocusing gel gives better resolution.

Plate 1*B*. Tubulin region of two-dimensional gels of diploid (*ben*A22/wt (FGSC no. 5) protein extract: (*a*), *ben*A22 (2⁺); (*b*), Diploid *ben*A22, FGSC no. 5, (2⁺, wild-type); (*c*), FGSC no. 5 (wild-type). From Sheir-Neiss *et al.* (1978). *Cell*, **15**, 639–48. © The MIT Press.

Plate 2*A*. Two-dimensional gel electrophoresis of the proteins of (*a*), Parental wild-type (B3); (*b*), *ben*A17R10 (1⁻ β-tubulin); (*c*), *ben*A21R7 (2⁻ β-tubulin).

Plate 2*B*. Two-dimensional gel electrophoresis of the proteins of: (*a*), *ben*A11 (wild-type α-tubulin); (*b*), *tub*A1 (*ben*A11R7) (1⁺ α-tubulin).

Plate 2*C*. Two-dimensional gel electrophoresis of the proteins of: (*a*), R153 (wild-type α-tubulin); (*b*), *tub*A1 (1⁺ α-tubulin); (*c*), R153/A1 diploid (wild type and 1⁺ α-tubulin). From Morris *et al.* (1979). *Cell*, **16**, 437–42. © The MIT Press.

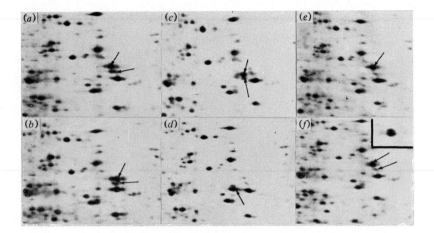

PLATE 1A

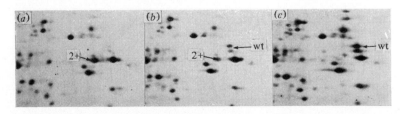

PLATE 1B

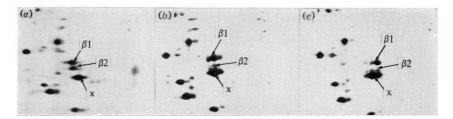

Plate 2*A*

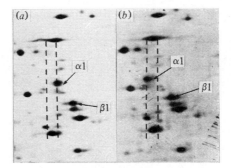

Plate 2*B*

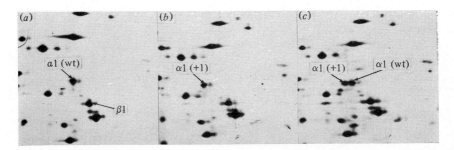

Plate 2*C*

MITOSIS AND MEIOSIS: NUCLEAR DIVISION MECHANISMS

J. D. DODGE* AND K. VICKERMAN†

*Department of Botany, Royal Holloway College, Egham, UK

†Department of Zoology, University of Glasgow, Glasgow, UK

INTRODUCTION

Mitosis takes many forms and exhibits a wide range of variation in eukaryotic micro-organisms, although it achieves the same basic objective in all of them: the segregation of the sister chromatids. This is in great contrast to the uniformity found in higher eukaryotes where there are but two basic types; that found in animals where centrioles are associated with spindle organisation, and the acentric mitosis of higher plants where there are no obvious pole-associated structures. In the fungi, algae and protozoa there is a veritable 'living museum' filled with examples of nuclear division, some involving elaboration of one structure or aspect of the process at the expense of the others. Here, then, is fruitful material for studying the basic mechanisms of nuclear division and at the same time the possibility of plotting some stages in the evolution of the more or less standard mitoses of higher organisms. Perhaps, because the nucleus is such a fundamental part of the cell, phylogeny based on nuclear type may be more reliable than that based on chloroplasts, flagella or aspects of the morphology of the cells (Oakley, 1978a).

Numerous reviews of various aspects of mitosis and meiosis in these groups have appeared in the past few years. From these further information may be obtained of processes and structures which cannot be described in great detail in the present review. The fungi have been reviewed by Fuller (1976), Wells (1977) and Heath (1978), the algae by Leedale (1970) and Dodge (1973), certain protozoa by Hollande (1972, 1974), and all the lower eukaryotes by Heath (1980). The spindle, including the various types observed in lower organisms, has been treated in great detail by Fuge (1974, 1977) and Kubai (1975), and division in general by Pickett-Heaps (1969, 1974, 1975) and Ris (1975). Meiosis in eukaryotic protists has been reviewed by Heywood & Magee (1976).

For the first 100 years or so the study of nuclear division in eukaryotic protists relied entirely on light microscopy, in conjunction with various chromosome stains. Bearing in mind the smallness of many of the nuclei, an amazing amount was discovered as is witnessed particularly by the remarkable work of Bĕlǎr (1926) which showed something of the variety of types of mitosis. In recent years advances in light microscopy have greatly aided studies on nuclear division processes; examples of such developments include the use of rectified polarised light to give improved image quality in observing spindle changes, and the use of differential interference contrast optics to display chromosome boundaries in relief with little interference from out-of-focus objects. The single-sideband edge-enhancement microscope recently described by Ellis (1978) promises to overcome the difficulties raised by chamber birefringence in following mitosis in living cells in culture or perfusion chambers.

The past twenty years, however, belong largely to the electron microscopists who have put details into such formerly ill-defined structures as spindle fibres and kinetochores, and have enabled us to visualise the nuclear envelope with its comings and goings, whereas previously its presence or absence could only be deduced. This chapter will mainly be concerned with ultrastructural aspects of the principal types of mitosis revealed in eukaryotic protists. We shall, in addition, discuss some contributions to our understanding of nuclear division mechanisms in general, arising from recent studies on unicellular eukaryotes.

THE BASIC MECHANISMS OF MITOSIS

A nuclear envelope composed of two membranes, and microtubules composed of polymerised tubulin subunits, are distinctive eukaryotic cell features. Nuclear division in most eukaryotes is associated with the appearance of a new microtubular configuration – the mitotic spindle – which ensures accurate bi-partitioning of the replicated genome between daughter cells. The spindle is composed of two major classes of microtubule-containing fibres, (1) the chromosomal fibres which usually link a specific region of the chromosomes, the kinetochores, with the spindle pole, and (2) polar or continuous fibres which pass from pole to pole. Spindle fibre birefringence gives a direct measurement of the packing density of the parallel microtubules composing the fibres visualised by polarised light microscopy (Sato, Ellis & Inoué, 1975).

In prokaryotes, separation of the daughter genomes is achieved by movement apart of the plasma membrane-attachment sites of the chromosomes (reviewed by Leibowitz & Schaechter, 1975); it is presumed that local membrane synthesis plays a part in the process but the mechanics are not understood. In higher eukaryotes, the inner nuclear membrane bears attachment sites for specific regions of the chromosomes; such sites are believed to represent a means for the ordered arrangement of chromosomes during interphase and in prophase when the nuclear envelope is breaking down (reviewed by Franke, 1974). With the onset of metaphase, however, the chromosomes are clearly attached to the formed spindle apparatus and their motion during mitosis is generally believed to be a consequence of changes taking place in the length and position of spindle fibre microtubules. Spindle microtubules are labile structures that are sequentially assembled and disassembled during mitosis. These microtubules are now known to be in dynamic equilibrium with free tubulin molecules. Free tubulins appear to enter a steady state microtubule from one end at about the same rate that the molecules leave from the other (Margolis & Wilson, 1978). Daughter chromosome separation at anaphase is accompanied by shortening of the chromosomal fibres and the rate of chromosomal movement is compatible with a process of disassembly of the chromosomal fibre microtubules taking place at their poleward ends (reviewed by Inoué, 1976; Inoué & Kiehart, 1978). Whether or not the force for chromosome movement is provided by such disassembly is uncertain; actin has been found in association with chromosomal fibres (Sanger & Sanger, 1976; Forer, 1976) and the force for movement may be generated by an actomyosin system. The continuous fibres do not contain actin and their role at anaphase is thought to be one of spindle elongation, pushing the daughter chromosome complements further apart. The microtubules composing the fibres may arise from microtubule organising centres at the spindle poles and spindle elongation is achieved first by growth of individual microtubules in the direction of the other pole and then by sliding of the two sets of overlapping microtubules with respect to one another (McIntosh, Hepler & van Wie, 1969; see Holwill, this volume).

Kubai (1975) has suggested that in the evolution of eukaryotes there has been a gradual shift from membranes to microtubules as producers of the primary forces for chromosome movement, and that during evolution microtubules might have acquired the capacity for force production as an adjunct to their static skeletal properties. As microtubules assume importance in cell division, so do the microtubule organising centres which initiate microtubule assembly (Pickett-Heaps, 1969). The kinetochore is widely regarded as the organising centre for the microtubules composing chromosomal fibres, and in animal cells the centriole was long presumed to be the organising centre for polar (including continuous) fibres, though a more plausible interpretation is that the polar organising centres are associated with the centriole for purposes of mitosis. During nuclear division in the majority of eukaryotic protists the nuclear envelope remains intact. This extraordinary difference from nuclear division in higher eukaryotes raises several problems of location of microtubule organising centres, as pointed out in the following survey of the major types of mitosis looked at from this point of view.

TYPES OF MITOSIS

For convenience of description four types of mitosis may be distinguished with regard to the condition of the nuclear envelope and the origin of the spindle.

Closed mitosis with internal spindle

In this type of division the nuclear envelope remains intact throughout and any spindle apparatus is formed within the nucleus. Such divisions are found in all three groups of organisms under consideration but they show a number of variations, particularly in the organisation of the spindle.

Possibly the most primitive of this type is found in members of the Euglenophyceae (Sommer & Blum, 1965, Leedale, 1968; Chaly, Lord & Lafontaine, 1977; Pickett-Heaps & Weik, 1977; Gillott & Triemer, 1978). Here, the process has mainly been studied in *Euglena* but representatives of *Astasia* and *Phacus* have also been examined. The start of division is signalled by the nucleus moving towards the anterior end of the cell until it is closely pressed against the reservoir. As this is happening the flagellar bases duplicate and the two pairs separate across the anterior end of the cell. Prophase commences with the appearance in the nucleus of microtubules which are orientated across the cell. The nucleolus begins to elongate in the same orientation as the microtubules and the chromosomes may become more condensed. At metaphase the nucleus is rather cylindrical and the nucleolus more drawn out; the chromosomes are aggregated around the centre of the nucleolus. Many microtubules can now be seen, some of which are grouped into bundles which may contact a kinetochore region of a chromosome (Pickett-Heaps & Weik, 1977). In anaphase, which takes a relatively long time, the nucleus extends considerably, becoming first fusiform and then dumb-bell shaped as the chromosomes separate into two groups. The nucleolus, which has microtubules lying along its surface, becomes stretched and finally breaks into two portions late in telophase. The end of division sees the equatorial region, which is now devoid of chromosomes, becoming constricted around the extended remains of the spindle. Finally, the nuclear envelope breaks and rejoins to give two daughter nuclei and a large length of interzonal spindle which may be abandoned in the cytoplasm (Pickett-Heaps & Weik, 1977).

Opinions differ over the organisation and connections of the spindle. Two groups of workers (Sommer & Blum, 1965; Gillott & Triemer, 1978) regard the basal bodies as having a role in the organisation of the spindle whereas others (Leedale, 1968; Pickett-Heaps & Weik, 1977) do not. Kinetochores were thought not to be present (Leedale, 1968) but they have been fairly clearly demonstrated in recent studies (Pickett-Heaps & Weik, 1977; Gillott & Triemer, 1978). However, there appear to be no connections between microtubules and the nuclear envelope or any microtubule organising centres so it is not clear how the 'spindle' can play an active part in the segregation of the chromatids, which is said to take place autonomously (Leedale, 1968).

A similar type of mitosis is found in the trypanosomes (Vickerman & Preston, 1970; Heywood & Weinman, 1978). In some species (*Trypanosoma brucei, Trypanosoma cyclops*) the spindle appears to consist solely of pole-to-pole microtubules which pass through or to one side of kinetochore-like plaques; chromatin-like material remains attached to the nuclear envelope throughout division. In *Trypanosoma raiae*, however, the microtubules converge on the kinetochore-like plaques suggesting a chromosomal fibre function, though condensed chromatin is not discernible. Neither kinetochores nor condensed chromatin have been observed during mitosis of the amoeboid stage of the amoeboflagellate *Naegleria gruberi* (Schuster, 1975), though division appears to resemble that of *Euglena* and some trypanosomes in that the intranuclear spindle surrounds the persistent nucleolus. In both *N. gruberi* and the trypanosomes it has been suggested that nuclear envelope growth may play the most important part in the extension of the nucleus and thus the separation of the two sets of chromosomes.

The micronucleus of many ciliates also has a fairly similar mitosis to that described for *Euglena*. In *Paramecium aurelia* (Jurand & Selman, 1970) and *Loxodes magnus* (Raikov, 1973) the nuclear envelope remains intact throughout division. Numerous microtubules form within the nucleus and some of these may be distinctly associated into bundles. There are no spindle polar structures of any sort and at metaphase the nucleus is clearly ovoid, with the microtubules more or less parallel rather than converging at the spindle poles. Each chromosome is attached by a kinetochore to several microtubules but kinetochores are absent from micronuclear mitosis in most species. At late anaphase there is a considerable extension of the continuous microtubules leaving a long interzonal strand when the two daughter nuclei form. In the dividing micronucleus of *Paramecium bursaria*, elaborate kinetochore-like structures ('micro-lamellae') are associated with microfilaments as well as microtubules in the spindle apparatus (Lewis, Witkus & Vernon, 1976); the significance of this will be discussed later.

Several organisms have mitotic divisions of the general type described above but with the addition of recognisable spindle-organising structures or, to use the current terminology, 'spindle pole bodies'. These are plate or disc shaped electron dense plaques inserted into the nuclear envelope or applied to one or other of its membranes; they serve as microtubule organising centres generating spindle microtubules from the inner side. They have been most extensively investigated for Ascomycetes and their role in mitosis will be returned to later (p. 91). Earlier reports suggesting that spindle tubules in these organisms consist entirely of the pole-to-pole variety (e.g. *Saccharomyces cerevisiae*, Robinow & Marak, 1966; *Podospora anserina*, Zickler, 1970) may be in need of revision in the light of more recent work (Peterson & Ris, 1976).

Mitosis in the malaria parasite *Plasmodium yoellii* is intranuclear with approximately 30 microtubules radiating from each spindle pole body, of which 16 to 20 are attached to kinetochores on uncondensed chromosomes (Schrevel, Asfaux-Fouchet & Bafort, 1976). When the spindle pole bodies and kinetochores divide, separation of the resulting hemispindles is achieved by expansion of the nuclear envelope which becomes evaginated to bring about apposition of the hemispindles and form the mitotic apparatus. Several spindles may form within the lobulated oocyst nucleus undergoing sporogony.

A further variation on closed nuclear division is found in the green (prasinophycean) flagellate *Pedinomonas* (Pickett-Heaps & Ott, 1974), the yellow-green filamentous alga *Vaucheria litorea* (Ott & Brown, 1972), the water fungi *Saprolegnia ferax* (Heath & Greenwood, 1970) and *Catenaria anguillulae* (Ichida & Fuller, 1968), the oomycete *Thraustotheca clavata* (Heath, 1974) and the radiolarian *Collozoum* (Hollande, Cachon & Cachon, 1969). In all of these organisms a pair of centrioles becomes associated with each pole of the nucleus, but outside the nuclear envelope, prior to the formation of the spindle. The spindles consist of both pole-to-pole microtubules and microtubules that are attached to chromosomes. The membraneous barrier provided by the nuclear envelope makes it unlikely that the centrioles play a part as microtubule organising centres in spindle organisation, though Heath's (1974) micrographs of *T. clavata* strongly suggest this. Chromosome separation is thought to take place by the extension of the continuous microtubules and in *V. litorea* and *C. anguillulae* a long interzonal spindle develops between the daughter nuclei, the separation of which is accomplished by the invagination of the nuclear envelope.

The existence of closed mitosis in all these organisms is of interest in that microtubules cannot be detected inside the prophase nucleus of dividing cells of higher plants and animals before the nuclear envelope has fragmented (see Fuge, 1977).

Closed mitosis with external spindle

This type of division is found in two groups of organisms, the dinoflagellates (both free-living and parasitic) and the hypermastigid flagellates. The most simple example is found in the free-living dinoflagellates *Gymnodinium micrum* (Leadbeater & Dodge, 1967), *Crypthecodinium cohnii* (Kubai & Ris, 1969), and *Amphidinium carterae* (Oakley & Dodge, 1974b, 1977) and the parasite *Haplozoon axiothellae* (Siebert & West, 1974). At the start of division, bundles of microtubules form at one side of the nucleus and then sink into it, becoming surrounded by nuclear envelope to form channels or tunnels. The chromosomes which are attached to the nuclear envelope surrounding the tunnels now replicate, commencing at their free ends. The number of tunnels is variable and there are no centrioles or spindle pole bodies associated with their 'poles'; in *A. carterae* microbodies are often seen at the ends of the bundles of microtubules. At 'metaphase' the chromosomes are rather loosely arranged around the tunnels and, by now, where they are attached to the nuclear envelope, a small pad has formed on the cytoplasmic side to which a pair of microtubules is attached, one directed towards each pole (Plate 2a). During anaphase the nucleus extends and the chromosomes segregate without any breakdown of the nuclear envelope. Finally, constriction of the envelope separates the daughter nuclei leaving a long interzonal spindle (Plate 1).

Early interpretations of this mitosis (Kubai & Ris, 1969; Ris, 1975) suggested that the division was entirely mediated by flowing of the nuclear membrane, the microtubules merely forming the framework along which separation takes place. However, it now seems that as some microtubules are continuous and others are chromosomal, and as the bundles of microtubules show distinct cross-bridges between adjacent tubules (Oakley & Dodge, 1977), then microtubule sliding and extension may be important factors in the anaphase movement.

A similar type of division has been found in a number of parasitic dinoflagellates but here there is another major difference; the presence of centrioles at the poles of the spindle. Several parasites have been studied; *Syndinium* (Ris & Kubai, 1974) and *Solenodinium* (Hollande, 1974), both endoparasites in radiolaria, and *Oodinium*, an ectoparasite (Cachon & Cachon, 1974) of appendicularians (urochordates). In all cases the chromosomes are attached to a double structure (kinetochore) consisting of two dense discs which are inserted into the nuclear envelope. The outer side of the disc has several microtubules attached to it and these run to centrioles which at prophase are situated at one side of the nucleus and then migrate such that as the single tunnel forms through the nucleus the chromosomes become segregated. This process is thought by Ris & Kubai (1974) to be brought about by the formation and growth of continuous microtubules between the separating centrioles. It is of interest that in all of these parasites there is a very small number of chromosomes (only four in *Syndinium*) and these contain a detectable amount of histone protein (Hollande, 1974) unlike the chromosomes of free-living dinoflagellates which normally consist essentially of DNA.

In one dinoflagellate, division does not exactly fit the categories described above; *Oxyrrhis marina* apparently has a cytoplasmic channel through the dividing nucleus but this contains no microtubules (Cachon, Cachon & Salvano, 1978). The chromosomes are nonetheless attached to the nuclear envelope and it is the extension of this which is said to be responsible for segregation.

In the hypermastigid flagellates such as *Barbulanympha ufalula* and *Trichonympha agilis* (Hollande & Valentin, 1967a, 1967b, 1971; Kubai, 1973) the spindle develops between the two halves of a structure called the rostrum which is associated with the flagellar roots. As the hemirostra separate, microtubule bundles radiate from the tips of giant 'centrioles' attached to each half rostrum. Some of these fibres overlap to form a paranuclear spindle while others approach the nucleus itself. Kinetochores form in the nucleus and become associated, in pairs, with the nuclear envelope but at this stage they are still inside the nucleus. Next, pores form in the nuclear envelope and, as the kinetochores become inserted, a connection is established between a fibrillar component of the kinetochore and spindle microtubules. The kinetochores are seen to be segregated into two groups in the nuclear envelope, one near each pole of the spindle. Finally there is active spindle elongation which completely separates the two groups of chromosomes (see p. 93). One point of considerable interest is that in *Trichonympha agilis*, according to Kubai (1973), the groups of sister chromatids are segregated apart whilst the kinetochores are attached to the nuclear envelope but before the spindle tubules are connected to them. This appears to be a clear case of membrane-mediated chromosome movement and supports the suggestion of Pickett-Heaps (1975) that the kinetochores of higher eukaryotes may have evolved from chromosome attachment sites on the nuclear envelope.

Semi-open mitosis

In representatives of all three groups of eukaryotic micro-organisms there are examples of mitosis in which the nuclear envelope remains substantially intact but gaps or polar fenestrae develop and make possible the intervention of cytoplasmic microtubule organising centres in the nuclear division. This type of division is characteristic of many green algae (Chlorophyceae) so it will first be described for the flagellates *Chlamydomonas reinhardii* and *Chlamydomonas moewusii* (Johnson & Porter, 1968; Triemer & Brown, 1974).

The first indication of mitosis is loss of the free parts of the flagella and the replication of the two basal bodies. The two pairs of flagellar bases then move apart across the anterior end of the cell and at this stage the nucleus moves near to the basal bodies which come to be situated just anterior to the spindle poles. The chromosomes condense, the nucleolus breaks down, and the nucleus becomes spindle-shaped as microtubules appear within it. Polar gaps form in the nuclear envelope and it may be that some microtubules enter the nucleus from the cytoplasm. There is no obvious connection between the basal bodies and the spindle, although Triemer & Brown (1974) have suggested that they act as centrioles; neither are any spindle pole bodies present. During anaphase the nucleus elongates and two daughter nuclei are formed.

Numerous other green algae have been studied by electron microscopy and the following are examples which have an essentially similar division to that described for *Chlamydomonas*: *Oedogonium cardiacum* (Pickett-Heaps & Fowke, 1970), *Kirchneriella lunaris* (Pickett-Heaps, 1970), *Tetraspora* (Pickett-Heaps, 1973) and *Volvox* (Deason & Darden, 1971). Some of these have centrioles at the spindle poles whilst others have only rudimentary structures.

A similar division is found in many flagellates of the Prasinophyceae, although here the rhizoplasts (flagellar roots) may play some part in the organisation of the spindle which is essentially cytoplasmic (Stewart, Mattox & Chandler, 1974; Pearson & Norris, 1975; Mattox & Stewart,

1977). During anaphase when the spindle elongates considerably, the pole-to-chromosome distance remains constant suggesting that there must be something connecting microtubules to the chromosome, although only Oakley & Dodge (1974a), working with *Tetraselmis tetrathele* and Floyd (1978) with *Asteromonas gracilis* have reported the presence of small kinetochores to which one to three microtubules are attached (Plate 2c). In *A. gracilis* the basal bodies lie at one side of the spindle and the nuclear envelope remains substantially intact throughout mitosis. Here the pole-to-chromosome distance shortens considerably in anaphase.

In the algal flagellates of the Chloromonadophyceae the mitosis of *Vacuolaria virescens*, as described by Heywood (1978), also fits this general pattern. In early prophase numerous microtubules are seen radiating from the flagellar bases over the anterior surface of the nucleus. The chromosomes become condensed and the nucleolus disperses. At the poles microtubules enter the nucleoplasm through gaps in the nuclear envelope. The nucleus now becomes elliptical and the chromosomes form into a fairly distinct metaphase plate. Many microtubules are continuous but others are attached to the chromosomes by distinct kinetochores (Heywood & Godward, 1972).

At anaphase the groups of chromosomes separate and the nucleus becomes elongated. Finally the old nuclear envelope either breaks down or may become incorporated into the new envelopes. In *Cyanophora paradoxa*, a flagellate of uncertain systematic position, the mitosis is similar to that described for *Vacuolaria* (Pickett-Heaps, 1972a). Preliminary reports for the unicellular red alga *Porphyridium purpureum* (Bronchart & Demoulin, 1977) show a mitosis with open ends but other workers find definite spindle pole bodies associated with several microbodies (Schornstein & Scott, 1978).

There are several fungi which have a semi-open mitosis and this is perhaps the main type in the myxomycetes (Mycetozoa). However, there is a distinct difference compared to the examples already quoted because the nuclear envelope remains substantially intact until it breaks down just before telophase. In *Physarum polycephalum* (Ryser, 1970) the spindle forms within the nucleus and consists of both continuous microtubules and those attached to chromosomes by disc-shaped kinetochores. Following the anaphase elongation the nuclear envelope disappears, first at the polar regions and later from the interzone, before the new envelope forms. In the fungus *Sorosphaera veronicae* (Braselton, Miller & Pechak, 1975) centrioles are situated at the poles and the spindle microtubules seem to be partly cytoplasmic in origin.

There are clearly defined kinetochores and, following the extensive anaphase elongation, there appear to be no microtubules between the two groups of chromosomes. In this organism the nucleolus is persistent.

Open mitosis

This is, of course, the main type of mitosis found in higher eukaryotes, and just as there is a difference between its nature in animals and plants, so we find a number of variations in the lower organisms. These differences mainly concern the spindle-organising structures.

In the giant amoeba *Pelomyxa carolinensis* (Roth & Daniels, 1962) there are no centrioles or spindle pole bodies and the metaphase nucleus is broad with parallel microtubules. Much the same situation seems to exist in the flagellate *Prymnesium parvum* (Manton, 1964) although the mitotic figure is not so wide and there is a suspicion of convergence of the microtubules towards the 'poles'. Mitosis is also acentric in the green filamentous alga *Raphidonema longiseta* (Pickett-Heaps, 1976) and rather surprisingly Golgi bodies are generally present at the spindle poles. However, in another member of the order Ulotrichales *Klebsormidium subtillissimum* (Pickett-Heaps, 1972b), centrioles are present at the poles. In the desmid *Cosmarium botrytis* the spindle has no obvious pole bodies but appears multipolar and the metaphase plate is unusually wide (Pickett-Heaps, 1972c). In all of these examples the nucleolus breaks down during division.

A fairly similar situation is found in the Cryptophyceae (Oakley & Dodge, 1973, 1076; Oakley & Bisalputra, 1977; Oakley, 1978b; Oakley & Heath, 1978). Here, the spindle microtubules are first formed in association with the separating flagellar bases and the anterior-moving nucleus. They then become reorientated and when the nuclear envelope breaks down a rather rectangular spindle is formed (Plate 3) with both continuous microtubules running through the chromatin plate and others attached to chromosomes by small kinetochores (Plate 7b). Anaphase separation, which is very rapid, results in the chromosome mass separating into two clumps as the spindle elongates. This is followed by the reconstitution of the nuclear envelope which often seems to involve the utilisation of adjacent portions of endoplasmic reticulum.

Another variation is found in the chrysophycean flagellates where *Ochromonas danica* has rather unusual spindle pole bodies in the form of flagellar roots or rhizoplasts (Slankis & Gibbs, 1972; Bouck & Brown, 1973). Following replication of the basal bodies a rhizoplast extends from one of each pair to the surface of the nucleus. In prophase the basal bodies and associated rhizoplasts separate and at this time there is a proliferation of microtubules which then enter the nucleus, via polar gaps, where they form the spindle. The rhizoplasts form very broad spindle poles and some microtubules extend from pole to pole whilst others appear to be attached to the chromosomes. The nuclear envelope now breaks down completely and anaphase separation takes place with a marked lengthening of the spindle.

A type of mitosis which may have developed from that of the Chrysophyceae is found in the diatoms (Bacillariophyceae) (Manton, Kowallik & von Stosch, 1969a; Tippit, McDonald & Pickett-Heaps, 1975; Picket-Heaps, McDonald & Tippit, 1975; Pickett-Heaps, Tippit & Andreozzi, 1978a, b; McDonald, Pickett-Heaps, McIntosh & Tippitt, 1977). Here, there are no basal bodies or flagellar roots to act as spindle poles but instead an unique body, termed the persistent polar complex, is situated at one side of the interphase nucleus. At prophase this body separates into two plates and between these, numerous microtubules form to give rise to a spindle. This sinks into the nucleus and comes to lie across the middle as the nuclear envelope breaks down. This unusual central spindle consists, in fact, of two half-spindles which overlap (see p. 94) and in addition some bundles of microtubules run from the poles to clumps of chromatin. It has not been possible to distinguish individual chromosomes or kinetochores.

In a number of basidiomycete fungi the nuclear envelope partly breaks down during mitosis but as this occurs not only at the poles, they are included here as open mitoses. As described by Girbardt (1968, 1971) for *Polystictus versicolor* and McCully & Robinow (1972) for *Leucosporidium scottii* there is a dense structure adjacent to the interphase nucleus which was called a kinetochore-equivalent or microtubule-organising centre, now a spindle pole body. This consists of two globular ends joined by a dense strand. The nuclear envelope partly breaks down and the nucleolus seems to be dissolved. Following this the two globular ends of the body move apart across the nucleus and microtubules form between them to give rise to the spindle. The two daughter sets of chromosomes move apart as the spindle extends and there are probably simple kinetochore attachments connecting some of the microtubules with the chromosomes (McCully & Robinow, 1972, Fig. 12). When the nuclear envelope forms around the daughter nuclei there is a very long spindle body between them.

In the slime mould *Physarum flavicomum* (Aldrich, 1969) the rather strange situation exists in which the haploid myxamoebae have centriolar division which is completely open whilst the diploid plasmodial stage lacks centrioles and has a closed division.

Amitosis

Division of the macronucleus of ciliates during binary fission is frequently dubbed 'amitotic' in that an organised spindle and discrete chromosomes are not usually visible, and daughter cells with unequal macronuclear DNA contents are frequently produced. The nuclear envelope remains intact throughout division with chromatin attached to its inner membrane. Elongation of the macronucleus is associated with intranuclear microtubules and often juxtanuclear microtubules paralleling the axis of elongation. In *Blepharisma* only juxtanuclear microtubules are found (Jenkins, 1977). Because of their position, orientation and time of appearance, these microtubules are believed to be either karyoskeletal, giving appropriate prefission shape to the macronucleus, or perhaps kinetic, providing force for separation of genomes or stretching of the macronucleus (see review by Raikov, 1976). This inexact distribution of genetic material is presumably tolerated because the macronucleus is polygenomic and contains a large number of redundant functional units. Although each subunit (chromosome, gene) replicates completely, these units may be partitioned unequally between the daughter macronuclei at fission. Analysis of variance within cell lines of *Paramecium aurelia* has shown that macronuclear DNA content is regulated so that a constant variance is maintained from one cell generation to the next. Half of the variation in macronuclear DNA content introduced into the population at a particular fission by inequality of division is compensated for during the subsequent period of DNA synthesis; half of the remaining variation is removed during the subsequent cell cycle (Berger & Schmidt, 1978). If mitosis is defined as nuclear division in which each daughter nucleus contains at least one full genome, then ciliate macronuclear division falls within this definition and should not be described as amitosis.

Amitotic division has been described from one other situation, the eukaryotic symbiont nucleus found within the cells of the dinoflagellate *Peridinium balticum* (Tippit & Pickett-Heaps, 1976). In this symbiont, which is thought to be related to the chrysophyte algae because of its type of chloroplasts, the nucleus becomes drawn out and then simply constricts into two. This happens in synchrony with the division of the dinoflagellate nucleus also present in the cell.

MEIOSIS

Reduction division has now been studied in many fungi, several Protozoa and in members of some of the algal groups. This area has been recently reviewed in detail (Heywood & Magee, 1976). In general it is difficult to distinguish meiotic stages from those of mitosis apart from the unique prophase. Here, the presence of a synaptonemal complex has been demonstrated in zygotes and pachytene stages in *Chlamydomonas moewusii* (Triemer & Brown, 1977) and the diatom *Lithodesmium undulatum* (Manton *et al.*, 1969*b*), amongst the micro-algae, in *Ceratiomyxa fruticulosa* (Furtado & Olive, 1971) and *Blepharisma* (Jenkins, 1977) amongst the Protozoa, and in numerous fungi from the Phycomycetes, Ascomycetes and Basidiomycetes (see Heywood & Magee, 1976; Wells, 1977).

In general the synaptonemal complex of protists consists of a central region 90–120 nm wide which includes a dark-staining stripe 10–30 nm wide. Either side of the central region there are lateral striped components 30–50 nm wide and beyond these there is dense-staining chromatin which presumably represents the homologous chromosomes. Synaptonemal complexes normally seem to end at the nuclear envelope which prompts the suggestion that they may be attached to it (Setliff, Hoch & Patton, 1974; Triemer & Brown, 1977). In the ascomycete *Ascophanus* synaptonemal complexes have been found in a haploid nucleus (Zickler, 1973).

The later stages of meiosis follow the same types of division, regarding nuclear envelope state and spindle organisation, as found at mitosis. This has been well documented for the diatom *Lithodesmium* where the only real difference is that at Meiosis I the spindle is much larger than at Meiosis II or mitosis. Microtubules at the equator number approximately 325 in Meiosis I but only 200 in the other divisions (Manton *et al.*, 1969*b*, 1970*a*, *b*).

THE ROLE OF MICROTUBULE ORGANISING CENTRES IN SPINDLE FORMATION

From the above accounts it can be seen that although the structural details of mitosis may vary from one eukaryote to another or even from one stage in the life-cycle to another, a bipolar spindle apparatus composed of microtubules is almost invariably present. Attempts to understand the role of spindle microtubules in chromosome segregation, however, have been frustrated by the complexity of spindle structure in higher eukaryotes. Eukaryotic protists provide simpler systems for studying the fundamentals of mitosis; the spindle is often composed of relatively few microtubules (Table 1) and in some cases a combined ultrastructural, biochemical and genetical approach can be made to specific problems (see Morris, this volume), one such problem is the nature of microtubule organising centres.

Table 1. *Number of microtubules composing half spindle in various eukaryotes*

Organism/cell	Microtubule number	Reference
Uromyces phaseoli	108	Heath & Heath, 1976
Thruastotheca clavata	41	Heath, 1974
Saccharomyces cerevisiae	21	Peterson & Ris, 1976
Fragilaria capucino	21	Tippit, Schulz & Pickett-Heaps, 1978
Human (HeLa cells)	2400 ($\pm$ 20%)	McIntosh & Landis, 1971
Haemanthus	1700	Jensen & Bajer, 1973

Data from Tippit *et al.* (1978).

Particularly illuminating in this respect are studies on the spindle of *Saccharomyces cerevisiae* (Peterson & Ris, 1976; Byers, Schriver & Goetsch, 1978; Hyams & Borisy, 1978). The yeast interphase nucleus has a single plaque-like spindle pole body inserted into the nuclear envelope and this divides into two as budding commences. Microtubules form at the intranuclear surface of each spindle pole body as the bud enlarges and the two daughter bodies migrate around the nucleus to occupy polar positions; the longer microtubules from opposite poles then interdigitate to form the spindle apparatus. The number of microtubules composing the spindle depends on cell ploidy. High voltage electron microscopy of serial thick sections has enabled the tracking of individual microtubules through complete spindles and indicated a one-to-one correlation between the number of linkage groups per nucleus (17 in the haploid strain) and the number of chromosomal (short) microtubules. The number of pole-to-pole (long) microtubules is small (five to ten) and does not vary with chromosome number. Identifiable kinetochores are not present, moreover the chromosomes remain uncondensed throughout nuclear division so are not individually visible. Apart from this, mitosis in yeast is comparable to that observed in typical eukaryotes (Peterson & Ris, 1976).

Spindle pole bodies have been isolated by osmotic lysis of yeast spheroplasts and shown to act as foci for microtubule polymerisation *in vitro* when challenged with chick or pig neurotubulin (Borisy, Peterson, Hyams & Ris, 1975; Byers *et al.*, 1978; Hyams & Borisy, 1978). Microtubule organising activity is sensitive to tryptic digestion but not to DNAase, RNAase or phospholipase (Hyams & Borisy, 1978). Microtubule growth is primarily on to the intranuclear face of the spindle pole body and there appears to be a limited number of sites for initiation, depending on the ploidy of the cell and the stage in the cell cycle at which the spindle pole bodies are isolated. Thus spindle pole bodies from exponentially growing cells nucleate a sub-set of microtubules equal in number to the interpolar microtubules seen *in vivo*, while those from stationary phase cultures (arrested in G1) initiate a number of microtubules equal to the total chromosomal plus interpolar microtubules of the strain from which they were derived. It appears that the two classes of microtubules composing the spindle are each subject to strict temporal control.

In yeast both chromosomal and pole-to-pole microtubules appear to arise from the spindle pole bodies in the absence of differentiated chromosome kinetochores. In higher eukaryotes the role of the kinetochore as a microtubule organising centre has been demonstrated most strikingly by inducing microtubule assembly at the kinetochores of isolated HeLa cell chromosomes *in vitro* (Telzer, Moses & Rosenbaum, 1975). Recent work on *Barbulanympha*, one of the giant flagellates from the hindgut of the wood roach (*Cryptocercus punctulatus*) has confirmed that kinetochores do not necessarily generate the chromosomal fibres in the mitotic spindle of unicellular eukaryotes (Ritter, Inoué & Kubai, 1978).

Mitosis in *Barbulanympha* and related hypermastigid flagellates was the subject of classic studies by Cleveland and his collaborators (see Cleveland, 1963, for references). The synchronous mitoses induced in these flagellates by the host's moulting hormone, together with the large size of the nucleus and chromosomes, make them ideal subjects for such investigations The giant centrioles – each with its proximal reproducing end and its distal fibrogenic centrosphere – were long regarded by cytologists as models for their smaller counterparts in animal cells, until electron microscopists (Grimstone & Gibbons, 1966; Hollande & Valentin, 1967a, b) showed that these centrioles lacked the characteristic nine microtubule-triplet ultrastructure and Hollande suggested the term 'atractophore' for them. Each fibrous rod-like atractophore descends from a half-rostrum at the anterior end of the organism and bears at its free end a spherical centrosphere, 8.5 μm in diameter, from which microtubule bundles radiate in astral arrays. Some of these astral bundles pass to the other centrosphere to form a birefringent central spindle, while others extend back towards the nucleus where they make contact with the already divided chromosome kinetochores inserted in the nuclear envelope. In this way chromosomal fibres are formed and they shorten to hoist the nucleus up towards the central spindle. How shortening is effected is unknown, but as the kinetochores are drawn towards the centrospheres, the nuclear envelope is locally drawn out as point protrusions. This poleward movement of chromosomes results in the segregation of daughter sets to the two poles of the nucleus and constitutes the first part of anaphase (anaphase A): in the second part (anaphase B) the central spindle elongates to as much as five times its original length, pushing the daughter sets apart and causing the nucleus to constrict into two.

FORCE PRODUCTION AND TRANSMISSION IN THE MITOTIC SPINDLE

The segregation of chromosome complements in anaphase is achieved by movement of chromosomes towards the spindle poles (anaphase A) and movement apart of the poles themselves (anaphase B). In *Barbulanympha* and the hypermastigids, these two components of anaphase are sequential, chromosomal fibre shortening being completed before the central spindle begins to elongate (Inoué & Ritter, 1978). In the parasitic dinoflagellate *Syndinium*, anaphase A is missing as the chromosome kinetochores on the nuclear envelope remain linked to the spindle poles by short microtubular fibres which do not shorten further as spindle elongation separates sister kinetochores.

We have mentioned earlier the possibility that the motile forces for the two movements of anaphase may derive from different molecular mechanisms. The velocity of chromosome movement is strictly proportional to the rate of loss of birefringence and slow disassembly of microtubules could provide the motive force in anaphase A; another possibility is that chromosomal fibres progressively shorten through activity of an actomyosin system as in muscle contraction. Microtubule assembly and disassembly have also been implicated in anaphase B but the sliding of polar microtubules relative to one another as in ciliary axoneme bending (Holwill, this volume) is currently the favoured explanation of spindle elongation (see Inoué & Kiehart, 1978). We present here evidence from unicellular eukaryotes supporting some of these ideas.

Several authors have reported the presence of actin and myosin in the half-spindle (chromosomal fibre) region. Anti-actin immunofluorescence was found at the poles and along the chromosomal fibres of rat kangaroo cells in culture (Cande, Lazarides & McIntosh, 1977) and the anti-actin unstained fibres were observed to shorten in anaphase. The interzone between the separating chromosomes could be stained with anti-tubulin but not anti-actin sera, showing that the anti-actin staining did not simply reflect microtubule distribution. Fluorescein-conjugated anti-myosin serum also stains the chromosomal fibre region of the spindle (Fujiwara & Pollard, 1976). Griffith & Pollard (1978) have presented evidence for actin-microtubule interaction mediated by microtubule-associated proteins. The characteristic 6 nm thick filaments of F-actin have rarely been observed in electron micrographs of the spindle but one notable exception is the dividing micronucleus of *Paramecium bursaria* (Lewis *et al.*, 1976). Here at metaphase the chromosomes are attached by way of elaborate cup-shaped kinetochores ('microlamellae') to a meshwork of microfilaments. Towards the polar regions the microfilaments join with microtubules that converge and terminate on the nuclear envelope. During metaphase and anaphase the chromosomes are apparently moved by the microfilaments pulling on their microlamellar attachments. No actin-specific staining was attempted on the microfilamentous region which disappeared at telophase. Inoué & Kiehart (1978) cast doubt on the role of an actomysin system in chromosome movement as microinjection of homologous anti-myosin antibody into *Asterias* eggs prevented several successive cleavage divisions without affecting nuclear division in their experiments.

McIntosh's sliding filament model of mitosis (McIntosh, Heppler & van Wie, 1969) attempted to account for the movements of both anaphase A and anaphase B by postulating sliding interaction between chromosomal microtubules and polar microtubules of antipolar polarity. Electron microscopists have observed lateral filaments projecting from spindle microtubules and sometimes forming bridges between them in the same way that the dynein arms on the A microtubules bridge peripheral doublets in the cilium or flagellum. Lateral interactions between the two types of spindle microtubules would be impossible in the spindles of many unicellular eukaryotes but not between polar microtubules of antipolar polarity, and some of the best evidence for this type of mechanism has come from serial reconstruction of microtubule distribution in the central spindle of diatoms (Pickett-Heaps *et al.*, 1975, 1978*a*, *b*; Tippit *et al.*, 1975; McDonald *et al.*, 1977; Tippit & Pickett-Heaps, 1977; Pickett-Heaps & Tippit, 1978).

One interesting feature of the diatom spindle as revealed by microtubule tracking is the presence of truly continuous microtubules stretching from pole to pole in prophase (*Fragilaria*), and occasionally persisting into metaphase (*Lithodesmium undulatum*; Manton *et al.*, 1970*a*, *b*) and even anaphase (*Diatoma vulgare*; McDonald *et al.*, 1977). An analogous situation occurs in certain fungal spindles, e.g. *Mucor hiemalis* (McCully & Robinow, 1973), *Phycomyces blakesleeanus*

(Franke & Dean, 1973), *Saccharomyces cerevisiae* (Peterson & Ris, 1976), and *Polysphondylium violaceum* (Roos, 1975); in these fungi it is possible that spindle elongation is largely by microtubule growth. Later events in *Fragilaria* mitosis suggest that such continuous spindles consist of two sets of oppositely polarised microtubules which grow by assembly, presumably at opposite ends. The distinction made by Fuge (1977) between spindles that have continuous microtubules (some fungi, diatoms) and those that do not (higher eukaryotes) may be quite artificial if continuous microtubules are converted into polar microtubules merely by sliding.

Microtubule sliding in cilia and flagella depends upon the mechano-chemically active inter-microtubule dynein bridges and as yet direct evidence that the cross bridges between spindle polar fibres are composed of dynein is lacking. Indirect evidence is available, however, in the demonstration that chromosome movement in the isolated sea urchin mitotic apparatus or lysed mitotic cell is inhibited by anti-dynein serum (Sakai *et al.*, 1976) but not by anti-myosin serum, as explained above. Also sodium orthovanadate, an effective inhibitor of dynein (but not myosin) ATPase activity, blocks spindle elongation in lysed kangaroo rat cells (Cande & Wolniak, 1978).

The morphological diversity of mitotic spindles described above provokes the question 'Are we also dealing with an equally wide variety of spindle movement mechanisms?'. The somewhat piecemeal understanding of mitotic mechanisms that we have today suggests that functional diversity may be much less marked than morphological diversity may lead us to believe, and that similar mechanisms may operate in both higher and lower eukaryotes.

REFERENCES

ALDRICH, H. C. (1969). The ultrastructure of mitosis in myxamoeba and plasmodia of *Physarum flavicomum*. *American Journal of Botany*, **56**, 290–9.
BĚLĂR, K. (1926). Der Formwechsel der Protistenkerne. Jena-Gustav Fischer, 1–420.
BERGER, J. D. & SCHMIDT, H. J. (1978). Regulation of macronuclear DNA content in *Paramecium tetraaurelia*. *Journal of Cell Biology*, **76**, 116–26.
BORISY, G. G., PETERSON, J. B., HYAMS, J. S. & RIS, H. (1975). Polymerisation of microtubules onto the spindle pole body of yeast. *Journal of Cell Biology*, **67**, 38a.
BOUCK, G. B. & BROWN, D. L. (1973). Microtubule biogenesis and cell shape in *Ochromonas*. I. The distribution of cytoplasmic and mitotic microtubules. *Journal of Cell Biology*, **56**, 350–59.
BRASELTON, J. P., MILLER, C. E. & PECHAK, D. G. (1975). The ultra-structure of cruciform nuclear division in *Sorosphaera veronicae* (Plasmodiophoromycete). *American Journal of Botany*, **62**, 349–58.

BRONCHART, R. & DEMOULIN, V. (1977). Unusual mitosis in the red alga *Porphyridium purpurem*. *Nature, London*, **268**, 80–1.

BYERS, B., SCHRIVER, K. & GOETSCH, L. (1978). The role of spindle pole bodies and modified microtubule ends in the initiation of microtubule assembly in *Saccharomyces cerevisiae*. *Journal of Cell Science*, **30**, 331–52.

CACHON, J. & CACHON, M. (1974). Comparaison de la mitose des Peridiniens libres et parasites à propos de celle des *Oodinium*. *Comptes rendus hebdomadaire des séances de l'Académie des sciences*, **278D**, 1735–7.

CACHON, J., CACHON, M. & SALVANO, P. (1978). Chromosome segregation without any microtubule involvement. *Journal of Protozoology*, **25**, 49A.

CANDE, W. Z. & WOLNIAK, S. M. (1978). Chromosome movement in lysed mitotic cells in inhibited by vanadate. *Journal of Cell Biology*, **79**, 573–80.

CANDE, W. Z., LAZARIDES, F. & McINTOSH, J. R. (1977). Comparison of the distribution of actin and tubulin in the mammalian mitotic spindle as seen by indirect immunofluorescence. *Journal of Cell Biology*, **72**, 552–67.

CHALY, N., LORD, A. & LAFONTAINE, J. G. (1977). A light and electron microscope study of nuclear structure throughout the cell cycle in the euglenoid *Astasia longa*. *Journal of Cell Science*, **27**, 23–45.

CLEVELAND, L. R. (1963). Functions of flagellate and other centrioles in cell reproduction. In *The Cell in Mitosis*, ed. L. Levine, pp. 3–31. New York: Academic Press.

DEASON, T. R. & DARDEN, W. H. (1971). The male initial and mitosis in *Volvox*. In *Contributions in Phycology*, ed. B. C. Parker & R. M. Brown, pp. 67–79. Lawrence, Kansas: Allen Press.

DODGE, J. D. (1973). *The Fine Structure of Algal Cells*. 261 pp. London & New York: Academic Press.

ELLIS, G. W. (1978). Advances in visualization of mitosis *in vivo*. In *Cell Reproduction*. ICN-UCLA Symposia on Molecular and Cellular Biology, vol. XII, ed. E. R. Dirksen, D. M. Prescott, & C. F. Fox, pp. 465–76. New York: Academic Press.

FLOYD, G. (1978). Mitosis and cytokinesis in *Asteromonas gracilis*, a wall-less green monad. *Journal of Phycology*, **14**, 440–5.

FORER, A. (1976). Actin filaments and birefringent spindle fibres during chromosomal movements. In *Cell Motility*, ed. R. Goldman, T. Pollard & J. Rosenbaum, pp. 1273–93. Cold Spring Harbor Laboratory.

FRANKE, W. W. (1974). Structure, biochemistry and functions of the nuclear envelope. *International Review of Cytology, Supplement* 4, 72–236.

FRANKE, W. W. & DEAN, P. (1973). Mitosis in *Phycomyces blakesleeanus*. *Archiv für Mikrobiologie*, **96**, 121–9.

FUGE, H. (1974). Ultrastructure and function of the spindle apparatus. Microtubules and chromosomes during nuclear division. *Protoplasma*, **82**, 289–320.

FUGE, H. (1977). Ultrastructure of the mitotic spindle. *International Review of Cytology, Supplement* 6, 1–58.

FUJIWARA, K. & POLLARD, T. D. (1976). Fluorescent antibody localization of myosin in the cytoplasm, cleavage furrow, and mitotic spindle of human cells. *Journal of Cell Biology*, **71**, 848–75.

FULLER, M. S. (1976). Mitosis in fungi. *International Review of Cytology*, **45**, 113–53.

FURTADO, J. S. & OLIVE, L. S. (1971). Ultrastructural evidence of meiosis in *Ceratiomyxa fruticulosa*. *Mycologia*, **63**, 413–16.

GILLOTT, M. A. & TRIEMER, R. E. (1978). The ultrastructure of cell division in *Euglena gracilis*. *Journal of Cell Science*, **31**, 25–35.

GIRBARDT, M. (1968). Ultrastructure and dynamics of the moving nucleus. In *Symposia of the Society for Experimental Biology*, **22**, ed. H. Smith & J. E. Pearce, pp. 249–59. Cambridge University Press.

GIRBARDT, M. (1971). Ultrastructure of the fungal nucleus. II. The kinetochore equivalent. *Journal of Cell Science*, **9**, 453–73.

GRIFFITH, L. M. & POLLARD, T. D. (1978). Evidence for actin filament-microtubule interaction mediated by microtubule-associated proteins. *Journal of Cell Biology*, **78**, 958–65.

GRIMSTONE, A. V. & GIBBONS, I. R. (1966). The fine structure of the centriolar apparatus and associated structures in the complex flagellates *Trichonympha* and *Pseudotrichonympha*. *Philosophical Transactions of the Royal Society, Series B*, **250**, 215–42.

HEATH, I. B. (1974). Mitosis in the fungus *Thraustotheca clavata*. *Journal of Cell Biology*, **60**, 204–20.

HEATH, I. B. (1978). Experimental studies of mitosis in fungi. In *Nuclear Division in the Fungi*, ed. I. B. Heath. New York: Academic Press.

HEATH, I. B. (1980). Variant mitoses in lower eukaryotes: Indicators of the evolution of mitosis? *International Review of Cytology*, **64**, in press.

HEATH, I. B. & GREENWOOD, A. D. (1970). Centriole replication and nuclear division in *Saprolegnia*. *Journal of General Microbiology*, **62**, 139–48.

HEATH, I. B. & HEATH, M. C. (1976). Ultrastructure of mitosis in cowpea rust fungus *Uromyces phaseoli* var. *vignae*. *Journal of Cell Biology*, **70**, 592–607.

HEYWOOD, P. (1978). Ultrastructure of mitosis in the chloromonadophycean alga *Vacuolaria virescens*. *Journal of Cell Science*, **31**, 37–51.

HEYWOOD, P. & GODWARD, M.B.E. (1972). Centromeric organization in the chloromonadophycean alga *Vacuolaria virescens*. *Chromosoma*, **39**, 333–9.

HEYWOOD, P. & MAGEE, P. T. (1976). Meiosis in protists. Some structural and physiological aspects of meiosis in algae, fungi and protozoa. *Bacteriological Reviews*, **40**, 190–240.

HEYWOOD, P. & WEINMAN, D. (1978). Mitosis in the haemoflagellate *Trypanosoma cyclops*. *Journal of Protozoology*, **25**, 287–92.

HOLLANDE, A. (1972). Le déroulement de la cryptomitose et les modalités de la ségrégation des chromatides dans quelques groupes de Protozoaires. *Annales Biologique*, **11**, (fasc. 9–10), 427–66.

HOLLANDE, A. (1974). Etude comparée de la mitose syndinienne et de celle des Péridiniens libres et des Hypermastigines. Infrastructure et cycle évolutif des Syndinides parasites de Radiolaires. *Protistologica*, **10**, 413–51.

HOLLANDE, A. & VALENTIN, J. (1967a). Interpretation des structures dites centriolaires chez les Hypermastigines symbiontes des termites et du *Cryptocercus*. *Comptes rendus hebdomadaire des séances de l'Académie des sciences*, **264D**, 1868–71.

HOLLANDE, A. & VALENTIN, J. (1967b). Morphologie et infrastructure du Genre *Barbulanympha*, hypermastigine symbiontique de *Cryptocercus punctulatus* Scudden. *Protistologica*, **3**, 257–87.

HOLLANDE, A. & VALENTIN, J. (1971). Les attractophores, l'induction du fuseau et la division cellulaire chez les Hypermastigines. *Protistologica*, **7**, 5–100.

HOLLANDE, A., CACHON, J. & CACHON, M. (1969). La Dinomitose attractophorienne à fuseau endonucléaire ches les Radiolaires *Thallassophysidae*: son homologie avec la mitose des Foraminifères et avec celle des levures. *Comptes rendus hebdomadaire des séances de l'Académie des sciences*, **269D**, 172–82.

HYAMS, J. S. & BORISY, G. G. (1978). Nucleation of microtubules *in vitro* by isolated spindle pole bodies of the yeast *Saccharomyces cerevisiae*. *Journal of Cell Biology*, **78**, 401–14.

ICHIDA, A. A. & FULLER, M. S. (1968). Ultrastructure of mitosis in the aquatic fungus *Catenaria anguillulae*. *Mycologia*, **60**, 141–55.

INOUÉ, S. (1976). Chromosome movement by reversible assembly of microtubules. In *Cell Motility*, ed. R. Goldman, T. Pollard & J. Rosenbaum, pp. 1317–28. Cold Spring Harbor Laboratory.

INOUÉ, S. & KIEHART, D. P. (1978). In-vivo analysis of mitotic spindle dynamics. In *Cell Reproduction*. ICN-UCLA Symposia on Molecular and Cellular Biology, vol. XII, ed. E. R. Dirksen, D. M. Prescott, & C. F. Fox, pp. 433–44. New York: Academic Press.

INOUÉ, S. & RITTER, J. JR. (1978). Mitosis in *Barbulanympha*. II. Dynamics of a two-stage anaphase, nuclear morphogenesis and cytokinesis. *Journal of Cell Biology*, **77**, 655–84.

JENKINS, R. A. (1977). The role of microtubules in macronuclear division of *Blepharisma*. *Journal of Protozoology*, **24**, 264–75.

JENSEN, C. & BAJER, A. (1973). Spindle dynamics of the arrangement of microtubules *Chromosoma, Berlin*, **44**, 73–89.

JOHNSON, U. G. & PORTER, K. R. (1968). Fine structure of cell division in *Chlamydomonas reinhardi*. Basal bodies and microtubules. *Journal of Cell Biology*, **38**, 403–25.

JURAND, A. & SELMAN, G. G. (1970). Ultrastructure of the nuclei and intranuclear microtubules of *Paramecium aurelia*. *Journal of General Microbiology*, **60**, 357–64.

KUBAI, D. F. (1973). Unorthodox mitosis in *Trichonympha agilis*: kinetochore differentiation and chromosome movement. *Journal of Cell Science*, **13**, 511–52.

KUBAI, D. F. (1975). The evolution of the mitotic spindle. *International Review of Cytology*, **43**, 167–227.

KUBAI, D. F. & RIS, H. (1969). Division in the dinoflagellate *Gyrodinium cohnii* (Schiller). *Journal of Cell Biology*, **40**, 508–28.

LEADBEATER, B. & DODGE, J. D. (1967). An electron microscope study of nuclear and cell division in a dinoflagellate. *Archiv für Mikrobiologie*, **57**, 239–54.

LEEDALE, G. F. (1968). The nucleus in *Euglena*. In *The Biology of Euglena*, vol. 1, ed. D. E. Buetow, pp. 185–242. New York: Academic Press.

LEEDALE, G. F. (1968). The nucleus in *Euglena*. In *The Biology of Euglena*, vol. 1, *of the New York Academy of Sciences*, **175**, 429–53.

LEIBOWITZ, P. J. & SCHAECHTER, M. (1975). The attachment of the bacterial chromosome to the cell membrane. *International Review of Cytology*, **41**, 1–28.

LEWIS, L. M., WITKUS, E. R. & VERNON, G. M. (1976). The role of microtubules and microfilaments in the micronucleus of *Paramecium bursaria* during mitosis. *Protoplasma*, **89**, 203–19.

McCULLY, E. K. & ROBINOW, C. F. (1972). Mitosis in heterobasidiomycetous yeasts. I. *Leucosporidium scottii* (*Candida scotti*). *Journal of Cell Science*, **10**, 857–81.

McCULLY, E. K. & ROBINOW, C. F. (1973). Mitosis in *Mucor hiemalis*: a comparative light and electron microscope study. *Archiv für Mikrobiologie*, **94**, 133–48.

McDONALD, K. D., PICKETT-HEAPS, J. D., McINTOSH, J. R. & TIPPIT, D. H. (1977). On the mechanism of anaphase elongation in *Diatoma vulgare*. *Journal of Cell Biology*, **74**, 377–88.

McINTOSH, J. R. & LANDIS, S. C. (1971). The distribution of spindle microtubules during mitosis in cultured human cells. *Journal of Cell Biology*, **49**, 468–92.

McINTOSH, J. R., HEPLER, P. K. & VAN WIE, D. G. (1969). Model for mitosis. *Nature, London*, **224**, 638–63.

MANTON, I. (1964). Observations with the electron microscope on the division cycle in the flagellate *Prymnesium parvum* Carter. *Journal of the Royal Microscopical Society*, **83**, 317–25.

MANTON, I., KOWALLIK, K. & VON STOSCH, H. A. (1969a). Observations on the fine structure and development of the spindle at mitosis and meiosis in a marine centric diatom (*Lithodesmium undulatum*). I. Preliminary survey of mitosis in spermatogonia. *Journal of Microscopy*, **89**, 295–320.

MANTON, I., KOWALLIK, K. & VON STOSCH, H. A. (1969b). Observations on the fine structure and development of the spindle at mitosis and meiosis in a marine centric diatom (*Lithodesmium undulatum*). II. The early meiotic stages in male gametogenesis. *Journal of Cell Science*, **5**, 271–98.

MANTON, I., KOWALLIK, K. & VON STOSCH, H. A. (1970a). Observations on the fine structure and development of the spindle at mitosis and meiosis in a marine centric diatom (*Lithodesmium undulatum*). III. The later stages of meiosis I in male gametes. *Journal of Cell Science*, **6**, 131–57.

MANTON, I., KOWALLIK, K. & VON STOSCH, H. A. (1970b). Observations on the fine structure and development of the spindle at mitosis and meiosis in a marine centric diatom (*Lithodesmium undulatum*). IV. The second meiotic division and conclusion. *Journal of Cell Science*, **7**, 407–43.

MARGOLIS, R. L. & WILSON, L. (1978). Opposite end assembly and disassembly of microtubules at steady state *in vitro*. *Cell*, **13**, 1–18.

MATTOX, K. R. & STEWART, K. D. (1977). Cell division in the scaly green flagellate *Heteromastix angulata* and its bearing on the origin of the Chlorophyceae. *American Journal of Botany*, **64**, 931–45.

OAKLEY, B. R. (1978a). Some advantages and limitations of mitosis as a phylogenetic criterion. *Biosystems*, **10**, 59–64.

OAKLEY, B. R. (1978b). Mitotic spindle formation in *Cryptomonas* and *Chroomonas* (Cryptophyceae). *Protoplasma*, **95**, 333–46.

OAKLEY, B. R. & BISALPUTRA, T. (1977). Mitosis and cell division in *Cryptomonas* (Cryptophyceae). *Canadian Journal of Botany*, **22**, 2789–800.

OAKLEY, B. R. & DODGE, J. D. (1973). Mitosis in the Cryptophyceae. *Nature, London*, **244**, 521–2.

OAKLEY, B. R. & DODGE, J. D. (1974a). Mitosis and cell division in *Tetraselmis*, a member of the Prasinophyceae. *British Phycological Journal*, **9**, 222.

OAKLEY, B. R. & DODGE, J. D. (1974b). Kinetochores associated with the nuclear envelope in the mitosis of a dinoflagellate. *Journal of Cell Biology*, **63**, 322–5.

OAKLEY, B. R. & DODGE, J. D. (1976). The ultrastructure of mitosis in *Chroomonas salina* (Cryptophyceae). *Protoplasma*. **88**, 241–54.

OAKLEY, B. R. & DODGE, J. D. (1977). Mitosis and cytokinesis in the dinoflagellate *Amphidinium carterae*. *Cytobios*, **17**, 35–46.

OAKLEY, B. R. & HEATH, I. B. (1978). The arrangement of microtubules in serially sectioned spindles of the alga *Cryptomonas*. *Journal of Cell Science*, **31**, 53–70.

OTT, D. W. & BROWN, R. M. (1972). Light and electron microscopical observations on mitosis in *Vaucheria litorea*. *British Phycological Journal*, **7**, 361–74.

PEARSON, B. A. & NORRIS, R. E. (1975). Fine structure of cell division in *Pyramimonas parkeae* Norris & Pearson. *Journal of Phycology*, **11**, 113–24.

PETERSON, J. B. & RIS, H. (1976). Electron microscopic study of the spindle and chromosome movement in the yeast *Saccharomyces cerevisiae*. *Journal of Cell Science*, **22**, 219–42.

PICKETT-HEAPS, J. D. (1969). The evolution of the mitotic apparatus: an attempt at comparative ultrastructural cytology in dividing plant cells. *Cytobios*, **1**, 257–80.

PICKETT,HEAPS, J. D. (1970). Mitosis and autospore formation in the green alga *Kirchneriella lunaris*. Protoplasma, **70**, 325–47.

PICKETT-HEAPS, J. D. (1972*a*). Cell division in *Cyanophora paradoxa*. New Phytologist, **71**, 561–7.

PICKETT-HEAPS, J. D. (1972*b*). Cell division in *Klebsormidium subtillissimum* (formerly *Ulothrix subtilissima*) and its possible phylogenetic significance. *Cytobios*, **6**, 167–83.

PICKETT-HEAPS, J. D. (1972*c*). Cell division in *Cosmarium botrytis*. *Journal of Phycology*, **8**, 343–60.

PICKETT-HEAPS, J. D. (1973). Cell division in *Tetraspora*. *Annals of Botany*, **37**, 1017–25.

PICKETT-HEAPS, J. D. (1974). The evolution of mitosis and the eukaryotic condition. *Biosystems*, **6**, 36–48.

PICKETT-HEAPS, J. D. (1975). Aspects of spindle evolution. *Annals of the New York Academy of Sciences*, **253**, 352–61.

PICKETT-HEAPS, J. D. (1976). Cell division in *Raphidonema longiseta*. *Archiv für Protistenkunde*, **118**, 209–14.

PICKETT-HEAPS, J. D. & FOWKE, L. C. (1970). Cell division in *Oedogonium*. II. Nuclear division in *O. cardiacum*. *Australian Journal of Biological Sciences*, **23**, 71–92.

PICKETT-HEAPS, J. D. & OTT, D. W. (1974). Ultrastructural morphology and cell division in *Pedinomonas*. *Cytobios*, **11**, 41–58.

PICKETT-HEAPS, J. D. & TIPPIT, D. H. (1978). The diatom spindle in perspective. *Cell*, **14**, 455–67.

PICKETT-HEAPS, J. D., MCDONALD, K. L. & TIPPIT, D. H. (1975). Cell division in the pennate diatom *Diatoma vulgare*. *Protoplasma*, **86**, 205–42.

PICKETT-HEAPS, J. D., TIPPIT, D. H. & ANDREOZZI, J. (1978*a*). Cell division in the pennate diatom *Pinnularia*. I. Early stages of mitosis. *Biologie Cellulaire*, **33**, 71–8.

PICKETT-HEAPS, J. D., TIPPIT, D. H. & ANDREOZZI, J. (1978*b*). Cell division in the pennate diatom *Pinnularia*. II. Later stages in mitosis. *Biologie Cellulaire*, **33**, 79–84.

PICKETT-HEAPS, J. D. & WEIK, K. L. (1977). Cell division in *Euglena* and *Phacus*. I. Mitosis. In *Mechanisms and Control of Cell Division*, ed. T. L. Rost & E. M. Gifford, pp. 308–36. Dowden, Hutchinson and Ross.

RAIKOV, I. (1973). Mitose intranucléaire acentrique du micronoyau de *Loxodes magnus*. Cilié Holotriche. Etude ultrastructurale. *Comptes rendus hebdomadaire des séances de l'Académie des sciences*, **276D**, 2385–8.

RAIKOV, I. B. (1976). Evolution of macronuclear organization. *Annual Review of Genetics*, **10**, 413–40.

RIS, H. (1975). Primitive mitotic mechanisms. *Biosystems*, **7**, 298–304.

RIS, H. & KUBAI, D. F. (1974). An unusual mitotic mechanism in the parasitic protozoan *Syndinium* Sp. *Journal of Cell Biology*, **60**, 702–20.

RITTER, H. JR., INOUÉ, S. & KUBAI, D. (1978). Mitosis in *Barbulanympha*. I. Spindle. structure, formation and kinetochore engagement. *Journal of Cell Biology*, **77**, 638–54.

ROBINOW, C. F. & MARAK, J. (1966). A fiber apparatus in the nucleus of the yeast cell. *Journal of Cell Biology*, **29**, 129–51.

ROOS, U. P. (1975). Mitosis in the cellular slime mold *Polysphondylium violaceum*. *Journal of Cell Biology*, **64**, 480–91.

ROTH, L. E. & DANIELS, E. W. (1962). Electron microscopic studies of mitosis in amoebae. II. The giant amoeba *Pelomyxa carolinensis*. *Journal of Cell Biology*, **12**, 57–78.

RYSER, U. (1970). Die Ultrastruktur der Mitosekerne in den Plasmodien von *Physarum poylcephalum*. *Zeitschrift für Zellforschung und mikroskopische Anatomie*, **110**, 108–30.

SAKAI, H., MABUSHI, I., SHIMODA, S., KURIYAMA, R., OGAWA, K. & MOHRI, H. (1976). Induction of chromosome motion in the glycerol isolated mitotic apparatus: nucleotide specificity and effects of anti-dynein and myosin sera on the motion. *Development, Growth & Differentiation*, **18**, 211–19.

SANGER, J. W. & SANGER, J. M. (1976). Actin localization during cell division. In *Cell Motility*, ed. R. Goldman, T. Pollard & J. Rosenbaum, pp. 1295–316. Cold Spring Harbor Laboratory.

SATO, H., ELLIS, G. W. & INOUÉ, S. (1975). Microtubular origin of mitotic spindle form birefringence. Demonstration of applicability of Wiener's equation. *Journal of Cell Biology*, **67**, 501–17.

SCHORNSTEIN, K. & SCOTT, J. (1978). Ultrastructure of cell division in *Porphyridium*. *Journal of Phycology*, **14**, (Supplement), 30.

SCHREVEL, J., ASFAUX-FOUCHET, F. & BAFORT, M. (1976). Étude ultrastructurale des mitoses multiples au cours de la sporogony du *Plasmodium berghei*. *Journal of Ultrastructure Research*, **59**, 332–50.

SCHUSTER, F. L. (1975). Ultrastructure of mitosis in the amoeboflagellate *Naegleria gruberi*. *Tissue and Cell*, **7**, 1–12.

SETLIFF, E. C., HOCH, H. C. & PATTON, R. F. (1974). Studies on nuclear division in basidia of *Poria latemarginata*. *Canadian Journal of Botany*, **52**, 2323–33.

SIEBERT, A. E. & WEST, J. A. (1974). The fine structure of the parasitic dinoflagellate *Haplozoon axiothellae*. *Protoplasma*, **87**, 17–35.

SLANKIS, T. & GIBBS, S. P. (1972). The fine structure of mitosis and cell division in the Chrysophycean alga *Ochromonas danica*. *Journal of Phycology*, **8**, 243–56.

SOMMER, J. R. & BLUM, J. J. (1965). Cell division in *Astasia longa*. *Experimental Cell Research*, **39**, 504–27.

STEWART, K. D., MATTOX, K. R. & CHANDLER, C. D. (1974). Mitosis and cytokinesis in *Platymonas subcordiformis*, a scaly green monad. *Journal of Phycology*, **10**, 65–79.

TELZER, B. R., MOSES, J. & ROSENBAUM, J. L. (1975). Assembly of microtubules onto kinetochores of isolated mitotic chromosomes of HeLa cells. *Proceedings of the National Academy of Sciences, USA*, **72**, 4023–7.

TIPPIT, D. H. & PICKETT-HEAPS, J. D. (1976). Apparent amitosis in the binucleate dinoflagellate *Peridinium balticum*. *Journal of Cell Science*, **21**, 273–89.

TIPPIT, D. H. & PICKETT-HEAPS, J. D. (1977). Cell division in the pennate diatom *Surirella ovalis*. *Journal of Cell Biology*, **73**, 702–27.

TIPPIT, D. H., MCDONALD, K. L. & PICKETT-HEAPS, J. D. (1975). Cell division in the centric diatom *Melosira varians*. *Cytobiologie*, **12**, 28–51.

TIPPIT, D. H., SCHULZ, D. & PICKETT-HEAPS, J. D. (1978). Analysis of the distribution of spindle microtubules in the diatom *Fragilaria*. *Journal of Cell Biology*, **79**, 737–63.

TRIEMER, R. E. & BROWN, R. M. (1974). Cell division in *Chlamydomonas moewusii*. *Journal of Phycology*, **10**, 419–33.

TRIEMER, R. E. & BROWN, R. M. (1977). Ultrastructure of meiosis in *Chlamydomonas reinhardtii*. *British Phycological Journal*, **12**, 23–44.

VICKERMAN, K. & PRESTON, T. M. (1970). Spindle microtubules in the dividing nuclei of trypanosomes. *Journal of Cell Science*, **6**, 365–83.

WELLS, K. (1977). Meiotic and mitotic divisions in the Basidiomycotina. In *Mechanisms and Control of Cell Division*, ed. T. L. Rost & E. M. Gifford, pp. 337–74. Dowden, Hutchinson & Ross.

ZICKLER, D. (1970). Division spindle and centrosomal plaques during mitosis and meiosis in some Ascomycetes. *Chromosoma, Berlin*, **30**, 287–304.

ZICKLER, D. (1973). Fine structure of chromosome pairing in ten Ascomycetes: meiotic and premiotic (mitotic) synaptonemal complexes. *Chromosoma, Berlin*, **40**, 401–16.

EXPLANATION OF PLATES

PLATE 1

A section through a late anaphase stage of mitosis in the dinoflagellate *Amphidinium carterae*. Note the remains of the spindle interzonal microtubules between the two daughter nuclei and running into the division tunnels through the nuclei. Chromosomes and a nucleolus are present in each nucleus. Magnification × 16 000. (After Oakley & Dodge, 1977.)

PLATE 2

Kinetochores in three algal groups. (*a*) A longitudinal section through a mitotic tunnel at metaphase in the dinoflagellate *Glenodinium hallii* showing microtubules, some of which are attached to the nuclear envelope (arrow) at depressions to which chromosomes are also attached. Magnification × 70 000. (*b*) Metaphase in *Cryptomonas* sp. (Cryptophyceae) showing a section of a chromosome to which microtubules are attached from either side. The kinetochores (arrow) appear very simple in structure. Magnification × 70 000. (After Oakley, 1978*b*.). (*c*) Part of a metaphase in *Tetraselmis tetrathele* (Prasinophyceae) showing what appears to be a pair of chromatids to one of which a distinct kinetochore can be seen to be attached (arrow). Part of the 'broken down' envelope is visible at the bottom of the figure. Magnification × 37 000. (Courtesy of B. R. Oakley.)

PLATE 3

A median section through a metaphase plate in the crynptomonad *Cryptomonas* sp. showing the condensed chromatin and spindle tubules some of which appear to end in the chromatin whilst others pass through it. The mitotic figure here is rather broad and the nuclear envelope is entirely broken down. Magnification × 48 000. (From Oakley & Bisalputra, 1977.)

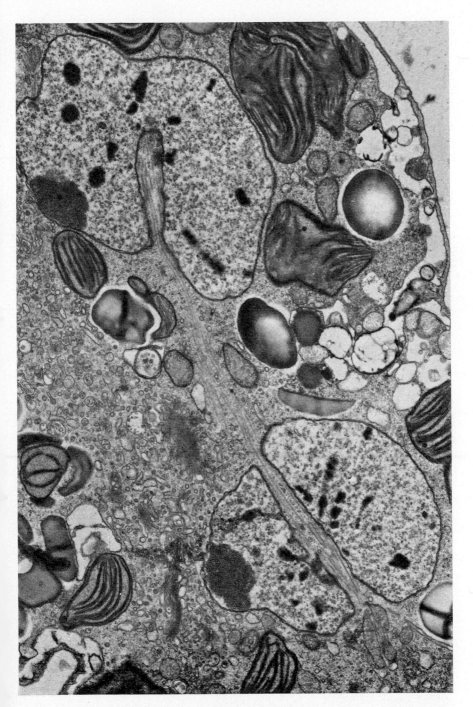

PLATE 1

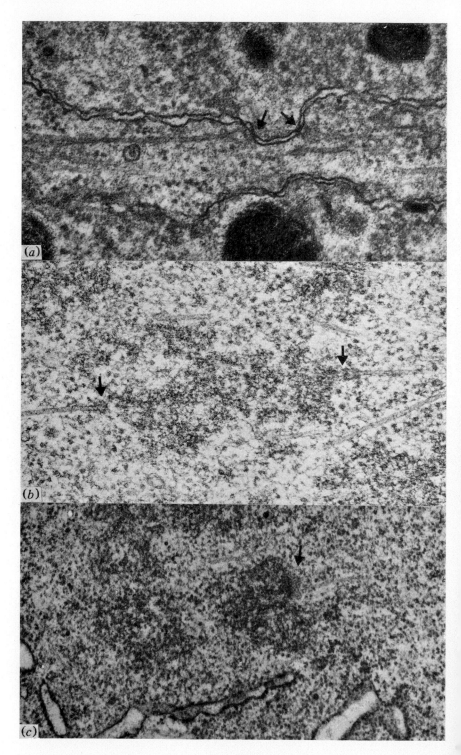

PLATE 2

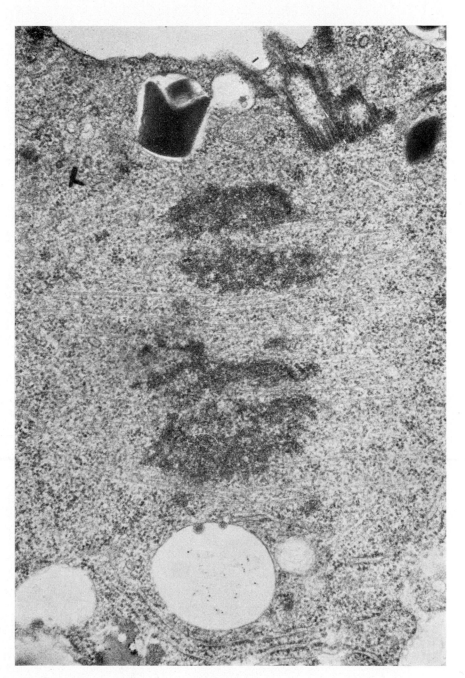

PLATE 3

THE PLASMA MEMBRANE OF
CANDIDA ALBICANS AND ITS ROLE
IN THE ACTION OF ANTI-FUNGAL DRUGS

DAVID KERRIDGE

Sub-department of Chemical Microbiology,
Department of Biochemistry,
University of Cambridge,
Tennis Court Road,
Cambridge CB2 1QW, UK

INTRODUCTION

The fundamental role of the plasma membrane in all living cells is to separate the cellular contents from the external environment. In prokaryotic organisms which lack specialised membraneous organelles, it provides for the transport of nutrients into, and metabolites out of the cell; it is involved in cell wall biosynthesis, and in aerobic bacteria it is the site of oxidative phosphorylation. Certain of these functions are associated with intracellular organelles in eukaryotic cells but the plasma membrane of eukaryotic cells possesses one additional property: exocytosis, a process by which vesicles derived from the Golgi apparatus fuse with the plasma membrane and discharge their contents to the exterior of the cell. This process is involved in both secretion and cell envelope biosynthesis. In addition, eukaryotic cells, which are not completely enclosed by a cell wall, can engulf soluble and particulate material by endocytosis (Stanier, 1970; Salton, 1978).

A number of antimicrobial agents interact specifically with the plasma membranes of both prokaryotic and eukaryotic cells but since selective toxicity depends upon differences between the pathogen and its host few can be used in the treatment of clinical infections. This is certainly so in the treatment of bacterial infections where the majority of antibiotics used clinically inhibit either mucopeptide or protein synthesis. (Gale *et al.*, 1972). It is perhaps surprising, therefore, that in the treatment of mycotic infections where we are dealing with a eukaryotic pathogen of a eukaryotic host, the two clinically important groups of antimycotic drugs both interact with the fungal plasma membrane causing impairment of function, cessation of growth and ultimately cell death (Cartwright 1975). The antibiotics used in the treatment of systemic fungal infections are shown in Table 1. The polyene antibiotics are produced by a variety of *Streptomyces* spp. (Hamilton-Miller, 1973) and these include amphotericin B, which is still the most effective drug for the treatment of systemic infections. Unfortunately its effectiveness is limited by serious side effects for the patient and the drug cannot be given prophylactically to patients who are at risk from fungal infections. There are similar problems with the synthetic imidazole antimycotic drugs in treating systemic infections. 5-Fluorocytosine is well tolerated, however, and high serum levels can be attained, but unfortunately selection of resistant strains of fungi is a clinical problem.

Table 1. *Clinically important antifungal drugs*

Antifungal drug	Pathogenic fungus	Target site
Polyene macrolide		Plasma membrane
Amphotericin B	*Aspergillus fumigatus*	
	Blastomyces dermatitidis	
	Candida albicans	
	Cryptococcus neoformans	
	Histoplasma capsulatum	
Candicidin ⎱	*Candida albicans*	
Nystatin ⎰	(superficial infections)	
Imidazole drugs		Plasma membrane
Clotrimazole	*Aspergillus fumigatus*	
Econazole	*Candida albicans*	
Miconazole		
5-Fluorocytosine	*Candida albicans*	RNA metabolism
	Cryptococcus neoformans	

Systemic fungal infections have assumed a greater relative importance since the introduction of an effective antibacterial therapy and the problem is further exacerbated by the absence of any really satisfactory antifungal drug. Apart from certain dermatophytes that are obligate pathogens, the pathogenic fungi have a saprobic existence and their pathogenic activities are incidental to their normal life cycle. The pathogenic fungi can be further divided into those fungi, for example *Histoplasma capsulatum* and *Coccidioides immitis*, which can infect a normal healthy host, and the other 'opportunistic pathogens', for example *Aspergillus fumigatus* and *Candida albicans*, which normally infect only those patients whose normal defence mechanisms are impaired. Unlike the other fungi, *C. albicans* occurs as a commensal associated with man and is important in that not only can it cause serious infections in a compromised host, but also it exhibits structural dimorphism and can exist in both a yeast and mycelial form (Kerridge 1978. Odds, 1979).

The role of the plasma membrane is not restricted to that of a target site for the antimycotic drug; it can provide a barrier and so prevent the drug reaching its target within the cell and even in some instances provide an active transport mechanism whereby the inhibitor is accumulated within the cell. This article is restricted to a consideration of the interaction of the clinically important (or potentially clinically important) antifungal drugs with the plasma membrane of *Candida albicans*.

ISOLATION AND CHARACTERISATION OF THE PLASMA MEMBRANE

In studying the mechanism of action of any antimicrobial drug it is essential to correlate the observations on growth and metabolism *in vivo* with the molecular interaction of the drug and its target *in vitro*. When the target is part of a subcellular organelle then clearly the first task after its identification as the primary site of action is its isolation and characterisation. In yeasts, isolation and characterisation of the plasma membrane is more complicated than from prokaryotic cells (Salton, 1978). Not only is there a complex cell wall to remove, but the cell also contains a number of membranous inclusions some of which (the Golgi apparatus and vesicles derived from it) are involved in the synthesis of the plasma membrane. An effective separation of the different membrane constituents present in the cell homogenate depends upon relatively small differences in size and density of the different components.

The three basic techniques used for the isolation of plasma membranes from yeasts are:

(1) enzymic removal of the cell wall using preparations ranging from the digestive juices of the snail *Helix pomatia* to purified β1–3 glucanases, followed by osmotic lysis and subsequent purification of the plasma membranes by differential and density gradient centrifugation (Longley, Rose & Knights, 1968; Schibeci, Rattray & Kidby, 1973a; Marriott, 1975; Santos, Villaneuva & Sentandreu, 1978);

(2) mechanical disruption of the cells followed by differential or density gradient centrifugation to purify the plasma membranes (Matile, Moor & Mühlethaler, 1967);

(3) sonic or mechanical disruption of the cells followed by centrifugal separation of the cell envelope from the other cellular constituents and finally enzymic digestion of the cell wall to leave the plasma membrane intact (Nurminen, Oura & Soumalainen 1970).

Different criteria have been used to measure the purity of the preparations; these range from the appearance of the plasma membranes in the electron microscope to the presence of specific enzymes (Matile *et al.*, 1967) and the specific isotopic labelling of the protoplast membrane (Schibeci *et al.*, 1973*b*; Marriott, 1975). These together with the different isolation procedures make it difficult to compare published analytical data. There is one further problem in *Candida albicans* in that it exhibits an environmentally controlled yeast mycelial dimorphism and at present it is not possible to distinguish between those differences in the composition of the plasma membrane which result from phenotypic variation and those which are primarily associated with the yeast mycelial transformation.

Table 2. *Chemical composition of plasma membranes*
of Candida albicans

	% composition[a]		
	Yeast		Mycelia
	Exponential phase	Stationary phase	Exponential phase
Protein	52 ± 2	43 ± 3	45 ± 3
Lipid	43 ± 3	31 ± 2.5	31 ± 3
Carbohydrates	9 ± 0.5	26 ± 3	25 ± 4
Nucleic acid	0.3 ± 0.1	0.2 ± 0.2	0.5 ± 0.1
Phospholipids			
Phosphatidyl-ethanolamine	70 ± 4	34 ± 2	50 ± 4
Phosphatidyl-serine Phosphatidyl-inositol	11 ± 2	11 ± 3	
Phosphatidyl-choline	4 ± 1	44 ± 3.5	50 ± 4
Sphingolipid	15 ± 4	10 ± 1	
Neutral lipids			
Sterol ester	40 ± 4	47 ± 4	28 ± 1
Triglyceride	24 ± 3	29 ± 2.5	36 ± 2
Free fatty acid	17 ± 3	12 ± 1	27 ± 4
Free sterol	19 ± 2	13 ± 1	9 ± 1

From Kerridge *et al.* (1976*b*). [a]Values ± standard deviation.

There have been no systematic analyses of the effect of the environment on the composition of the plasma membrane of *Candida albicans* but there are quantitative and qualitative differences in the chemical composition of membranes isolated from exponentially growing and stationary phase yeast cells and the exponentially growing mycelia (Marriott, 1975; Kerridge *et al.*, 1976*b*). These are shown in Table 2.

One major difference lies in the high carbohydrate content of the plasma membranes prepared from stationary phase yeast cells and the exponentially growing mycelia of *C. albicans*. This could reflect either a failure to remove all traces of the cell wall during protoplast preparations or an increased content of membrane glycoproteins (or both). Microscopic observations of protoplast production from the mycelial phase of *C. albicans* would support the latter hypothesis. However, given the difficulties of preparing protoplasts from stationary phase cultures of *C. albicans* the high carbohydrate content of plasma membranes prepared from these cells may reflect retention of some cell wall material. The other major differences reside in the content and type of phospholipids present. There are considerable differences in the relative proportions of the different types of phospholipids present in plasma membranes isolated from organisms harvested during the exponential and stationary phase of growth; phosphatidylserine, phosphatidylinositol and sphingolipid are absent from the plasma membranes prepared from exponentially growing mycelia.

PLASMA MEMBRANE – DRUG INTERACTIONS

The uptake of antimycotic drugs into Candida albicans

One of the primary functions of the plasma membrane is the selective transport of nutrients from the environment into the cell and a number of antifungal drugs are sufficiently similar in structure to the normal metabolites to be transported into the cell by the same mechanisms. Although the primary target of these antimycotics is far removed from the plasma membrane, this organelle plays an essential role in the inhibitory action of these compounds. So far only a limited number of such compounds have been studied, and of these, only 5-fluorocytosine is used clinically in the treatment of systemic candidiasis.

5-Fluorocytosine (Fig. 1) was originally developed as an antileukaemic drug but it has proved valuable in the treatment of certain mycotic infections. The mode of action of this compound is well understood (see Polak & Scholer, 1975); it is transported into *Candida albicans* by a cytosine permease which is also responsible for the uptake of adenine and hypoxanthine (Polak & Grenson, 1973). The uptake of each of the substrates adenine, cytosine, hypoxanthine and 5-fluorocytosine is competitively inhibited by the others and the apparent K_m for the uptake of each substrate is the same as the inhibition constant K_i when it is used to inhibit the uptake of the other compounds. Once inside the cell, 5-fluorocytosine is deaminated to 5-fluorouracil, which is subsequently incorporated into cellular RNA resulting in growth inhibition. The development of resistance by *C. albicans* to 5-fluorocytosine during therapy is common, but although it might be expected that this would frequently result from the loss of the cytosine permease, this is not necessarily so; Polak & Scholer (1975) examined 29 clinical isolates resistant to 5-fluorocytosine and found that none lacked the cytosine permease.

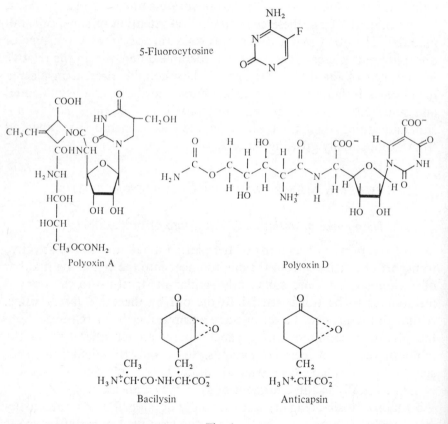

Fig. 1

The polyoxins are widely used in Japan to control *Alternaria kikuchiana*, the causative agent of pear black spot. Polyoxin D (Fig. 1) is a structural analogue of UDP-*N*-acetylglucosamine and competitively inhibits chitin synthetase (Gooday, 1977) which in *Saccharomyces cerevisiae* (Cabib, 1975) and *Candida albicans* (Braun & Calderone, 1978) is associated with the plasma membrane. However, it is not the interaction of the antibiotic with a specific membrane bound enzyme I wish to consider here, but rather the role of the plasma membrane in transporting polyoxins into the cell. Although it might be expected that a specific inhibitor of chitin synthetase would prove an ideal chemo-therapeutic agent for treating mycotic infections (since chitin is absent from the host) this has not been the case. For many fungi, including *C. albicans*, the minimum growth inhibitory concentration (MGIC) is greatly in excess of the concentration of polyoxin D required to inhibit chitin synthetase *in vitro* (Table 3) and it would appear that the antibiotic

Table 3. *The inhibition of fungal growth and chitin synthetase by polyoxin D*

	Aspergillus fumigatus[a]	Coprinus cinereus[b]	Mucor rouxii[c]	Neurospora crassa[d]	Saccharomyces cerevisiae[e]
MGIC (μM)	–	0.5	19	190	2000
K_1 chitin synthetase (μM)	5.2	3	0.6	1.4	0.5

[a]Archer (1977); [b]Gooday & De Rousset-Hall (1975); [c]McMurrough & Bartnicki-Garcia (1971); [d]Glaser & Brown (1957); [e]Keller & Cabib (1971), Bowers, Levin & Cabib (1974).

is not reaching its target site within the cell. In *A. kikuchiana* (Hori, Kakiki & Misato, 1977), polyoxin A inhibits both germ tube formation and chitin synthesis *in vitro* and the inhibitory effects on growth but not chitin synthetase *in vitro* are reversed by the addition of dipeptides. These dipeptides also inhibit the uptake of ^{14}C-labelled polyoxin A into the fungus and, for glycylglycine, kinetic analysis demonstrated that it competitively inhibits the uptake of polyoxin A. Similar results on the effect of peptides on the antifungal action of polyoxin D have been reported by Mitani & Inoué (1968) in *Pellicularia sasakii*.

Another example of an inhibitor being transported into the cell on a transport system normally reserved for peptides is bacilysin (Fig. 1). This compound is a dipeptide antibiotic produced by *Bacillus subtilis* (Abraham & Florey, 1949) and is active against both bacteria and *Candida albicans* (Kenig & Abraham, 1976). The activity of bacilysin is antagonised by a variety of peptides and by glucosamine and *N*-acetyl glucosamine. Bacilysin is broken down by extracts of *Staphylococcus aureus* into its constituent amino acids and the anticapsin (Fig. 1) produced inhibits glucosamine synthetase. The results are consistent with the hypothesis that bacilysin is transported into *C. albicans* by a peptide transport system and growth inhibition results from an inhibition of glucosamine synthetase by the anticapsin produced by intra-cellular hydrolytic cleavage of bacilysin.

Peptide transport into bacteria has recently been exploited by Allen *et al.* (1978) in the development of the phosphonopeptides, a new class of synthetic antibacterial agents. Polyoxin D and bacilysin are transported in fungi by a peptide transport system; it is possible that studies on the uptake mechanism for, and subsequent metabolism of these compounds could provide the basis for the development of a new range of clinically important antimycotic drugs.

The plasma membrane as a barrier to the entry of antibiotics

The intrinsic resistance of Gram negative bacteria to a variety of antibiotics has long been known to be associated with the failure of the antibiotic to penetrate the outer membrane barrier (Costerton, Ingram & Cheng, 1974). Rough mutants of *Salmonella typhimurium* are more susceptible to hydrophobic antibiotics than the parent strain (Roantree, Kuo & MacPhee, 1977) and Nikaido (1976) proposed that penetration of hydrophobic antibiotics occurs when the normally hydrophilic surface of the outer membrane is altered, so exposing the phospholipid bilayer and allowing the inhibitors to penetrate by dissolving in the hydrocarbon interior of the outer membrane. A more detailed genetic and chemical analysis of the role of the cell envelope in drug resistance in *Neisseria gonorrhoeae* has demonstrated the importance of both peptidoglycan structure and the outer membrane proteins (Guymon, Walstad & Sparling, 1978).

In fungi the role of the cell envelope in determining intrinsic resistance has not been so thoroughly examined. There is evidence that the cell wall of *Saccharomyces cerevisiae* has a limited porosity and will exclude polyethylene glycols of molecular weight greater than 800 (Scherrer, Louden & Gerhardt, 1974). In spite of this, polyene antimycotics with molecular weights in the region of 1000 can penetrate to the plasma membrane.

Phenotypic changes do occur in the structure and organisation of the cell wall of *Candida albicans* after the cessation of growth which result in increased resistance to both the polyene antibiotics and the imidazole drugs (Gale, 1974; Kerridge *et al.*, 1976*b*; Cassone, Kerridge & Gale, 1979) but there have been no studies on the role of the plasma membrane of *C. albicans* in determining resistance to antimycotic drugs comparable to those of Rose and his colleagues on alcohol tolerance in *Saccharomyces cerevisiae* (e.g. Thomas, Hossack & Rose, 1978).

In *Saccharomyces cerevisiae* the role of the plasma membrane in determining resistance to a number of inhibitors of mitochondrial function has been examined by Rank, Robertson & Phillips (1975), and

Rank, Robertson & Bussey (1978). In the multiple resistant strains they isolated, resistance did not result from either enzymic modification of the drugs or from modification of the cell wall, since protoplasts were also resistant. They examined the possibility that this generalised decrease in permeability to a number of inhibitors resulted from an alteration in either the lipid composition or viscosity of the plasma membrane. Experimental artefacts were eliminated by using four different methods to prepare plasma membranes from sensitive and multiple drug-resistant isolates of two isogenic groups of *S. cerevisiae* but no differences were detected in either lipid composition or membrane viscosity which could account for multiple drug resistance in these strains. However, given the results of Guymon *et al.* (1978) on drug resistance in bacteria, it would certainly be of interest to examine the plasma membrane proteins in these strains of yeast.

The plasma membrane as a target for antibiotic action

In this section I wish to consider the two groups of clinically important antimyotic drugs which interact with the plasma membrane of *Candida albicans* causing cessation of growth and ultimately cell death. Although both the polyene antibiotics and the synthetic imidazole drugs impair membrane function, the molecular basis of action is quite distinct and the groups will be discussed separately. Historically, the polyenes predate the imidazoles and the first polyene, nystatin, was isolated in 1950 by Hazen & Brown. Since then some 200 different polyenes have been isolated, but only a few are sufficiently non-toxic to be used clinically. These antibiotics are characterised by a carbon ring containing both a conjugated double bond system and an hydrophilic region and closed by lactonisation. The different polyenes differ in the number of carbon atoms in the ring, the number of conjugated double bonds, the number of hydroxyl groups and the presence or absence of a glycosidically linked carbohydrate (Hamilton-Miller, 1973; Cartwright, 1975; Thomas, 1976; Norman, Spielvogel & Wong, 1976; Kobayashi & Medoff, 1977; Kerridge, 1979). The complete three dimensional structure of amphotericin B has been elucidated by Mechlinski *et al.* (1970) and it is an elongated rigid molecule with opposing hydrophobic and hydrophilic faces (Fig. 2).

The specificity of the interaction of the polyene macrolide antibiotics with plasma membranes of sensitive organisms and the extent of the resulting damage depends both on the antibiotic and the lipid composition, in particular the sterols, of the membrane (Norman *et al.*, 1976). The molecular mechanisms by which the polyenes interact with the plasma membranes of sensitive cells differ from one polyene to another. In the case of members of the filipin complex the association of the antibiotic with the membrane sterol is thought to result in the formation of complexes some 20 nm in diameter within the hydrophobic region of the lipid bilayer causing gross membrane damage (Verkleij *et al.*, 1973).

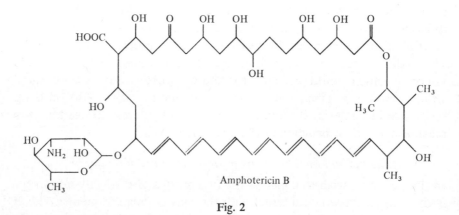

Fig. 2

The incorporation of the larger polyenes (amphotericin B and nystatin) into the plasma membrane of sensitive organisms would appear to result in the production of an aqueous pore consisting of an annulus of eight amphotericin molecules linked hydrophobically to the membrane sterol. This gives rise to an aqueous pore in which the hydroxyl residues of the polyene face inwards to give an effective pore diameter of from 0.4–1 nm. The length of the annulus is such that two half pores are required to span the plasma membrane (de Kruijff & Demel, 1974; Andreoli, 1973; Holz, 1974). Marty & Finkelstein (1975) have proposed a somewhat modified model in which either a single or double polyene annulus can span the plasma membrane in each case distorting the lipid bilayer in the vicinity of the aqueous pore.

However, the role of the sterol in mediating the interaction of the polyene molecule with the plasma membrane requires revision in the light of recent experimental findings. In-vitro studies with artificial membranes (Hsu-Chen & Feingold, 1973) and in-vivo studies with *Mycoplasma mycoides* subspecies *Capri* (Archer, 1976) have provided evidence that the fluidity of the lipid phase of the plasma membrane is also of primary importance in the polyene interaction. Under certain conditions sterols, far from enhancing the disruptive effect of polyenes on liposomal membranes, can have the opposite effect. Liposomes prepared from lecithin with saturated fatty acid side chains are sensitive to polyenes but if a sterol is incorporated into the bilayer they are resistant. It has been suggested that these results can be explained by an osmotic swelling and breakage of the liposomes (de Kruijff *et al.*, 1974) but since *Mycoplasma mycoides* subspecies *Capri* is not osmotically fragile nor does it leak K^+ at 2 °C (Archer, 1976) it is unlikely that all the results can be explained in this manner. It would appear that both the membrane sterol and the overall state of the membrane lipid organisation determine its sensitivity or resistance to the polyenes and the disruptive interaction demands that the plasma membrane lipids are in an ordered state. The relatively simple sterol polyene models so far proposed need to be modified to take these findings into account.

The molecular basis of the interaction of the polyenes with the plasma membrane of *Candida albicans* is still uncertain but there are two further questions to be considered: firstly, how does this interaction result in the cessation of growth and ultimately cell death, and secondly, what environmental and genetic factors affect the interaction of the antibiotics with the plasma membrane? The latter question has some considerable clinical importance since the maximum serum levels of amphotericin B that can be attained are low and even minor changes in the sensitivity of *C. albicans* could result in a failure of therapy (Hamilton-Miller, 1972*a*). One of the first observable effects after the addition of amphotericin B to a culture of *C. albicans* is a release of intracellular K^+ ions; a feature that has been used by a number of workers to monitor the interaction of the polyenes with the plasma membrane (Hammond, Lambert & Kliger, 1974; Gale, 1974; Johnson, White & Williamson, 1978). It has been assumed that cessation of growth results from the loss of K^+ ions and other low molecular weight constituents of the cell and that death results from the concomitant uptake of protons into the cell (Lampen, 1966). Support for this hypothesis comes from the finding by Liras & Lampen (1974) that the growth inhibitory effects of candicidin on *Saccharomyces cerevisiae* are prevented by the addition of 85mM-KCl and 45mM-$MgCl_2$. Similarly with *C. albicans* the inhibitory effects of amphotericin B methyl ester on both cell growth and the incorporation of ^{14}C-labelled phenylalanine into protoplasts are reversed by the addition of both K^+ and Mg^{2+}, but not when each ion is added separately (Kerridge, Koh & Johnson, 1976*a*). Unlike the protection afforded by the addition of sterols to the growth medium (Archer & Gale, 1976) the protection by cations does not result from the prevention of polyene binding to the plasma membrane since the inhibitory effects are observed immediately the protoplasts are harvested and resuspended

in a fresh medium lacking both polyene and protecting ions. The interaction of amphotericin B methyl ester with the protoplast membrane is
reversible provided the protecting ions are present after the removal of
the antibiotic from the suspending medium. The inhibition of protoplast
synthesis by filipin cannot be prevented by the addition of K^+ and Mg^{2+}
giving further support to the hypothesis that although the target
organelle is the same, the molecular mechanisms are different. The
presence of both cations is required to reverse the fungistatic effect of
amphotericin B on *C. albicans* but either ion alone prevented both the
fungicidal effects of amphotericin B and the massive uptake of protons
which occurred when the polyene was added to a suspension of the
organism in an unbuffered medium at pH 3.7 (Table 4). Proton uptake
did not occur when amphotericin B was added to a suspension of *C.
albicans* at pH 5.8 (Kerridge & Ingram, unpublished observations).

Table 4. *The protective effects of ions on the fungicidal action of
amphotericin B on* Candida albicans

Additions	pH of the suspending medium		Survival (%)
	Initial	Final (10 min)	
None	3.6	3.6	100
Amphotericin B (1μg ml^{-1})	3.6	5.7	0.2
Amphotericin B (1μg ml^{-1}) 45mM-MgCl$_2$	3.75	3.9	100
Amphotericin B (1μg ml^{-1}) 85mM-KCl	3.8	3.8	100
Amphotericin B (1μg ml^{-1}) 45mM-MgCl$_2$ 85mM-KCl	3.8	3.8	100

C. albicans was incubated at a suspension density of 1mg dry wt per ml
in an unbuffered medium; salts and amphotericin B were added as
required. The pH value was monitored using a recording pH meter and
the viability determined by serial dilution and plating onto a nutrient
medium.

The relationship between amphotericin-induced membrane permeability and its fungicidal action has been questioned by Hsu-Chen &
Feingold (1974). In examining the properties of a number of mutagen-
induced polyene-resistant strains of *Candida albicans* they observed that
in two of their isolates polyene-induced release of phosphate or ultra-
violet-absorbing material did not result in, or parallel, extensive killing.
This finding was confirmed by examining polyene-induced leakage from
liposomal vesicles prepared from total lipid extracts of both the parent
strain and the resistant mutants. Subsequently Chen, Chou & Feingold

(1978) examined both polyene-induced K$^+$ release and the fungicidal effects of these antibiotics on *C. albicans* and found that these effects could be dissociated, with nystatin proving to be rather more effective than amphotericin B in causing leakage of K$^+$ rather than cell death. The hypothesis that K$^+$ release is the primary event after addition of antibiotic has been questioned in an investigation of amphotericin B inhibition of maltose fermentation in *Saccharomyces cerevisiae* by Palacios & Serrano (1978), who demonstrated a polyene-induced increase in the permeability of the plasma membrane to protons. Since K$^+$ accumulation in yeast is linked to a proton gradient (Peña, 1975), this could account for the polyene-induced K$^+$ release. De Kruijff *et al.* (1974) have reported that pimaricin, although an effective antifungal antibiotic, binds to *Acholeplasma laidlawii* cells and egg lecithin liposomes without affecting their permeability and it would appear that for this polyene antibiotic impairment of membrane functions other than permeability are involved in its activity. Amphotericin B, nystatin and levorin have been shown to inhibit ATPase and lactate dehydrogenase in a membrane fraction derived from lysed protoplasts of *C. albicans* (Solov'eva, Belousova & Tereshin, 1976). The concentration of polyene required to inhibit the activity of these enzymes *in vitro* was greatly in excess of the minimum growth inhibitory concentration, but the authors suggested that the fungistatic effects of these antibiotics could result in part from an inhibition of membrane-bound enzymes. So in spite of all the evidence that has been adduced in favour of a polyene-induced loss of low molecular weight cellular constituents being responsible for their fungicidal and fungistatic effects, it may be that the production of aqueous pores in the plasma membrane is not the only event responsible for growth inhibition and we must look for an inhibition of other plasma membrane functions.

I wish now to consider the effect of both phenotypic and genotypic changes in the composition of the plasma membrane of *Candida albicans* on sensitivity of the organism to amphotericin B. There has unfortunately not been a systematic analysis of phenotypic variation in the composition of the plasma membrane and a correlation of such changes with polyene sensitivity. Johnson *et al.* (1978) examined the factors affecting the susceptibility of *C. albicans* to polyene-induced K$^+$ release using continuous culture techniques and found that it was influenced by the nature of the carbon source, growth rate, growth temperature and pH value of the growth medium. There were changes in the internal concentration of K$^+$ which could affect the polyene-induced K$^+$ release but the authors did not investigate the composition of the plasma membrane under the different environmental conditions – a daunting task but an important one, since environmentally induced changes in plasma membrane composition can affect sensitivity of yeasts to polyene antibiotics. In *Saccharomyces cerevisiae* growth temperature affects both the sterol content of the cells and their ability to bind nystatin (Venables & Russell, 1975). After growth at 20 °C the sterol content was 0.7% of the dry weight and 2.4×10^6 molecules of nystatin were bound per cell whereas growth at 40 °C resulted in a reduction of the total sterol to 0.37% of the dry weight and 9.8×10^5 molecules of nystatin were bound per cell. The interaction of amphotericin B methyl ester (AME) with the plasma membrane of *C. albicans* is also influenced by the fatty acid content, Koh *et al.* (1977). The growth requirements of a fatty acid auxotroph of *C. albicans* can be satisfied by a variety of unsaturated fatty acids and the sensitivity to AME (measured by K^+ release) ranged from 0.08 ± 0.02 μg ml^{-1} for organisms grown in the presence of palmitoleic acid ($C_{16:1}$) to 1.2 ± 0.3 μg ml^{-1} when grown in the presence of oleic acid ($C_{18:1}$). The sterol contents of the organism were little affected by the nature of the added unsaturated fatty acid. As in the parent strain, stationary phase cells from the mutant were less sensitive to AME and for the organisms grown in the presence of oleic acid, protoplasts were also resistant and here the resistance was associated with a change in the plasma membrane rather than in the cell wall (cf. Gale, 1974). This is a result which clearly supports the hypothesis that the sterol phospholipid interactions as well as the sterol *per se* are important in determining the interaction of the polyene with the plasma membrane of sensitive cells. Not all changes in the lipid composition of the plasma membrane of *C. albicans* affect polyene sensitivity. There are major differences between the lipid compositions of the plasma membranes isolated from the yeast and mycelial phases of *C. albicans* yet protoplasts derived from these cultures show the same sensitivity to AME-induced K^+ release (Table 1, Marriott, 1974).

The occurrence of polyene resistant mutants of *Candida albicans* is not a clinical problem. In a study of 2015 clinical isolates of *C. albicans*, Athar & Winner (1971) found none resistant to either nystatin or amphotericin B. Polyene resistant mutants of *C. albicans* have been isolated in a number of laboratories and used to study both sterol metabolism and the interaction of the polyene with the plasma membrane. It is clear from the literature that in the majority of cases the sterol metabolism of the mutant strain differs from that of the parent but the results are conflicting and it is impossible to derive a unified hypothesis for the mechanism of resistance. Polyene resistance in *C. albicans* has been associated with a decreased level of ergosterol in mutants selected by continuous subculture in the presence of the antibiotic (Athar & Winner, 1971) and with an increased level of total ergosterol in resistant strains isolated after mutagenesis with N-methyl-N'-nitro-N-nitroso-guanidine (Hamilton-Miller, 1972b). In neither case were the plasma membranes isolated and characterised nor was any account taken of possible phenotypic variation in lipid composition. Virina et al. (1976) isolated a strain of C. albicans resistant to amphotericin B and found that although ergosterol was apparently absent and replaced by other sterols, the organism bound as much amphotericin B as the parent sensitive strain. The sterols isolated from both the sensitive and resistant strains were capable of protecting the sensitive strain from the action of amphotericin B. The authors suggest that although the sterols of the resistant strain retain the ability to interact with amphotericin B they do not form specific complexes with the antibiotic within the plasma membrane and so membrane permeability is not impaired.

More detailed analyses of the lipid composition of polyene resistant strains of Candida albicans produced by mutagen treatment have been carried out by Subden et al. (1977) and Pierce et al. (1978) but the results have been of more value in elucidating the pathways of sterol bio-synthesis in this organism than the mechanism of resistance. It is clear from the results of Pierce et al. (1978) that alterations in the lipid alone are insufficient to explain the resistance. Their results are summarised in Table 5. There are significant differences in the total amount of sterol

Table 5. *Properties of polyene-resistant strains of* Candida albicans

Strain	MGIC (μg ml^{-1})		Total sterol		Major sterol (%)
	Nystatin	Amphotericin B	% Dry wt	% Esterified	
Parent	8	0.23	0.15	27	Ergosterol (62.5)
C7	25	1.36	0.189	19	Ergosterol (70.4)
E4	102	6.9	0.367	34	ergosta-5, 8, 22 -trienol (46.5)
C4	250	500	0.722	26	24-methyl, 24, 25 -dihydrolanosterol
D10	1000	500	0.499	26	24-methyl, 24, 25 -dihydrolanosterol

Data taken from Pierce et al. (1978).

present with the resistant strains having in all cases more than the parent and apart from strain C7, where ergosterol was still the major sterol, there are differences in the sterol composition. In these mutants polyene resistance can best be correlated with the total quantity and structure of

the sterols present but it is clear from a comparison of the parent strain
and strain C7, and strains C4 and D10, which have similar sterol and
lipid compositions, that additional factors must be involved in the
development of resistance to the polyenes. These results, like the others
on polyene resistance in *C. albicans*, were obtained by analysis of the
total lipid fraction, not of purified plasma membranes, and although
there is evidence of phenotypic variation in the lipid composition of
these strains no attempt was made to study this further.

The evidence that the plasma membrane of *Candida albicans* is the
primary target for the imidazole antimycotics is less convincing than for
the polyenes. It is possible from the data available that the fungistatic
and fungicidal properties of these compounds result from the inhibition
of a number of cellular functions one of which is the permeability of the

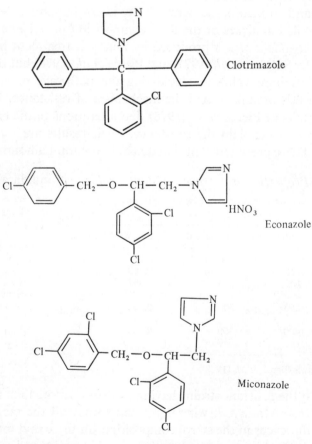

Fig. 3

plasma membrane. The imidazoles in current clinical use are: clotrimazole (Plempel *et al.*, 1969, 1970); miconazole (van Cutsem & Thienpont, 1972); and econazole (Thienpont *et al.*, 1975). They are all hydrophobic molecules with an essential imidazole moiety (Fig. 3) and it is perhaps not surprising that the plasma membrane is a potential target for their action and that apart from minor differences their modes of action are similar. The evidence for their interaction with the plasma membrane of *C. albicans* can be grouped under three headings; (i) effects on membrane permeability, (ii) binding studies, and (iii) effects on cellular morphology.

The first indication that these compounds interacted with the plasma membrane of *Candida albicans* came from observations that at fungicidal (50 μM) but not fungistatic (13 μM) concentrations, clotrimazole caused a massive release of radioactivity from organisms previously labelled with either ^{32}P or ^{42}K (Iwata, Yamaguchi & Hiratani, 1973). Similar results were obtained for miconazole which at the MGIC (20 μM) caused a release of Na^+- and ^{32}P-labelled material but not K^+ from *C. albicans*, at concentrations greater than 50 μM there was an appreciable leakage of all cellular material presumably associated with gross membrane damage (Swamy, Sirsi & Rao, 1974). Addition of either Mg^{2+} or Ca^{2+} at concentrations above 1 mM antagonised the miconazole-induced leakage of ultraviolet-light absorbing material, presumably by preventing its association with the plasma membrane. Miconazole at concentrations below the MGIC had a selective effect on the uptake of nutrients into *C. albicans* (van den Bossche, 1974). At 1.04 nM, miconazole had no effect on the uptake of glucose, glycine or leucine, reduced the uptake of adenine, guanine and hypoxanthine and enhanced the uptake of adenosine and guanosine.

The studies of the binding of [^{3}H]miconazole to the cellular constituents of *Candida albicans* do not provide conclusive evidence for a specific association with the plasma membrane (van den Bossche, 1974). 75% of the bound radioactivity was associated with the 16 000 g pellet (cell walls and plasma membranes) after 7 h incubation. If however the incubation was continued for 24 h, 57% of the bound radioactivity was found in the 100 000*g* pellet (microsomal fraction) and only 19% in the 16 000*g* pellet.

The morphological changes in *Candida albicans* resulting from incubation in the presence of the imidazole drugs have been reported by Iwata *et al.* (1973) for clotrimazole; De Nollin & Borgers (1974, 1975) for miconazole; and Preusser (1976) for econazole. In all cases the presence of the antimycotic drug at fungistatic concentrations results in changes in the cell envelope, primarily in the plasma membrane and in the structure and organisation of the cellular organelles; at fungicidal concentrations the changes in the cellular membranes are more pronounced and associated with major degradative changes in the cell.

Further support for the hypothesis that the plasma membrane is a target for the imidazole antimycotics comes from the finding that miconazole will lyse mammalian erythrocytes (Swamy, Sirsi & Rao, 1976) and that the antifungal activity of both clotrimazole and miconazole is antagonised by several classes of lipids containing unsaturated fatty acids (Yamaguchi, 1977, 1978). Although Iwata et al. (1973) were unable to detect any effect of clotrimazole on either respiration of Candida albicans or on mitochondria isolated from C. albicans these results were not confirmed by Dickinson (1977) who found that miconazole (100 μM) caused an uncoupling and, at higher concentrations, an inhibition of oxidative phosphorylation in isolated rat liver mitochondria, with concomitant damage to the mitochondrial membrane.

Recent studies have suggested that impairment of plasma membrane functions may not be the complete explanation of the mode of action of the imidazole drugs. The first was by de Nollin et al., 1977, who observed a decreased cytochrome c oxidase and cytochrome c peroxidase activity and an increased catalase activity in Candida albicans when incubated in the presence of miconazole at a fungistatic concentration. At a fungicidal concentration these enzymes disappeared and these authors put forward an hypothesis involving these enzymes and cellular hydrogen peroxide to account for both the fungistatic and fungicidal action of miconazole. There has been a more interesting development recently with the finding by van den Bossche et al. (1978) that miconazole at concentrations as low as 0.1 μM, causes a reduction in the synthesis of ergosterol and an accumulation of methylated sterols (14-methyl fecosterol and obtusifoliol, a 4,14,-dimethyl sterol) in C. albicans. A result of great significance in the light of the findings of Nes et al. (1978) is that sterols with methyl groups on C-4 and C-14 will not support growth of Saccharomyces cerevisiae under strictly anaerobic conditions. It could be that the imidazole drugs exert their inhibitory effect on C. albicans not only by impairing the functioning of the existing plasma membrane but also by inhibiting the formation of a functional membrane.

FUTURE PROSPECTS

What of the future? Clearly the most important task is a systematic study of phenotypic variation in the composition of the plasma membrane of Candida albicans. There is already evidence that such changes can not only influence the molecular interactions between antimycotic drugs and the plasma membrane but also are important in the structural dimorphism of this organism. These studies should not be limited to chemically defined media, valuable though such studies are, but should be extended to complex media. After all *C. albicans* is not growing *in vivo* in a glucose ammonium salts medium. Will it be possible to exploit further the fungal plasma membrane as a specific target in the development of new antimycotic drugs to control both superficial and systemic infections? One way ahead is the chemical modification of existing compounds and already amphotericin B, in spite of its complexity, has been chemically modified to produce the methyl ester (Mechlinski & Schaffner, 1972). This has resulted in an increased apparent solubility and reduced toxicity without a reduction in the antifungal activity (Chen *et al.*, 1977). The development of inhibitors specific for enzymes involved in the synthesis of the cell envelope of *C. albicans* is an exciting possibility. Already the polyoxins have been shown to inhibit the membrane bound chitin synthetase and the imidazole drugs to inhibit a sterol demethylase and if it is possible to exploit the cell's own specific transport systems (as for 5-fluorocytosine) to accumulate such inhibitors within the cell then we will be entering a new era in the chemotherapy of mycotic infections.

REFERENCES

ABRAHAM, E. P. & FLOREY H. W. (1949). Antibiotics from the genus *Bacillus*. In *Antibiotics*, vol. 1, ed. H. W. Florey, E. Chain, N. G. Heatley, M. A. Jennings, A. G. Sanders, E. P. Abraham & M. E. Florey, p. 457. Oxford: Oxford University Press.

ALLEN, J. G., ATHERTON, F. R., HALL, M. J., HASSALL, C. H., HOLMES, S. W., LAMBERT, R. W., NISBET, L. J. & RINGROSE, P. S. (1978). Phosphonopeptides, a new class of synthetic antibacterial agents. *Nature, London*, **272**, 56–8.

ANDREOLI, T. E. (1973). On the anatomy of amphotericin B–cholesterol pores in lipid membranes. *Kidney International*, **4**, 337–45.

ARCHER, D. B. (1976). Effect of the lipid composition of *Mycoplasma mycoides* subspecies *Capri* and phosphatidylcholine vesicles upon the action of polyene antibiotics. *Biochimica et Biophysica Acta*, **436**, 68–76.

ARCHER, D. B. (1977). Chitin biosynthesis in protoplasts and subcellular fractions of *Aspergillus fumigatus*. *Biochemical Journal*, **164**, 653–8.

ARCHER, D. B. & GALE, E. F. (1976). Antagonism by sterols of the action of amphotericin and filipin on the release of potassium ions from *Candida albicans* and *Mycoplasma mycoides* subsp. *Capri*. *Journal of General Microbiology*, **90**, 187–90.

ATHAR, M. A. & WINNER, H. T. (1971). The development of resistance by *Candida* species to polyene antibiotics *in vitro*. *Journal of Medical Microbiology*, **4**, 505–17.

BOWERS, B., LEVIN, G. & CABIB, E. (1974). Effect of polyoxin D on chitin synthesis and septum formation in *Saccharomyces cerevisiae*. *Journal of Bacteriology*, **119**, 564–75.

BRAUN, P. C. & CALDERONE, R. A. (1978). Chitin synthesis in *Candida albicans*: Comparison of yeast and hyphal forms. *Journal of Bacteriology*, **135**, 1472–7.

CABIB, E. (1975). Molecular aspects of yeast morphogenesis. *Annual Review of Microbiology*, **29**, 191–214.

CARTWRIGHT, R. Y. (1975). Antifungal drugs. *Journal of Antimicrobial Chemotherapy*, **1**, 141–62.

CASSONE, A., KERRIDGE, D. & GALE, E. F. (1979). Ultrastructural changes in the cell wall of *Candida albicans* following cessation of growth and their possible relationship to the development of polyene resistance. *Journal of General Microbiology*, **110**, 339–47.

CHEN, W. C., CHOU, D-L. & FEINGOLD, D. S. (1978). Dissociation between ion permeability and the lethal action of polyene antibiotics in *Candida albicans*. *Antimicrobial Agents and Chemotherapy*, **13**, 914–17.

CHEN, W. C., SUD, I. J., CHOU, D-L. & FEINGOLD, D. S. (1977). Selective toxicity of the polyene antibiotics and their methyl ester derivatives. *Biochemical and Biophysical Research Communications*, **74**, 480–7.

COSTERTON, J. W., INGRAM, J. M. & CHENG, K. J. (1974). Structure and function of the cell envelope of Gram negative bacteria. *Bacterial Reviews*, **38**, 87–110.

DE KRUIJFF, B. & DEMEL, R. A. (1974). Polyene antibiotic-sterol interactions in membranes of *Acholeplasma laidlawii* cells and lecithin liposomes. III. Molecular structure of the polyene antibiotic–cholesterol complexes, *Biochimica et Biophysica Acta*, **339**, 57–70.

DE KRUIJFF, B., GERRITSEN, W. J., OERLEMANS, A., VAN DIJCK, P. W. M., DEMEL, R. A. & VAN DEENEN, L. L. M. (1974). Polyene antibiotic sterol interactions in membranes of *Acholeplasma laidlawii* cells and lecithin liposomes. II. Temperature dependence of the polyene antibiotic–sterol complex formation. *Biochimica et Biophysica Acta*, **339**, 44–56.

DE NOLLIN, S. & BORGERS, M. (1974). The ultrastructure of *Candida albicans* after *in vitro* treatment with miconazole. *Sabouraudia*, **12**, 341–51.

DE NOLLIN, S. & BORGERS, M. (1975). Scanning electron microscopy of *Candida albicans* after in-vitro treatment with miconazole. *Antimicrobial Agents and Chemotherapy*, **7**, 704–11.

DE NOLLIN, S., VAN BELLE, H., GOOSSENS, F., THONE, F. & BORGERS, M. (1977). Cytochemical and biochemical studies of yeasts after in-vitro exposure to miconazole. *Antimicrobial Agents and Chemotherapy*, **11**, 500–13.

DICKINSON, D. P. (1977). The effects of miconazole on rat liver mitochondria. *Biochemical Pharmacology*, **26**, 541–2.

GALE, E. F. (1974). The release of potassium ions from *Candida albicans* in the presence of polyene antibiotics. *Journal of General Microbiology*, **80**, 451–65.

GALE, E. F., CUNDLIFFE, E., REYNOLDS, P. E., RICHMOND, M. H. & WARING, M. J. (1972). *The Molecular Basis of Antibiotic Action*. New York: John Wiley & Sons.

GLASER, L. & BROWN, D. H. (1957). The synthesis of chitin in cell-free extracts of *Neurospora crassa*. *Journal of Biological Chemistry*, **228**, 729–42.

GOODAY, G. W. (1977). Biosynthesis of the fungal wall-mechanisms and implications. *Journal of General Microbiology*, **99**, 1–11.

GOODAY, G. W. & DE ROUSSET-HALL, A. (1975). Properties of chitin synthetase from *Coprinus cinereus*. *Journal of General Microbiology*, **89**, 137–45.

GUYMON, L. F., WALSTAD, D. L. & SPARLING, P. F. (1978). Cell envelope alterations in antibiotic-sensitive and -resistant strains of *Neisseria gonorrhoeae*. *Journal of Bacteriology*, **136**, 391–401.

HAMILTON-MILLER, J. M. T. (1972a). A comparative in-vitro study of amphotericin B, clotrimazole and 5-fluorocytosine against clinically isolated yeasts. *Sabouraudia*, **10**, 276–83.

HAMILTON-MILLER, J. M. T. (1972b). Sterols from polyene resistant mutants of *Candida albicans*. *Journal of General Microbiology*, **73**, 201–3.

HAMILTON-MILLER, J. M. T. (1973). Chemistry and biology of the polyene macrolide antibiotics. *Bacteriological Reviews*, **37**, 166–96.

HAMMOND, S. M., LAMBERT, P. A. & KLIGER, B. N. (1974). The mode of action of polyene antibiotics; induced potassium leakage in *Candida albicans*. *Journal of General Microbiology*, **81**, 325–36.

HAZEN, E. L. & BROWN, R. (1950). Two antifungal agents produced by a soil actinomycete. *Science*, **112**, 423.

HOLZ, R. W. (1974). The effects of the polyene antibiotics nystatin and amphotericin B on thin lipid membranes. *Annals of the New York Academy of Sciences*, **235**, 469–79.

HORI, M., KAKIKI, K. & MISATO, T. (1977). Antagonistic effects of dipeptides on the uptake of Polyoxin A by *Alternaria kikuchiana*. *Journal of Pesticide Science*, **2**, 129–49.

HSU-CHEN, C. C. & FEINGOLD, D. S. (1973). Polyene action on lecithin liposomes; effect of cholesterol and fatty acyl chains. *Biochemical and Biophysical Research Communications*, **51**, 972–8.

HSU-CHEN, C. C. & FEINGOLD, D. S. (1974). Two types of resistance to polyene antibiotics in *Candida albicans*. *Nature, London*, **251**, 656–9.

IWATA, K., KANDA, Y., YAMAGUCHI, H. & OSUMI, M. (1973). Electron microscopic studies on the mecanism of action of clotrimazole on *Candida albicans*. *Sabouraudia*, **11**, 205–9.

IWATA, K., YAMAGUCHI, H. & HIRATANI, T. (1973). Mode of action of clotrimazole. *Sabouraudia*, **11**, 158–66.

JOHNSON, B., WHITE, R. J. & WILLIAMSON, G. M. (1978). Factors influencing the susceptibility of *Candida albicans* to the polyenoic antibiotics nystatin and amphotericin B. *Journal of General Microbiology*, **104**, 325–33.

KELLER, F. A. & CABIB, E. (1971). Chitin and yeast budding: properties of chitin synthetase from *Saccharomyces carlsbergensis*. *Journal of Biological Chemistry*, **246**, 160–6.

KENIG, M. & ABRAHAM, E. P. (1976). Antimicrobial activities and antagonists of bacilysin and anticapsin. *Journal of General Microbiology*, **94**, 37–45.

KERRIDGE, D. (1978). Current problems and developments in medical mycology. In *Companion to Microbiology*, ed. A. T. Bull & P. M. Meadow, pp. 215–49. London & New York Longman.

KERRIDGE, D. (1979). The Polyene Macrolide antibiotics. *Postgraduate Medical Journal*, **55**, 30–3.

KERRIDGE, D., KOH, T. Y. & JOHNSON, A. M. (1976a). The interaction of amphotericin B methyl ester with protoplasts of *Candida albicans*. *Journal of General Microbiology*, **96**, 117–23.

KERRIDGE, D., KOH, T. Y., MARRIOTT, M. S. & GALE, E. F. (1976b). The production and properties of protoplasts from the dimorphic yeast *Candida albicans*. In *Microbial and Plant Protoplasts*, ed. J. F. Peberdy, A. H. Rose, H. J. Rogers & E. C. Cocking, pp. 23–38. London, New York, San Francisco: Academic Press.

KOBAYASHI, G. S. & MEDOFF, G. (1977). Antifungal agents: recent developments. *Annual Review of Microbiology*, **31**, 291–308.

KOH, T. Y., MARRIOTT, M. S., TAYLOR, J. & GALE, E. F. (1977). Growth characteristics and polyene sensitivity of a fatty acid auxotroph of *Candida albicans*. *Journal of General Microbiology*, **102**, 105–10.

LAMPEN, J. O. (1966). Interference by polyene antibiotics (especially nystatin & filipin) with specific membrane functions. *Symposium of the Society for General Microbiology*, **16**, ed. B. A. Newton & P. E. Reynolds, pp. 111–30. Cambridge University Press.

LIRAS, P. & LAMPEN, J. O. (1974). Protection by K+ and Mg²+ of growth and macromolecular synthesis in candicidin treated yeast. *Biochimica et Biophysica Acta*, **374**, 159–63.

LONGLEY, R. P., ROSE, A. H. & KNIGHTS, B. A. (1968). Composition of the protoplast membrane of *Saccharomyces cerevisiae*. *Biochemical Journal*, **108**, 401–12.

MARRIOTT, M. S. (1974). Studies on the cell envelope of *Candida albicans*. PhD Thesis, University of Cambridge.

MARRIOTT, M. S. (1975). Isolation and characterisation of plasma membranes from the yeast and mycelial forms of *Candida albicans*. *Journal of General Microbiology*, **86**, 115–32.

MARTY, A. & FINKELSTEIN, A. (1975). Pores formed in lipid bilayer membranes by nystatin, differences in one sided and two sided action. *Journal of General Physiology*, **65**, 515–26.

MATILE, P., MOOR, H. & MUHLETHALER, K. (1967). Isolation and properties of plasmalemma in yeast. *Archiv für Mikrobiologie*, **58**, 201–11.

MCMURROUGH, I. & BARTNICKI-GARCIA, S. (1971). Properties of a particulate chitin synthetase from *Mucor rouxii*. *Journal of Biological Chemistry*, **246**, 4008–16.

MECHLINSKI, W. & SCHAFFNER, C. B. (1972). Polyene macrolide derivatives. 1. *N*-acylation and esterification reactions with amphotericin B. *Journal of Antibiotics* (*Tokyo*), **25**, 256–8.

MECHLINSKI, W., SCHAFFNER, C. P., GANIS, P. & AVITABILE G. (1970). Structure and absolute configuration of the polyene macrolide antibiotic amphotericin B. *Tetrahedron Letters*, **44**, 3873–6.

MITANI, M. & INOUE, Y. (1968). Antagonists of antifungal substance polyoxin. *Journal of Antibiotics*, **21**, 492–6.

NES, W. R., SEKULA, B. C., NES, D. & ADLER, J. H. (1978). The functional importance of structural features of ergosterol in yeast. *Journal of Biological Chemistry*, **253**, 6218–25.

NIKAIDO, H. (1976). Outer membrane of *Salmonella*. Transmembrane diffusion of some hydrophobic molecules. *Biochimica et Biophysica Acta*, **433**, 118–32.

NORMAN, A. W., SPIELVOGEL, A. M. & WONG, R. C. (1976). Polyene antibiotic-sterol interactions. *Advances in Lipid Research*, **14**, 127–70.

NURMINEN, T., OURA, E. & SUOMALAINEN, H. (1970). The enzymic composition of the isolated cell wall and plasma membrane of baker's yeast. *Biochemical Journal*, **116**, 61–9.

ODDS, F. C. (1979). Candida *Candidiosis*. Leicester University Press.

PALACIOS, J. & SERRANO, R. (1978). Proton permeability induced by polyene antibiotics. A plausible mechanism for their inhibition of maltose fermenation in yeast. *FEBS Letters*, **91**, 198–201.

PEÑA, A. (1975). Studies on the mechanism of K+ transport in yeast. *Archives of Biochemistry and Biophysics*, **167**, 397–419.

PIERCE, A. M., PIERCE, H. D., UNRAU, A. M. & OEHLSCHLAGER, A. C. (1978). Lipid composition and polyene antibiotic resistance of *Candida albicans* mutants. *Canadian Journal of Biochemistry*, **56**, 135–42.

PLEMPEL, M., BARTMANN, K., BUECHEL, K. H. & REGEL, E. (1969). Experimentelle Befunde uber ein neues, oral wirksames Antimykoticum mit breiten Wirkungspekrum. *Deutsche Medizinische Wochenschraft*, **94**, 1356–64.

PLEMPEL, M., BARTMANN, K., BRUECHEL, K. H. & REGEL, E. (1970). Bay b5097, a new orally applicable antifungal substance with broad spectrum activity. *Antimicrobial Agents and Chemotherapy* – 1969. 270–4.

POLAK, A. & GRENSON, M. (1973). Evidence for a common transport system for cytosine, adenine and hypoxanthine in *Saccharomyces cerevisiae* and *Candida albicans*. *European Journal of Biochemistry*, **32**, 276–82.

POLAK, A. & SCHOLER, H. J. (1975). Mode of action of 5-fluorocytosine and mechanisms of resistance. *Chemotherapy*, **21**, 113–30.

PREUSSER, H.-J. (1976). Effects of in vitro treatment with econazole on the ultrastructure of *Candida albicans*. *Mykosen*, **19**, 304–16.

RANK, G. H., ROBERTSON, A. J. & BUSSEY, H. (1978). The viscosity and lipid composition of the plasma membrane of multiple drug resistant and sensitive yeast strains. *Canadian Journal of Biochemistry*, **56**, 1036–41.

RANK, G. H., ROBERTSON, A. & PHILLIPS, K. (1975). Reduced plasma membrane permeability in a multiple cross resistant strain of *Saccharomyces cerevisiae*. *Journal of Bacteriology*, **122**, 359–66.

ROANTREE, R. J., KUO, T-T. & MACPHEE, D. G. (1977). The effect of defined lipopolysaccharide core defects upon antibiotic resistances of *Salmonella typhimurium*. *Journal of General Microbiology*, **103**, 223–39.

SANTOS, E., VILLANUEVA, J. R. & SENTANDREU, R. (1978). The plasma membrane of *Saccharomyces cerevisiae*, isolation and some properties. *Biochimica et Biophysica Acta*, **508**, 39–54.

SALTON, M. R. J. (1978). Structure and function of bacterial plasma membranes. *Symposium of the Society for General Microbiology*, **28**, ed. R. Y. Stanier, H. J. Rogers & J. B. Ward, pp. 201–23. Cambridge University Press.

SCHERRER, R., LOUDEN, L. & GERHARDT, P. (1974). Porosity of the yeast cell wall and membrane. *Journal of Bacteriology*, **118**, 534–40.

SCHIBECI, A., RATTRAY, J. B. M. & KIDBY, D. K. (1973a). Isolation and identification of yeast plasma-membrane. *Biochimica et Biophysica Acta*, **311**, 15–25.

SCHIBECI, A., RATTRAY, J. B. M. & KIDBY, D. K. (1973b). Electron microscope autoradiography of labelled yeast plasma-membrane. *Biochimica et Biophysica Acta*, **323**, 532–38.

SOLOV'EVA, N. N., BELOUSOVA, I. I. & TERESHIN, I. M. (1976). Influence of polyene antibiotics on the lactate dehydrogenase and ATPase activity of the membrane fraction of *Candida albicans*. *Khimiko-Farmatsevticheskii Zhurnal*, **10**, 18–20.

STANIER, R. Y. (1970). Some aspects of the biology of cells and their possible evolutionary significance. *Symposium of the Society for General Microbiology*, **20**, ed. H. P. Charles & B. C. J. G. Knight, pp. 1–38. Cambridge University Press.

SUBDEN, R. E., SAFE, L., MORRIS, D. C., BROWN, R. G. & SAFE, S. (1977). Eburicol, lichesterol, ergosterol and obtusifoliol from polyene antibiotic-resistant mutants of *Candida albicans*. *Canadian Journal of Microbiology*, **23**, 751–4.

SWAMY, K. H. S., SIRSI, M. & RAO, G. R. (1974). Studies on the mechanism of action of miconazole: effect of miconazole on respiration and cell permeability of *Candida albicans*. *Antimicrobial Agents and Chemotherapy*, **5**, 420–5.

SWAMY, K. H. S., SIRSI, M. & RAO, G. R. (1976). Studies on the mechanism of action of miconazole – II interaction of miconazole with mammalian erythrocytes. *Biochemical Pharmacology*, **25**, 1145–50.

THIENPONT, D., VAN CUTSEM, J., VAN NUETEN, J. M., NIEMEGEERS, C. J. E. & MARSBOOM, R. (1975). Biological and toxicological properties of econazole – a broadspectrum antimycotic. *Arzneimittel-Forschung (Drug Res..)*, **25**, 224–30.

THOMAS, A. H. (1976). Analysis and Assay of polyene antifungal antibiotics. *The Analyst*, **101**, 321–40.

THOMAS, D. S., HOSSACK, J. A. & ROSE, A. H. (1978). Plasma-membrane lipid composition and ethanol tolerance in *Saccharomyces cerevisiae*. *Archives of Microbiology*, **117**, 239–45.

VAN CUTSEM, J. & THIENPONT, D. (1972). Miconazole a broad spectrum antimycotic agent with antibacterial activity. *Chemotherapy*, **17**, 392–404.

VAN DEN BOSSCHE, H. (1974). Biochemical effects of miconazole on fungi – 1. Effects on the uptake and/or utilization of purines, pyrimidines, nucleosides, amino acids and glucose by *Candida albicans*. *Biochemical Pharmacology*, **23**, 887–99.

VAN DEN BOSSCHE, H., WILLEMSENS, G., COOLS, W., LAUWERS, W. F. J. & LE JEUNE, L. (1978). Biochemical effects of miconazole on fungi. II. Inhibition of ergosterol biosynthesis in *Candida albicans*. *Chemical and Biological Interactions*, **21**, 59–78.

VENABLES, P. & RUSSELL, A. D. (1975). Nystatin induced changes in *Saccharomyces cerevisiae*. *Antimicrobial Agents and Chemotherapy*, **7**, 121–7.

VERKLEIJ, A. J., DE KRUIJFF, B., GERRITSEN, W. F., DEMEL, K. A., VAN DEENEN, L. L. M. & VERVERGAERT, P. H. J. (1973). Freeze etch electron microscopy of erythrocytes, *Acholeplasma laidlawii* cells and liposomal membranes after the action of filipin and amphotericin B. *Biochimica et Biophysica Acta*, **291**, 577–81.

VIRINA, A. M., FEISIN, A. H., FATEEVA, L. I., KASUMOV, KH.H., BELOUSOVA, I. I. & TERESHIN, I. M. (1976). Interaction of polyene antibiotics with sensitive and resistant strains of *Candida albicans*. *Khimiko-Farmatsevticheskii Zhurnal*, **10**, 12–16.

YAMAGUCHI, H. (1977). Antagonistic action of lipid components of membranes from *Candida albicans* and various other lipids on two imidazole antimycotics, clotrimazole and miconazole. *Antimicrobial Agents and Chemotherapy*, **12**, 16–25.

YAMAGUCHI, H. (1978). Protection by unsaturated lecithin against the imidazole antimycotics, clotrimazole and miconazole. *Antimicrobial Agents and Chemotherapy*, **13**, 423–6.

EXPLANATION OF PLATE

Space filling model of amphotericin B, showing the molecule as a rigid rod rather than as a circular structure with a hole in the centre as suggested by the structural formula presented in Fig. 2.

PLATE 1

THE HYDROGENOSOME

MIKLÓS MÜLLER

The Rockefeller University, New York, N.Y. 10021, USA

INTRODUCTION

Eukaryotic cells represent a high level of subcellular differentiation characterised by the presence of various membrane-bounded organelles and are regarded as relatively late products in the evolution of life. It is likely that the structural and biochemical organisation of all present day eukaryotic cells was established in that period of evolution when the change of the atmosphere from anaerobic to aerobic had already occurred. Thus the typical eukaryotic cell is well-adapted to an aerobic mode of life. In this regard the most important structural and biochemical adaptation is the presence of mitochondria (Broda, 1975).

Mitochondria represent the most efficient biological machinery for aerobic conversion of the energy of organic nutrients into the phosphate bond energy of ATP which is used as a common energy source in all biological processes. These properties give a major energetic advantage to those cells that contain mitochondria.

Electron microscopy discloses marvellous structural differences in protozoa, showing that unicellularity permits far-reaching morphological adaptations in this group. It is to be expected that morphological diversity is accompanied by biochemical differences, and in fact the evidence for the biochemical diversity of protozoa is constantly increasing. Unfortunately, the relatively small number of species that are available in axenic cultures, necessary to study biochemical properties, sets a limit to the information easily obtained.

Functional mitochondria are regarded as primary characteristics of eukaryotic cells, and their absence is usually taken as sign of excessive irreversible differentiation (mammalian red blood cells) or as a result of transient metabolic adaptations (anaerobically-grown yeasts). It is therefore of some interest that, according to electron microscopic findings, certain protozoan species seem to lack mitochondria. Such morphological evidence, however suggestive, is clearly not sufficient to demonstrate the absence of mitochondrial functions and their structural basis from these organisms. Fortunately a few of the species apparently lacking mitochondria are available in axenic culture and their biochemical study gives data which clearly suggest the absence of mitochondrial functions.

The aim of this chapter is to summarise and put into a general, often hypothetical and provocative perspective, recent results on trichomonads, in which mitochondria are replaced by hydrogenosomes, an unusual type of redox organelle. This topic has been discussed in earlier reviews (Müller, 1975, 1976; Müller, Lindmark & Mack, 1978). Biochemical properties of prokaryotic micro-organisms and those of cells and cell organelles of higher animals will be mentioned frequently, but to save space, no specific references will be given for each statement made. Standard texts of biochemistry and microbial physiology cover these topics adequately.

MORPHOLOGY OF TRICHOMONADS

Trichomonad flagellates are parasitic protozoa inhabiting the alimentary tract or the genito-urinary tract of vertebrates. Several species are important in human or veterinary medicine (Honigberg, 1978a, b). Although it is likely that most of the subsequent presentation is valid for all trichomonads, the discussion will be limited to those few species whose biochemistry has been studied in some detail. These include two pathogenic species, *Tritrichomonas foetus* of the genito-urinary tract of cattle and *Trichomonas vaginalis* of the human genito-urinary tract, as well as a more primitive, possibly ancestral form, *Monocercomonas* sp. from reptile intestine. The most extensive studies concern *T. foetus* and therefore this species will be central to this chapter.

Trichomonad protoza are typical zooflagellates: their flagellar apparatus consists of three (*Tritrichomonas foetus*) or four (*Trichomonas vaginalis*) anterior flagella and a recurrent flagellum forming an undulating membrane. *Monocercomonas* has no undulating membrane. The presence of skeletal organelles is highly characteristic of trichomonads. The axostyle consists of a sheet of longitudinally attached microtubules which run in the axis of the cell from the basal bodies of the flagella to the posterior end of the cell. The costa is a cross-striated band, similar to flagellar rootlets found in other protozoa, and follows the undulating membrane under the cell surface.

The presence of granules 0.5–2.0 μm diameter in the cytoplasm of trichomonads has been well known for decades. These organelles show a close topographical relationship to the axostyle and costa, and often are arranged in rows along them (Plate 1a). It is this arrangement which gave them the name paraxostylar and paracostal granules. Because these organelles stain strongly with certain dyes used in light microscopy, they are also called chromatic granules. Their recent biochemical characterisation led to the introduction of a term reflecting their metabolic role (Lindmark & Müller, 1973a). This term is the 'hydrogenosome', to be used extensively below.

In earlier papers these granules were often described as microbody-like structures. Although mammalian microbodies (i.e. peroxisomes) and hydrogenosomes show certain morphological similarities, careful examination shows the latter to have characteristic features (Plate 1b–d; Filadoro, 1970; Müller, 1973). The organelles are spherical or elongated and are surrounded by a membrane. The finer structure of this membrane is not yet satisfactorily established. It is usually described as a single unit membrane, but recent electron micrographs indicate that it might consist of two closely opposed unit membranes (Plate1d; Kulda, Králová & Vávra, 1977). The matrix of the organelles is finely granular, often containing a denser 'core'. Lying inside the membrane, at the periphery of the organelle, is an elongated structure which turns out to be a flattened membrane-limited vesicle (Filadoro, 1970; Kulda et al., 1977). In the matrix there are a few, rather dense granules which might be ribosomes.

The hydrogenosomes occupy a considerable portion of the cell volume, suggesting that they play an important function in the cell. Morphological data give little information on the biogenesis of hydro-genosomes. The presence of some dumb-bell-shaped structures (Plate 1b) and a detailed morphometric analysis suggest that binary division occurs which is not synchronised with cytokinesis (Nielsen & Diemer, 1976).

ENERGY METABOLISM OF TRICHOMONADS

Pyruvate metabolism

Trichomonad flagellates are aerotolerant anaerobes (Mack & Müller, 1978) which have a fermentative metabolism. They show intense respiration if air is present, i.e. they are able to use O_2 as terminal acceptor for some of the metabolically-generated electrons. Under anaerobic conditions they are able to eliminate reducing equivalents in the form of H_2, a rather unusual property among eukaryotic organisms although widespread among prokaryotes (Gray & Gest, 1965).

The nature of the mechanism of H_2 formation is of considerable interest. In fermentative prokaryotes, one of two major enzymatic systems can be involved in this process. In the facultative anaerobic Enterobacteriaceae, formate is the proximal substrate of the 'formate hydrogen lyase' system producing H_2. In Clostridia and many other

strict anaerobes it is the oxidation of pyruvate which produces the
reducing equivalents leading to H_2, a process formerly designated as
'phosphoroclastic' pyruvate metabolism. This is a multi-step pathway
consisting of the oxidative decarboxylation of pyruvate resulting in the
formation of acetyl-CoA and the reduction of an electron carrier
protein with a low redox potential: ferredoxin or flavodoxin (Fig.1*b*).

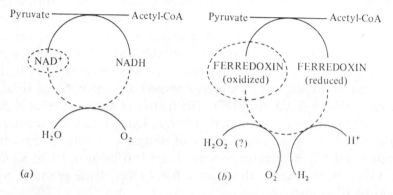

Fig. 1. Oxidative decarboxylation of pyruvate (*a*) in mitochondria and many aerobic
micro-organisms and (*b*) in hydrogenosomes and several strict anaerobes.

The reaction is catalysed by pyruvate:ferredoxin oxidoreductase.
Pyridine nucleotides cannot serve as primary electron acceptors for this
enzyme. Since its activity is reversible, it is also called pyruvate synthase. Hydrogenase mediates the re-oxidation of the electron carrier by
protons, a process which results in the formation of H_2. The carrier is
auto-oxidisable in the presence of O_2. In prokaryotes the acetyl-CoA
formed is converted to acetyl phosphate which in turn can yield acetate
with the conservation of the energy of the phosphate bond by substrate
level phosphorylation. Since acetyl phosphate is not the primary product
of pyruvate oxidation, the term 'phosphoroclastic' does not describe the
process accurately.

The above pathway differs in several significant points from the
oxidative decarboxylation of pyruvate in mitochondria and in prokaryotes other than strict anaerobes (Fig. 1*a*). Although the final product of
the reaction is acetyl-CoA in both cases, in the latter the reaction is
catalysed in several steps by the pyruvate dehydrogenase complex. In
this reaction, NAD^+ is the final electron acceptor and the enzyme cannot
donate electrons to ferredoxin or similar electron transport proteins.
The reaction is practically irreversible, in contrast to the reaction catalysed by pyruvate:ferredoxin oxidoreductase. The re-oxidation of
NADH occurs usually in the electron transport chain with O_2 as the

terminal electron acceptor, a process accompanied by oxidative phosphorylation.

The nature of the H_2-producing system in trichomonads has been
under consideration for some time. Although some components of the
facultative anaerobic system were reported to exist in trichomonads
(Lindblom, 1961), it is now established that the mechanism of H_2
formation is similar to that observed in strict anaerobes (Bauchop,
1971; Edwards & Mathison, 1970; Lindmark & Müller, 1973a). The
metabolic map on Fig. 2 gives a schematic representation of the reac-

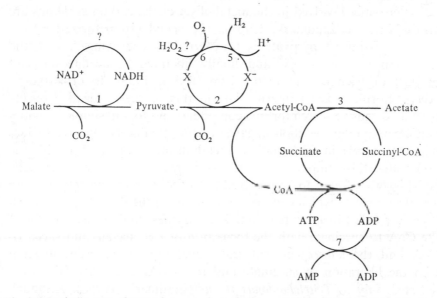

Fig. 2. Metabolic map of *Tritrichomonas foetus* hydrogenosomes. Step 1 Malate dehydrogenase (decarboxylating); step 2 pyruvate: ferredoxin oxidoreductase; step 3, acetate-
succinate CoA transferase; step 4, succinate thiokinase; step 5, hydrogenase; step 6,
terminal oxidase of unknown nature; step 7, adenylate kinase. X, unknown electron transport protein(s), possibly akin to ferredoxin. Reproduced with permission from Müller &
Lindmark (1978).

tions involved. All trichomonads studied (Lindmark & Müller, 1973a,
1974a; Lindmark, Müller & Shio, 1975) contain pyruvate:ferredoxin
oxidoreductase and hydrogenase. Both activities function with exo-
geneous clostridial ferredoxin and with free flavin nucleotides as electron
acceptors. These compounds can be replaced by methyl viologen, a low
redox potential electron acceptor which easily substitutes the natural
acceptor in ferredoxin-linked reactions. Pyridine nucleotides were
ineffective with either enzyme.

The low redox potential electron transport protein linked to the above processes has not been yet identified in trichomonads. The reactivity of the enzymes with clostridial ferredoxin and their overall similarity to the corresponding enzymes of clostridia suggest that a ferredoxin-like protein does participate in the process *in vivo*. Trichomonads contain relatively high amounts of flavins (Čerkasovová, 1970; Lindmark & Müller, 1973b). Preliminary e.p.r. studies indicate the presence in *Tritrichomonas foetus* of a few iron–sulphur proteins and flavoproteins with different mid-level potentials (D. Lloyd, T. Ohnishi, D. G. Lindmark & M. Müller, unpublished), one of which might represent the natural carrier.

The enzymes involved in the metabolism of the carbon skeletons are also of interest (Lindmark, 1976a). The acetyl group formed by the decarboxylation of pyruvate is transferred to coenzyme A with the formation of acetyl-CoA, and finally liberated as acetate, an end product of trichomonad metabolism (Ryley, 1955). In prokaryotes, acetyl phosphate is an intermediate in this process. Negative results from an extensive search for this compound and for enzymatic activities participating in its metabolism gave evidence that acetyl phosphate does not play a role in trichomonad metabolism, in agreement with the notion that this compound occurs only in prokaryotes. Two enzymatic activities were found to be involved in the conversion of acetyl-CoA to acetate, a process which conserves the energy of the thioester bond in an ATP- (or GTP)-forming reaction. The enzymes catalyse the transfer of the CoA to succinate with the formation of free acetate and succinyl-CoA and the subsequent substrate level phosphorylation of ADP with the liberation of succinate and free CoA. These activities were detected only in *Tritrichomonas foetus* (Lindmark, 1976a). Although *Trichomonas vaginalis* also produces acetate, possibly by a similar mechanism, attempts to detect the enzymes responsible for its production gave no positive results (D. G. Lindmark, personal communication).

Localisation of enzymes in hydrogenosomes

Cell fractionation studies using differential and isopycnic centrifugation demonstrated that in trichomonads all the steps of the pathway from pyruvate to acetate and H_2 are localised in large cytoplasmic granules (Lindmark & Müller, 1973a, 1974a; Lindmark et al., 1975). Most enzymes in these organelles exhibit structure-bound latency suggesting that the granules are surrounded by a membrane. The equilibrium density of these organelles in sucrose gradients is about 1.25 g ml^{-1}, higher than that of mitochondria or peroxisomes. Electron microscopy showed that the dominant components in fractions enriched significantly in the characteristic enzymatic activities were the paracostal and paraxostylar granules. These data established that the enzymes of H_2 production are localised in the well-known microbody-like organelles of trichomonads. On this basis we named these structures 'hydrogenosomes' (Lindmark & Müller, 1973a).

In addition to the enzymes of pyruvate metabolism, decarboxylating malate dehydrogenase (Brugerolle & Metenier, 1973; Lindmark & Müller, 1973a) and adenylate kinase (Čerkasov et al., 1978) are also present in the hydrogenosomes.

Further characterisation of the hydrogenosomes was attempted by searching for enzymes that are characteristic components of known organelles of other eukaryotic cells. The following components of mitochondria could not be found in Tritrichomonas foetus hydrogenosomes: cytochromes (Lloyd, Lindmark & Müller, 1979a), magnesium activated and N,N'-dicyclohexylcarbodiimide-sensitive ATPase (Lloyd, Lindmark & Müller, 1979b). Although hydrogenosomes are able to oxidise NADH at a low rate (Čerkasov et al., 1978), the predominant NADH oxidase is localised in the cytoplasm (Brugerolle & Metenier, 1973; Čerkasovová & Čerkasov, 1974). Components of peroxisomes are also absent from hydrogenosomes: catalase is absent from Trichomonas vaginalis (Ryley, 1955), and although present in T. foetus, it is localised in the cytoplasm. Peroxisomal direct oxidases could not be detected (Müller, 1973).

Respiratory properties of hydrogenosomes

Although under anaerobic conditions the main function of hydrogenosomes is pyruvate oxidation coupled with the formation of H_2, in the presence of O_2 they show intense respiration with pyruvate or malate if certain cofactors are also present (Čerkasov et al., 1978; Müller & Lindmark, 1978). The nature of this respiration is not clear to date. It is resistant to most inhibitors and uncouplers of mitochondrial respiration (Čerkasov et al., 1978), a finding in agreement with the absence of cytochromes and ATPase from the hydrogenosomes (Lloyd et al., 1979a, b). The composition of the system transporting electrons from pyruvate to O_2 is under study and e.p.r. spectroscopy indicates that several of the detected iron–sulphur proteins and flavoproteins might participate in the process (D. Lloyd, T. Ohnishi, D. G. Lindmark & M. Müller, unpublished). The terminal oxidase is clearly not a cytochrome oxidase.

Integration of the hydrogenosome into cellular metabolism

Figs. 3 and 4 show our present concepts of the carbohydrate metabolism of *Tritrichomonas foetus* and *Trichomonas vaginalis*. Pyruvate and

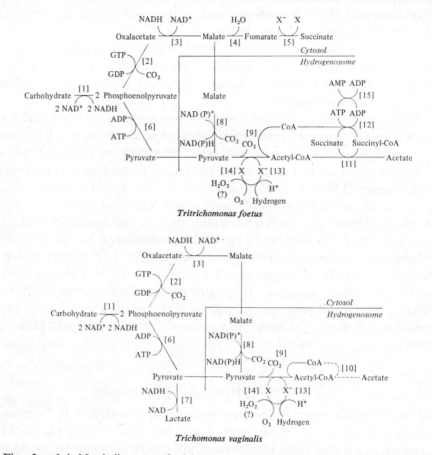

Figs. 3 and 4. Metabolic maps of trichomonad flagellates. [1], Glycolytic enzymes; [2], *P*-enolpyruvate carboxykinase (GDP); [3], malate dehydrogenase (NAD); [4], fumarate hydratase; [5], fumarate reductase; [6], pyruvate kinase; [7], lactate dehydrogenase; [8], malate dehydrogenase (decarboxylating, NAD(P)); [9], pyruvate: ferredoxin oxidoreductase; [10], deacylation of CoA; [11], acetate:succinate CoA-transferase; [12], succinate thiokinase; [13], hydrogenase; [14], aerobic reoxidation of ferredoxin; [15], adenylate kinase. Modified from Müller (1976) and Müller *et al.* (1978).

malate, the main substrates of hydrogenosomal metabolism are, produced in the cytoplasm by glycolysis and by CO_2 fixation into *P*-enolpyruvate and the reduction of the resulting oxaloacetate (Ryley, 1955; Arese & Cappuccinelli, 1974). After entering the hydrogenosome, these substrates are oxidised. Thus the hydrogenosomal activity represents an oxidative extension of the cytoplasmic pathways, made possible by the introduction of a supplementary mechanism of removal of reducing equivalents that are generated in this oxidative step. The energy yield of the additional oxidation can be utilised in a substrate level phosphorylation step. Thus, at least in theory, the organisms derive extra benefits from the presence of hydrogenosomal pathways (Čerkasovová, 1970).

The in-vivo metabolic flow through these organelles is considered next. Anaerobic metabolic balances established by Ryley (1955) for *Tritrichomonas foetus* show that all the carbon of carbohydrate utilised can be accounted for in acetate and succinate, produced in equimolar quantities. This finding implies that one half of the carbon entering glycolysis will pass through the hydrogenosome. Under aerobic conditions, which clearly favour the oxidative pathway, the ratio of acetate to succinate is higher, indicating a still higher carbon flow through the organelle.

The anaerobic oxidative processes occurring in *Tritrichomonas foetus* hydrogenosomes seem not to be involved in the re-oxidation of glycolytically formed NADH, a process indispensable for the continuous operation of the glycolytic pathway. In fact, recently a strain of *T. foetus* which seems to have lost its hydrogenosomal metabolic capacity was developed under drug pressure (Čerkasovová & Čerkasov, 1978). Thus the functioning of the hydrogenosomal metabolic machinery may not be absolutely essential for *T. foetus*.

The case of *Trichomonas vaginalis* is less clearly understood. Available information on the anaerobic metabolic balance suggests that the organism can thrive well on simple homolactic fermentation (Gobert, Chaigneau & Savel, 1971) which can proceed without the participation of hydrogenosomes in the carbohydrate metabolism. In our work significant acetate production, a possible indicator of hydrogenosomal metabolism, was demonstrated although lactate was also always present (S. R. Mack & M. Müller, unpublished). It seems therefore that the hydrogenosome can play a significant role also in *T. vaginalis*.

The participation of hydrogenosomal respiration in the aerobic gas exchange of trichomonads is an open question.

ANTI-TRICHOMONAD AGENTS

Trichomonas infections of humans and of farm animals are usually treated with nitroimidazole derivatives. Metronidazole was the first such compound introduced in human medicine and is still extensively used (Fig. 5). In recent years several of its analogues with similar activity were put on the market. In addition to trichomonads, these drugs are highly effective against certain other eukaryotic and also prokaryotic micro-organisms, all of which are strict anaerobes. These findings suggest that

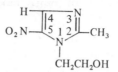

Fig. 5. Metronidazole, an antitrichomonad drug.

similarities in the metabolism of these organisms are responsible for their susceptibility to nitroimidazole derivatives.

Experiments showing a cessation of H_2 production when metronidazole or other nitroimidazoles were added to a suspension of *Trichomonas vaginalis* (Edwards & Mathison, 1970; Edwards, Dye & Carne, 1973; Müller *et al.*, 1978) indicated that the action of these compounds could be related to metabolic processes dependent on ferredoxin-linked reactions, i.e. on reactions localised in hydrogenosomes of trichomonads. Recent findings strongly indicate that the reduction of the nitro group of the drug is a crucial step in its action (Lindmark & Müller, 1976). It is assumed that this reduction is achieved by ferredoxin-like proteins with low redox potentials which are of great significance in anaerobes and play a minor or no role in other organisms. Metronidazole can easily be reduced by reduced ferredoxin in a non-enzymatic reaction (Lindmark & Müller, 1976). It is now assumed that in the trichomonad cell metronidazole acts as an acceptor for the electrons that give rise to H_2 in the hydrogenosomes under anaerobic conditions, thus inhibiting its production.

It is not the interference with H_2 production that is responsible for the toxicity of nitroimidazoles for trichomonads. Other nitroderivatives (e.g. 2,4-dinitrophenol, Müller *et al.*, 1979; nitrofurans, Edwards *et al.*, 1973; M. Müller, unpublished) also can act as electron acceptors without having toxic effects. The reduction of metronidazole, however, produces certain short lived derivatives or free radicals which exert a toxic action (Müller, Lindmark & McLaughlin, 1977). Although little is known about this toxic action and it is likely that the reactive products have multiple targets in the cell, it is worth mentioning that electron micrographs of *Trichomonas vaginalis* exposed to high concentrations of nitroimidazoles show, among other effects, structural damage of the hydrogenosomes (Buchner & Edwards, 1975; Carosi *et al.*, 1977).

BIOLOGICAL NATURE OF THE HYDROGENOSOME

The above discussion of the morphology and enzyme content of the hydrogenosomes shows that they differ significantly from mitochondria. Although the major substrate for both is pyruvate, its metabolism in each organelle differs. On the basis of its metabolic properties, especially the presence of substrate level phosphorylation only and the absence of cytochrome-mediated electron transport, the hydrogenosomes are closer to primitive anaerobic bacteria of a clostridial nature than to mitochondria.

It is suggested that the hydrogenosome might be an anaerobic equivalent of the mitochondrion (Müller, 1975; Čerkasovová et al., 1976; Whatley, John & Whatley, 1979). Although this suggestion is only a hypothesis at present, we can compare the known properties of the two organelles. Unfortunately, the available data are rather scanty and some of them need further confirmation. Clearly the most important point would be the unequivocal demonstration in hydrogenosomes of the presence or absence of genetic and protein-synthetic machinery. A single electron micrograph has been published depicting a circular DNA molecule observed in a hydrogenosomal preparation from *Tritricho-monas foetus* (Čerkasovová et al., 1976). There are no further data on this point and especially none on the presence of ribonucleic acids or of the enzymes of a genetic apparatus. Cell fractionation studies showed the presence of two components, not directly related to the energy metabolism, in the hydrogenosomes. These are cardiolipin (Čerkasovová et al., 1976), also a characteristic component of the mitochondrial inner membrane, and a cyanide-insensitive superoxide dismutase (Lindmark & Müller, 1974b). Similar types of superoxide dismutase are often found in the matrix of mitochondria and in the cytoplasm of bacteria. It has to be added here, however, that the extra-organellar cytoplasm also contains superoxide dismutase similar in its properties to the organellar enzyme. The morphology of the two organelles differs in several aspects. It might be significant, however, that, similarly to mitochondria, the hydrogenosomes seem to be surrounded not by one, but by two unit membranes. Further data are necessary to verify or disprove the above hypothesis, but the available evidence is not incompatible with it.

Although the evolutionary origin of mitochondria and chloroplasts is still an open question it is possible that the hydrogenosome:clostridium pair represents a further example in the list of organelle and 'its free living prokaryotic ancestor' pairs which includes the mitochondrion:

aerobic bacterium and the chloroplast:cyanobacterium. In the latter two cases the comparative consideration of the members of each pair provided a great impetus and sometimes also help in the study of both the organelle and of the prokaryote. We feel that it could do so also in the practically unexplored area of the organelles of anaerobic protozoa, and especially as it concerns the hydrogenosomes.

OTHER PROTOZOA

In addition to trichomonads there are other groups of protozoa in which electron microscopic evidence suggests the absence of mitochondria. In some of these cases the morphological data are corroborated by biochemical findings. On this basis two questions deserve brief consideration. Firstly, whether protozoa other than trichomonads are likely to contain hydrogenosomes, and secondly, whether anaerobic pyruvate decarboxylation linked to a low redox potential electron acceptor is necessarily localised in a membrane-bounded organelle.

The biochemical characteristics of hydrogenosomes indicate that, if they are present anywhere else than in trichomonads, they will most likely occur in protozoa from anaerobic habitats. For parasitic protozoa, such habitats are usually parts of the alimentary tract of their hosts. In fact morphological evidence shows the absence of typical mitochondria from many, but certainly not from all, parasitic protozoa in such habitats. Among these, species belonging to two major systematic groups contain microbody-like organelles with morphological similarities to hydrogenosomes. One group is the Hypermastiginida of termites and wood-eating roaches. Since trichomonad affinities of the Hypermastiginida are striking, although their cellular organisation is much more advanced, the presence of hydrogenosomes is a possibility (Hollande & Valentin, 1969). The other group consists of the holotrich ciliates and possibly also of the oligotrich ciliates of the rumen and similar habitats (Anderson & Dumont, 1966). The strictly anaerobic nature of these organisms makes it likely that their microbody-like organelles are hydrogenosomes. The absence of typical mitochondria and the presence of microbody-like organelles were also reported in free-living ciliates from anaerobic bottom sediments of oceans (Fenchel, Perry & Thane, 1977). A hydrogenosomal identification would be tempting in this case, too. In none of these cases do we have the biochemical evidence which is indispensable for the identification of the organelles as hydrogenosomes. The above data give, however, some suggestions for future studies. The difficulty of obtaining the necessary material and especially the lack of appropriate methods of axenic cultivation might keep open for a long time the question of the systematic distribution of hydrogenosomes among various protozoa.

In certain protozoa, not only mitochondria but also structures resembling hydrogenosomes are lacking. Such organisms occur in several major taxonomic groups. Of those found in vertebrate intestine we mention certain *Entamoeba* species, especially *Entamoeba histolytica*, and a flagellate, *Giardia lamblia* because recent studies permit their biochemical comparison with trichomonads (Lindmark, 1976b; Weinbach, Diamond & Claggett, 1976; Reeves *et al.*, 1977; Weinbach *et al.*, 1978; Lindmark, 1979). These organisms are aero-tolerant anaerobes and their respiration is not cytochrome-mediated. More importantly, they contain a pyruvate decarboxylating enzyme which, similar to trichomonads, is not linked to NAD but functions with low redox potential electron acceptors, including clostridial ferredoxin. Acetate was observed among the end products of carbohydrate metabolism and preliminary evidence suggests that it is formed by mechanisms similar to those found in trichomonads. However, no production of hydrogen and no hydrogenase could be detected. Cell fractionation studies, however gentle the homogenisation procedure employed, showed the pyruvate decarboxylating enzyme to be localised in the non-sedimentable portion of the cytoplasm (Lindmark, 1976b; Reeves *et al.*, 1977; D. G. Lindmark, personal communication) and gave us no evidence for the existence of hydrogenosomes in these species. It remains to be seen whether the concomitant lack of hydrogenase reflects any causal relationship. It is clear, however, that the low redox potential pathway of pyruvate oxidation is not obligately connected in protozoa with a specialised, membrane-limited subcellular compartment.

REFERENCES

ANDERSON, E. & DUMONT, J. N. (1966). A comparative study of the concrement vacuole of certain endocommensal ciliates – a so-called mechanoreceptor. *Journal of Ultrastructural Research*, **15**, 414–50.

ARESE, P. & CAPPUCCINELLI, P. (1974). Glycolysis and pentose phosphate cycle in *Trichomonas vaginalis*: I. Enzyme activity pattern and the constant proportion quintet. *International Journal of Biochemistry*, **5**, 859–65.

BAUCHOP, T. (1971). Mechanism of hydrogen formation in *Trichomonas foetus*. *Journal of General Microbiology*, **68**, 27–33.

BRODA, E. (1975). *The Evolution of Bioenergetic Processes*, pp. 1–168. Oxford: Pergamon Press.

BRUGEROLLE, G. & METENIER, G. (1973). Localisation intracellulaire et caracterisation de deux types de malate deshydrogénase chez *Trichomonas vaginalis* Donné, 1836. *Journal of Protozoology*, **20**, 320–7.

BUCHNER, Y. & EDWARDS, D. I. (1975). The effect of metronidazole and nitrofurans on the morphology of *Trichomonas vaginalis*. *Journal of Antimicrobial Chemotherapy*, **1**, 229–34.

CAROSI, G., FILICE, G., SUTER, F. & DEI CAS, A. (1977). *Trichomonas vaginalis*: Effect of tinidazole on ultrastructure *in vitro*. *Experimental Parasitology*, **43**, 315–25.

ČERKASOV, J., ČERKASOVOVÁ, A., KULDA, J. & VILHELMOVÁ, D. (1978). Respiration of hydrogenosomes of *T. foetus*. I. ADP-dependent oxidation of malate and pyruvate. *Journal of Biological Chemistry*, **253**, 1207–14.

ČERKASOVOVÁ, A. (1970). Energy-producing metabolism of *T. foetus*. I. Evidence for control of intensity and the contribution of aerobiosis to total energy production. *Experimental Parasitology*, **27**, 165–78.

ČERKASOVOVÁ, A. & ČERKASOV, J. (1974). Location of the NADH oxidase activity in fractions of *Tritrichomonas foetus* homogenate. *Folia Parasitologica (Praha)*, **21**, 193–203.

ČERKASOVOVÁ, A. & ČERKASOV, J. (1978). Absence of pyruvate splitting processes in *Tritrichomonas foetus* resistant to metronidazole. *Journal of Protozoology* **25**, 34A–5A.

ČERKASOVOVÁ, A., ČERKASOV, J., KULDA, J., & REISCHIG, J. (1976). Circular DNA and cardiolipin in hydrogenosomes, microbody-like organelles of trichomonads. *Folia Parasitologica (Praha)*, **23**, 33–7.

EDWARDS, D. I., DYE, M. & CARNE, H. (1973). The selective toxicity of antimicrobial nitroheterocyclic drugs. *Journal of General Microbiology*, **76**, 135–45.

EDWARDS, D. I. & MATHISON, G. E. (1970). The mode of action of metronidazole against *Trichomonas vaginalis*. *Journal of General Microbiology*, **63**, 297–302.

FENCHEL, T., PERRY, T. & THANE, A. (1977). Anaerobiosis and symbiosis with bacteria in free living ciliates. *Journal of Protozoology*, **24**, 154–63.

FILADORO, F. (1970). Fine structure of chromatic granules in *Trichomonas vaginalis* Donné. *Experientia*, **26**, 213–14.

GOBERT, N., CHAIGNEAU, M. & SAVEL, J. (1971). Etude des gaz libérés au cours de la culture en anaérobiose de *Trichomonas vaginalis*. *Comptes Rendus des Séances de la Société de Biologie*, **165**, 276–82.

GRAY, C. T. & GEST, H. (1965). Biological formation of molecular hydrogen. *Science*, **148**, 186–92.

HOLLANDE, A. & VALENTIN, J. (1969). Appareil de Golgi, pinocytose, lysosomes, mitochondries, bactéries symbiontiques, atractophores et pleuromitose chez les hypermastigines du genre *Joenia*. Affinités entre joeniides et trichomonadines. *Protistologica*, **5**, 39–86.

HONIGBERG, B. M. (1978a). Trichomonads of veterinary importance. In *Parasitic Protozoa*, vol. 2, ed. J. P. Kreier, pp. 163–273. New York: Academic Press.

HONIGBERG, B. M. (1978b). Trichomonads of importance in human medicine. In *Parasitic Protozoa*, vol. 2, ed. J. P. Kreier, pp. 275–454. New York: Academic Press.

KULDA, J., KRÁLOVÁ, J. & VÁVRA, J. (1977). The ultrastructure and 3,3′-diaminobenzidine cytochemistry of hydrogenosomes of trichomonads. *Journal of Protozoology*, **24**, 51A–2A.

LINDBLOM, G. P. (1961). Carbohydrate metabolism of trichomonads: Growth, respiration, and enzyme activity in four species. *Journal of Protozoology*, **8**, 135–50.

LINDMARK, D. G. (1976a). Acetate production by *Tritrichomonas foetus*. In *Biochemistry of Parasites and Host-Parasite Relationships*, ed. H. van den Bossche, pp. 15–21. Amsterdam: North-Holland.

LINDMARK, D. G. (1976b). Certain enzymes of the energy metabolism of *Entamoeba invadens* and their subcellular localization. In *Proceedings of the International Conference on Amebiasis*, ed. B. Sepulveda & L. S. Diamond, pp. 185–9. Mexico: Instituto Mexicano de Siguro Social.

LINDMARK, D. G. (1979). Metabolism of *Giardia lamblia*. *Journal of Protozoology*, **26**, 11A.

LINDMARK, D. G. & MÜLLER, M. (1973a). Hydrogenosome, a cytoplasmic organelle of the anaerobic flagellate, *Tritrichomonas foetus*, and its role in pyruvate metabolism. *Journal of Biological Chemistry*, **248**, 7724–8.

LINDMARK, D. G. & MÜLLER, M. (1973b). Subcellular distribution of flavins in two trichomonad species. *Journal of Protozoology*, **20**, 500.

LINDMARK, D. G. & MÜLLER, M. (1974a). Biochemical cytology of trichomonad flagellates. II. Subcellular distribution of oxidoreductases and hydrolases in *Monocercomonas* sp. *Journal of Protozoology*, **21**, 374–8.

LINDMARK, D. G. & MÜLLER, M. (1974b). Superoxide dismutase in the anaerobic flagellates, *Tritrichomonas foetus* and *Monocercomonas* sp. *Journal of Biological Chemistry*, **249**, 4634–7.

LINDMARK, D. G. & MÜLLER, M. (1976). Antitrichomonad action, mutagenicity, and reduction of metronidazole and other nitroimidazoles. *Antimicrobial Agents and Chemotherapy*, **10**, 476–82.

LINDMARK, D. G., MÜLLER, M. & SHIO, H. (1975). Hydrogenosomes in *Trichomonas vaginalis*. *Journal of Parasitology*, **61**, 552–4.

LLOYD, D., LINDMARK, D. G. & MÜLLER, M. (1979a). Respiration of *Tritrichomonas foetus*: absence of detectable cytochromes. *Journal of Parasitology*, **65**, 466–9.

LLOYD, D., LINDMARK, D. G. & MÜLLER, M. (1979b). Adenosine triphosphatase of *Tritrichomonas foetus*. *Journal of General Microbiology*, **115**, 301–7.

MACK, S. R. & MÜLLER, M. (1978). Effect of oxygen and carbon dioxide on the growth of *Trichomonas vaginalis* and *Tritrichomonas foetus*. *Journal of Parasitology*, **64**, 927–9.

MÜLLER, M. (1973). Biochemical cytology of trichomonad flagellates. I. Subcellular localization of hydrolases, dehydrogenases, and catalase in *Tritrichomonas foetus*. *Journal of Cell Biology*, **57**, 453–74.

MÜLLER, M. (1975). Biochemistry of protozoan microbodies (peroxisomes, α-glycerophosphate oxidase bodies and hydrogenosomes). *Annual Review of Microbiology*, **29**, 467–83.

MÜLLER, M. (1976). Carbohydrate and energy metabolism of *Tritrichomonas foetus*. In *Biochemistry of Parasites and Host–Parasite Relationships*, ed. H. van den Bossche, pp. 3–14. Amsterdam: North-Holland.

MÜLLER, M. & LINDMARK, D. G. (1978). Respiration of hydrogenosomes of *Tritrichomonas foetus*. II. Effect of CoA on pyruvate oxidation. *Journal of Biological Chemistry*, **253**, 1215–18.

MÜLLER, M., LINDMARK, D. G. & MCLAUGHLIN, J. (1977). Mode of action of metronidazole on anaerobic microorganisms. *Metronidazole. Proceedings of the International Metronidazole Conference, Montreal*, ed. S. M. Feingold, J. McFadzean, & F. J. C. Roe, pp. 12–19. Amsterdam: Excerpta Medica.

MÜLLER, M., LINDMARK, D. G. & MACK, S. R. (1978). *Trichomonas vaginalis*: Metabolism and the mode of action of antitrichomonad nitroimidazoles. *Bulletins et Mémoires de la Société de Médecine de Paris*, **181**, 141–5.

MÜLLER, M., NSEKA, V., MACK, S. R. & LINDMARK, D. G. (1979). The effects of 2, 4-dinitrophenol on trichomonads and *Entamoeba invadens*. *Comparative Biochemistry and Physiology*, **64B**, 97–100.

NIELSEN, M. H. & DIEMER, N. H. (1976). The size, density, and relative area of chromatic granules ('hydrogenosomes') in *Trichomonas vaginalis* Donné from cultures in logarithmic and stationary growth. *Cell and Tissue Research*, **167**, 461–5.

REEVES, R. E., WARREN, L. G., SUSSKIND, B. & LO, H.-S. (1977). An energy-conserving pyruvate-to-acetate pathway in *Entamoeba histolytica*. Pyruvate synthase and a new acetate thiokinase. *Journal of Biochemical Chemistry*, **252**, 726–31.

RYLEY, J. F. (1955). Studies on the metabolism of the protozoa. 5. Metabolism of the parasitic flagellate *Trichomonas foetus*. *Biochemical Journal*, **59**, 361–9.

WEINBACH, E. C., CLAGGETT, C. E., TAKEUCHI, T. & DIAMOND, L. S. (1978). Biological oxidations and flavoprotein catalysis in *Entamoeba histolytica*. *Archivos de Investigacion Médica* (México), **9** (Supl. 1), 89–98.

WEINBACH, E. C., DIAMOND, L. S. & CLAGGETT, C. E. (1976). Iron–sulfur proteins of *Entamoeba histolytica*. *Journal of Parasitology*, **62**, 127–8.

WHATLEY, J. M., JOHN, P. & WHATLEY, F. R. (1979). From extracellular to intracellular: the establishment of mitochondria and chloroplasts. *Proceedings of the Royal Society of London, Series B*, **204**, 165–87.

EXPLANATION OF PLATE

PLATE 1

Hydrogenosomes of trichomonad flagellates. (*a*) Row of paracostal hydrogenosomes (arrow) along the costa in *Tritrichomonas foetus*. Phase contrast image of cell fixed under the coverslip by glutaraldehyde diffusing from one side. um, undulating membrane. (*b*) Elongated hydrogenosomes of *Monocercomonas* sp. n, nucleus; G, Golgi complex. (*c*) Hydrogenosomes of *Tritrichomonas foetus* showing 'marginal plate' (m) and dense core (c). ax, axostylar microtubules; g, glycogen granules. (*d*) Hydrogenosome of *Monocercomonas* sp. with indications of a double membrane.

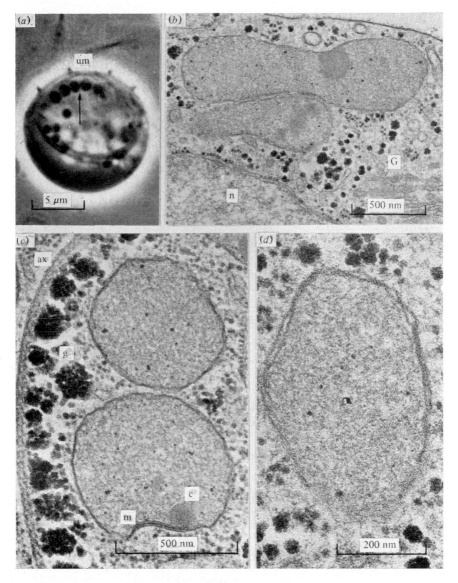

PLATE 1

STRUCTURE, FUNCTION, BIOGENESIS AND GENETICS OF MITOCHONDRIA

DAVID LLOYD* AND GEOFFREY TURNER†

*Department of Microbiology, University College, Newport Road, Cardiff, UK

†Department of Bacteriology, The Medical School, University Walk, Bristol, UK

INTRODUCTION

Recent advances in our appreciation of the bioenergetics and biogenesis of mitochondria have provided a growing awareness of the complexity and dynamic organisation of biochemical events that occur in these membrane-bounded organelles which are found in all aerobically-growing eukaryotes. Almost all studies focus either on the inner mitochondrial membrane as a permeability barrier (to H^+, adenine and pyridine nucleotides and other metabolites) and as the site of the respiratory chain and energy conservation, or on the subcellular systems responsible for the synthesis of this membrane (nuclear DNA and cyto-ribosomes, mitochondrial DNA and mitochondrial ribosomes, and the two systems of protein synthesis). Here we highlight major recent developments in understanding the molecular organisation of the inner membrane, its functional mechanisms and dynamics from the sub-molecular to the cellular level, its mode of synthesis and assembly, and its genetics and inheritance. For fuller accounts of mitochondrial bio-chemistry and genetics, especially in micro-organisms, the reader is referred to Lloyd (1974) and Bandlow et al. (1977), and for introductory accounts, to previous articles in this series (Lloyd, 1969; Hughes, Lloyd & Brightwell, 1970).

MITOCHONDRIAL POPULATION DYNAMICS AND MOTILITY

Until recently it was accepted that most cells contain a large number of mitochondria, and that those which possess one were exceptional. But following the observation that some small marine flagellates sometimes have a single mitochondrion (Manton, 1959; Manton & Parke, 1960), this was shown also for Chlorella sp. (Atkinson, John & Gunning, 1974), trypanosomes (Paulin, 1975) and (exceptionally) for a yeast (Hoffman & Avers, 1973). More commonly, a few highly-branched mitochondria have been identified, either by reconstruction of images from serial sections, or by the use of high voltage electron microscopy with thick sections. Thus several unicellular algae contain a mitochondrial reticulum (Leedale & Buetow, 1970; Arnold et al., 1972; Burton & Moore, 1974; Grobe & Arnold, 1975; Osafune et al., 1975). In other organisms, many small unbranched mitochondria are present (Heywood, 1977). Fragmentation of a large branched mitochondrial structure to give many smaller mitochondria occurs at a specific time in the cell cycle of *Euglena gracilis* (Calvayrac et al., 1974), and immediately after zygote formation in *Chlamydomonas reinhardii* (Grobe & Arnold, 1977). In yeast the transformation from a few mitochondria in rapidly-growing cells to a population of numerous small mitochondria in stationary phase cultures has been demonstrated (Stevens, 1977). In all stages of growth, yeast cells contain a mitochondrion which is ramified and considerably larger in size than the others. A deeper appreciation of the population dynamics of mitochondria is needed to understand the mechanism of segregation of mitochondrial DNA at cell division and the processes of mitochondrial mutation and recombination. In this context it is interesting that mitochondrial DNA in yeast rarely occurs as isolated molecules, but is usually distributed in aggregates, 'chondrolites' (Williamson et al., 1977).

Mitochondrial locomotion can occur independently of cytoplasmic flow; the fine-structural events underlying mitochondrial motility indicate that these active movements are controlled by metabolic state (Bereiter-Hahn, 1978).

ELECTRON TRANSPORT AND ENERGY CONSERVATION

The components of the respiratory chain can be grouped together as four complexes (I–IV) which correspond to the units that remain associated through the classical sequence of biochemical purification steps (Fig. 1). Rapid advances in e.p.r. studies have enabled the characterisation of the iron–sulphur centres (Ohnishi, 1979). The relationship of the respiratory chain to energy-conserving reactions is more evident when the components associated with four levels of mid-point oxidation–reduction potential are grouped together (Fig. 2); these groups then constitute isopotential pools of redox components. Energy-transducing redox carriers operate between the pools; at Site I, $Fe-S_{N-2}$ may serve this function, and at Sites II and III cytochromes b_T and cytochrome c

oxidase, respectively, are responsible for energy conservation (Chance, 1977). The nature of the 'energy pool' (sometimes referred to as the 'energised state') is still enigmatic, but the following observations provide strong support for the basic postulates of the chemiosmotic

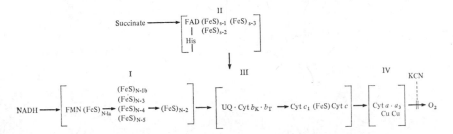

Fig. 1. The mitochondrial electron transport chain. Components are indicated in four groups corresponding to complexes I–IV as separated by chemical purification procedures. FeS, iron–sulphur centre; UQ, ubiquinone; Cyt, cytochrome. (Chance 1977).

hypothesis (Mitchell, 1966). (1) Most of the energy-transducing units are able to generate a H^+ gradient and/or membrane potential across the 'coupling membranes' in which they are located. These processes are reversible. (2) Energy transfer between the energy-transducing units can occur by way of a H^+ gradient and/or membrane potential. Agents that abolish the H^+ gradient and/or membrane potential uncouple these

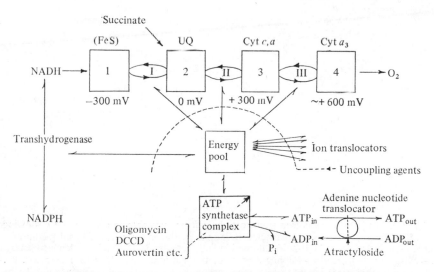

Fig. 2. The mitochondrial electron-transport chain and associated reactions of energy conservation. The components of the respiratory chain are grouped in isopotential pools and the three sites of energy conservation (I, II, III) occur at energy-transducing redox components which span the gaps in redox potential between these pools.

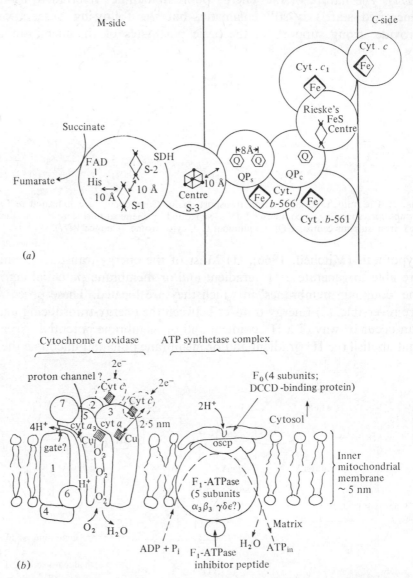

Fig. 3(*a*). Model of the electron transport chain from succinate to cytochrome *c* showing relative orientations and intercarrier separations in the inner mitochondrial membrane. Q, ubiquinone; QP$_s$ and QP$_c$, ubiquinone-binding proteins; SDH, succinate dehydrogenase. S-1, S-2 and S-3 are iron–sulphur centres of the succinate dehydrogenase complex. (T. Ohnishi, unpublished). (*b*) Models of cytochrome oxidase and ATP synthetase in relation to the sidedness of the inner mitochondrial membrane. Numbers 1 to 7 indicate subunits of cytochrome *c* oxidase (Chance, Leigh & Waring, 1977; Dockter, Steinemann & Schatz, 1978).

processes of energy transfer. Although the essential features of the chemiosmotic hypothesis are now experimentally established, the precise mechanisms involved in the generation of energy are still a mystery. 'Localised' or 'micro' chemiosmosis also provides a hypothesis for ATP formation by energised H^+, but in membrane micro-environments rather than on a transmembrane basis (Williams, 1978).

Areas of investigation include the layout of molecular components with special reference to the sidedness of the inner mitochondrial membrane, the mechanism and stoichiometry of H^+ translocation, the precise pathway and mechanisms of electron transport, and the mechanism of the ATP synthesis reaction itself (Boyer *et al.*, 1977). Models of the topographical relationships of complexes II and III, and cytochrome *c* oxidase and ATP synthetase within the inner mitochondrial membrane are shown in Fig. 3.

Mitochondrial ATP synthetase complex

A striking feature of ATP synthetases isolated from membranes engaged in energy conversion (bacterial, chloroplast or mitochondrial) is that they all share common features; suggesting that this enzyme arose early in evolution, long before the divergence of the animal and plant kingdoms. The mitochondrial ATP synthetase complex (oligomycin-sensitive ATPase) represents a highly perfected energy-converting molecule, and is the most complicated of all known enzymes. In yeast it consists of at least eleven different subunits and some phospholipids, and has a mol. wt. of 468 000 (Tzagoloff, Rubin & Sierra, 1973; Ryrie & Gallagher, 1979). A model of its four functional parts (Fig. 3) indicates that F_1-ATPase (which has five subunits) contains the catalytic site(s) responsible for ATP synthesis and hydrolysis, and reversibly binds the F_1-ATPase inhibitor peptide (mol. wt. 6800). Although F_1-ATPase can itself catalyse the hydrolysis of ATP, ATP synthesis requires also the participation of the F_0 component (the membrane factor with four subunits), phospholipids, and the oligomycin-sensitivity conferring protein (OSCP). The stoichiometry of the various subunits is not established with certainty. ADP competes with ATP for binding to F_1-ATPase and the kinetics of inhibition indicate strong co-operativity; titration with thiol-blocking reagents reveals the presence of 12 thiol groups per molecule of unfolded, reduced F_1-ATPase, but these reagents do not inhibit ATP hydrolysis. Experiments with carboxypyridine disulphide (Gautheron & Godinot, 1977) and dibutylchloromethyltin chloride (Griffiths & Hyams, 1977) confirm earlier results which indicated the presence of vicinal thiols; inhibition by 2,3-butanediol implicates arginyl groups (Marcus, Schuster & Lardy, 1976), and inhibition by 4-chloro-7-nitrobenzofurazan provided evidence for the presence of a tyrosyl group on the β-subunit of F_1-ATPase, at or near the active centre of the enzyme (Ferguson $et\ al.$, 1975). The mechanism by which F_0 functions during the synthesis of ATP is uncertain; the chemiosmotic hypothesis of Mitchell (1966) visualises this segment of the complex acting as a channel to direct H^+ to the active centre in F_1; inhibitors binding to F_0 must then act by preventing H^+-translocation through this 'pore'. One of the subunits of F_0 is particularly hydrophobic, having a low content of polar amino acids, and a mol. wt. of approximately 8000 (Cattell $et\ al.$, 1971; Wachter, Sebald & Tzagoloff, 1977; Graf & Sebald, 1978). Dicyclohexylcarbodiimide (DCCD) binds irreversibly to this subunit, preventing H^+ translocation across the membrane, and there is evidence that this subunit might be the 'protonophore' of the ATPase complex (Nelson $et\ al.$, 1977). The chemical hypothesis of energy conservation would require the mediation of a mobile carrier between the electron transport chain and the ATP synthetase complex.

Control of the hydrolytic activity of ATP synthetase (and hence of all ATP-driven reactions including reversed electron flow, transhydrogenase and ion transport) is mediated by the F_1-ATPase inhibitor peptide (Pullman & Monroy, 1963) which reversibly associates at specific binding sites on F_1-ATPase. This peptide is present, presumably in the bound state under conditions where ATP synthesis proceeds apace (e.g. in aerobically-grown catabolite-derepressed yeast). Table 1 compares the sensitivities to inhibitors of ATPases in cell-free extracts of some yeasts and protozoa. The increased sensitivities (especially to those inhibitors which bind to F_0) after catabolite derepression in $Schizosaccharomyces\ pombe$ are due to elevated levels of bound F_1-ATPase inhibitor peptide (Lloyd & Edwards, 1976). Even in $Saccharomyces\ carlsbergensis$ grown anaerobically in the presence of 10% glucose, the ATP synthetase complex still shows a characteristic spectrum of inhibitor sensitivities, and has thus not lost any of the inhibitor binding sites (T. G. Cartledge & D. Lloyd, unpublished), although under these conditions it must function solely as an ATP phosphohydrolase, possibly involved in Ca^{2+} or K^+ transport across the inner membrane of the 'promitochondrion'. While some protozoal mitochondrial ATP synthetases possess a full complement of inhibitor binding sites (e.g. $Acanthamoeba\ castellanii$, S. W. Edwards & D. Lloyd, unpublished), others appear to lack sites commonly assumed to act at or near the active catalytic site of F_1-ATPase. Thus the mitochondrial ATPase of $Crithidia\ fasciculata$ is insensitive to inhibition by spegazzinine, efa-

Table I. I_{50} values for inhibitors of mitochondrial ATPases in cell-free extracts of yeasts and protozoa

Organism ▶ Inhibitor	[a]Schizosaccharomyces pombe (Glucose-repressed)	(Glucose-derepressed)	[b]Saccharomyces carlsbergensis (Anaerobically-grown, glucose-repressed)	[c]Acanthamoeba castellanii	[d]Crithidia fasciculata	[e]Tetrahymena pyriformis ST
Oligomycin	3.4	0.6	0.2	0.45	4.6	
N,N'-Dicyclohexylcarbodiimide	8.5	2.5	9.0	4.1	1.9	80.0
Triethyltin sulphate	0.058	0.02	NT	0.01	1.24	20.0
Dibutylchloromethyltin chloride	NT	NT	0.27	0.002	NT	3.5
Venturicidin	0.028	0.008	1.4	1.4	3.0	10.0
Leucinostatin	NT	NT	0.53	0.19	11.0	50.0
Dio-9	5.2	1.2	0.4	0.02	6.6	
Efrapeptin	0.018	0.011	0.09	0.07	NI	150
Aurovertin	0.75	0.45	NT	NT	NI	
Citreoviridin	NT	NT	NT	0.26	NI	
4-Chloro-7-nitrobenzofurazan	2.8	2.0	0.75	0.86	NI	
Quercetin	4.9	44	0.4	0.66	2.6	
Spegazzinine	18.5	12.5	0.19	5.0	NI	

I_{50} values measured at [a]pH 8.6 (Lloyd & Edwards, 1976); [b]pH 9.3 (T. G. Cartledge & D. Lloyd, unpublished; [c]pH 7.8 (S. W. Edwards & D. Lloyd, unpublished); [d]pH 7.8 N. Yarlett and D. Lloyd, unpublished; [e]pH 7.2. NI, not inhibitory; NT, not tested.

peptin, aurovertin, citreoviridin and 4-chloro-7-nitrobenzofurazan (N. Yarlett & D. Lloyd, unpublished). The ATPase activity of mitochondria from *Tetrahymena pyriformis* is insensitive to many of the classical inhibitors (M. D. Unitt and D. Lloyd, unpublished).

Alternative pathways of electron transport

Some capacity for respiration remains in many eukaryotes when inhibitors of the main respiratory chain are present at concentrations which completely prevent the operation of cytochrome c oxidase. Several alternative terminal oxidases have been identified (e.g. the salicylhydroxamic acid-sensitive oxidases present in a wide variety of eukaryotes, and the α-glycerophosphate oxidase and cytochrome o of trypanosomes; Degn, Lloyd & Hill, 1978). Alternative terminal oxidases serve an obvious function in circumstances where the main respiratory chain is impaired so that it has a lowered or negligible capacity for electron transport (e.g. in many mutant organisms, or in organisms growing with electron transport inhibitors), but the necessity for their operation alongside fully-functional main chains is not understood. Cytochrome c peroxidase in yeast and trypanosome mitochondria provides an alternative electron transport pathway which by-passes energy conservation at Site III (Erecińska et al., 1973; Kusel, Boveris & Storey, 1973). Peroxide generation occurs at the ubiquinone region of the chain during respiration and its formation is believed to occur via superoxide anion (Boveris, 1978).

MITOCHONDRIAL CYTOSOLIC INTER-RELATIONSHIPS

A number of specific transport systems (translocators) are located in the inner mitochondrial membrane and act as regulators during the controlled passage of low molecular weight solutes across it. Anion transport systems include those specific for malate, succinate, oxoglutarate, citrate, glutamate and pyruvate, phosphate and adenine nucleotides. Cation transport systems include the high affinity Ca^{2+}-carrier; Ca^{2+} uptake is energy-requiring in the sense that the movement of the ion occurs in response to a membrane potential, negative inside, generated across the inner membrane as a result of respiratory chain or ATPase activity.

The adenine nucleotide translocator permits the exchange of the ADP and ATP located in the matrix space with ATP or ADP of the cytosol; stoichiometry of this exchange is 1:1. The affinity of the translocase for ADP is higher than the affinity of phosphotransferases located outside the inner mitochondrial membrane (such as adenylate kinase) (Vignais, 1976). Therefore ADP formed in the cytosol as a consequence of ATP-consuming reactions diffuses freely through the outer mitochondrial membrane into the intermembrane space where it is preferentially trapped by the adenine nucleotide translocator to be phosphorylated by the oxidative phosphorylation system. ATP translocation *into* mitochondria is of vital importance in organisms when oxidative phosphorylation is not occurring (Kovác, Kolarov & Subik, 1977). Evidence against a primary role for the adenine nucleotide translocase in the regulation of energy metabolism has been presented (Wilson, Owen & Holian, 1977). The intra-mitochondrial phosphate potential ($[ATP]/[ADP][P_i]$) is between ten- and 100-fold lower than that in the cytosol, and the rate of mitochondrial respiration is dependent on the extra-mitochondrial phosphate potential. In both intact cells and isolated mitochondria, energy conservation Sites I and II are in near equilibrium with the extra-mitochondrial phosphate potential, and this implies that the mechanism for respiratory control is exerted at Site III (cytochrome c oxidase). Experiments with *Candida utilis* have suggested that the cellular respiratory rate and the NADH generation rate are regulated by the ratio $[ATP]/[ADP][P_i]$ and not by the adenylate energy charge $([ATP] + \frac{1}{2}[ADP]/[ATP] + [ADP] + [AMP])$ (Atkinson, 1977). Thus the intracellular $[P_i]$ plays an important part in the regulatory mechanisms of energy metabolism (Erecińska *et al.*, 1977).

The inner mitochondrial membrane is impermeable to NAD^+ and NADH, and the entry of reducing equivalents is mediated by specific shuttle systems; control is also exerted by this membrane on the processes of compartmentation of fatty acids (Greville, 1966).

Control of mitochondrial function in growing organisms

A plethora of information on the flexibility of mitochondrial function and the variability of mitochondrial structure and composition has come from experiments in which micro-organisms have been grown under conditions of environmental stress. Examples of this approach include growth of yeasts under anaerobic or catabolite-repressing conditions, growth with limiting concentrations of iron, copper, sulphate, haem, choline, pantothenate or inositol, growth with inhibitors of electron transport (such as antimycin A or cyanide) or inhibitors or uncouplers of energy conservation, or growth in the presence of inhibitors of macromolecular synthesis such as ethidium bromide and chloramphenicol (Lloyd, 1974). The environmentally-modified mitochondria that are developed under these conditions, and that in all but the most extreme cases can still perform a modicum of energy generation, indicate the versatility of mitochondrial organisation and its limits.

A different question can then be posed: is it possible to study the function of mitochondria *in vivo* during 'normal' undisturbed cellular growth and division? To do this evidently requires the use of cultures in which organisms divide synchronously in the absence of any measurable metabolic perturbation. The methods of choice for obtaining synchronous cultures are based on selection methods in which the population used is of a homogeneous size-class. Provided that the size selection procedure is carried out rapidly and avoids transient anaerobiosis, nutrient starvation, temperature changes and high centrifugal fields, then perturbation is minimal (Lloyd *et al.*, 1975; Lloyd & Ball, 1979). The oscillatory energy metabolism observed in the fission yeast, *Schizosaccharomyces pombe*, and in the trypanosome, *Crithidia fasciculata*, after size-selection by velocity sedimentation through density gradients (Poole, Lloyd & Kemp, 1973; Poole & Lloyd, 1973,1974; Edwards, Statham & Lloyd, 1975) has more recently been observed in other protozoa and yeasts in synchronous cultures prepared under conditions of minimum perturbation (Phillips & Lloyd, 1978; Lloyd, Phillips & Statham, 1978; Edwards & Lloyd, 1978; Kader & Lloyd, 1979; C. L. Bashford, B. Chance, D. Lloyd & R. K. Poole, unpublished). In all these organisms we observed large-amplitude respiratory oscillations (with periods of about 1 h for the protozoa and 30 min in *Candida utilis*) which were only attenuated as synchrony decayed. The sustained character and low frequency of these oscillations distinguished them from oscillatory mitochondrial functions previously observed *in vitro* (Gooch & Parker, 1974). Indeed the cell-cycle-dependent respiratory oscillations have time constants which place them in the epigenetic time domain (Goodwin, 1963; 1976), rather than in the domain of metabolic oscillations, readily produced by perturbation of steady-state conditions (Hess & Chance, 1978; Lloyd & Ball, 1979). A partial explanation for the oscillatory nature of energy generation during the cell cycle has resulted from work with *Acanthamoeba castellanii* (Edwards & Lloyd, 1978; S. W. Edwards & D. Lloyd, unpublished). This organism has a cell cycle time of about 8 h, and its respiration oscillates with a periodicity of about one-seventh of this time. Maximal uncoupling of electron transport from the reactions of energy conservation (i.e. maximum acceleration of the respiratory rate) was achieved by adding an uncoupler at respiratory minima; as organisms progress through phases of high respiratory activity little stimulation of respiration was produced by this treatment. Conversely, organisms exhibited greater sensitivity to inhibition by cyanide when respiration was fast rather when it was slow. It was also shown that the intracellular level of ADP oscillated in phase with changes in respiration rates. From these results it may be concluded that the phenomenon we observe is an example of mitochondrial respiratory control *in vivo* (in which the rate of mitochondrial electron transport is controlled by the availability of ADP; Chance & Williams, 1956). But why should cellular growth necessitate *slow* state transitions from near the active respiratory State 3 (where ADP pools are large) into almost the 'resting' State 4 (when the intra-mitochondrial concentration of ADP becomes the rate-limiting factor that determines overall respiratory activity)? The answer to this question cannot reside in the reactions of energy metabolism which display extremely fast state-changes (State 4 to State 3 transitions produced by adding ADP to mitochondrial suspensions take less than 1s), but in epigenetic control loops involving the processes of transcription and translation. When we determined carefully the time course of accumulation in synchronous cultures of total cellular RNA and protein, we observed that these major products of biosynthesis did not show continuous increases, but also oscillated with a period of one-seventh of the cell cycle (S. W. Edwards

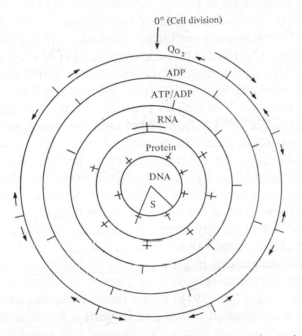

Fig. 4. Map of the cell cycle of *Acanthamoeba castellanii* represented as a circle from 0° to 360°. 0° represents time zero in a synchronous culture and 360° the time when all the cells have completed division. Maxima of cell cycle events are thus shown as phase angles with respect to their timings in synchronous cultures (Edwards & Lloyd 1978). Results for respiratory maxima represent the means ($\pm$ SD) of five determinations; those for RNA and protein are each from two experiments. S, S-phase of DNA synthesis; Qo_2, respiration rate of organisms in growth medium.

& D. Lloyd, unpublished). The phase relationships between the observed variations (Fig. 4) indicate that respiratory rate, ADP, and total protein content are all approximately in phase, but out of phase with cellular RNA content and the intracellular ATP/ADP ratio. We propose therefore that the cell cycle of *A. castellanii* is temporally ordered as a sequence of seven sub-cycles, in which discrete phases of protein synthesis and degradation alternate; close coupling of energy demand with mitochondrial energy generation via the respiratory control mechanism requires that as the organism progresses through its cell cycle it under-goes transitions almost from State 3 to State 4. Similar close coupling of energy demand to its supply as indicated by three sub-cycles of protein synthesis and degradation coupled to respiratory control *in vivo* in an unrelated organism, *Candida utilis* (D. Lloyd, S. W. Edwards & J. Ball, unpublished) tempts the suggestion that, at least in the cell cycles of eukaryotes, such a process may be widespread, or perhaps even univer-sal. This concept of 'periodic turnover' has far-reaching consequences. It suggests that the view of life as a 'dynamic equilibrium' in which the maintenance of steady-state levels of cellular components occurs by a balancing of simltaneously-occurring synthetic and degradative pro-cesses (Hopkins, 1913) is, like 'the apparent stability of adenylate charge' (Atkinson, 1977), an oversimplification which stems from time-averaged observations of asynchronous cultures (Lloyd *et al.*, 1978). Time resolu-tion of the temporal hierarchy in synchronous cultures (and perhaps eventually in single cells) indicates that even in cultures growing at or near their maximum rates, the process of turnover is active and in-escapable, and has a time structure, which although at present only poorly understood, must have a deep significance. The observation of periodic turnover in growing cells also has repercussions in the inter-pretation of calculations based on aerobic growth yields (Tempest, 1978) and the quantitative validity of measurements of 'maintenance energy' (Pirt, 1975). An indication that the previously reported half-life deter-minations for the proteins in growing cells do not adequately allow for post-incorporation and recycling of isotope label, comes from the studies of Bakalkin *et al.* (1978) which indicate that the half-life of the products of mitochondrial translation in yeast is less than one hour.

Periodic expression of mitochondrial enzyme activities during the cell cycle

Studies of the cell cycle of *Schizosaccharomyces pombe* have revealed that the activities of several mitochondrial enzymes are expressed dis-continuously during synchronous growth and division. Fig. 5 summar-ises the findings of Poole & Lloyd (1973), Poole, Lloyd & Chance (1974) and Edwards & Lloyd (1977) with glucose-repressed cultures of *S. pombe*. The second and third maxima of ATPase activity approxi-mately coincide with the two maxima for cytochromes $a + a_3$ and b-563 (b_T), and the third maximum of ATPase (at 0.65 cycle) is close to the single maximum observed for cytochrome c oxidase activity. That the observed oscillations in enzyme activity are a consequence of oscilla-tions in enzyme protein is evident in the case of cytochromes from data from difference spectra, and for ATPase from independent immunological assays using antibody raised to purified F_1-ATPase (B. Evans, G. El'Khayat, S. W. Edwards & D. Lloyd, unpublished). That different enzymes of the inner mitochondrial membrane are not expressed at identical times in the cell cycle suggests that the stoichiometry of

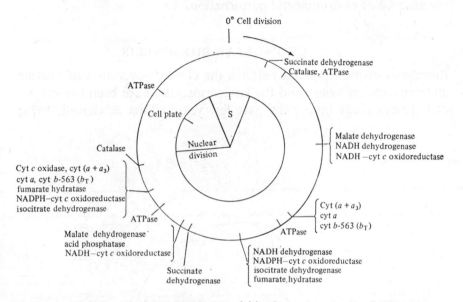

Fig. 5. Map of timings of maxima in enzyme activities during the cell cycle of glucose-grown *Schizosaccharomyces pombe*. 0° indicates the mid-point of doubling of cell numbers in synchronous cultures, S represents the duration of the S phase of DNA synthesis (Poole & Lloyd, 1973; Poole *et al.*, 1974; Edwards & Lloyd, 1977).

individual enzyme components in the membrane (e.g. succinate dehydrogenase and cytochrome *c* oxidase) may vary at different stages. Discontinuous expression of the inhibitor-binding sites of mitochondrial ATPase (Lloyd & Edwards, 1977) may have several different possible explanations: (*a*) periodic synthesis of subunits bearing specific inhibitor-binding sites; (*b*) periodic binding and assembly of subunits from precursor pools; (*c*) periodic modification of existing membrane-bound ATPase (by restricted proteolysis or changes in enzyme conformation) so as to expose latent inhibitor-binding sites; (*d*) periodic modification of neighbouring membrane components (fatty acid or sterol components); (*e*) periodic synthesis of compounds that adsorb or inactivate added inhibitors; or (*f*) modulation of the ATP synthetase complex by binding of F_1-inhibitor peptide (possibly in response to cycles of energisation and de-energisation of the membrane). Although we cannot yet completely assess the relative importance of these various parameters in the expression of mitochondrial ATPase some evidence for the first

alternative has been obtained. This experimental approach provides
information on the pathway and temporal organisation of assembly,
integration, and functional modulation of this most important and
complicated enzyme *in vivo* during the growth and division of cells in
the absence of environmental perturbation.

MITOCHONDRIAL BIOGENESIS

Biogenesis of mitochondria requires the close co-operation of nuclear
and extranuclear genes, and the latter especially have been the subject
of intensive study during the past few years (Borst & Grivell, 1978;

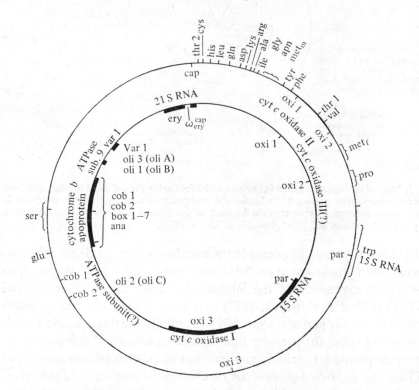

Fig. 6. Map of yeast mtDNA. Inner circle shows approximate location of genetic markers,
and probable gene products (Borst & Grivell, 1978; Heyting & Menke, 1979). Outer circle
shows location of mitochondrial tRNA species in relation to other genetic markers (Wesolowski & Fukuhara, 1979). Exact map positions vary somewhat according to the strain and
mapping method used. ery, erythromycin resistance; cap, chloramphenicol resistance; oxi,
defective in cytochrome oxidase synthesis (I, II, III refer to the mitochondrially-synthesised
subunits of the cytochrome oxidase complex); par, paromomycin resistance; oli, oligomycin
resistance; ana, antimycin A resistance; cob, box, defective in cytochrome *b* synthesis, in
some cases affecting also cytochrome oxidase (numbers refer to loci). For further details see
text.

Linnane & Nagley, 1978). Although a wide range of organisms has been used in these studies, by far the most productive has been the yeast *Saccharomyces cerevisiae*, primarily because of the possibility of deleting some or all of the mitochondrial genome without loss of viability. Such mutants, defective in essential mitochondrial functions, may then grow by fermentation. It is this unique feature which has permitted the isolation of a large variety of mutants, and the construction of a map of the genome (Fig. 6). Further detailed analysis of the genome is continuing, using a combination of genetical and physical techniques, including restriction endonuclease mapping and DNA sequencing. Cloning of mitochondrial DNA (mt DNA) from yeast in bacteria in order to facilitate the study of particular areas of the genome has not been necessary, since cytoplasmic 'petite' mutants provide a convenient way of amplifying such areas (Faye *et al.*, 1973). Since yeast has a number of features which are not typical of eukaryotes, studies on other organisms have helped to give a more complete picture.

Mitochondrial DNA and its coding capacity

The mtDNA of micro-organisms is circular except for that of some ciliates, and ranges in mol. wt. from 10×10^6 in *Chlamydomonas reinhardii* (Ryan *et al.*, 1978) to 62×10^6 in *Podospora anserina* (Cummings, Belcour & Grandchamp, 1979; Table 2). Each cell contains many copies of the mtDNA, though the actual number varies with

Table 2. *Size of mitochondrial DNA for different eukaryotes*

Organism	Mol. wt. $\times 10^{-6}$	Reference
Animals (Echinodermata, Arthropoda, Chordata)	9–12	Borst, 1970
Chlamydomonas reinhardii	9.8	Ryan *et al.*, 1978
Acanthamoeba castellanii	26	Bohnert, 1973
Paramecium aurelia	28[a]	Goddard & Cummings, 1975
Tetrahymena pyriformis	29[a]	Goldbach *et al.*, 1977
Kluyveromyces lactis	22	Sanders *et al.*, 1974
Saccharomyces cerevisiae *Saccharomyces carlsbergensis*	45–51	Hollenberg *et al.*, 1969; Sanders *et al.*, 1977
Neurospora crassa	40	Agsteribbe, Kroon & van Bruggen, 1972; Bernard, Bade & Küntzel, 1975
Aspergillus nidulans	21	López-Pérez & Turner, 1975; Stępień *et al.*, 1978
Aspergillus amstelodamii	27	A. Earl, G. Turner & C. Lazarus, (unpublished)
Podospora anserina	62	Cummings *et al.*, 1979
Pea	70	Kolodner & Tewari, 1972

[a]Detected only as linear DNA.

growth conditions and ploidy (Williamson, 1970). It is uncertain whether all of the DNA is transcribed. For instance, about one half of *Saccharomyces cerevisiae* mitochondrial genome consists of A–T rich ($< 5\%$ G–C) stretches, which are unlikely to code for protein (Bernardi *et al.*, 1976) though they could be involved in the processing of primary transcripts (Bos *et al.*, 1978a). Despite this large variation in size, little is known about variations in coding capacity of these different DNAs, since only in *S. cerevisiae* is much information available on gene/ polypeptide relationships. Nevertheless, it is likely that there will be differences, since it has already been demonstrated that the dicyclohexylcarbodiimide-binding protein, a subunit of mitochondrial ATPase involved in proton translocation (Fig. 3), is coded by the nucleus in *Neurospora crassa* and synthesised on cytoplasmic ribosomes (Jackl & Sebald, 1975; Sebald, Sebald-Althaus & Wachter, 1977); but mitochondrially coded in *S. cerevisiae* and synthesised on mitochondrial ribosomes (Tzagoloff & Meagher, 1972; Wachter *et al.*, 1977). *Aspergillus nidulans* is similar to *Neurospora crassa* in this respect (Turner, Imam & Küntzel, 1979).

Small variations in DNA composition are readily seen following analysis of the DNA using restriction endonucleases, and differences in restriction sites, small insertions and deletions are seen in mtDNAs of closely-related species within the same genus in yeast (Bernardi *et al.*, 1976; Sanders *et al.*, 1977), *Paramecium aurelia* (Maki & Cummings, 1977), *Neurospora crassa* (Bernard & Küntzel, 1976), *Tetrahymena pyriformis* (Goldbach *et al.*, 1977) and *Aspergillus nidulans* (Plate 1).

Mapping of mitochondrial genes in yeast (Fig. 6) has been carried out by a variety of techniques, but all have involved the use of deletions occurring in cytoplasmic petite mutants. These spontaneously occurring mutants, which may also be induced at high frequency by a variety of drugs, including ethidium bromide, have lost variable amounts of wild-type genome. In most cases, the DNA remaining consists of a reiterated, contiguous stretch of DNA representing part of the wild-type genome (Faye *et al.*, 1973), though internal deletions and rearrangements have been observed (Morimoto *et al.*, 1975; Heyting *et al.*, 1979). The DNA of petites may be compared to that of the wild-type strain either by hybridisation techniques (Sriprakash *et al.*, 1976) or by the use of restriction endonucleases (Sanders *et al.*, 1976). If petites are made from wild-type strains that contain appropriate mitochondrial mutations as markers, the presence or absence of these markers in the derived petites can be correlated with the retention or loss of DNA, and a map constructed which locates particular mitochondrial mutations on the physical map. A map has also been constructed genetically by observing the frequency of co-retention of markers in petites, and the results are in good agreement with those obtained by physical mapping (Schweyen *et al.*, 1976).

Correlation of a particular mutation with the gene for a mitochondrial product has usually been indirect, for example, the loss of a polypeptide or alteration in its electrophoretic mobility. A mutation conferring oligomycin resistance, *oli-1*, was shown to affect the aggregation properties of subunit 9 (the dicyclohexylcarbodiimide-binding protein) of the ATPase complex (Tzagoloff, Akai & Foury, 1976), while other mutations, originally referred to as *mit⁻*, lacked cytochrome oxidase and/or cytochrome *b* (Slonimski & Tzagoloff, 1976). A temperature-sensitive mutant, unable to assemble cytochrome oxidase at a non-permissive temperature, and having an altered electrophoretic mobility for one of the cytochrome oxidase subunits at the permissive temperature, suggested that the *oxi-3* locus codes for Subunit I of the enzyme (Eccleshall *et al.*, 1978). Similarly, mutants at the *oxi-1* locus, also defective in cytochrome oxidase synthesis, lacked Subunit II, but had instead smaller fragments which reacted with antiserum raised against Subunit II (Cabral *et al.*, 1978). Only in the case of the dicyclohexylcarbodiimide-binding protein has there been a direct demonstration of a mutation (*oli-1*), altering the amino acid sequence of a polypeptide, thus demonstrating its location in a structural gene (Wachter *et al.*, 1977). Complete or partial loss of a particular subunit could result from a mutation in a regulatory gene. An oligomycin-resistant, cold-sensitive mutant at the *oliA* (*oli-3*) locus, which is tightly linked to the *oliB* (*oli-1*) locus, was affected in the synthesis of the 20 000 mol. wt. subunit of the ATPase complex, and it was suggested that *oliB* might be in the structural gene for this subunit (Groot Obbink *et al.*, 1976). However, further studies on other mutants mapping at the same locus have shown that they affect the 8000 mol. wt. dicyclohexylcarbodiimide-binding protein (Murphy *et al.*, 1978). Whether the loci *oli-1* and *oli-3* are in the same gene or not remains unclear, and DNA and/or polypeptide sequencing of the mutants will be necessary.

In view of the difficulties inherent in the indirect approach, attempts have been made to obtain direct transcription and translation of mtDNA *in vitro*. So far, only small polypeptides (<10 000 mol. wt.) have been obtained using an *Escherichia coli* cell-free system, though antigenic determinants for cytochrome oxidase have been detected amongst the products using appropriate antisera (Moorman *et al.*, 1978). By using petites retaining different segments of the genome, it should be possible to use this approach for mapping (Grivell & Moorman, 1977). An alternative approach has been made to isolate and physically map mRNA species for particular polypeptides (Moorman, van Ommen & Grivell, 1978).

MtDNAs of all organisms so far examined contain genes for large and small rRNAs and for tRNAs. Mapping of the genes for these RNAs is fairly straightforward, since it can be carried out by direct hybridisation of labelled RNA to restriction fragments. Because of this, it is not confined to yeast.

In mammalian tissues (e.g. HeLa cells, Ojala & Attardi, 1977), *Neurospora crassa* (Terpstra, Holtrop & Kroon, 1977; Hahn *et al.*, 1979) and *Aspergillus nidulans* (C. M. Lazarus, H. Lünsdorf, U. Hahn & H. Küntzel, unpublished results), the genes for large and small rRNA lie close together, though in *Saccharomyces cerevisiae* (Sanders, Heyting & Borst, 1975; Fig. 6) and *Tetrahymena pyriformis* (Goldbach *et al.*, 1978) they are far apart. *T. pyriformis* is unusual in that its mtDNA, always seen as linear molecules, has a terminal duplication-inversion, and the 21S rRNA gene is located in this duplicated segment. Only one copy of the 14S rRNA gene is present. The large and small rRNA genes of *N. crassa* are transcribed together to form a 32S precursor, which is then processed to yield the 25S and 19S species (Kuriyama & Luck, 1973), though no precursor containing both large and small rRNA has been detected in yeast.

Genes for tRNA have been mapped in yeast (van Ommen, Groot & Borst, 1977; Fig. 6) and *Neurospora crassa* (Kroon *et al.*, 1976) by hybridisation of total labelled 4S mtRNA with mtDNA. In both organisms, most of the genes seem to be in a limited sector of the genome, though a few are scattered over the remainder of the genome. It is not yet certain whether there are sufficient genes to code for all the tRNA species which might be required by the mitochondrial translation system. Data from yeast (van Ommen *et al.*, 1977) indicate that up to 26 tRNA genes are present, though the authors point out that the wobble hypothesis (Crick, 1966) requires 32 tRNA species. Approximately 25 mitochondrial tRNA species have been detected in *Neurospora crassa* (de Vries *et al.*, 1978). In contrast to yeast, where all tRNAs detected within the mitochondria seem to be coded on the mitochondrial genome, there is evidence that *Tetrahymena pyriformis* imports the majority of tRNA species, and only seven species (for four amino acids) have been detected which are transcribed from the mitochondrial genome (Chiu *et al.*, 1975; Suyama & Hamada, 1976).

In addition to the functionally-defined components, a mitochondrially synthesised polypeptide 'var 1' has been detected by gel electrophoresis following specific labelling of mitochondrially-synthesized protein using $^{35}SO_4^{2-}$ (Douglas & Butow, 1976). The molecular weight of this component is strain dependent in the range of 40–44 000, and the extranuclear location of its gene (Fig. 6) was first demonstrated by conventional genetic tests taking advantage of the variability as a genetic marker. It has since been shown that a stretch of DNA at the *var 1* locus also shows inter-strain variation which correlates with the polypeptide variation (Strausberg *et al.*, 1978). The function of this polypeptide is as yet unknown, but its discovery highlights the possibility that not all mitochondrial gene products may be detected simply by examination of known enzyme complexes such as cytochrome oxidase, cytochrome *b* and ATPase. Approximately 83% of the yeast mitochondrial genome is unaccounted for in terms of gene products, and work is now in progress to elucidate the function of this DNA.

Gene organisation

Fine mapping of the large rRNA gene in certain strains of *Saccharomyces cerevisiae* by restriction endonuclease analysis and electron microscopy has shown that the gene is not continuous, but contains an intervening sequence of 1000 base pairs which does not code for the mature 21S RNA (Bos *et al.*, 1978*b*; Heyting *et al.*, 1979). Similar observations have now been made in *Neurospora crassa* (Hahn *et al.*, 1979) and *Aspergillus nidulans* (C. M. Lazarus, H. Lünsdorf, U. Hahn & H. Küntzel, personal communication). There is also indirect evidence that the genes for cytochrome *b* apoprotein and cytochrome oxidase Subunit I contain intervening sequences (Slonimski *et al.*, 1978; Grivell & Moorman, 1977). 'Split genes' of this type appear to be a feature of eukaryotic and not prokaryotic gene organisation (Flavell, Glover & Jeffreys, 1978), and their discovery in the mitochondrial genome raises questions about the endosymbiont hypothesis, which stresses the similarity between mitochondria and bacteria, and sees mitochondria as the descendents of engulfed bacteria (e.g. Stanier, 1970; and for counter argument see Raff & Mahler, 1972). Doolittle (1978), however, has proposed that split genes represent the more primitive form of gene organisation, once common to all prokaryotes, and that some organisms evolved to lose the splits. In doing so, they gained some advantages of a more streamlined and efficient gene organisation, but lost evolutionary potential, leaving us with today's prokaryotes. The evolutionary advantages conferred by split genes are discussed by Gilbert (1978). Mutations at the boundary regions between introns (intervening sequences) and exons (coding sequences) could, by affecting the splicing of the primary transcript, result in loss or gain of whole amino acid sequences, and if the splicing mechanism were not 100% efficient, the old product might still be made alongside the newly-modified polypeptide. In addition, recombination would occur at a higher frequency because of the presence of introns. Thus the split genes in mitochondria might represent the gene organisation in primitive prokaryotes.

Since the possibility exists of making mutations in both introns and exons (Slonimski et al., 1978), the mitochondrial genes of yeast may provide valuable information about the function of introns and the expression of eukaryotic genes in a way which cannot be studied in higher organisms, in a system which is simpler than any nuclear genome.

A large number of mutants affecting cytochrome *b* (or cytochrome reductase activity) including *mit⁻* mutants (Slonimski & Tzagoloff, 1976) and antimycin A resistant mutants (Lang et al., 1976) have been mapped in a region which includes the structural gene for apocytochrome *b* (Slonimski et al., 1978). The apoprotein of cytochrome *b* in both *Neurospora crassa* (Weiss & Ziganke, 1977) and yeast (Lin & Beattie, 1978) seems to be a single polypeptide with a mol. wt. of approximately 30 000, which is synthesised on the mitochondrial ribosomes. However, the genetic loci affecting the synthesis of cytochrome *b* are spread over a region of approximately 6000 base pairs, while only 900 would be required for a polypeptide of 30 000 mol. wt., suggesting the presence of non-translated sequences in the gene. Slonimski et al. (1978) suggest that some of the mutant loci may be located in the coding regions for the polypeptide (exons), and include possible termination mutations which result in the synthesis of incomplete peptides, while other loci may be located in the intervening sequences (introns). Mutants at some of these *box* loci also affect the synthesis of cytochrome $a + a_3$, cytochrome oxidase activity and Subunit I of cytochrome oxidase.

Sequencing of the cytochrome *b* apoprotein and the corresponding DNA region should test this hypothesis, and give a more precise location for the mutations. The way will then be open for a study on the role of intervening sequences in the control of transcription and translation.

Assembly of the inner mitochondrial membrane

Much was already known about the sites of translation of mitochondrial proteins before detailed mapping of the genes began (Schatz & Mason, 1974). Protein synthesis was studied both in isolated mitochondria and in whole cells in which cytoplasmic protein synthesis had been blocked by the presence of cycloheximide, an inhibitor specific for 80S ribosomes. Similarly, protein synthesis on the smaller, mitochondrial ribosomes could be blocked by chloramphenicol. All the evidence to date indicates that mitochondrially-coded proteins are translated on the mitochondrial ribosomes. It is also clear that biogenesis of cytochrome oxidase, ATP synthetase (ATPase) and mitochondrial ribosomes themselves (whose proteins are mostly imported) requires the co-operation of mitochondrial and cytoplasmic protein synthesis. However, little is known about the transport of cytoplasmically-synthesised mitochondrial components into mitochondria, their integration with mitochondrially-synthesised components, or about how these processes are regulated, although many observations have been made on the effect of blocking either of the systems (see below).

Following the discovery that some cytoplasmic ribosomes were firmly bound to the mitochondrial outer membrane in yeast (Kellems & Butow, 1972), it was suggested that they might be involved in the vectorial translation of mitochondrial proteins directed by nuclear mRNA (Kellems, Allison & Butow, 1974, 1975). Recent studies with *Neurospora crassa* (which also has mitochondrially-bound 80S ribosomes, Michel *et al.*, 1976) on the apoprotein of cytochrome *c*, a cytoplasmic product, have shown that the protein is liberated into a cytoplasmic pool, an observation which tends to contradict the vectorial-translation hypothesis (Hallermayer, Zimmerman & Neupert, 1977). Further studies using a cell-free system have shown that the apoprotein is converted into the holoprotein only when incubated with mitochondria (Korb & Neuport, 1978). The authors suggest that *in vivo*, apocytochrome *c* synthesised on the cytoplasmic ribosomes is first released into the cytoplasm, and is then taken up by the mitochondria, a process which is not directly dependent on either cytoplasmic or mitochondrial protein synthesis (Harmey *et al.*, 1977). The haem group is then added at the inner membrane, and a conformational change may prevent the cytochrome *c* from leaving the mitochondrion, since the holocytochrome *c* is found mostly in the mitochondrial fraction, and the apocytochrome *c* in the cytoplasm.

Cytochrome oxidase in yeast and *Neurospora crassa* is assembled from mitochondrially-synthesised Subunits I–III and cytoplasmically-synthesised Subunits IV–VII (Sebald, Machleidt & Otto, 1973; Schatz & Mason, 1974; Figs. 3 and 6). Ebner, Mason & Schatz (1973) found that cytoplasmic petite mutants, lacking cytochrome oxidase activity and mitochondrial protein synthesis, still produced three of the four cytoplasmic subunits, and transferred these into the mitochondria. There is immunological evidence that the four cytoplasmic subunits may be synthesised as a 55 000-mol. wt. precursor, which is detectable in the cytoplasm (Poyton & McKemmie, 1976).

Coupling between the two protein-synthesising systems does not appear to be very tight. Mitochondrial protein synthesis is able to continue for some time in isolated mitochondria, and it has been shown that this depends on the presence of an intra-mitochondrial pool of cytoplasmically-synthesised precursors present at the time of isolation (Poyton & Kavanagh, 1976). Once this pool is exhausted, the mito-chondrial protein synthesis can be re-stimulated by addition of a cyto-plasmic extract (post-polysomal supernatant). It has been further demonstrated, using specific antisera, that the stimulating factor(s) in the extract are antigenically related to Subunits IV and VI of the cyto-chrome oxidase (Poyton & McKemmie, 1976). The same phenomenon is observed *in vivo*, and has been used successfully as an approach to labelling mitochondrially-synthesised proteins (Sebald *et al.*, 1973). If cells are first incubated with chloramphenicol, cytoplasmically synthe-sised polypeptides accumulate. The cells are then washed free of chlor-amphenicol, and cycloheximide is added to block further cytoplasmic synthesis. Under such conditions, mitochondrial protein synthesis re-starts, and the accumulated cytoplasmic precursors act as a reservoir for assembly of complete enzyme complexes. Overall control of the synthesis of cytochrome oxidase is rather complex, since nuclear mutants have been isolated which contain all the cytoplasmically-synthesised subunits of the enzyme, but lack one or more of the mito-chondrially-synthesised subunits (Cabral & Schatz, 1978).

Although there is no direct evidence, it has been proposed that mtDNA codes for a repressor(s) which acts on certain nuclear genes and prevents their transcription, thus keeping the rate of cytoplasmic protein synthesis in phase with mitochondrial protein synthesis (Barath & Küntzel, 1972a, b). Similar models had already been proposed to account for control of mtDNA replication during cell growth (William-son, 1970; Lloyd *et al.*, 1971). Inhibition of mitochondrial protein synthesis with chloramphenicol or ethidium bromide can result in the 'overproduction' of cytoplasmically-synthesised components such as cytochrome *c* (von Jagow & Klingenberg, 1972), mtRNA polymerase (Barath & Küntzel, 1972a), elongation factors (Barath & Küntzel, 1972b) and tRNA synthetases (Beauchamp & Gross, 1976).

INHERITANCE AND RECOMBINATION OF
MITOCHONDRIAL DNA

Strict uniparental inheritance of the mitochondrial genome is observed in a wide variety of organisms, including *Neurospora crassa* (Mitchell, M. B. & Mitchell, H. K., 1952), *Podospora anserina* (Belcour, 1975), *Xenopus* (Dawid & Blackler, 1972), *Equus* (Hutchison *et al.*, 1974), *Aspergillus nidulans* (Rowlands & Turner, 1976) and rat (Hayashi *et al.*, 1978). In yeast, however, most zygotes usually contain mitochondrial genomes from both parents, which may be replicated and expressed in both vegetative diploid progeny and ascospores. Some zygotes are found in which the mitochondrial genome comes from one parent only, and the number of these uniparental zygotes can be increased by biasing the cross in favour of one parent (Birky *et al.*, 1978). The authors suggest that maternal inheritance observed in many organisms may depend upon a biased input of organelle DNA molecules which usually favours the maternal parent, followed by failure of the minority (paternal) molecules to replicate in many or all zygotes.

Mitochondrial recombination has been studied in yeast by analysis of random diploid cells obtained following mating between two haploid strains, and by zygote cell lineage analysis for individual zygotes to observe the segregation of markers at each division (Thomas & Wilkie, 1968; Coen *et al.*, 1970; Wilkie & Thomas, 1973). These studies have shown that a number of generations are required before all the diploid cells become homoplasmic, giving opportunity for a number of rounds of recombination to occur. Since mitochondria present in mixtures in the same cytoplasm may recombine as long as the heteroplasmic state persists, quantification of recombination data has been difficult in most cases. Nevertheless, recombination has been useful in tests for allelism, and to give some idea of the degree of linkage by observation of low or high recombination frequencies, and as a result to identify particular loci in yeast (Slonimski & Tzagoloff, 1976), *Aspergillus nidulans* (Lazarus & Turner, 1977) and *Podospora anserina* (Belcour & Begel, 1977). The fact that recombination of mtDNA does occur has also been invaluable in yeast mapping studies in the 'marker rescue' technique (Gingold *et al.*, 1969). To test whether a certain marker is still present in a petite DNA following deletion, the petite strain must first be crossed with a respiratory competent strain before the marker can be expressed, a process which requires mitochondrial recombination to occur.

How widespread is mitochondrial recombination?

Whether mtDNA of an organism undergoes recombination can be tested either by use of mitochondrially-located mutations, or by examination of restriction endonuclease digests (Plate I). Using the former method, recombination has been demonstrated in *Aspergillus nidulans* (Rowlands & Turner, 1974) and *Podospora anserina* (Belcour & Begel, 1977) in addition to yeast. Although mitochondrial markers are known in *Paramecium aurelia* and *Neurospora crassa*, no recombination between them has been demonstrable. This may be because the markers are allelic, especially in the case of *N. crassa*, where the extranuclear mutants are all slow-growing with cytochrome defects (Bertrand & Pittenger, 1972a). One of these, *poky* (Mitchell, M. B. & Mitchell, H. K., 1952), has a defect in the mitochondrial ribosomes (Lambowitz, Chua & Luck, 1976), presumably leading to a defect in mitochondrial protein synthesis. Recombination between markers of different strains cannot occur during the sexual cycle in *A. nidulans* and *N. crassa*, since strict maternal inheritance is observed (Rowlands & Turner, 1976; Bertrand & Pittenger, 1972a) but mitochondria of different strains may be brought together in the same cytoplasm by heterokaryon formation provided that the strains are heterokaryon compatible (Rowlands & Turner, 1974, 1976; Bertrand & Pittenger, 1972b).

Because of the existence of restriction endonuclease site differences, even amongst fairly closely related strains or sub-species, recombination of DNA may be observed directly as a result of the formation of 'hybrid' mitochondria. Such recombined DNA has been seen in *Saccharomyces cerevisiae* (Fonty et al., 1978) and in *Aspergillus* spp. (Plate I). Recombination of mtDNA in *Aspergillus* spp. can be studied only by the use of protoplast fusion, since strains with observable differences in mtDNA tend to be heterokaryon-incompatible as a result of differences at one or more nuclear loci (A. J. Earl, R. G. B. Dales, J. H. Croft & G. Turner, unpublished). Protoplast fusion thus offers a way of overcoming natural barriers to cytoplasmic mixing (Kevei & Perberdy, 1977; Dales & Croft, 1977) and once the mtDNAs are present in the same cytoplasm, even a difference in size of 10% between mitochondrial genomes does not prevent recombination (Plate 1). Such hybrids may offer possibilities for studying the significance of mtDNA size variation, and for identifying species-specific genes.

Interspecies transfer of mitochondria has been carried out in *Paramecium aurelia* by micro-injection (Beale & Knowles, 1976), a technique which permits mitochondria of two different species to be brought together in the same cytoplasm. Erythromycin-resistant mitochondria were injected into different species, and under appropriate growth conditions, replaced the original, sensitive mitochondria. It has also been shown by restriction pattern analysis that the mtDNA of the resistant species replaces that of the sensitive species (Cummings, Goddard & Maki, 1976). However, to date no recombination of the mtDNAs to generate a new restriction pattern has been observed, and it is possible that there is no mechanism in *Paramecium aurelia* for such recombination.

The actual mechanism(s) by which mtDNA does recombine in yeast and *Aspergillus* spp. is now under investigation. Some interesting phenomena have been observed in yeast, though whether they are generally applicable to mitochondrial recombination has yet to be investigated. Fonty *et al.* (1978) examined the restriction patterns of parental and hybrid strains of yeast following recombination between strains with pattern differences, and found evidence for the occurrence of unequal recombination, which most likely occurs in the spacers between genes, and leaves the genes themselves unaffected.

There is also evidence for gene conversion. For some time, it has been known that polarity is observed in some extranuclear crosses in yeast; that is, certain alleles from one of the parents predominate in the progeny. These alleles which showed polarity were said to be linked to a locus designated ω^+, and when an ω^+ strain was crossed with an ω^- strain, a type of gene conversion occurred ensuring that the majority of the progeny were ω^+ (Dujon, Slonimski & Weill, 1974). Recently, molecular studies have identified ω^+ as a stretch of DNA approximately 1000 base pairs long (Heyting & Menke, 1979) which occurs in the split in the 21S RNA gene. Polar markers, such as chloramphenicol and erythromycin resistance, are located in this region (Fig. 6).

Gene conversion has also been observed at relatively high frequency at the *var 1* locus (Strausberg *et al.*, 1978), which codes for a variable polypeptide of unknown function (Douglas & Butow, 1976). It was shown that different molecular weight forms of the polypeptide product of the gene *var 1* could be accounted for by two DNA segments, of about 36 and 57 base pairs. These segments could be inserted independently and at different frequencies into other *var 1* alleles during crosses, and the conversions could be detected by examining the polypeptides produced in the progeny.

In most organisms so far studied, mitochondrial recombination is severely restricted in a number of different ways, and it seems likely that this will slow down the rate of mitochondrial evolution (Sager, 1972).

It will be interesting to learn more about the mechanism(s) of recombination in organisms in which it does occur once the barriers to cytoplasmic mixing have been overcome.

REFERENCES

AGSTERIBBE, E., KROON, A. M. & VAN BRUGGEN, E. F. J. (1972). Circular DNA from mitochondria of *Neurospora crassa*. *Biochimica et Biophysica Acta*, **269**, 299–303.

ARNOLD, C. G., SCHIMMER, O., SCHÖTZ, F. & BATHELET, H. (1972). Die Mitochondrien von *Chlamydomonas reinhardii*. *Archiv für Mikrobiologie*, **81**, 50–67.

ATKINSON, A. W., JR., JOHN, P. C. L. & GUNNING, B. E. S. (1974). The growth and division of the single mitochondrion and other organelles during the cell cycle of *Chlorella*, studied by quantitative stereology and three-dimensional reconstruction. *Protoplasma*, **81**, 77–109.

ATKINSON, D. E. (1977). *Cellular Energy Metabolism and its Regulation*, 293 pp. New York: Academic Press.

BAKALKIN, G. Y., KALNOV, S. L., GALKIN, A. V., ZUBATOV, A. S. & LUZIKOV, V. N. (1978). The lability of the products of mitochondrial protein synthesis in *Saccharomyces cerevisiae*. A novel method for protein half-life determination. *Biochemical Journal*, **170**, 569–76.

BANDLOW, W., SCHWEYEN, R. J., WOLF, K. & KAUDEWITZ, F. (1977). *Mitochondria 1977, Genetics and Biogenesis of Mitochondria*, 669 pp. Berlin & New York: Walter de Gruyter.

BARATH, Z. & KÜNTZEL, H. (1972a). Induction of mitochondrial RNA polymerase in *Neurospora crassa*. *Nature New Biology*, **240**, 195–7.

BARATH, Z. & KÜNTZEL, H. (1972b). Co-operation of mitochondrial and nuclear genes specifying the mitochondrial genetic apparatus in *Neurospora crassa*. *Proceedings of the National Academy of Sciences, USA*, **69**, 1371–4.

BEALE, G. H. & KNOWLES, J. K. C. (1976). Interspecies transfer of mitochondria in *Paramecium aurelia*. *Molecular and General Genetics*, **143**, 197–201.

BEAUCHAMP, P. M. & GROSS, S. R. (1976). Increased mitochondrial leucyl- and phenylalanyl tRNA synthetase activity as a result of inhibition of mitochondrial protein-synthesis. *Nature, London*, **261**, 338–40.

BELCOUR, L. (1975). Cytoplasmic mutations isolated from protoplasts of *Podospora anserina*. *Genetical Research*, **25**, 155–61.

BELCOUR, L. & BEGEL, O. (1977). Mitochondrial genes in *Podospora anserina*: Recombination and linkage. *Molecular and General Genetics*, **153**, 11–21.

BEREITER-HAHN, J. (1978). Intracellular motility of mitochondria: role of the inner compartment in migration and shape changes of mitochondria in XTH-cells. *Journal of Cell Science*, **30**, 99–115.

BERNARD, U. & KÜNTZEL, H. (1976). Physical mapping of mitochondrial DNA from *Neurospora crassa*. In *The Genetic Function of Mitochondrial DNA*, ed. C. Saccone & A. M. Kroon, pp. 105–9. Amsterdam, Oxford & New York: North Holland.

BERNARD, U., BADE, E. & KÜNTZEL, H. (1975). Specific fragmentation of mitochondrial DNA from *Neurospora crassa* by restriction endonuclease Eco RI. *Biochemical and Biophysical Research Communications*, **64**, 783–9.

BERNARDI, G., PRUNELL, A., FONTY, G., KOPECKA, H. & STRAUSS, F. (1976). The mitochondrial genome of yeast: Organization, evolution and the petite mutation. In *Genetic Function of Mitochondrial DNA*, ed. C. Saccone & A. M. Kroon, pp. 185–98. Amsterdam, Oxford & New York: North Holland.

BERTRAND, H. & PITTENGER, T. H. (1972a). Isolation and classification of extranuclear mutants of *Neurospora crassa*. *Genetics*, **71**, 521–33.

BERTRAND, H. & PITTENGER, T. H. (1972b). Complementation among cytoplasmic mutants of *Neurospora crassa*. *Molecular and General Genetics*, **117**, 82–90.

BIRKY, C. W., DEMKO, C. A., PERLMAN, P. S. & STRAUSBERG, R. L. (1978). Uniparental inheritance of mitochondrial genes in yeast. Dependence on input bias of mitochondrial DNA and preliminary investigation of mechanism. *Genetics*, **89**, 615–51.

BOHNERT, H. J. (1973). Circular mitochondrial DNA from *Acanthamoeba castellanii* (Neff-strain). *Biochimica et Biophysica Acta*, **324**, 199–205.

BORST, P. (1970). Mitochondrial DNA: structure, information content, replication and transcription. In *Control of Organelle Development. Symposia of the Society for Experimental Biology*, 24, ed. P. L. Miller, pp. 201–26. Cambridge University Press.

BORST, P. & GRIVELL, L. A. (1978). The mitochondrial genome of yeast. *Cell*, **15**, 705–23.

BOS, J. L., VAN KREIJL, C. F., PLOEGAERT, F. H., MOL, J. N. M. & BORST, P. (1978a). A conserved and unique (AT)-rich segment in yeast mitochondrial DNA. *Nucleic Acids Research*, **5**, 4563–78.

BOS, J. L., HEYTING, C., BORST, P., ARNBERG, A. C. & VAN BRUGGEN, E. F. J. (1978b). An insert in the single gene for the large ribosomal RNA in yeast mitochondrial DNA. *Nature, London*, **275**, 336–8.

BOVERIS, A. (1978). Production of superoxide anion and hydrogen peroxide in yeast mitochondria. In *Biochemistry and Genetics of Yeast, Pure and Applied Aspects*, ed. M. Bacila, B. L. Horecker & A. O. M. Stoppani, pp. 65–80. New York, San Francisco & London: Academic Press.

BOYER, P. D., CHANCE, B., ERNSTER, L., MITCHELL, P., RACKER, E. & SLATER, E. C. (1977). Oxidative phosphorylation and photo-phosphorylation. *Annual Review of Biochemistry*, **46**, 955–1026.

BURTON, M. D. & MOORE, J. (1974). The mitochondrion of the flagellate, *Polytomella agilis*. *Journal of Ultrastructural Research*, **48**, 414–19.

CABRAL, F. & SCHATZ, G. (1978). Identification of cytochrome *c* oxidase subunits in nuclear yeast mutants lacking functional enzyme. *Journal of Biological Chemistry*, **253**, 4396–401.

CABRAL, F., SOLIOS, M., RUDON, Y., SCHATZ, G., CLAVILIER, L. & SLONIMSKI, P. P. (1978). Identification of the structural gene for yeast cytochrome *c* oxidase subunit II on mitochondrial DNA. *Journal of Biological Chemistry*, **253**, 279–304.

CALVAYRAC, R., BERTAUX, O., LEFORT-TRAN, M. & VALENCIA, R. (1974). Généralisation du cycle mitochondrial chez *Euglena gracilis Z.* en cultures synchrones, heterotrophe et phototrophe. *Protoplasma*, **80**, 355–70.

CATTELL, K. J., LINDOP, C. R., KNIGHT, I. G. & BEECHEY, R. B. (1971). The identification of the site of action of *N, N'*-dicyclohexylcarbodiimide as a proteolipid in mitochondrial membranes. *Biochemical Journal*, **125**, 169–77.

CHANCE, B. (1977). Electron transfer: pathways, mechanisms, and controls. *Annual Review of Biochemistry*, **46**, 967–80.

CHANCE, B., LEIGH, J. S. JR. & WARING, E. (1977). The reaction of oxygen with cytochrome oxidase. In *Structure and Function of Energy-transducing Membranes. BBA Library* 14, ed. K. van Dam & B. F. van Gelder, pp. 1–10. Amsterdam: Elsevier/North Holland Biomedical Press.

CHANCE, B. & WILLIAMS, G. R. (1956). The respiratory chain and oxidative phosphorylation. *Advances in Enzymology*, **17**, 65–134.

CHIU, N., CHIU, A. & SUYAMA, Y. (1975). Native and imported transfer RNA in mitochondria. *Journal of Molecular Biology*, **99**, 37–50.

COEN, D., DEUTSCH, J., NETTER, P., PETROCHILO, E. & SLONIMSKI, P. P. (1970). Mitochondrial genetics I – Methodology and phenomenology. In *Control of Organelle Development*, ed. P. L. Miller, pp. 449–96. 24th Symposium of the Society for Experimental Biology, Cambridge University Press.

CRICK, F. H. C. (1966). Codon–anticodon pairing: The wobble hypothesis. *Journal of Molecular Biology*, **19**, 548–55.

CUMMINGS, D. J., BELCOUR, L. & GRANDCHAMP, C. (1979). Mitochondrial DNA from *Podospora anserina*. I. Isolation and characterisation. *Molecular and General Genetics*, **171**, 229–38.

CUMMINGS, D. J., GODDARD, J. M. & MAKI, R. A. (1976). Mitochondrial DNA from *Paramecium aurelia*. In *The Genetic Function of Mitochondrial DNA*, ed. C. Saccone & A. M. Kroon, pp. 119–30. Amsterdam, Oxford & New York: North Holland.

DALES, R. G. B. & CROFT, J. H. (1977). Protoplast fusion and the isolation of heterokaryons and diploids from vegetatively incompatible strains of *Aspergillus nidulans*. *FEMS Microbiology Letters*, **1**, 201–3.

DAWID, I. B. & BLACKLER, A. W. (1972). Maternal and cytoplasmic inheritance of mitochondrial DNA in *Xenopus*. *Developmental Biology*, **29**, 152–61.

DEGN, H., LLOYD, D. & HILL, G. C. (1978). *Functions of Alternative Terminal Oxidases*, 196 pp. Oxford: Pergamon Press.

DE VRIES, H., DE JONGE, J. C., SCHNELLER, J.-M., MARTIN, R. P., DIRHEIMER, G. & STAHL, A. J. C. (1978). *Neurospora crassa* mitochondrial transfer RNAs. *Biochimica et Biophysica Acta*, **520**, 419–27.

DOCKTER, M. E., STEINEMANN, A. & SCHATZ, G. (1978). Mapping of yeast cytochrome *c* oxidase by fluorescence resonance energy transfer. *Journal of Biological Chemistry*, **253**, 311–17.

DOOLITTLE, W. F. (1978). Genes in pieces: were they ever together? *Nature, London*, **272**, 581–2.

DOUGLAS, M. G. & BUTOW, R. A. (1976). Variant forms of mitochondrial translation products in yeast: Evidence for location of determinants on mitochondrial DNA. *Proceedings of the National Academy of Sciences, USA*, **73**, 1083–6.

DUJON, B., SLONIMSKI, P. P. & WEILL, L. (1974). Mitochondrial genetics IX: A model for recombination and segregation of mitochondrial genomes in *Saccharomyces cerevisiae*. *Genetics*, **78**, 415–37.

EBNER, E., MASON, T. L. & SCHATZ, G. (1973). Mitochondrial assembly in respiratory-deficient mutants of *Saccharomyces cerevisiae*. II. Effect of nuclear and extrachromosomal mutants on the formation of cytochrome *c* oxidase. *Journal of Biological Chemistry*, **248**, 5369–78.

ECCLESHALL, T. R., NEEDLEMAN, R. B., STORM, E. M., BUCHFERE, B. & MARMUR, J. (1978). Temperature sensitive yeast mitochondrial mutant with altered cytochrome c oxidase subunit. *Nature, London*, **273**, 67–70.

EDWARDS, S. W. & LLOYD, D. (1977). Mitochondrial adenosine triphosphatase of the fission yeast, *Schizosaccharomyces pombe* 972 h⁻. Changes in activity and oligomycin-sensitivity during the cell cycle of catabolite-repressed and -derepressed cells. *Biochemical Journal*, **162**, 39–46.

EDWARDS, S. W. & LLOYD, D. (1978). Oscillations of respiration and adenine nucleotides in synchronous cultures of *Acanthamoeba castellanii*: mitochondrial respiratory control *in vivo*. *Journal of General Microbiology*, **108**, 197–204.

EDWARDS, C., STATHAM, M. & LLOYD, D. (1975). The preparation of large-scale synchronous cultures of the trypanosomatid, *Crithidia fasciculata* by cell-size selection: changes in respiration and adenylate charge through the cell-cycle. *Journal of General Microbiology*, **88**, 141–52.

ERECIŃSKA, M., OSHINO, N., LOH, P. & BROCKLEHURST, E. (1973). *In vitro* studies on yeast cytochrome c peroxidase and its possible function in the electron transfer and energy coupling reactions. *Biochimica et Biophysica Acta*, **292**, 1–12.

ERECIŃSKA, M., STUBBS, M., MIYATA, Y., DITRE, C. M. & WILSON, D. F. (1977). Regulation of cellular metabolism by intracellular phosphate. *Biochimica et Biophysica Acta*, **462**, 20–35.

FAYE, G., FUKUHARA, H., GRANDCHAMP, C., LAZOWSKA, J., MICHEL, F., CASEY, J., GETZ, G. S., LOCKER, J., RABINOWITZ, M., BOLOTIN-FUKUHARA, M., COEN, D., DEUTSCH, J., DUJON, B., NETTER, P. & SLONIMSKI, P. P. (1973). Mitochondrial nucleic acids in the petite colonie mutants: deletions and repetitions of genes. *Biochemie*, **55**, 779–92.

FERGUSON, S. J., LLOYD, W. J., LYONS, M. H. & RADDA, G. (1975). The mitochondrial ATPase. Evidence for a single essential tyrosine residue. *European Journal of Biochemistry*, **54**, 117–26.

FLAVELL, R. A., GLOVER, D. M. & JEFFREYS, A. J. (1978). Discontinuous genes. *Trends in Biochemical Sciences*, **3**, 241–4.

FONTY, G., GOURSOT, R., WILKIE, D. & BERNARDI, G. (1978). Mitochondrial genome of wild-type yeast cells. VII. Recombination in crosses. *Journal of Molecular Biology*, **119**, 213–35.

GAUTHERON, D. C. & GODINOT, C. (1977). Structure and function of ATP synthase. In *Living Systems as Energy Converters*, ed. R. Buvet, M. J. Allen & J.-P. Masué, pp. 89–102. Amsterdam: Elsevier–North Holland Biomedical Press.

GILBERT, W. (1978). Why genes in pieces? *Nature, London*, **271**, 501.

GINGOLD, E. B., SAUNDERS, G. W., LUKINS, H. B. & LINNANE, A. W. (1969). Biogenesis of mitochondria. X. Reassortment of the cytoplasmic genetic determinants for respiratory competence and erythromycin resistance in *Saccharomyces cerevisiae*. *Genetics*, **62**, 735–44.

GODDARD, J. M. & CUMMINGS, D. J. (1975). Structure and replication of mitochondrial DNA from *Paramecium aurelia*. *Journal of Molecular Biology*, **97**, 593–609.

GOLDBACH, R. W., ARNBERG, A. C., VAN BRUGGEN, E. F. J., DEFIZE, J. & BORST, P. (1977). The structure of *Tetrahymena pyriformis* mitochondrial DNA I. Strain differences and occurrence of inverted repetitions. *Biochimica et Biophysica Acta*, **477**, 37–50.

GOLDBACH, R. W., BORST, P., BOLLEN-DE BOER, J. E. & VAN BRUGGEN, E. F. J. (1978). The organisation of ribosomal RNA genes in the mitochondrial DNA of *Tetrahymena pyriformis* strain ST. *Biochimica et Biophysica Acta*, **521**, 169–86.

GOOCH, V. D. & PACKER, L. (1974). Oscillatory systems in mitochondria. *Biochimica et Biophysica Acta*, **346**, 245–60.

GOODWIN, B. C. (1963). *Temporal Organisation in Cells: A Dynamic Theory of Cellular Control Processes*, 164 pp. London: Academic Press.

GOODWIN, B. C. (1976). *Analytical Physiology of Cells and Developing Organisms*, 249 pp. London: Academic Press.

GRAF, T. & SEBALD, W. (1978). The dicyclohexylcarbodiimide-binding protein of the mitochondrial ATPase complex from beef heart. Isolation and amino acid composition. *FEBS Letters*, **94**, 218–22.

GREVILLE, G. D. (1966). Factors affecting the utilization of substrates by mitochondria. In *Regulation of Metabolic Processes in Mitochondria. BBA Library* 7, ed. J. H. Tager, S. Papa, E. Quagliariello & E. C. Slater, pp. 86–107. Amsterdam: Elsevier.

GRIFFITHS, D. E. & HYAMS, R. L. (1977). Oxidative phosphorylation: a role for lipoic acid and unsaturated fatty acids. *Biochemical Society Transactions*, **5**, 207–8.

GRIVELL, L. A. & MOORMAN, A. F. M. (1977). A structural analysis of the oxi 3 region on yeast mitochondria DNA. In *Mitochondria 1977, Genetics and Biogenesis of Mitochondria*, ed. W. Bandlow, R. J. Schweyen, K. Wolf & F. Kaudewitz, pp. 371–84. Berlin & New York: Walter de Gruyter.

GROBE, B. & ARNOLD, C.-G. (1975). Evidence of a large ramified mitochondrion in *Chlamydomonas reinhardii*. *Protoplasma*, **86**, 291–4.

GROBE, B. & ARNOLD, C.-G. (1977). The behaviour of mitochondria in the zyogote of *Chlamydomonas reinhardii*. *Protoplasma*, **93**, 357–61.

GROOT OBBINK, D. J., HALL, R. M., LINNANE, A. W., LUKINS, H. B., MONK, B. C., SPITHILL, T. W. & TREMBATH, M. K. (1976). In *The Genetic Function of Mitochondrial DNA*, ed. C. Saccone & A. M. Kroon, pp. 163–73. Amsterdam, Oxford & New York: North Holland.

HAHN, U., LAZARUS, C. M., LÜNSDORF, H. & KÜNTZEL, H. (1979). Split gene for mitochondrial 24S ribosomal RNA of *Neurospora crassa*. *Cell*, **17**, 191–200.

HALLERMAYER, G., ZIMMERMAN, R. & NEUPERT, W. (1977). Kinetic studies of cytoplasmically synthesized proteins into the mitochondria in intact cells of *Neurospora crassa*. *European Journal of Biochemistry*, **81**, 523–32.

HARMEY, M. A., HALLERMAYER, G., KORB, H. & NEUPERT, W. (1977). Transport of cytoplasmically synthesized proteins into the mitochondria in a cell free system from *Neurospora crassa*. *European Journal of Biochemistry*, **81**, 533–44.

HAYASHI, J. I., YONEKAWA, H., GOTON, O., WATANABE, J. & TAGASHIR, Y. (1978). Strictly maternal inheritance of rat mitochondrial DNA. *Biochemical and Biophysical Research Communications*, **83**, 1032–8.

HESS, B. & CHANCE, B. (1978). Oscillating enzyme reactions. *Theoretical Chemistry*, **4**, 159–79.

HEYTING, C., MEIJLINK, F. C. P. W., VERBEET, M. Ph., SANDERS, J. P. M., Bos, J. L. & BORST, P. (1979). Fine structure of the 21S ribosomal RNA region on yeast mitochondrial DNA I. Construction of the physical map and location of the cistron for the 21S mitochondrial ribosomal RNA. *Molecular and General Genetics*, **168**, 231–50.

HEYTING, C. & MENKE, H. H. (1979). Fine structure of the 21S ribosomal RNA region on yeast mitochondrial DNA III. Physical location of mitochondrial genetic markers and the molecular nature of ω. *Molecular and General Genetics*, **168**, 279–91.

HEYWOOD, P. (1977). Evidence from serial sections that some cells contain large numbers of mitochondria. *Journal of Cell Science*, **26**, 1–8.

HOFFMAN, H. P. & AVERS, C. J. (1973). Mitochondrion of yeast: ultrastructural evidence for one giant branched organelle per cell. *Science*, **181**, 749–51.

HOLLENBERG, C. P., BORST, P., THURING, R. W. J. & VAN BRUGGEN, E. F. J. (1969). Size, structure and genetic complexity of yeast mitochondrial DNA. *Biochimica et Biophysica Acta*, **186**, 417–19.

HOPKINS, F. G. (1913). The dynamic side of biochemistry. *British Medical Journal*, **2**, 713–24.

HUGHES, D. E., LLOYD, D. & BRIGHTWELL, R. (1970). Structure, function and distribution of organelles in prokaryotic and eukaryotic microbes. In *Organization and Control in Prokaryotic and Eukaryotic Cells. Symposium of the Society for General Microbiology*, 20, ed. H. P. Charles & B. C. J. G. Knight, pp. 295–322. Cambridge University Press.

HUTCHISON, III, C. A., NEWBOLD, J. E., POTTER, S. S. & EDGELL, M. H. (1974). Maternal inheritance of mammalian mitochondrial DNA. *Nature, London*, **251**, 536–8.

JACKL, G. & SEBALD, W. (1975). Identification of two products of mitochondrial protein synthesis associated with mitochondrial adenosine triphosphatase. *European Journal of Biochemistry*, **54**, 97–106.

KADER, J. & LLOYD, D. (1979). Respiratory oscillations and heat evolution in synchronous cultures of *Candida utilis* NCYC 193. *Journal of General Microbiology*, **114**, 455–61.

KELLEMS, R. E., ALLISON, V. F. & BUTOW, R. A. (1974). Cytoplasmic-type 80S ribosomes associated with yeast mitochondria II. Evidence for the association of cytoplasmic ribosomes with the outer mitochondrial membrane *in situ*. *Journal of Biological Chemistry*, **249**, 3297–303.

KELLEMS, R. E., ALLISON, V. F. & BUTOW, R. A. (1975). Cytoplasmic-type 80S ribosomes associated with yeast mitochondria. IV. Attachment of ribosomes to the outer membrane of isolated mitochondria. *Journal of Cell Biology*, **65**, 1–14.

KELLEMS, R. E. & BUTOW, R. A. (1972). Cytoplasmic-type 80S ribosomes associated with yeast mitochondria. I. Evidence for ribosome binding sites on yeast mitochondria. *Journal of Biological Chemistry*, **247**, 8043–50.

KEVEI, F. & PERBERDY, J. F. (1977). Interspecific hybridization between *Aspergillus nidulans* and *Aspergillus rugulosus* by fusion of somatic protoplasts. *Journal of General Microbiology*, **102**, 255–62.

KOLODNER, R. & TEWARI, K. K. (1972). Physicochemical characterization of mitochondrial DNA from pea leaves. *Proceedings of the National Academy of Sciences, USA*, **69**, 1830–4.

KORB, H. & NEUPERT, W. (1978). Biogenesis of cytochrome *c* in *Neurospora crassa*: Synthesis of apocytochrome *c*, transfer to mitochondria and conversion to holocytochrome *c*. *European Journal of Biochemistry*, **91**, 609–20.

KOVÁČ, L., KOLAROV, J. & SUBIK, J. (1977). Genetic determination of the mitochondrial adenine nucleotide translocation system and its role in the eukaryotic cell. *Molecular and Cellular Biochemistry*, **14**, 11–14.

KROON, A. M., TERPSTRA, P., HOLTROP, N. DE VRIES, H., VAN DER BOGERT, C., DE JONGE, J. & AGSTERIBBE, E. (1976). The mitochondrial RNAs of *Neurospora crassa*: Their function in translation and their relation to the mitochondrial genome. In *Genetics and Biogenesis of Mitochondria*, ed. W. Bandlow, R. J. Schweyen, K. Wolf & F. Kaudewitz, pp. 685–96. Berlin & New York: Walter de Gruyter.

KURIYAMA, Y. & LUCK, D. J. L. (1973). Ribosomal RNA synthesis in mitochondria of *Neurospora crassa*. *Journal of Molecular Biology*, **73**, 425–37.

KUSEL, J. P., BOVERIS, A. & STOREY, B. T. (1973). H_2O_2 production and cytochrome *c* peroxidase activity in mitochondria isolated from the trypanosomatid haemoflagellate *Crithidia fasciculata*. *Archives of Biochemistry and Biophysics*, **158**, 799–805.

LAMBOWITZ, A. M., CHUA, N.-H. & LUCK, D. J. L. (1976). Mitochondrial ribosome assembly in *Neurospora*. Preparation of mitochondrial ribosome precursor particles, sites of synthesis of mitochondrial ribosomal proteins and studies on the *poky* mutant. *Journal of Molecular Biology*, **107**, 223–53.

LANG, B., BURGER, G., BANDLOW, W., KAUDEWITZ, F. & SCHWEYEN, R. J. (1976). Antimycin- and funiculosin-resistant mutants in *Saccharomyces cerevisiae*: New markers on the mitochondrial DNA. In *Genetics and Biogenesis of Mitochondria*, ed. W. Bandlow, R. J. Schweyen, K. Wolf & F. Kaudewitz, pp. 461–5. Berlin & New York: Walter de Gruyter.

LAZARUS, C. M. & TURNER, G. (1977). Extranuclear recombination in *Aspergillus nidulans*: Closely-linked multiple chloramphenicol- and oligomycin-resistance loci. *Molecular and General Genetics*, **156**, 303–11.

LEEDALE, G. F. & BUETOW, D. (1970). Observations on the mitochondrial reticulum in living *Euglena gracilis*. *Cytobiologie*, **1**, 195–202.

LIN, L. F. H. & BEATTIE, D. S. (1978). Formation of yeast mitochondrial membrane 6. Purification and properties of a major cytochrome *b* peptide from bakers yeast. *Journal of Biological Chemistry*, **253**, 2412–18.

LINNANE, A. W. & NAGLEY, P. (1978). Structural mapping of mitochondrial DNA. *Archives of Biochemistry and Biophysics*, **187**, 277–89.

LLOYD, D. (1969). The development of organelles concerned with energy production. In *Microbial Growth. Symposium of the Society for General Microbiology*, 19, ed. P. Meadow & S. J. Pirt, pp. 299–332, Cambridge University Press.

LLOYD, D. (1974). *The Mitochondria of Micro-organisms*, 553 pp. London: Academic Press.

LLOYD, D. & BALL, J. (1979). Perturbation of respiration in *Candida utilis*: induction of metabolic oscillations. *Journal of General Microbiology*, **114**, 463–6.

LLOYD, D. & EDWARDS, S. W. (1976). Mitochondrial adenosine triphosphatase of the fission yeast, *Schizosaccharomyces pombe* 972 h⁻. Changes in activity and inhibitor-sensitivity in response to catabolite repression. *Biochemical Journal*, **160**, 335–42.

LLOYD, D. & EDWARDS, S. W. (1977). Mitochondrial adenosine triphosphatase of the fission yeast *Schizosaccharomyces pombe* 972 h⁻. Changes in inhibitor sensitivities during the cell cycle indicate similarities and differences in binding sites. *Biochemical Journal*, **162**, 581–90.

LLOYD, D., JOHN, L., EDWARDS, C. & CHAGLA, A. H. (1975). Synchronous cultures of micro-organisms: large-scale preparation by continuous-flow size selection. *Journal of General Microbiology*, **88**, 153–8.

LLOYD, D., EDWARDS, C., EDWARDS, S. W., EL'KHAYAT, G., JENKINS, S. J., JOHN, L., PHILLIPS, C. A. & STATHAM, M. (1978). The stability of adenylate energy charge values. *Trends in Biochemical Sciences*, **3**, N138–9.

LLOYD, D., PHILLIPS, C. A. & STATHAM, M. (1978). Oscillations of respiration, adenine nucleotide levels and heat evolution in synchronous cultures of *Tetrahymena pyriformis ST* prepared by continuous-flow selection. *Journal of General Microbiology*, **106**, 19–26.

LLOYD, D., TURNER, G., POOLE, R. K., NICHOLL, W. G. & ROACH, G. I. (1971). A hypothesis of nuclear-mitochondrial interactions for the control of mitochondrial biogenesis based on experiments with *Tetrahymena pyriformis*. *Sub-cellular Biochemistry*, **1**, 93–7.

LÓPEZ-PÉREZ, M. J. & TURNER, G. (1975). Mitochondrial DNA from *Aspergillus nidulans*. *FEBS Letters*, **58**, 159–63.

MAKI, R. A. & CUMMINGS, D. J. (1977). Characterisation of mitochondrial DNA from *Paramecium aurelia* with EcoRI and HaeII restriction endonucleases. *Plasmid*, **1**, 106–14.

MANTON, I. (1959). Electron microscopical observations on a very small flagellate: the problem of *Chromulina pusilla* Butcher. *Journal of the Marine Biological Association, UK*, **38**, 319–33.

MANTON, I. & PARKE, M. (1960). Further observations on small green flagellates with special reference to possible relatives of *Chromulina pusilla* Butcher. *Journal of the Marine Biological Association, UK*, **39**, 275–98.

MARCUS, F., SCHUSTER, S. M. & LARDY, H. A. (1976). Essential arginyl residues in mitochondrial adenosine triphosphatase. *Journal of Biological Chemistry*, **251**, 1775–80.

MICHEL, R., HALLERMAYER, G., HARMEY, M. A., MILLER, F. & NEUPERT, W. (1976). Significance of 80S ribosomes associated with *Neurospora crassa* mitochondria. In *Genetics and Biogenesis of Chloroplasts and Mitochondria*, ed. Th. Bücher, W. Neupert, W. Sebald & S. Werner, pp. 725–30. Amsterdam, New York & Oxford: North-Holland.

MITCHELL, M. B. & MITCHELL, H. K. (1952). A case of 'maternal inheritance' in *Neurospora crassa*. *Proceedings of the National Academy of Sciences, USA*, **38**, 442–9.

MITCHELL, P. (1966). *Chemiosmotic Coupling in Oxidative and Photosynthetic Phosphorylation*, 192 pp. Bodmin: Glynn Research Ltd.

MOORMAN, A. F. M., GRIVELL, L. A., LAMIE, F. & SMITS, H. L. (1978). Identification of mitochondrial gene products by DNA-directed protein synthesis *in vitro*. *Biochimica et Biophysica Acta*, **518**, 351–65.

MOORMAN, A. F. M., VAN OMMEN, G. J. & GRIVELL, L. A. (1978). Transcription in yeast mitochondria. Isolation and physical mapping of messenger RNAs for subunits of cytochrome *c* oxidase. *Molecular and General Genetics*, **160**, 13–24.

MORIMOTO, R., LEWIN, A., HSU, H.-J., RABINOWITZ, M. & FUKUHARA, H. (1975). Restriction endonuclease analysis of mitochondrial DNA from grande and genetically characterised cytoplasmic petite clones of *Saccharomyces cerevisiae*. *Proceedings of the National Academy of Sciences, USA*, **72**, 3868–72.

MURPHY, M., GUTOWSKI, S. J., MARZUKI, S., LUKINS, H. B. & LINNANE, A. W. (1978). Mitochondrial oligomycin-resistance mutations affecting the proteolipid subunit of the mitochondrial ATPase. *Biochemical and Biophysical Research Communications*, **85**, 1283–90.

NELSON, N., EYTAN, E., NOTSANI, B., SIGRIST, H., SIGRIST-NELSON, K. & GITLER, C. (1977). Isolation of a chloroplast *N, N'*-dicyclohexylcarbodiimide-binding proteolipid active in proton translocation. *Proceedings of the National Academy of Sciences, USA*, **74**, 2375–8.

OHNISHI, T. (1979). Mitochondrial iron–sulphur flavodehydrogenases. In *Membrane Proteins in Energy Transduction*, ed. R. A. Capaldi, pp. 1–87. New York & Basel: Marcel Dekker.

OJALA, D. & ATTARDI, G. (1977). Detailed physical map of HeLa cell mitochondrial DNA and its alignment with positions of known genetic markers. *Plasmid*, **1**, 78–105.

OSAFUNE, T., MIHARA, S., HASE, E. & OHKURO, I. (1975). Formation and division of giant mitochondria during the cell cycle of *Euglena gracilis* Z. in synchronous culture. I. Some characteristics of changes in the morphology of mitochondria and oxygen-uptake activity of cells. *Plant and Cell Physiology*, **16**, 313–26.

PAULIN, J. J. (1975). The chondriome of selected trypanosomatids. A three-dimensional study based on serial thick sections and high voltage electron microscopy. *Journal of Cell Biology*, **66**, 404–13.

PHILLIPS, C. A. & LLOYD, D. (1978). Continuous-flow size selection of *Tetrahymena pyriformis* strain ST: changes in volume, DNA, RNA and protein during synchronous growth. *Journal of General Microbiology*, **105**, 95–103.

PIRT, S. J. (1975). *Principles of Microbe and Cell Cultivation*, 274 pp. Oxford: Blackwell Scientific Publications.

POOLE, R. K. & LLOYD, D. (1973). Oscillations of enzyme activities during the cell cycle of a glucose-repressed fission-yeast *Schizosaccharomyces pombe* 972 h⁻. *Biochemical Journal*, **136**, 195–207.

POOLE, R. K. & LLOYD, D. (1974). Changes in respiratory activities during the cell-cycle of the fission yeast *Schizosaccharomyces pombe* 972 h⁻ growing in the presence of glycerol. *Biochemical Journal*, **144**, 141–8.

POOLE, R. K., LLOYD, D. & CHANCE, B. (1974). The development of cytochromes during the cell cycle of a glucose-repressed fission yeast *Schizosaccharomyces pombe* 972 h⁻. *Biochemical Journal*, **138**, 201–10.

POOLE, R. K., LLOYD, D. & KEMP, R. B. (1973). Respiratory oscillations and heat evolution in synchronously dividing cultures of the fission yeast *Schizosaccharomyces pombe* 972 h⁻. *Journal of General Microbiology*, **77**, 209–20.

POYTON, R. O. & KAVANAGH, J. (1976). Regulation of mitochondrial protein synthesis by cytoplasmic proteins. *Proceedings of the National Academy of Sciences, USA*, **73**, 3947–51.

POYTON, R. O. & MCKEMMIE, E. (1976). The assembly of cytochrome *c* oxidase from *Saccharomyces cerevisiae*. In *Genetics and Biogenesis of Mitochondria*, ed. W. Bandlow, R. J. Schweyen, K. Wolf & F. Kaudewitz, pp. 207–14. Berlin & New York: Walter de Gruyter.

PULLMAN, M. E. & MONROY, G. C. (1963). A naturally occurring inhibitor of mitochondrial adenosine triphosphatase. *Journal of Biological Chemistry*, **238**, 3762–9.

RAFF, R. A. & MAHLER, H. R. (1972). The non-symbiotic origin of mitochondria. *Science*, **177**, 575–82.

RAPER, K. B. & FENNELL, D. I. (1965). *The Genus Aspergillus*, 686 pp. Baltimore: Williams & Wilkins.

ROWLANDS, R. T. & TURNER, G. (1974). Recombination between extranuclear genes conferring oligomycin-resistance and cold sensitivity in *Aspergillus nidulans*. *Molecular and General Genetics*, **133**, 151–61.

ROWLANDS, R. T. & TURNER, G. (1976). Maternal inheritance in *Aspergillus nidulans*. *Genetical Research*, **28**, 281–90.

RYAN, R., GRANT, D., CHIANG, K. S. & SWIFT, T. H. (1978). Isolation and characterisation of mitochondrial DNA from *Chlamydomonas reinhardtii*. *Proceedings of the National Academy of Sciences, USA*, **75**, 3268–72.

RYRIE, I. J. & GALLAGHER, A. (1979). The yeast mitochondrial ATPase complex. Subunit composition and evidence for a latent protease contaminant. *Biochimica et Biophysica Acta*, **545**, 1–14.

SAGER, R. (1972). *Cytoplasmic Genes and Organelles*. New York & London: Academic Press.

SANDERS, J. P. M., WEIJERS, P. J., GROOT, G. S. P. & BORST, P. (1974). Properties of mitochondrial DNA from *Kluyveromyces lactis*. *Biochimica et Biophysica Acta*, **374**, 136–44.

SANDERS, J. P. M., HEYTING, C. & BORST, P. (1975). The organization of genes in yeast mitochondrial DNA I. The genes for large and small ribosomal RNA are far apart. *Biochemical and Biophysical Research Communications*, **65**, 699–707.

SANDERS, J. P. M., HEYTING, C., DIFRANCO, A., BORST, P. & SLONIMSKI, P. P. (1976). The organisation of genes in yeast mitochondrial DNA. In *The Genetic Function of Mitochondrial DNA*, ed. C. Saccone & A. M. Kroon, pp. 259–72. Amsterdam, Oxford & New York: North Holland.

SANDERS, J. P. M., HEYTING, C., VERBEET, M. Ph., MEIJLINK, F. C. P. W. & BORST, P. (1977). The organisation of the genes in yeast mitochondrial DNA III. Comparison of physical maps of the mitochondrial DNAs from three wild-type *Saccharomyces* strains. *Molecular and General Genetics*, **157**, 239–61.

SCHATZ, G. & MASON, T. L. (1974). The biosynthesis of mitochondrial proteins. *Annual Review of Biochemistry*, **43**, 51–87.

SCHWEYEN, R. J., WEISS-BRUMMER, B., BACKHAUS, B. & KAUDEWITZ, F. (1976). Localization of seven gene loci on a circular map of the mitochondrial genome of *Saccharomyces cerevisiae*. In *The Genetic Function of Mitochondrial DNA*, ed. C. Saccone & A. M. Kroon, pp. 251–8. Amsterdam, Oxford & New York: North Holland.

SEBALD, W., MACHLEIDT, W. & OTTO, J. (1973). Products of mitochondrial protein synthesis in *Neurospora crassa*. Determination of equimolar amounts of three products in cytochrome oxidase on the basis of amino acid analysis. *European Journal of Biochemistry*, **38**, 311–24.

SEBALD, W., SEBALD-ALTHAUS, M. & WACHTER, E. (1977). Altered amino acid sequence of the DCCD-binding protein of the nuclear oligomycin-resistant mutant AP-2 from *Neurospora crassa*. In *Mitochondria 1977*, ed. W. Bandlow, R. J. Schweyen, K. Wolf & F. Kaudewitz, pp. 433–40. Berlin: Walter de Gruyter.

SLONIMSKI, P. P. & TZAGOLOFF, A. (1976). Localization in yeast mitochondrial DNA of mutations expressed in a deficiency of cytochrome oxidase and/or coenzyme QH_2-cytochrome c reductase. *European Journal of Biochemistry*, **61**, 27–41.

SLONIMSKI, P. P., CLAISSE, M. L., FOUCHER, M., JACQ, C., KOCHKO, A., LAMOUROUX, A., PAJOT, P., PERRODIN, G., SPYRIDAKIS, A. & WAMBIER-KLUPPEL, M. L. (1978). Mosaic organization and expression of the mitochondrial DNA region controlling cytochrome c reductase and oxidase. III. A model of structure and function. In *Biochemistry and Genetics of Yeasts, Pure and Applied Aspects*, ed. M. Bacila, B. L. Horecker & A. O. M. Stoppani, pp. 391–401. New York, San Francisco & London: Academic Press.

SRIPRAKASH, K. S., MOLLOY, P. L., NAGLEY, P., LUKINS, H. B. & LINNANE, A. W. (1976). Biogenesis of mitochondria XLI. Physical mapping of mitochondrial genetic markers in yeast. *Journal of Molecular Biology*, **104**, 485–503.

STANIER, R. Y. (1970). Some aspects of the biology of cells and their possible evolutionary significance. In *Organization and Control in Prokaryotic and Eukaryotic Cells. Symposium of the Society for General Microbiology*, 20, ed. H. P. Charles & B. C. J. G. Knight, pp. 1–38. Cambridge University Press.

STĘPIEŃ, P. P., BERNARD, U., COOKE, H. J. & KÜNTZEL, H. (1978). Restriction endonuclease cleavage map of mitochondrial DNA from *Aspergillus nidulans*. *Nucleic Acids Research*, **5**, 317–30.

STEVENS, B. J. (1977). Variation in number and volume of the mitochondria in yeast according to growth conditions. A study based on serial sectioning and computor graphics reconstruction. *Biologie Cellulaire*, **28**, 37–56.

STRAUSBERG, R. L., VINCENT, R. D., PERLMAN, P. S. & BUTOW, R. A. (1978). Asymmetric gene conversion at inserted segments on yeast mitochondrial. DNA. *Nature, London*, **276**, 577–83.

SUYAMA, Y. & HAMADA, J. (1976). Imported tRNA: its synthetase as a possible transport protein. In Genetics and Biogenesis of Chloroplasts and Mitochondria, ed. Th. Bücher, W. Neupert, W. Sebald & S. Werner, pp. 763–70. Amsterdam, Oxford & New York: North Holland.

TEMPEST, D. W. (1978). The biochemical significance of growth yields: a reassessment. *Trends in Biochemical Sciences*, **3**, 180–4.

TERPSTRA, P., HOLTROP, M. & KROON, A. M. (1977). The ribosomal RNA genes on *Neurospora crassa* mitochondrial DNA are adjacent. *Biochimica et Biophysica Acta*, **478**, 146–55.

THOMAS, D. Y. & WILKIE, D. (1968). Recombination of mitochondrial drug-resistance factors in *Saccharomyces cerevisiae*. *Biochemical and Biophysical Research Communications*, **30**, 368–72.

TURNER, G., IMAM, G. & KÜNTZEL, H. (1979). Mitochondrial ATPase complex of *Aspergillus nidulans* and the dicyclohexylcarbodiimide-binding protein. *European Journal of Biochemistry*, **97**, 565–71.

TZAGOLOFF, A., AKAI, A., & FOURY, F. (1976). Assembly of the mitochondrial membrane system XVI. Modified form of the ATPase proteolipid in oligomycin resistant mutants of *Saccharomyces cerevisiae*. *FEBS Letters*, **65**, 391–5.

TZAGOLOFF, A. & MEAGHER, P. (1972). Assembly of the mitochondrial membrane system. VII. Mitchondrial synthesis of subunit proteins of the rutamycin-sensitive adenosine triphosphatase. *Journal of Biological Chemistry*, **247**, 594–603.

TZAGOLOFF, A., RUBIN, M. S. & SIERRA, M. F. (1973). Biosynthesis of mitochondrial enzymes. *Biochimica et Biophysica Acta*, **301**, 71–104.

VAN OMMEN, G. J. B., GROOT, G. S. P. & BORST, P. (1977). Fine-structure physical mapping of 4S RNA genes on mitochondrial DNA of *Saccharomyces cerevisiae*. *Molecular and General Genetics*, **154**, 255–62.

VIGNAIS, P. V. (1976). Molecular and physiological aspects of adenine nucleotide transport in mitochondria. *Biochimica et Biophysica Acta*, **456**, 1–38.

VON JAGOW, G. & KLINGENBERG, M. (1972). Close correlation between antimycin titre and cytochrome *b* content in mitochondria of chloramphenicol-treated *Neurospora crassa*. *FEBS Letters*, **24**, 278–82.

WACHTER, E., SEBALD, W. & TZAGOLOFF, A. (1977). Altered amino-acid sequence of the DCCD-binding protein in the oli 1-resistant mutant D273–10b/A-21 of *Saccharomyces cerevisiae*. In *Mitochondria 1977, Genetics and Biogenesis of Mitochondria*, ed. W. Bandlow, R. J. Schweyen, K. Wolf & F. Kaudewitz, pp. 441–50. Berlin & New York: Walter de Gruyter.

WEISS, H. & ZIGANKE, B. (1977). Partial identification, stoichiometry and site of translation of the subunits of ubiquinone:cytochrome c oxidoreductase. In *Mitochondria 1977, Genetics and Biogenesis of Mitochondria*, ed. W. Bandlow, R. J. Schweyen, K. Wolf & F. Kaudewitz, pp. 463–72. Berlin & New York: Walter de Gruyter.

WESOLOWSKI, M. & FUKUHARA, H. (1979). The genetic map of transfer RNA genes of yeast mitochondria: correction and extension. *Molecular and General Genetics*, **170**, 261–75.

WILKIE, D. & THOMAS, D. Y. (1973). Mitochondrial genetic analysis by zygote cell lineages in *Saccharomyces cerevisiae*. *Genetics*, **73**, 367–77.

WILLIAMS, R. J. P. (1978). The history and the hypotheses concerning ATP-formation by energised protons. *FEBS Letters*, **85**, 9–19.

WILLIAMSON, D. H. (1970). The effect of environmental and genetic factors on the replication of mitochondrial DNA in yeast. In *Control of Organelle Development. Symposia of the Society for Experimental Biology*, 24, ed. P. L. Miller, pp. 247–76. Cambridge University Press.

WILLIAMSON, D. H., JOHNSTON, L. H., RICHMOND, K. M. V. & GAME, J. C. (1977). Mitochondrial DNA and the heritable unit of the yeast mitochondrial genome: a review. In *Mitochondria 1977, Genetics and Biogenesis of Mitochondria*, ed. W. Bandlow, R. J. Schweyen, K. Wolf & F. Kaudewitz, pp. 1–24. Berlin & New York: Walter de Gruyter.

WILSON, D. F., OWEN, C. S. & HOLIAN, A. (1977). Control of mitochondrial respiration: a quantitative evaluation of the roles of cytochrome *c* and oxygen. *Archives of Biochemistry and Biophysics*, **182**, 749–62.

EXPLANATION OF PLATE

PLATE 1

Recombination between the mtDNAs of *Aspergillus nidulans* (*Eidam*) Wint. and *Aspergillus nidulans* var. *echinulatus* Fennell & Raper (Raper & Fennell, 1965). *A. nidulans* (*Eidam*) Wint. carried the extra-nuclear oligomycin-resistance marker (*oliA1*) (Rowlands & Turner, 1974). The hybrid strain was obtained following protoplast fusion between the two parental strains by selecting for oligomycin-resistance in the *A. nidulans* var. *echinulatus* nuclear background. DNA isolated from the mitochondria of each strain (Stępień *et al.*, 1978) was digested with HindIII restriction endonuclease, and analysed by electrophoresis on 1.1 % agarose. (*a*) *A. nidulans* (*Eidam*) Wint., mol. wt. 20.3×10^6; (*b*) hybrid, mol. wt. 22.2×10^6; (*c*) *A. nidulans* var. *echinulatus*, mol. wt. 23.6×10^6. Bands of equal mol. wt. are joined by dashed lines. Mol. wts. are given for bands in lane (*a*), (A. J. Earl, R. G. B. Dales, J. H. Croft & G. Turner, unpublished).

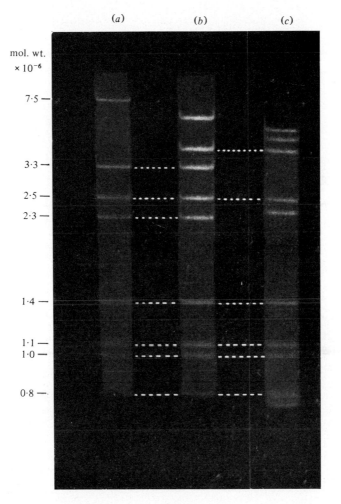

PLATE 1

CHLOROPLASTS OF EUKARYOTIC MICRO-ORGANISMS

JOHN A. RAVEN

*Department of Biological Sciences, University of Dundee,
Dundee DD1 4HN, UK*

INTRODUCTION

All eukaryotic micro-organisms that possess chloroplasts are, to an unrepentant botanist, classified as algae. These chloroplasts, like those of macrophytes, enable the organisms which contain them to carry out O_2-evolving photosynthesis and to live phototrophically. Among the prokaryotes, these metabolic and ecological characteristics are only found in cyanobacteria. There is thus considerable overlap in the habitats exploited by cyanobacteria and the various classes of eukaryotic micro-algae. However, the really largest habitat for photolithotrophs, the euphotic zone of the open sea, has a very small biomass and productivity of cyanobacteria relative to that of eukaryotic algae. Thus the 0.4 or so of world primary productivity that is carried out by marine phytoplankton is largely due to the chloroplasts contained in eukaryotic micro-algae (Cooper, 1975).

Eukaryotic micro-algae are also found in benthic aquatic and terrestrial habitats, but they rarely achieve dominance there, the dominant plants being macrophytes (macro-algae, bryophytes and tracheophytes). However, in some benthic and terrestrial habitats the micro-algae compete well with dominant macrophytes by taking part in symbioses with macro-heterotrophs; examples are lichens (micro-alga fungus) competing with poikilohydric bryophytes on land, and corals (micro-alga coelenterate) competing with macro-algae in shallow tropical seas. The significance of such symbioses is eloquently argued by Lewis (1973). The micro-algae have thus come to dominate terrestrial and benthic aquatic habitats either by evolving into macrophytes, or by entering into symbiotic associations. In each case, the structure and function of the chloroplasts seems to be basically similar, although they form a part of very different structural organisations and heterotrophic metabolic patterns.

The very great ecological importance of eukaryotic micro-algae, as well as their value as experimentally amenable eukaryotic photolithotrophs, makes the study of their chloroplasts also important. A major difficulty with such studies is that of obtaining isolated chloroplast preparations structurally and functionally intact and free of contamination by other cell components. Such preparations can now be routinely obtained from leaves of angiosperms (Lilley & Walker, 1975) and from the thalli of coenocytic macro-algae (Cobb & Rott, 1978). The problems involved in extracting chloroplasts from micro-algae are discussed by contributors to Hellebust & Craigie (1978), who point out that functional sub-chloroplast fractions (thylakoids, soluble enzymes) are more easily obtained than intact chloroplasts. For this reason a basic premise of my subsequent discussion, i.e. that the chloroplast(s) of micro-algae

Table 1. *List of symbols and abbreviations*

CF_0	H+ uniporter component of the ATP synthetase complex
CF_1	ATPase component of the ATP synthetase complex
cyt *b*-559	*b*-type cytochrome existing in a number of 'states' with respect to mid-point oxidation–reduction potential, functioning close to Photosystem II
cyt *b*-563	*b*-type cytochrome probably involved in cyclic redox reactions
cyt *c*	*c*-type cytochrome: soluble, analogous to mitochondrial cytochrome *c*, interchangeable with Pcy
cyt *f*	*c*-type cytochrome, insoluble, analogous to mitochondrial cytochrome c_1
Fd	ferredoxin: a low potential, soluble iron–sulphur protein; in some algae, particularly under Fe deficiency, replaced by the flavoprotein, flavodoxin
Fd:NADP o-r	ferredoxin: NADP oxido-reductase; a flavoprotein involved in non-cyclic and possibly cyclic redox reactions
LHPPC	Light-harvesting pigment–protein complex. A pigment–protein complex containing shorter-wavelength absorbing forms of chlorophyll *a* and class-specific accessory pigments (e.g. chlorophyll *b* in the Chlorophyceae and Euglenophyceae; peridinin in the Dinophyceae), generally more closely associated with, and more ready to transfer excitation energy to, Photosystem II. In the Rhodophyceae (and cyanobacteria) the role of LHPPC is largely taken over by the phycobilin-containing phycobilisomes which are attached to, rather than being integral components of, the thylakoid.
Photosystem I	A pigment-protein complex containing longer-wavelength absorbing forms of chlorophyll *a*, including the photochemically active *P*-700. Catalyses transmembrane, energy-requiring e− transfer in Photoreaction I of cyclic and non-cyclic redox reactions
Photosystem II	A pigment-protein complex containing shorter-wavelength absorbing forms of chlorophyll *a*, including the photochemically active P-680. Catalyses transmembrane, energy-requiring e− transfer in Photoreaction II of non-cyclic redox reactions
Pcy	Plastocyanin, a copper-containing protein interchangeable with cyt *c*
PQ	Plastoquinone, a quinone which acts as a transmembrane (H) uniporter in cyclic and non-cyclic redox reactions
P-430	Insoluble iron–sulphur complex acting as the low-potential electron acceptor of Photosystem I.
P-680	Form of protein-associated chlorophyll *a* which carries out the primary photochemical reaction of Photosystem II
P-700	Form of protein-associated chlorophyll *a* which carries out the primary photochemical reaction in Photosystem I
Q	Primary acceptor of Photosystem II: probably a quinone
S	Donor to Photosystem II, probably involving Mn and acting as a 'charge-accumulating' part of the O_2 evolution mechanism

are capable of the complete sequence of photosynthetic reactions from light absorption to the production of reduced carbon compounds which can support the heterotrophic metabolism of the extrachloroplastic part of the cell, rests on less direct evidence than one would wish for.

Granted that the chloroplast is (in the sense just described) competent to carry out photosynthesis, I propose to analyse the structure–function relationships of chloroplasts from a quantitative point of view, paying particular attention to three 'design considerations'. One of these is the intrinsic efficiency of the process, i.e. the ratio of useful work output to energy input; another is the work output per unit of catalytic and structural material used. The balance between these two considerations in various metabolic systems is likely to be subject to considerable selection pressure with different outcomes in organisms adapted to different environments (Odum & Pinkerton, 1955; Grime, 1979). The third 'design consideration', and one which is particularly important in the photosynthetic apparatus, is that of safety: it is possible that other-wise advantageous mechanisms with a local or global optimal balance of the first two design features might, in the long term, be too risky. The risk of photosynthesis is related to the occurrence of light, a photo-catalyst (chlorophyll) and oxygen together in the photosynthetic apparatus.

STRUCTURE AND FUNCTION OF MICRO-ALGAL CHLOROPLASTS: UNITY AND DIVERSITY

There is a fundamental unity in the structure and composition of chloroplasts which applies to all photolithotrophic eukaryotes (Coombs & Greenwood, 1976); these common features will be outlined before the difference between the various classes of micro-algae are briefly mentioned.

Fig. 1 shows the basic structure of a chloroplast. The membrane components seem to be interpretable most readily as fluid mosaics of dispersed protein in a continuous phase of polar lipid bilayer (Singer & Nicolson, 1972). Such membranes act as general barriers to the transport of most metabolites, with catalysis of the transport of selected metabolites. The conversion of light energy into chemical energy (ATP and $NADPH_2$) occurs within (or across) the thylakoid membrane from the intra-thylakoid space. About half of the mass of the thylakoid membrane is polar lipid (chlorophylls, carotenols, glycolipids and phospholipids), carotenes and quinones; the remainder is protein, including the protein components of light-harvesting and photo-chemically active pigment–protein complexes, various redox catalysts and the ATP synthetase complex (CF_1–CF_0). Thus the continuous phase of the membrane consists of glyco- and phospholipids, with most of the pigments associated with proteins (Anderson, 1975). The orientation of the proteins in (or on) the membrane is discussed in the next section.

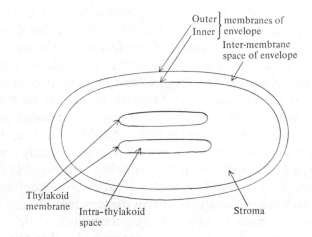

Fig. 1. Generalised chloroplast ultrastructure. A basic chloroplast illustrating its membrane-bound compartments. Additional membranes around the chloroplast are found in certain algal classes.

The other membranes of the typical chloroplast are the two concentric envelope membranes. Work on isolated chloroplasts of higher plants has shown that the outer membrane is permeable to molecules of molecular weight less than 500–1000, while the inner membrane is only 'passively' permeable to H_2O, O_2, CO_2, NH_3, etc. and contains transport catalysts for other metabolites (p. 193; Heber, 1974). Compositional data for envelopes are only available for the combined inner and outer membranes of higher plants: the lipids include glycolipids and caro-tenes resembling those of the thylakoid, but with no chlorophyll; the proteins include an ATPase, but there seem to be no redox catalysts in the envelope (Douce, Holtz & Benson, 1973; MacKender & Leech, 1974).

Qualitatively, the most significant aqueous phase in the chloroplast is the stroma which accounts for half of the total volume. This is topologically equivalent to prokaryote cytoplasm, with its own genome granting partial genetic and protein synthetic autonomy, the enzymes of the photosynthetic carbon reduction cycle (PCRC) and certain other enzymes (Givan & Leech, 1971; Leech & Murphy, 1976). The two other aqueous phases (the intra-thylakoid space and the inter-membrane space of the envelope) are much smaller (together less than one-tenth of the total volume). They are topologically equivalent to the extra-cellular space, and have a very limited diversity of enzymes (ten or so).

The unity of structure of all chloroplasts may be related to wider concepts of cell organisation (the 'sidedness' of membranes and the 'cytoplasmic' or 'external medium' nature of cytoplasmic phases) elaborated with varying emphases by Mitchell (1966, 1970), Heldt & Sauer (1971), Raven & Smith (1976), Smith & Raven (1978) and Raven (1979). It is important to note that the aqueous phase characteristics and membrane sidedness of chloroplasts and mitochondria can be explained by an endosymbiotic origin or by successive plasmalemma invaginations with equal facility (Smith & Raven, 1978).

Turning from unity to diversity of chloroplast structure and composition in micro-algae, much of the taxonomy of algae at the class level is based on such differences (Bold & Wynne, 1978). There are class-specific differences in thylakoid arrangement, and in the pigments they contain. All O_2-evolvers have chlorophyll a and β-carotene; the different classes have a diversity of other pigments as light-harvesting pigments in (chlorophylls, carotenoids) or on (phycobilins) the thylakoids. There are also differences (some with macrotaxonomic significance) in two important redox mediators, with ferredoxin and flavodoxin, and cytochrome c-552 and plastocyanin being interchangeable (Crofts & Wood, 1978). In the stroma, a common (but not universal) component of unicellular algae is the pyrenoid, readily identified by light or electron microscopy and consisting largely of the enzyme ribulosebisphosphate carboxylase (Kerby & Evans, 1978). The activity of this 'precipitated' enzyme, and the role of the pyrenoid in polysaccharide synthesis is not well understood. Finally, there are often class-specific variations in the number of membranes surrounding the chloroplast. The presence of two additional membranes is readily reconciled with general topological principles of cell construction (Coombs & Greenwood, 1976). The presence of one additional membrane is more perplexing, and has been related by Gibbs (1978) to the possible origin of such chloroplasts from endosymbiotic eukaryotes.

Lest the reader feels that the final message of this section is the topic with which it ends, i.e. diversity, it must be reiterated that the dominant impression gained from a survey of the structure, composition and function of algal chloroplasts is one of fundamental similarity.

QUANTITATIVE ASPECTS OF
STRUCTURE FUNCTION RELATIONS

Thylakoids

A widely accepted scheme for non-cyclic energy conversion in algal photosynthesis is shown in Fig. 2(*a*). In qualitative terms light energy is absorbed by pigment–protein complexes associated with two distinct sorts of photochemical reaction centres which catalyse two different photoreactions (denoted I and II); some of the light is absorbed by additional light-harvesting pigment–protein complexes which preferentially transmit excitation energy to Photoreaction II (Butler, 1978; Thornber, 1975; Prezelin & Alberte, 1978). The photochemical reactions are arranged vectorially such that an electron is transferred to an acceptor on the stroma side of the membrane, leaving an oxidised donor on the intra-thylakoid side of the membrane. The ultimate donor to Photoreaction II (H_2O) and the ultimate acceptor for Photoreaction I (CO_2), as well as the redox compound connecting the reaction II acceptor to the reaction I donor (PQ) are redox agents dealing in (H) rather than e^-. The result of these vectorial (H) and e^- movements is a transmembrane proton-motive force (p.m.f.) (Junge, 1977; Crofts & Wood, 1978). The H^+ circuit is completed via the ATP synthetase complex (Jagendorf, 1975; Avron, 1978). The end-products of the light energy conversion are O_2, $NADPH_2$ and ATP, the two latter being used mainly in CO_2 fixation. Studies on mutants of micro-algae have been useful in elucidating the pathways shown in Fig. 2 (Levine, 1969; Levine & Goodenough, 1970).

Quantifying first the running costs, the movement of an electron through each of the photoreactions covers an energy span of about 1 eV; the energy input is a light quantum with an effective energy content of about 1.76 eV for Photoreaction I, and 1.82 eV for Photoreaction II, i.e. each reaction stores some 0.56 of the total energy available. Subsequent reactions store some 1.15 eV as $NADPH_2$ and O_2, and 0.4 eV as $2H^+$ transported across the membrane, i.e. 1.55 eV out of the 2 eV in the primary charge separation products, or the 3.6 eV available in the two absorbed quanta. For the p.m.f., an H^+/ATP ratio of 3 (Avron, 1978) and a ΔG_p* of 0.5 V means storage of 0.33 V as ATP out of the available 0.4 V in the p.m.f. Overall, the energy stored per two quanta used and electron transferred is 1.5 eV out of the input of 3.58 eV (an

---

* ΔG_p, Free energy of ATP hydrolysis *in vivo* expressed per molecule of ATP in energetic equilibrium with a 1 e^- redox reaction.

efficiency of 42% which can only be attained in terms of *incident* irradiance when light absorption and transfer to reaction centres is perfectly efficient, and light is strictly limiting the overall rate of photosynthesis).

Another ATP-generating system is based on a cyclic redox system (Gimmler, 1977) using Photoreaction I, with a proton-motive Q cycle as envisaged by Mitchell (1976) (Fig. 2(*b*); cf. Crowther, Mills & Hind, 1979). The quantum/e^-/H^+ stoichiometry is $1:1:2$, so with an H^+/ATP ratio of 3 (Avron, 1978, Crofts & Wood, 1978) this reaction stores 0.33 eV out of the 1.8 eV input, an efficiency of 18%. An even lower efficiency of 9% is found for pseudocyclic ATP synthesis in which the Photoreaction I product of non-cyclic redox reactions is re-oxidised by O_2. Higher efficiencies can be computed if a proton-motive Q cycle is incorporated into non-cyclic redox reactions, or if an H^+/ATP ratio of 2 is assumed.

Turning to capital investments, it is possible to compute the power output of these redox and ATP synthesis reactions on the basis of the material or energy invested in structural and catalytic machinery of the thylakoid. The thylakoid area is a useful basis for such computations: the content of reaction centres in the thylakoid is some 0.5 pmol of each kind of reaction centre per cm^2 (J. A. Raven, unpublished), and the turnover is some 200 s^{-1}. This means that the rate of non-cyclic redox reactions is some 100 pmol e^- cm^{-2} s^{-1}. Since 1 eV $\simeq$ 100 kJ mol^{-1}, the energy storage per turnover of 1.5 eV means an energy storage rate (power output) of some 15 μJ cm^{-2} s^{-1} (15 μW cm^{-2}). On an inner membrane area basis the rate of ATP synthesis, and ΔG_p for the mitochondrial ATP synthesis mechanism is similar to that in thylakoids so the power output there is only some 3–4 μW cm^{-2}. On a protein basis the thylakoid shows up even better in that the thylakoid has only some two-thirds of the protein content per cm^2 of the inner mitochondrial membrane (J. A. Raven, unpublished).

We can make some tentative attempts to interpret the quantitative data in terms of membrane structure and optimise the rate and efficiency of energy conversion. The efficient use of limited irradiance requires efficient excitation energy transfer from the light-harvesting pigment molecules to the reaction centres, and the ability to balance the energy supplied to the two reaction centres in relation to different ratios of cyclic and non-cyclic energy transduction reactions and to variations in light quality (e.g. with depth in the sea) and thus in the relative absorption by the various pigment–protein complexes. These considerations can partly explain (in adaptive and mechanistic terms) the organisation of the various light-harvesting complexes in relation to the reaction

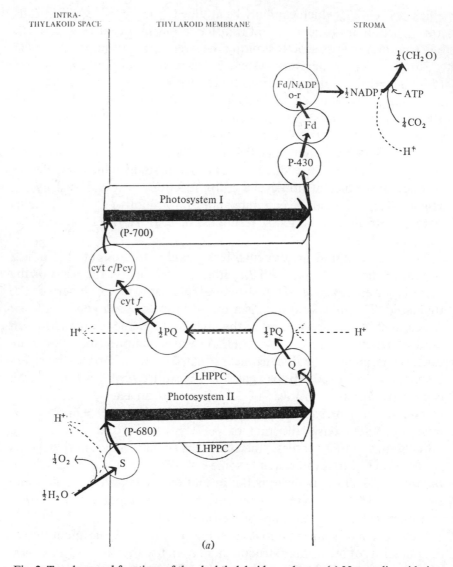

Fig. 2. Topology and functions of the algal thylakoid membrane. (*a*) Non-cyclic oxidation–reduction reactions generating a proton-motive force with a stoichiometry of 2H+ per e− (H) transported and per two quanta absorbed at low irradiances. These same two quanta also bring about the transfer of an e− (H) from H_2O to NADP+, generating $\frac{1}{4}O_2$ and $\frac{1}{2}NADPH_2$. The pathway of electron transport is indicated by thick continuous lines, with the light reactions (energy input) represented as broad arrows. H+ transport is represented by dashed lines. The approximate location of the various e− and H carriers in the membrane is indicated. Photosystem II operates between about +0.9V (P-680) and —0.1V (Q); Photosystem I operates between about +0.45V (P-700) and —0.55V (P-430). (*b*) Cyclic oxidation–reduction reactions generating a proton-motive force with a stoichiometry of 2H+ per e− (H) transported and per quantum absorbed by Photosystem I. The single light reaction (Photosystem I) used for cyclic redox reactions is the same as that used in non-cyclic redox reactions: cyclic and non-cyclic redox reactions can represent alternative functions of the same Photosystem I units, depending on the relative requirements of the

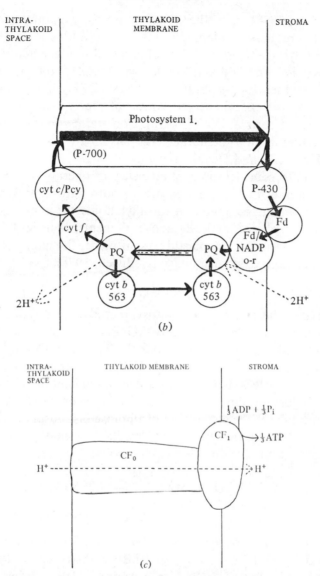

(b)

(c)

cell for reductant and ATP. The cyclic redox scheme shown involves a proton-motive Q cycle: twice as many H^+ and e^- are transferred from the stromal to the intra-thylakoid side of the membrane by PQ as e^- are transferred from the intra-thylakoid to the stromal side of the membrane by Photosystem I. The missing e^- is moved from the intra-thylakoid to the stromal side of the membrane by b-type cytochromes (including cyt b-563). The scheme for non-cyclic redox reactions shown in (a) above does not include a proton-motive Q cycle: any such Q cycle would probably involve cyt b-559. The cyclic oxidation–reduction reactions shown here do not involve the net generation of oxidant or reductant: the sole useful product is the proton-motive force. (c) ATP synthetase, catalysing the downhill transport of H^+ from the intra-thylakoid to the stromal side of the membrane coupled to the forma- tion of ATP from ADP and P_i in the stroma. The driving force for H^+ movement through the ATP synthetase is the proton-motive force generated by either cyclic or non-cyclic oxidation–reduction reactions. For explanation of abbreviations and symbols, see Table 1.

centres (Butler, 1978). A further requirement for economy of materials investment in organisms adapted to low light is a high ratio of pigment to redox catalysts; this is indeed what is found in algae adapted to low irradiances (Myers & Graham, 1971). Algae adapted to high irradiances probably have their light-saturated rate of $NADPH_2$ and ATP synthesis determined by the specific reaction rates of redox catalysts, especially plastoquinone (Crofts & Wood, 1978). Constraints on the reactivity of quinones acting as transmembrane (H) porters are discussed by Keck, Dilley & Ke (1970) and Hauska (1978).

In addition to considerations of efficiency of operation and economy of construction, a final 'design feature' of both natural and man-made energy transformers is safety. In the thylakoids the potentially lethal combination of photochemically active pigments, light and O_2 is constrained by a variety of plastid constituents, including carotenes, quinones and superoxide dismutase (Halliwell, 1978).

Stroma

The stroma is the site of the photosynthetic carbon reduction cycle (PCRC) which uses the ATP and $NADPH_2$ generated in the light-dependent thylakoid processes (p. 00) in the fixation of CO_2 into reduced carbon compounds (cf. Beardall et al., 1976). A scheme summarising the reactions of the cycle is shown in Fig. 3. The initial fixation of CO_2 is catalysed by ribulose bisphosphate carboxylase–oxygenase (E.C. 4.1.1.39), which also catalyses the conversion of ribulose bisphosphate (RuBP) to 3-P-glycerate (PGA) and P-glycolate by its oxygenase activity; these competing reactions are depicted in equations:

$$RuBP + CO_2 + H_2O \xrightarrow{\text{RuBP carboxylase}} 2\ PGA \qquad (1)$$

$$RuBP + O_2 \xrightarrow{\text{RuBP oxygenase}} PGA + P\text{-glycolate} \qquad (2)$$

In air-equilibrated solution at 25 °C the kinetics of the competing reactions are such that the ratio of RuBP carboxylase to RuBP oxygenase activity is about 5:1 (Badger & Andrews, 1974; Laing, Ogren & Hagemann, 1974) in both higher plants and in eukaryotic micro-algae (Jensen & Bahr, 1977; Lord & Brown, 1975; Berry et al., 1976). In most land plants (C_3 plants) and in algae grown at high CO_2 levels (1 % or more in the gas stream) the characteristics of photosynthesis in terms of O_2 inhibition of net CO_2 fixation and the rate of glycolate synthesis are consistent with the in-vitro kinetics of RuBP carboxylase–oxygenase (Laing et al., 1974; Berry & Bowes, 1973; Berry et al., 1976; Raven & Glidewell, 1979). The P-glycolate generated in this pathway is scavenged via the photorespiratory carbon oxidation cycle (PCOC) (Fig. 3) as well as, in the algae, excreted as such.

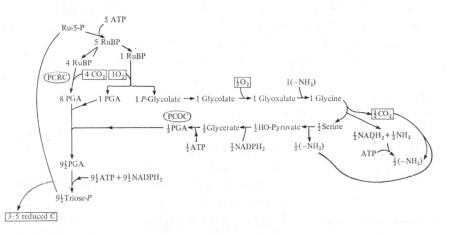

Fig. 3. Photosynthetic carbon reduction cycle (PCRC) and photorespiratory carbon oxidation cycle (PCOC). The stoichiometry of the PCRC and the PCOC shown here is based on a 4:1 ratio of the activities of the carboxylase–oxygenase activities of ribulose bisphosphate carboxylase–oxygenase. The scheme as drawn involves the removal of 3.4 C (as carbohydrate) for every turn of the cycles; the remaining 0.5 C of the 4 CO_2 fixed in gross photosynthesis is evolved in the glycine to serine step of the PCOC. The net evolution of $3.5O_2$ per $3.5CO_2$ taken up (net) is the resultant of $1.5O_2$ taken up (in RuBP oxygenase and glycolate oxidase activity) and $5O_2$ evolved (in the course of generating the $NADPH_2$ used in the PCOC and the PCRC). The $NADPH_2$ requirement is $10NADPH_2$ per net 3.5 C fixed, while the ATP requirement is $15\frac{1}{2}ATP$ per 3.5 C. The recycling of the NH_3 generated in the glycine to serine step is shown as involving the glutamine synthetase-glutamate synthetase sequence. Since it is unlikely that the $NADH_2$ also generated in the glycine to serine step can be used in the glutamate synthetase reaction in the manner shown in the Figure, it is likely that the $\frac{1}{2}NADH_2$ generated in the glycine to serine reaction is oxidised in the mitochondria (generating up to $1\frac{1}{2}ATP$) and that the glutamate synthetase reaction uses reduced ferredoxin generated in the light reactions. This would increase the O_2 uptake to $2O_2$ per net carbon fixed and the gross O_2 evolution to $5\frac{1}{2}$ per net carbon fixed into carbohydrate, but would decrease the ATP requirement from photophosphorylation from $15\frac{1}{2}$ to 14 since $1\frac{1}{2}ATP$ are made during the reoxidation of $NADH_2$. It should be noted that a number of intermediates in the PCRC between triose-P and Ru-5-P are not represented, and that stoichiometries make no allowances for biosynthetic utilisation of intermediates of the cycles (with the exception of 3.5C as carbohydrate from to PCRC). Removal of carbon other than as carbohydrate would involve a corresponding decrease in the carbon removed as carbohydrate.

However, when many algae are grown under more 'natural' conditions of inorganic carbon supply (i.e. solutions in equilibrium with 0.03% or less of CO_2 in the gas phase) the extent of the O_2 inhibition of photosynthesis, and the rate of glycolate synthesis, are both decreased (Raven & Glidewell, 1979). There is no evidence for a change in the V_{max}, $K_{\frac{1}{2}}$ or K_i of RuBP carboxylase–oxygenase for either CO_2 or O_2 during this adaptation, and the data are best explained by the occurrence of a 'CO_2 accumulating mechanism' based on an active HCO_3^- influx pump. This enhanced CO_2 concentration and CO_2/O_2 ratio at the site of RuBP carboxylase–oxygenase suppresses the oxygenase relative to the carboxylase activity and accounts for the reduced glycolate synthesis rate, PCOC activity and O_2 inhibition of net photosynthesis in a manner auxiliary carboxylation caralysed by P-enolpyruvate carboxylase micro-algae which have a 'CO_2 accumulation mechanism' based on an auxiliary carboxylation catalysed by P-enolpyruvate carboxylase (Berry *et al.*, 1976; Raven & Glidewell, 1978; Badger, Kaplan & Berry, 1977; cf. Findenegg, 1978).

The quantitative analysis of this phenomenon is difficult since it is difficult to quantify the possible running and capital costs of the 'CO_2 concentration' mechanisms. It seems likely that the 'CO_2 pump' involves active transport of HCO_3^- across the plasmalemma or the inner chloroplast envelope (Raven & Glidewell, 1978). The energy costs of such a pathway may well exceed 1ATP per CO_2 fixed if the CO_2 permeabilities of the chloroplast and plasmalemma membranes are at their usual high values, since this provides a large leak short-circuiting the HCO_3^- pump (Raven, 1980). Against this must be set savings in the energy costs of the glycolate-scavenging PCOC (Fig. 3) which deals with less glycolate per unit carbon fixed. Overall, the PCRC uses 3ATP and $2NADPH_2$ per carbon atom fixed. If we take 2ATP as the energy cost of 'CO_2 pumping' plus that of running the PCOC per net CO_2 fixed, the overall energy cost is that of 5ATP and $2NADPH_2$ or 710 kJ per mol C fixed. Since 500 kJ mol^{-1} are required to convert CO_2 to the carbohydrate level of reduction, the efficiency of this process is 70%. The overall efficiency of light energy conversion requires consideration of the efficiency of ATP production: assuming that the excess ATP is generated by cyclic photophosphorylation, 10.33 quanta are needed per molecule of carbon that is fixed, i.e. an overall efficiency of 27%.

Considerations of the capital costs of fixation are dominated by RuBP carboxylase–oxygenase, an enzyme with a high molecular weight (550 000) and a low specific reaction rate of RuBP oxygenase even when the carboxylase activity is repressed by saturating CO_2 and zero O_2, i.e. 10 mol CO_2 fixed per mol enzyme s^{-1} at 25 °C (Jensen & Bahr, 1977). The molecular weight is perhaps two to four times that of the 'average' soluble enzyme and its specific reaction rate perhaps ten times lower (see the analysis of glycolysis by Hess, 1973). Thus, even granted the more 'typical' molecular weight and specific reaction rate characteristics of the other PCRC enzymes, the protein and energy cost of the PCRC is higher (per unit substrate transformation rate) than that of many other pathways with a similar number of enzymic steps. This is illustrated by the quantity of soluble protein in the chloroplast (involved in the PCRC, together with certain other synthetic pathways) relative to that in the cytoplasm (which catalyses a much greater number of molecular transformations per unit time than does the chloroplast during balanced growth): this ratio is perhaps $1:2$ (Smillie, 1963; Rabinowitz et al., 1975). The molecular weight and turnover characteristics of RuBP carboxylase–oxygenase can be used, together with a reasonable upper limit on the quantity of dissolved, active enzyme per unit volume of stroma (1 nmol cm^{-3}), to estimate the volume of stroma required to support the fixation of 1 nmol CO_2 s^{-1} at about 0.1 cm^3. This computation assumes CO_2 saturation of the enzyme; this can be achieved either by supplying high external CO_2 concentrations for growth or by the activity of a 'CO_2 pump' at low exogenous CO_2 concentrations.

Space does not permit more than passing mention of the other synthetic activities of the stroma. With the exception of reduced C and (probably) of organic N (NO_2^- reductase, and the assimilation of NH_4^+ into organic combination via the glutamine synthetase–glutamate synthetase pathway, are plastid activities: Raven, 1980), the chloroplast is not a net supplier of intermediates to the 'heterotrophic' portions of the cell. Either the chloroplast is self-sufficient (e.g. the large subunit of the RuBP carboxylase–oxygenase gene, and its transcription and translation) or depends on the cytoplasm (e.g. many amino acids and other low molecular weight intermediates, and perhaps 80–90% of the protein species found in the chloroplast). These metabolic interactions involve complex controls and (see below) a number of selective transport mechanisms at the chloroplast envelope.

Chloroplast envelope

The data from higher plants mentioned in the previous section suggest that it is the inner of the two envelope membranes which is the major barrier to diffusion of hydrophilic solutes with a molecular weight below 1000, and is the site of the catalysts of their selective transport. The two best-characterised systems are those for P_i and various 3-C phosphate esters (capacity 140 pmol cm^{-2} s^{-1}) and for dicarboxylic acids (capacity 100 pmol cm^{-2} s^{-1}); Heber (1974), with the capacities referring to Spinacia chloroplasts at substrate saturation and 20 °C. Other porters transport inorganic ions, glucose, purines and pyrimidines (Gimmler, Schafer & Heber, 1974; Raven, 1976b; Schofer, Heber & Heldt, 1977; Barber & Thurman, 1978a, b). The available data are consistent with, but by no means prove, the occurrence of an ATP-powered H^+ efflux pump at the inner envelope membrane with a capacity of up to 100 pmol cm^{-2} s^{-1} (Gimmler *et al.*, 1974; Heber, 1974). Much further work is needed on these porters, particularly on the existence of primary active H^+ extrusion and its possible role in coupling other porters. Even less is known of the mechanism of macromolecule (mainly protein) transport across the envelope of higher plant chloroplasts: it seems likely that mediated transport is required at both envelope membranes rather than just the inner membrane (Highfield & Ellis, 1978; Borst & Grivell, 1978; cf. Gibbs, 1979).

A significant quantitative point relates to the comparison between the chloroplast and the other major energy-transducing organelle of eukaryotes, the mitochondrion. It was noted above that the thylakoid membrane can make more ATP per unit area and time than the inner mitochondrial membrane, although it is clear that the inner mitochondrial membrane has an additional burden of the transport of substrates and products. In the chloroplast such porters are in the envelope membrane, so a better comparison of the capital costs of ATP synthesis in the two organelles should compare the rate per unit area of inner mitochondrial membrane with that for thylakoid plus the associated envelope membrane. This correction reduces the ATP synthesis (cm^2 thylakoid:s)$^{-1}$ of the thylakoid by a factor of 0.8 (J. A. Raven, unpublished).

THE CHLOROPLAST IN THE CELL

Here we see how the quantitative data discussed in the previous section relate to the activities of the cell as a whole and, specifically, what fraction of the capacity for photosynthesis in the chloroplast(s) is actually expressed during photolithotrophic growth. This is significant in that up to 40% of the cell protein, dry mass and non-vacuolar volume is chloroplast (Smillie, 1963; Schotz, 1972; Atkinson, John & Gunning, 1974).

Taking cells growing photolithotrophically (e.g. many *Chlorella* strains growing at 25 °C) with a specific growth rate (μ_m) of 0.1 h^{-1}, with NO_3^- as N source and C as 50% of the dry weight, the rate of CO_2 fixation must be 1.17 μmol (g dry wt)$^{-1}$ s^{-1}, and the rate of $NADPH_2$ and ATP synthesis (for the PCRC, the CO_2 pump, and heterotrophic processes) must be 3.27 and 7 of those units, respectively. If all of the ATP and $NADPH_2$ is generated in the thylakoids, we can compute the thylakoid material required: taking an ATP synthesis rate of 0.1 nmol (cm^2 thylakoid·s)$^{-1}$, an area of 7 m^2 and a mass of 58 mg thylakoid per g dry wt (assuming 825 ng dry weight per cm^2 thylakoid). For *Chlorella*, a typical chlorophyll content is 30 mg (g dry wt)$^{-1}$, corresponding to 124 mg thylakoid (g dry wt)$^{-1}$ or 15 m^2 (g dry wt)$^{-1}$.

Taking RuBP carboxylase–oxygenase, fixation of 1.17 μmol (g dry wt)$^{-1}$s^{-1} CO_2 requires 0.117 μmol enzyme (g dry wt)$^{-1}$ if it is working at its maximum specific reaction rate of 10 μmol CO_2 fixed (μmol enzyme·s)$^{-1}$ at 25 °C (see previous Section). Thus 64 mg of RuBP carboxylase–oxygenase is required per g dry wt. If protein is half of the cell dry weight, RuBP carboxylase–oxygenase must be 13% of the cell protein, with at least 3% of the cell volume occupied by stroma (if the enzyme concentration is 1mM and the wet/dry weight ratio is 4). The measured RuBP carboxylase–oxygenase content of unicellular algae is about 10% of the total cell protein (Rabinowitz et al., 1975).

These figures refer to growth at 25 °C but some strains of *Chlorella* are able to grow with a specific growth rate of 0.3 h^{-1} at 35–40 °C (Eppley, 1972). This can be accounted for by increased specific reaction rate of enzymes without the need to increase the thylakoid area or the RuBP carboxylase–oxygenase content. The estimates of thylakoid and carboxylase–oxygenase content and activity thus suggest that the capacity for ATP and NADPH$_2$ synthesis, and CO_2 fixation, are one or two times that needed for the observed μm. This is supported by the finding that the short-term rate of CO_2 fixation uncoupled from growth in micro-algae can commonly exceed the rate used in balanced growth by about 1.5-fold (Myers, 1970; Raven, 1979).

A final point relates to the capacity for transport across the envelope. With an organic carbon efflux from the chloroplast of 0.8 of the rate of photosynthesis, and an average number of carbon atoms per molecule transported of 3, the carbon efflux from the chloroplast is 0.31 nmol (g dry wt)$^{-1}$s^{-1}. At a flux of 0.1 nmol (cm^2 envelope)$^{-1}$s^{-1}, we thus require 310 cm^2 of envelope area per g dry wt. *Chlorella fusca* has perhaps 4 m^2 of envelope area per g dry wt (Raven, 1980), so the required flux is only some 8 pmol cm^{-2} s^{-1}. A more serious test of the transport capacity of the inner envelope membrane is related to the cytoplasmic ATP requirement during photolithotrophic growth. If all of the ATP requirement of some 2 μmol (g dry wt)$^{-1}$ s^{-1} in the cytoplasm is met from photophosphorylation, the shuttle flux for dihydroxyacetone-*P* to *P*-glycerate (and for the reductant-recycling dicarboxylate porter; Heber, 1974) must be 50 pmol cm^{-2} s^{-1}.

CHLOROPLAST REGULATION

Chloroplast catalytic capacity as a function of trophic conditions

In the last section we saw that the chloroplast can comprise up to 40% of the cell dry mass, protein and lipid of a micro-algal cell. It is thus teleologically reasonable that cells with alternative, heterotrophic growth possibilities should repress the synthesis of plastids when heterotrophic growth is possible, thus relieving themselves of the synthetic liability of the temporarily redundant organelle (Raven, 1976a). The extent to which this repression occurs is variable; it does not occur at all in obligate photolithotrophs, and where it does occur the organism rarely shows an increased μ_m under heterotrophic as compared with photolithotrophic growth conditions. Facultative heterotrophs that do not show plastid repression in the presence of their heterotrophic growth substrates tend to have lower μ_m values for heterotrophic than for photolithotrophic growth (Raven, 1976a). Thus the majority of phototrophic micro-algae seem to 'prefer' growth on CO_2 and light, and can either not grow at all on organic compounds as their sole carbon and energy sources, or can only grow as fast on these compounds as they can photolithotrophically when plastid synthesis is repressed. This contrasts with the continuum of eukaryotic types noted by Raven (1976a) with the organisms occupying the 'no man's land' *not* confined to extreme environments *sensu* Brock (1969) and contrasts with the categories proposed by Whittenbury & Kelly (1977).

The variable extent of heterotrophic growth among eukaryotic microalgae, and the variable degree of repression of plastid development during heterotrophic growth, relates to the factor(s) determining the value of μ_m for phototrophic micro-algae. 'Specialist' heterotrophic micro-eukaryotes can achieve a faster μ_m (Griffin, Timberlake & Cherrey, 1974) than can the micro-algae growing either photolitho- or heterotrophically (Droop, 1974). The data just discussed and those in the last section suggest that μ_m in micro-algae is not *absolutely* determined by the photosynthetic capacity. The constraints on μ_m in eukaryotes may relate to some aspect of the translation mechanism (Maynard Smith, 1969); perhaps the greater minimum cell size in phototrophs, and the larger number of genes which they contain, are causally related to the lower value of μ_m.

A final point concerning repression of chloroplasts relates to the cyanobacteria: here the extent of compartmentation of the photosynthetic apparatus, the capacity for heterotrophic growth and the degree of repression of the photosynthetic apparatus under heterotrophic conditions are all less well developed than in most classes of eukaryotic algae (Carr, 1973; Evans & Carr, 1975). The extent to which the relative lack of co-ordinate repression of the various photosynthetic reactions is related to their lack of spatial co-ordination is not clear.

Regulation by the chloroplast

Another regulatory aspect of the occurrence of photosynthetic catalysts within chloroplasts is the possibility of differential regulation of the reactions occurring within the chloroplasts as opposed to those which occur outside the chloroplast. This differential distribution of catalysts between chloroplasts and cytosol involves both the occurrence of two separate genomes and their associated replication, transcription and translation apparatuses in the two compartments, and the selective transfer into chloroplasts of proteins that are nuclear-coded, but synthesised in the cytoplasm (Gray & Hollick, 1979; Givan, 1979; Freyssinett, 1978). It is important to reiterate that the catalysts we are referring to are membrane porters as well as water-soluble enzymes. Differential enzyme distribution, together with strictly regulated exchanges of many enzyme substrates products and modifiers, permit the chloroplast to bring about reactions under different regulatory controls from those in the remainder of the cell. This does not involve a complete exclusion from the chloroplast of the reactions specific to the 'heterotrophic' pathways (e.g. glycolysis and the oxidative pentose phosphate pathway). In green algae and higher plants at least, the reactions within the chloroplast are subject to controls which generally complement the regulation of photosynthesis (see Raven, 1976a). These controls allow 'heterotrophic' reactions to occur in darkened chloroplasts, but 'photosynthetic' reactions to predominate in the light (Kaiser & Bassham, 1979). Further, it allows the cytoplasm to carry out its reactions in light or dark without direct intervention of such chloroplast regulators as the stromal concentration of H^+ and Mg^{2+} or the reduction of vicinal -SH groups (Werdan, Heldt & Milovancev, 1975; Krause, 1977; Anderson & Avron, 1976). Other, putative messages from the chloroplast presumably regulate mitochondrial processes in the illuminated green cell (Raven, 1976a).

We may contrast these regulatory opportunities with those found in the cells of cyanobacteria in which the soluble enzymes are all in the cytoplasm where they are all subjected to the same regulatory signals (H^+ and Mg^{2+} activity, -SH groups etc.), and furthermore have the photosynthetic and oxidative phosphorylation pathways in the same (thylakoid) membranes. The widespread occurrence of obligate photolithotrophy in the cyanobacteria may relate to the problems of independent regulation of high-capacity systems for photosynthetic and respiratory metabolism.

Extrachloroplastic implications of photolithotrophy

(1) The nucleus contains the majority of the genes for plastid proteins in addition to the 'heterotrophic' genes in photolithotrophs. Possibly the increased number of genes in photolithotrophs which must be transcribed and translated per cell doubling is related to the lower specific growth rates associated with photolithotrophs when compared with specialist heterotrophs (with fewer genes) of otherwise similar organisation (cf. Maynard Smith, 1969). This may apply to prokaryotes as well where a similar difference in μ_m is found.

(2) The capacity for oxidative phosphorylation on a unit cell mass basis is lower than in comparable heterotrophs (Raven, 1976a), although the capacity on a unit mitochondrial basis is high (Raven, 1978). The possible savings in running costs and capital costs by carrying out most of the ATP synthesis with thylakoid membranes is discussed by Raven (1976a) who points out that there may be a decreased exploitation of 'heterotrophic' enzymes that use ATP if they are relatively inactive in the dark.

(3) There is frequently an active influx of HCO_3^- at the plasmalemma, and a corresponding pH-regulating H^+ influx, which is related to the ability of most micro-algae to photosynthesise rapidly at high external pH values. Also many micro-algae grown at low CO_2 levels partially suppress RuBP oxygenase activity. It is not clear whether a further HCO_3^- porter system is required at the inner chloroplast envelope membrane (Raven, 1980; Raven & Glidewell, 1979).

(4) Much of the 'scavenging' of glycolate (formed by unsuppressed RuBP oxygenase or in other ways) occurs outside the chloroplast: in a few microalgae the majority of the PCOC is in micro-bodies, as it is in higher plants; in most microalgae the mitochondria seem to play a much greater role (Raven, 1976a, 1979).

These four considerations generally involve the 'heterotrophic' extra-chloroplastic part of the cell in carrying out *extra* functions when compared with a heterotrophic cell; the oxidative capacity of mitochondria may, however, be *lower* in photolithotrophs than in comparable heterotrophs.

CONCLUSIONS

Can we draw any conclusions as to the way in which chloroplast design in general, and any variations which are found consistently in algae adopting particular ecological strategies, are related to the three design considerations mentioned in the first section (thermodynamic efficiency, useful work output per unit of chloroplast, and safety)? The data available suggest that the thermodynamic efficiency of the overall process (useful chemical products per unit light absorbed) approaches the same maximum value in algae of all taxonomic and ecological types. This apparent immutability of efficiency (as defined above) means that maximising the useful chemical products *per cell* in unit time when irradiance is limiting, is achieved by increasing the chlorophyll content per cell and by optimising the 'light-harvesting geometry' (Kirk, 1976) of the cell.

In algae adapted to higher irradiance in environments in which other factors (e.g. nutrient supply) permit high potential growth rates, the ratio of light-harvesting machinery to the catalysts of quantum conversion and subsequent reactions is decreased. Again, optimising useful chemical output per cell per unit time involves changes in the relative quantities of the various components of the chloroplasts, and the ratio of chloroplast(s) to the rest of the cell.

Thus we see that the great range of taxonomic and ecological categories of eukaryotic micro-algae is achieved largely by altering the quantity of the various components of the chloroplasts, and the ratio of chloroplast material to the rest of the cell, without apparently using changes in either the efficiency or the specific reaction rate of particular photosynthetic processes. This accords with the basic similarity of all eukaryotic chloroplasts.

Turning to the constraints on the molecular design of chloroplasts (which we have seen is so constant), the efficiency of the primary photochemical reactions is subject *inter alia* to an upper limit found for all photochemical reactions (Landsberg, 1977). The stoichiometry of the coupling of redox and adenylate reactions via H^+ is limited, *inter alia*, by the free energy of hydrolysis of ATP *in vivo* and the maximum steady-state transmembrane H^+ free energy difference. It would be very difficult to operate a $1H^+$ ATPase generating ATP with a free energy of hydrolysis *in vivo* of 55 kJ mol^{-1}, since the required transmembrane H^+ free energy difference would (even with a pH difference component of 4 units) involve an electrical potential difference of 310 mV: this potential difference across a lipid bilayer membrane is close to the 'punch through' threshold. We do not yet know enough about the molecular details of photosynthesis to follow the 'design' of chloroplasts to much greater detail than the broad generalisations already made; thus the less overtly adapative features of RuBP carboxylase–oxygenase (low specific reaction rate, O_2 inhibition) may be ascribed to natural selection pressures during the early evolution of the enzyme. By the time that O_2 inhibition became ecologically significant the enzyme had become so entrenched in phototrophic metabolism that its replacement by an O_2-insensitive enzyme was not possible (Lorimer & Andrews, 1973; Andrews & Lorimer, 1978). The 'CO$_2$ concentrating mechanism' (page 190) may represent an indirect attempt to overcome the deficiencies of RuBP carboxylase–oxygenase. Another deleterious effect of O_2 relates to the photo-oxidising tendency of the combination of light, O_2 and chlorophyll in aerobic photosynthesisers; such organisms all possess protective mechanisms which prevent photo-oxidation.

The considerations outlined in this section shows that at least some of the design characteristics of the chloroplasts of eukaryotic micro-algae can be interpreted in terms of the 'efficiency/rate/safety' trinity. However, these interpretations could apply almost equally to the cyanobacterial photosynthetic apparatus, and it is to the regulatory phenomena discussed earlier that the rationale for the eukaryotic organisation of photosynthesis into chloroplasts must be sought. To this appeal to 'regulation' can be added, if the endosymbiotic hypothesis of the origin of eukaryotic cells is accepted, the historical argument that the eukaryote cell received its photosynthetic apparatus 'pre-packaged' in its envelope membranes, one derived from the host and the other from its endosymbionts. By contrast, if a serial involution hypothesis of the origin of eukaryotic cells (e.g. Cavalier-Smith, 1976) is accepted, then the selective advantage of the improved regulatory characteristics of chloroplasts *versus* a dispersed photosynthetic apparatus must have been correspondingly greater.

Finally, it is hoped that the inevitable over-simplification imposed by limitations of space and the desire to present a coherent account of chloroplasts has not produced too glib a set of conclusions, thus obscuring the need for further research at all levels from the molecular to the ecological to reach a more complete understanding of the structure-function relationships of the chloroplasts of micro-organisms.

I wish to acknowledge Dr J. Beardall who has made valuable criticisms of the manuscript.

REFERENCES

ANDERSON, J. M. (1975). The molecular organisation of chloroplast thylakoids. *Biochimica et Biophysica Acta*, **416**, 191–235.

ANDERSON, L. E. & AVRON, M. (1976). Light modulation of enzyme activity in chloroplasts: generation of membrane-bound vicinal–dithiol groups by photosynthetic electron transport. *Plant Physiology*, **57**, 209–13.

ANDREWS, T. J. & LORIMER, G. H. (1978). Photorespiration – still unavoidable? *FEBS Letters*, **90**, 1–9.

ATKINSON, A. W., JR., JOHN, P. C. L. & GUNNING, B. E. S. (1974). The growth and division of the single mitochondrion and other organelles during the cell cycle of *Chlorella*, studies by quantitative sterology and three dimensional reconstruction. *Protoplasma*, **81**, 77–110.

AVRON, M. (1978). Energy transduction in photophosphorylation. *FEBS Letters*, **96**, 223–32.

BADGER, M. R. & ANDREWS, T. J. (1974). Effects of CO_2, O_2 and temperature on a high-affinity form of ribulose disphosphate carboxylase–oxygenase from spinach. *Biochemical and Biophysical Research Communications*, **60**, 204–10.

BADGER, M. R., KAPLAN, A. & BERRY, J. A. (1977). The internal CO_2 pool of *Chlamydomonas reinhardtii*: response to external CO_2. *Yearbook of the Carnegie Institution of Washington*, **76**, 362–6.

BARBER, D. J. & THURMAN, D. A. (1978*a*). Transport of glutamine into isolated pea chloroplasts. *Plant, Cell and Environment*, **1**, 297–303.

BARBER, D. J. & THURMAN, D. A. (1978*b*). Transport of purines and pyrimidines into isolated pea chloroplasts. *Plant, Cell and Environment*, **1**, 305–6.

BEARDALL, J., MUKERJI, D., GLOVER, H. E. & MORRIS, I. (1976). The path of carbon in photosynthesis by marine phytoplankton. *Journal of Phycology*, **12**, 409–17.

BERRY, J. A. & BOWES, G. (1973). Oxygen uptake *in vitro* by ribulose diphosphate carboxylase of *Chlamydomonas reinhardtii*. *Yearbook of the Carnegie Institution of Washington*, **72**, 405–7.

BERRY, J. A., BOYNTON, J., KAPLAN, A. & BADGER, M. R. (1976). Growth and photosynthesis of *Chalmydomonas reinhardtii* as a function of CO_2 concentration. *Yearbook of the Carnegie Institution of Washington*, **75**, 423–32.

BOLD, H. C. & WYNNE, M. J. (1978). *Introduction to the Algae: Structure and Reproduction*. pp. xiv + 706. Englewood Cliffs: Prentice-Hall.

BORST, P. & GRIVELL, L. A. (1978). The mitochondrial genome of yeast. *Cell*, **15**, 705–24.

BROCK, T. D. (1969). Microbial growth under extreme conditions. In *Microbial Growth. Symposium of the Society for General Microbiology*, **19**, ed. P. Meadow & S. J. Pirt, pp. 15–41. Cambridge University Press.

BUTLER, W. L. (1978). Energy distribution in the photochemical apparatus of photosynthesis. *Annual Review of Plant Physiology*, **29**, 345–78.

CARR, N. G. (1973). Metabolic control and autotrophic physiology. In *The Biology of Blue-Green Algae*, ed. N. G. Carr & B. A. Whitton, pp. 39–45. Oxford: Blackwell Scientific Publications.

CAVALIER-SMITH, T. (1976). The origin of nuclei and of eukaryotic cells. *Nature, London*, **256**, 463–8.

COBB, A. H. & ROTT, J. (1978). The carbon fixation characteristics of isolated *Codium fragile* chloroplasts. Chloroplast intactness, the effect of photosynthetic carbon reduction cycle intermediates, and the regulation of ribulose bisphosphate carboxylase *in vivo*. *New Phytologist*, **81**, 527–543.

COOMBS, J. & GREENWOOD, A. D. (1976). Compartmentation of the photosynthetic apparatus. In *The Intact Chloroplast*, ed. J. Barber, pp. 1–51. Amsterdam: Elsevier.

COOPER, J. P. (1975) (ed.). *Photosynthesis and Productivity in Different Environments*, pp. xxiv + 715. Cambridge: Cambridge University Press.

CROFTS, A. R. & WOOD, P. M. (1978). Photosynthetic electron-transport chains of plants and bacteria and their role as proton pumps. In *Current Topics in Bioenergetics*, vol. 7, *Photosynthesis, Part A*, ed. D. Rao Sanadi & L. P. Vernon, pp. 175–244. New York: Academic Press.

CROWTHER, D., MILLS, J. D. & HIND, G. (1979). Proton motive cyclic electron flow around photosystem one in intact chloroplasts. *FEBS Letters*, **98**, 386–90.

DROOP, M. R. (1974). Heterotrophy of carbon. In *Algal Physiology and Biochemistry*, ed. W. D. P. Stewart, pp. 530–59. Oxford: Blackwell Scientific Publications.

DOUCE, R., HOLTZ, R. B. & BENSON, A. A. (1973). Isolation and properties of the envelope of spinach chloroplasts. *Journal of Biological Chemistry*, **248**, 7215–22.

EPPLEY, R. W. (1972). Temperature and phytoplankton growth in the sea. *Fishery Bulletin*, **70**, 1063–85.

EVANS, E. H. & CARR, N. G. (1975). Dark-light transitions with a heterotrophic culture of a blue-green alga. *Biochemical Society Transactions*, **3**, 373–6.

FINDENEGG, G. R. (1978). Analysis of light-dark changes in internal pH of *Scenedesmus*. *Federation of European Societies of Plant Physiologists Meeting, Edinburgh, Abstracts*, pp. 192–3.

FREYSSINET, G. (1978). Determination of the site of synthesis of some *Euglena* cytoplasmic and chloroplast ribosomal proteins. *Experimental Cell Research*, **115**, 207–20.

GIBBS, S. P. (1978). The chloroplasts of *Euglena* may have evolved from symbiotic green algae. *Canadian Journal of Botany*, **56**, 2883–9.

GIBBS, S. P. (1979). The route of entry of cytoplasmically synthesised proteins into chloroplasts of algae possessing chloroplast endoplasmic reticulum. *Journal of Cell Science*, **35**, 253–66.

GIMMLER, H. (1977). Photophosphorylation *in vivo*. In *Encyclopedia of Plant Physiology, New Series*, vol. 5, part 1, *Photosynthetic electron transport and photophosphorylation*, ed. A. Trebst & M. Avron, pp. 448–72. Berlin: Springer-Verlag.

GIMMLER, H., SCHAFER, G. & HEBER, U. (1974). Low permeability of the chloroplast envelope toward cations. In *Proceedings of the Third International Congress on Photosynthesis*, ed. M. Avron, pp. 1381–92. Amsterdam: Elsevier.

GIVAN, A. L. (1979). Ribulose bisphosphate carboxylase from a mutant strain of *Chlamydomonas reinhardii* deficient in chloroplast ribosomes. The absence of both subunits and their pattern of synthesis during enzyme recovery. *Planta* **144**, 271–6.

GIVAN, C. V. & LEECH, R. M. (1971). Biochemical autonomy of higher plant chloroplasts and their synthesis of small molecules. *Biological Reviews*, **46**, 409–28.

GRAY, P. W. & HOLLICK, R. B. (1979). Base composition heterogeneity of *Euglena gracilis* chloroplast DNA. *Biochimica et Biophysica Acta*, **561**, 53–8.

GRIFFIN, D. H., TIMBERLAKE, W. E. & CHERRY, J. C. (1974). Regulation of macromolecular synthesis, colony development and specific growth rate of *Achlya bisexualis* during balanced growth. *Journal of General Microbiology*, **80**, 381–8.

GRIME, J. P. (1979). *Plant Strategies and Vegetation Processes*, pp. xi + 210. Chichester: John Wiley & Sons.

HALLIWELL, B. (1978). The chloroplast at work. A review of modern developments in our understanding of chloroplast metabolism. *Progress in Biophysics and Molecular Biology*, **33**, 1–54.

HAUSKA, G. (1978). Vectorial redox reactions of quinoid compounds and the topography of photosynthetic membranes. In *Photosynthesis 77*, ed. D. O. Hall, J. Coombs & T. W. Goodwin, pp. 185–96. London: The Biochemical Society.

HEBER, U. W. (1974). Metabolite exchange between chloroplast and cytoplasm. *Annual Review of Plant Physiology*, **25**, 393–421.

HELDT, H. W. & SAUER, F. (1971). The inner membrane of the chloroplast envelope as the site of specific metabolite transport. *Biochimica et Biophysica Acta*, **234**, 83–91.

HELLEBUST, J. A. & CRAIGIE, J. S. (1978). (eds.). *Handbook of Phycological Methods: Physiological and Biochemical Methods*. pp. xiv + 512. Cambridge: Cambridge University Press.

HESS, B. (1973). Organisation of glycolysis: oscillatory and stationary control. In *Symposia of the Society for Experimental Biology*, **27**, ed. D. D. Davies, pp. 105–31. Cambridge University Press.

HIGHFIELD, P. E. & ELLIS, R. J. (1978). Synthesis and transport of the small subunit of chloroplast ribulose bisphosphate carboxylase. *Nature, London*, **271**, 420–4.

JAGENDORF, A. T. (1975). Chloroplast membranes and coupling factor conformations. *Federation Proceedings*, **34**, 1718–22.

JENSEN, R. G. & BAHR, J. T. (1977). Ribulose 1,5-bisphosphate carboxylase–oxygenase. *Annual Review of Plant Physiology*, **28**, 379–400.

JUNGE, W. (1977). Membrane potentials in photosynthesis. *Annual Review of Plant Physiology*, **28**, 503–36.

KAISER, W. H. & BASSHAM, J. A. (1979). Carbon metabolism of chloroplasts in the dark: oxidative pentose phosphate pathway versus glycolytic pathway. *Planta*, **164**, 193–200.

KECK, R. W., DILLEY, R. A. & KE, B. (1970). Photochemical characteristics in a soybean mutant. *Plant Physiology*, **46**, 699–704.

KERBY, N. W. & EVANS, L. V. (1978). Isolation and partial characterization of pyrenoids from the brown alga *Pilayella littoralis*. *Planta*, **142**, 91–9.

KIRK, J. T. O. (1976). A theoretical analysis of the contribution of algal cells to the attenuation of light in natural waters. III. Cylindrical and spheroidal cells. *New Phytologist*, **77**, 341–58.

KRAUSE, G. H. (1977). Light-induced movement of magnesium ion in intact chloroplasts. Spectroscopic determination with eniochrome blue SE. *Biochimica et Biophysica Acta*, **460**, 500–10.

LAING, W. A., OGREN, W. L. & HAGEMANN, R. H. (1974). Regulation of soybean net photosynthetic CO_2 fixation by the interaction of CO_2, O_2 and ribulose 1,5 diphosphate carboxylase. *Plant Physiology*, **54**, 768–85.

LANDSBERG, P. T. (1977). A note on the thermodynamics of energy conversion in plants. *Photochemistry and Photobiology*, **26**, 313–14.

LEECH, R. M. & MURPHY, D. J. (1976). The co-operative function of chloroplasts in the biosynthesis of small molecules. In *The Intact Chloroplast*, ed. J. Barber, pp. 365–401. Amsterdam: Elsevier.

LEVINE, R. P. (1969). The analysis of photosynthesis using mutant strains of algae and higher plants. *Annual Review of Plant Physiology*, **20**, 523–40.

LEVINE, R. P. & GOODENOUGH, U. W. (1970). The genetics of photosynthesis and of the chloroplast in *Chlamydomonas reinhardtii*. *Annual Review of Genetics*, **4**, 397–408.

LEWIS, D. H. (1973). The relevance of symbiosis to taxonomy and ecology with particular reference to mutualistic symbioses and the exploitation of marginal habitats. In *Taxonomy and Ecology*, ed. V. H. Heywood, pp. 151–72. London: Academic Press.

LILLEY, R. McC. & WALKER, N. A. (1975). Carbon dioxide assimilation by leaves, isolated chloroplasts and ribulose bisphosphate carboxylase from spinach. *Plant Physiology*, **55**, 1087–92.

LORD, M. J. & BROWN, R. H. (1975). Purification and some properties of *Chlorella fusca* ribulose bisphosphate carboxylase *Plant Physiology*, **55**, 360–4.

LORIMER, G. H. & ANDREWS, T. J. (1973). Plant photorespiration – an inevitable consequence of the existence of atmospheric O_2. *Nature, London*, **243**, 359–60.

MAYNARD SMITH, J. (1969). Limitations on growth rate. in *Microbial Growth. Symposium of the Society for General Microbiology*, **19**, ed. P. Meadow & S. J. Pirt, pp. 1–13. Cambridge University Press.

MACKENDER, R. O. & LEECH, R. H. (1974). The galactolipid, phospholipid and fatty acid composition of the chloroplast envelope membranes of *Vicia faba*. *Plant Physiology*, **53**, 496–502.

MITCHELL, P. (1966). *Chemiosmotic Coupling and Energy Transduction*, pp. vi + 111. Bodmin, England: Glynn Research Limited.

MITCHELL, P. (1970). Membranes of cells and organelles: morphology, transport and metabolism. In *Organization and Control in Prokaryotic and Eukaryotic Cells. Symposium for the Society of General Microbiology*, **20**, ed. H. P. Charles & B. C. J. G. Knight, pp. 121–66. Cambridge University Press.

MITCHELL, P. (1976). Possible molecular mechanisms of the proton motive function of cytochrome systems. *Journal of Theoretical Biology*, **62**, 327–67.

MYERS, J. (1970). Genetic and adaptive physiological characteristics observed in the Chlorella. In *Prediction and Measurement of Photosynthetic Production*, pp. 447–54. Wageningen: PUDOC.

MYERS, J. & GRAHAM, J. R. (1971). The photosynthetic unit of *Chlorella* measured by repetitive short flashes. *Plant Physiology*, **48**, 282–6.

ODUM, H. T. & PINKERTON, R. L. (1955). Time's speed regulator: the optimum efficiency for maximum power output in physical and biological systems. *American Scientist*, **43**, 331–43.

PREZELIN, B. P. & ALBERTE, R. S. (1978). Photosynthetic characteristics and organisation of chlorophyll in marine dinoflagellates. *Proceedings of the National Academy of Sciences, USA*, **75**, 1801–4.

RABINOWITZ, H., REISFELD, A., SAGHER, D. & EDELMAN, W. (1975). Ribulose diphosphate carboxylase from autotrophic *Euglena gracilis*. *Plant Physiology*, **56**, 345–60.

RAVEN, J. A. (1976a). Division of labour between chloroplasts and cytoplasm. In *The Intact Chloroplast*, ed. J. Barber, pp. 403–43. Amsterdam: Elsevier.

RAVEN, J. A. (1976b). Glucose metabolism in *Hydrodictyon africanum* in relation to cell energetics. *New Phytologist*, **76**, 205–12.

RAVEN, J. A. (1978). Photosynthesis in cells and tissues. In *Photosynthesis 77*, ed. D. O. Hall, J. Coombs & T. W. Goodwin, pp. 147–55. London: Biochemical Society.

RAVEN, J. A. (1979). Photosynthesis in algae and cyanobacteria. In *Handbook of Food and Nutrition*, ed. M. Rechcigl Jr. Miami, Florida: CRC Press.

RAVEN, J. A. (1980). Nutrient transport in micro-algae. *Advances in Microbial Physiology*, in press.

RAVEN, J. A. & GLIDEWELL, S. M. (1978). C_4 characteristics of photosynthesis in the C_3 alga *Hydrodictyon africanum*. *Plant, Cell and Environment*, **1**, 185–97.

RAVEN, J. A. & GLIDEWELL, S. M. (1979). Processes limiting photosynthetic conductance. In *Physiological Processes Limiting Plant Productivity*, ed. C. B. Johnson. London: Butterworth, in press.

RAVEN, J. A. & SMITH, F. A. (1976). The evolution of chemiosmotic energy coupling. *Journal of Theoretical Biology*, **57**, 301–12.

SCHOFER, G., HEBER, U. & HELDT, H. W. (1977). Glucose transport into spinach chloroplasts. *Plant Physiology*, **60**, 286–9.

SCHOTZ, F., BATHELT, H., ARNOLD, C-G. & SCHIMMER, O. (1972). Die Architektur und Organization der *Chlamydomonas*-Zelle. Ergebniss der Elektronmikroscopie von Serienschnitten und der darous resultierenden driedimensenel Rekonstruktion. *Protoplasma*, **75**, 229–54.

SINGER, S. J. & NICOLSON, G. (1972). The fluid mosaic model of the structure of cell membranes. *Science*, **175**, 720–31.

SMILLIE, R. M. (1963). Formation and function of soluble proteins in chloroplasts. *Canadian Journal of Botany*, **41**, 123–54.

SMITH, F. A. & RAVEN, J. A. (1978). The evolution of H+ transport and its role in photosynthetic energy transduction. In *Light-transducing Membranes: Structure, Function and Evolution*. ed. D. W. Deamer, pp. 233–51. New York: Academic Press.

THORNBER, J. P. (1975). Chlorophyll-proteins: light-harvesting and reaction centre components in plants. *Annual Review of Plant Physiology*, **26**, 127–58.

WERDAN, K., HELDT, H. W. & MILOVANCEV, M. (1975). The role of pH in the regulation of carbon dioxide fixation in the chloroplast stroma. Studies on CO_2 fixation in the light and dark. *Biochimica et Biophysica Acta*, **396**, 276–92.

WHITTENBURY, R. & KELLY, D. P. (1977). Autotrophy: a conceptual phoenix. In *Symposium of the Society for General Microbiology* 27, ed. B. A. Haddock & W. A. Hamilton, pp. 121–49. Cambridge University Press.

WALL STRUCTURE AND BIOSYNTHESIS IN FUNGI

G. W. GOODAY* AND A. P. J. TRINCI†

*Department of Microbiology, Marischal College,
University of Aberdeen, Aberdeen, AB9 1AS, UK

†Department of Microbiology, Queen Elizabeth College,
Campden Hill, London W8 7AH, UK

INTRODUCTION

Some years ago Castle (1942) observed 'Thus, although the primary wall [of sporangiophores of *Phycomyces blakesleeanus*] is of the greatest physiological interest, we have the least satisfactory knowledge of its fine structure and mode of growth', and again (1953), 'It must be borne in mind that growth of the primary wall – growth in surface area, cell extension – is most likely a radically different process to thickening'. It is unfortunate that workers subsequently interested in wall growth and differentiation have not always taken sufficient note of Castle's perceptive observations. The literature (see Bartnicki-Garcia, 1968) abounds with analyses of the chemical composition of fungal walls, including many from workers who are primarily interested in hyphal morphogenesis. Unfortunately such analyses fail to distinguish between the primary and mature walls of hyphae and thus fail to provide information about the qualitative and quantitative changes which occur as a specific region of the wall is formed and matures. Even today we know comparatively little about the chemical structure or the physical and biochemical properties of the tip wall. It is salutary to recall how recently we appreciated that the tip wall always contains microfibrils (Hunsley & Burnett, 1970) and that vesicles are always associated with tip growth (McClure, Park & Robinson, 1968; Girbardt, 1969; Grove, Bracker & Morré, 1970).

Above all it was the work of Robertson (1959, 1965) which redirected attention to the hyphal tip. Like Castle he recognised that the crucial events involved in hyphal growth and morphogenesis occur in apical areas of primary wall growth rather than in regions of secondary wall growth far removed from the tip. Robertson suggested that hyphal shape is determined by a balance between extension of the tip wall and development of rigidity in the newly formed wall.

The inherent difficulties in studying tip growth should not be underestimated, the main problem being that the tip is only a very small proportion of the total length of a hypha. However, one successful approach to studying tip structure and differentiation is illustrated by the work of Hunsley & Burnett (1970) and Hunsley & Kay (1976). Hyphal growth and morphogenesis has been reviewed by Bartnicki-Garcia (1973), whilst Rosenberger (1976) has reviewed the chemical structure of fungal walls.

Vegetative hyphae increase in length as a result of tip extension. The tapered region of the tip involved in this growth is called the extension zone. The wall of the extension zone forms the primary wall of the hypha and contains chitin or cellulose microfibrils. Secondary wall growth occurs distal to the extension zone and consequently the wall increases in thickness with distance from the tip, eventually forming the mature wall of the hypha (Trinci & Collinge, 1975). Primary wall growth always involves the fusion of vesicles with the tip. Hyphal growth is highly polarised in that although precursors of wall polymers are synthesised in a large volume of cytoplasm they are only incorporated into the primary wall at the hyphal tip. Hyphal growth thus depends upon the ability of moulds to assemble and organise their primary walls very rapidly. Very high rates of linear growth may result from such a polarised system and because they possess it fungi are able to colonise solid substrates much more effectively than non-motile, unicellular micro-organisms (Bull & Trinci, 1977).

WALL EXTENSION

Wall expansion at the tip

In a hypha growing at a linear rate the shape of the extension zone (Fig. 1), its length and the radius of the hypha at the base of the zone are all constant. This constancy is maintained despite very rapid rates of primary wall synthesis. For example, a hypha of *Neurospora crassa* having an extension zone which is only 29 μm long extends at a rate of 38 μm min^{-1} (Steele & Trinci, 1975). The constant shape of the extension zone implies that both circumferential and longitudinal aspects of hyphal extension are patterned in some stable fashion.

Mycelial hyphae have extension zones which vary from about 2–30 μm (Steele & Trinci, 1975), whilst the extension zone of Stage I sporangiophores of *Phycomyces blakesleeanus* is about 2 mm long (Castle, 1958); a Stage I sporangiophore is a giant aerial hypha which has many if not all the characteristics of mycelial hyphae. For each species a direct relationship is observed between extension zone length and extension rate (Trinci & Halford, 1975; Steele & Trinci, 1975).

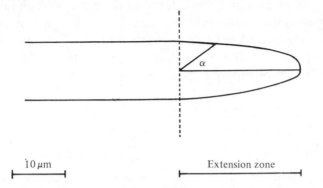

10 μm

Extension zone

Fig. 1. Extension zone of a leading hypha of *Neurospora crassa cot* 3 at the margin of a 'mature' colony grown on solid medium at 25 °C. Similar hyphae had a mean extension zone length of 15μm and a mean extension rate of 15 μm min⁻¹. See also p. 211.

Wall expansion is at its maximum at the hyphal apex and declines rapidly towards the base of the extension zone where the wall becomes 'rigidified', i.e. it no longer expands in surface area. Castle (1958) determined the specific growth rates in wall length and girth for the extension zone of State 1 sporangiophores of *Phycomyces blakesleeanus* and these data may be used to calculate the specific rates of area expansion along the tip (Fig. 2). Provided that the extension zone wall does not vary in thickness this Figure also gives the specific rates of wall synthesis at the tip. Castle's results showed that the allometric

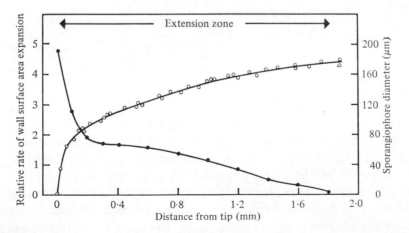

Fig. 2. Hyphal diameter (○) and specific rate of wall area expansion (●) of the extension zone of Stage I sporangiophores of *Phycomyces blakesleeanus*. Calculated from the data of Castle (1958).

coefficient (the ratio of the specific growth rates in the longitudinal and circumferential directions) varied between 1 and 3.5 in the distal half of the extension zone and was below 1 in the proximal half. Thus longitudinal growth was more rapid than circumferential growth in the distal half whilst the reverse was true in the proximal half. Saunders & Trinci (1979) showed that the allometric coefficient is determined by the variation in rate of supply of precursors to the extension zone wall.

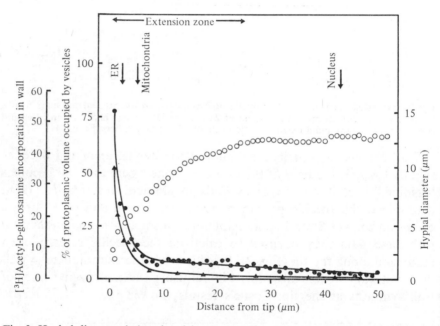

Fig. 3. Hyphal diameter (○) and vesicle concentration (●), expressed as a percentage of the volume occupied by vesicles, of *Neurospora crassa cot* 3 grown at 25 °C (adapted from Collinge & Trinci, 1974). Specific rate of incorporation of *N*-acetylglucosamine (▲) into hyphae of another strain of *N. crassa* grown at 20 °C; expressed as the number of silver grains per 10 μm² resulting from incorporation of *N*-[³H]acetyl-D-glucosamine into the wall for 1 min. The single arrows show where organelles were first observed; ER, endoplasmic reticulum. Redrawn from Gooday (1971).

Unfortunately, it is technically not possible to make direct measurements on mycelial hyphae similar to those made by Castle on sporangiophores. However, rates of wall synthesis at the tip of a vegetative hypha may be estimated by determining the rate of incorporation of a primary wall precursor such as *N*-acetylglucosamine (Fig. 3). The pattern of wall synthesis in mycelial hyphae determined by this method is similar (concave upwards) but not identical to that observed in tips of Stage I sporangiophores of *Phycomyces blakesleeanus* (Fig. 2).

Hyphal tips approximate to half ellipsoids of revolution rather than to hemispheres (Fig. 1; Trinci & Saunders, 1977). Trinci & Saunders suggest that the specific rate of area expansion of the tip wall (i.e. the rate of wall synthesis, since the extension zone wall is of constant thickness) is proportional to the cotangent of the angle (a) between the longitudinal axis of the hypha and the particular point on the tip wall (Fig. 1); it had previously been assumed that tips approximate to hemispheres and that consequently the specific rate of area expansion was proportional to the cosine of this angle (see Burnett, 1976). The cotangent model, which predicts a curve that is concave upwards, provides a better qualitative fit for the experimental data (Figs. 2 and 3) than does the cosine relationship which predicts a curve that is concave downwards (Trinci & Saunders, 1977).

Tip shape and wall elasticity

Saunders & Trinci (1979) showed that it is not possible to account for tip shape assuming the Reinhardt (1892) model of tip growth. In this hypothesis the wall is regarded as a rigid layer, with the expansion of each small area of the surface due to the intussusception of a suitable amount of material in that area itself. However, Saunders & Trinci (1979) showed that variation in the rate of supply of material to the tip wall determined the allometric coefficient, not the shape of the extension zone. Castle (1958) similarly concluded that his observations on tip growth in *Phycomyces blakesleeanus* 'did not support the idea that membrane [= wall] growth is based upon a quantal growth event'. The model proposed by Bartnicki-Garcia (1973) for tip growth of hyphae is based upon such an event but this was intended only to account for wall extension and not to explain how tip shape is determined.

Saunders & Trinci (1979) considered an alternative model of D'Arcy Thompson (1917) in which the wall is considered to be elastic, with the shape determined not by the addition of material but by the wall adjusting itself in response to turgor pressure so as to adopt a shape which minimises the surface energy. They concluded that tip shape is largely determined by wall elasticity. The model of Saunders & Trinci predicts a graph of wall elasticity versus distance from the hyphal apex to be concave upwards and it provides a reasonable qualitative fit to the experimental data of Ahlquist & Gamow (1973) and Ahlquist, Iverson & Jhasman (1975) on the mechanical and plastic extensibility of the intercalary extension zone wall of stage IVb sporangiophores of *Phycomyces blakesleeanus*. Ahlquist *et al.* (1975) found a good qualitative agreement between the observed extensibility of the wall and that predicted by a model which assumed that the wall was composed of stiff chitin microfibrils embedded in a Newtonian viscous fluid. Values in the model for wall thickness and microfibril spacing, overlap and orientation were estimated from experimental data. They observed a deviation between their predicted and experimental values at the tip of the extension zone and suggested that this was caused by some unknown parameters not included in the model.

Ortega (1976) modelled the behaviour of the intercalary extension zone of *Phycomyces blakesleeanus* using the Maxwell linear viscoelastic model which consists of a spring element and a viscous fluid element coupled in series. He found that walls of dead sporangiophores behaved like a solid but that during growth the sporangiophore wall could be modelled as a Newtonian viscous fluid and that the rate of flow (i.e. extension rate) was a function of wall viscosity and the turgor pressure of the protoplast. Like Saunders & Trinci (1979), Ortega suggests that the wall is a single regulatory parameter.

It is clear from Robertson's (1958) results that the elasticity of newly-formed primary wall material decreases rapidly with time, as the tip wall of *Fusarium oxysporum* became 'rigidified' within one minute after an osmotic shock. Unfortunately we do not know what causes this decline in wall elasticity.

Rosenberger (1976) has compared fungal walls, which essentially consist of a network of crystalline microfibrils embedded in an amorphous matrix, with such man-made composites as glass-fibre-reinforced plastic. He observed that such composites possess remarkable strength for their weight (clearly a desirable feature for a cell wall) and that a range of mechanical properties can be achieved with no change of materials. For example, change in the length and cross-section of fibres, in the area the fibres occupy, in the orientation of fibres and in the degree of interaction between fibres and matrix will all produce marked alterations in the behaviour of the composite (Holliday,1966; Kelly, 1966). Such observations may be significant in view of the changes known to occur in the microfibrillar component of fungal walls distal to the tip (see p. 214). Thus the elasticity of fungal walls may vary as a result of changes in the dimensions of the microfibrils. In particular, wall viscosity may increase, as in man-made composites, due to an increase in mean microfibril length. Burnett (1979) also suggests that wall elasticity may decrease during 'maturation' because of an increase in the diameter, number and packing of microfibrils in the wall. In addition he suggests that interlacing and aggregation of microfibrils may be involved in wall rigidification. Microfibril aggregation is observed during wall regeneration by protoplasts (Sietsma *et al.*, 1975; Peberdy,

1979) and during synthesis of chitin *in vitro* (Ruiz-Herrera *et al.*, 1975; Bracker, Ruiz-Herrera & Bartnicki-Garcia, 1976).

Wall extension in *Nitella axillaris* (Metraux & Taiz, 1977) and in higher plant cells (Davies, 1973) is stimulated by acid conditions and inhibited by Ca^{2+} or other ions (Ray & Baker, 1965). A correlation may be made between the relative effectiveness of ions in stimulating the extension of plant walls and their ability to reduce or increase the viscosity of pectin solutions (Smidsrød & Hang, 1972; Dow & Rubery, 1975).

Acid solutions cause the tips of fungal hyphae to burst (Bartnicki-Garcia & Lippman, 1972; Dow & Rubery, 1975). Bartnicki-Garcia & Lippman suggest that this bursting is a consequence of a disturbance in the balance between plasticisation of the tip wall by lytic enzymes and the simultaneous deposition of new wall material in the same zone. Dow & Rubery (1975) argue that the reversible and antagonistic effects of H^+ and Ca^{2+} on the integrity of the tip wall of *Mucor rouxii* are due to a modification of the physical properties of the matrix polymers, particularly mucoran, and not to a cleavage of covalent bonds. They suggest that H^+ ions cause the wall matrix to become less viscous, so that the wall becomes more extensible, whilst Ca^{2+} ions cause the wall matrix to become more viscous, so that the wall becomes more rigid.

The 'acid' effects described above may be significant in view of Turian's (1978) observation that the Spitzenkörper (a group of microvesicles, see p. 222) at the hyphal tip of *Neurospora crassa* has a lower pH than the distal cytoplasm. The 'rigidification' may be correlated with an increase in viscosity of the wall which results from changes in its microfibrillar and/or its matrix components. There is no evidence that the onset of wall rigidity is associated with any appreciable increase in wall thickness (Trinci & Collinge, 1975; Collinge, Miles & Trinci, 1978).

Spiral wall growth

Castle (1942) showed that the wall of the extension zone of Stage I sporangiophores of *Phycomyces blakesleeanus* spirals in a steep left-handed direction with a deviation of about 10° between the spiral axis and the long axis of the sporangiophore. Spiralling is even more pronounced in the intercalary extension zone of Stage IVb sporangiophores of this organism. The rate of rotation of the extension zone wall in Stage IVb sporangiophores is roughly proportional to its rate of elongation, except at the base of the extension zone where a significant amount of rotation occurs without measurable elongation (Ortega, Harris & Gamow, 1974). Spiral wall growth has also been observed in sporangiophores of *Mucor mucedo* (Castle, 1942) and conidiophores of *Aspergillus giganteus* (Trinci & Banbury, 1967). In these aerial hyphae, spiral growth may cause the tip to nutate (Trinci & Banbury, 1967).

Indirect evidence (Madelin, Toomer & Ryan, 1978; Trinci *et al.*, 1979) suggests that there is also a spiral element to wall extension in vegetative hyphae and under certain conditions hyphae may exhibit a helical morphology (Toaze-Soulet *et al.*, 1978). Thus tip rotation may be a universal feature of hyphal extension. This rotation may be related to some spiral structure (the microfibrils?) of the primary wall, but we know neither the nature of this connection nor the origin of the spiral structure.

Spiral wall growth is, of course, not confined to fungi. Even cells of bacteria which normally form straight rods may assume regular helical shapes when treated in certain ways, e.g. with penicillin G or after restarting growth from a stationary phase inoculum (Tilby, 1977). Thus there may be a helical component to wall growth in bacteria.

PRIMARY WALL

Wall structure

Relatively few attempts have been made to determine the structure, chemical composition and the physical and biochemical properties of the tip wall of hyphae. This is regrettable because it is clear that the properties and behaviour of the primary wall are crucial to hyphal morphogenesis (Robertson, 1965). The tips of all hyphae examined have been found to be bounded by a wall made up of microfibrils (Plate 1) embedded in an amorphous matrix; the microfibrils are of chitin or cellulose. They usually have a diameter of 10–25 nm (Table 1) and may be up to 2–3 μm long (Burnett, 1979). Since hyphae of *Neurospora crassa* extend at rates up to about 100 μm min^{-1}, it is obvious that microfibril assembly and deposition must occur extremely rapidly. Further, microfibrils must be synthesised throughout the growth zone since their numbers per unit volume of wall remain constant or increase slightly from tip to base of the extension zone (Burnett, 1979). As mentioned earlier, Burnett (1979) observed an increase in microfibril number, diameter and packing with distance from the tip of hyphae of *N. crassa*. It is clear that microfibrils form the major skeletal element of the wall. However, although microfibrils have a skeletal function in maintaining hyphal shape under normal conditions, they do not necessarily determine tip shape (Katz & Rosenberger, 1971*a*). Microfibrils (chitin or cellulose) also form part of the wall initially produced by regenerating protoplasts (Peberdy, 1979).

The extension zone wall of *Neurospora crassa* (Fig. 4) has a constant thickness of about 50 nm and is composed of an inner layer of chitin microfibrils embedded in protein and an outer layer of amorphous material which may be mainly protein but also contains small amounts of a laminarin-like glucan (R-glucan?) and a glycoprotein; the glycoprotein reticulum as a morphological entity has only been detected subapically. Thus although chitin accounts for only about 10% (w/w) of the gross composition of the walls of hyphae of *N. crassa*, it probably makes up a much larger proportion of the primary wall.

Table 1. *Diameters of microfibrils of chitin and cellulose*

Organism	Polymer	Microfibril diameter (nm)	Reference
Phytophthora palmivora	Cellulose	15–25	Hegnauer & Hohl (1973)
Phytophthora parasitica	Cellulose	11.5[a], 12.8[b]	Hunsley & Burnett (1968)
Neurospora crassa	Chitin	10.4[a], 18.9[b]	Hunsley & Burnett (1968)
Schizophyllum commune	Chitin	9.3[a], 12.0[b]	Hunsley & Burnett (1968)
Phycomyces blakesleeanus	Chitin	15–25	Bergman *et al.* (1969)
Polyporus myllitae	Chitin	10–22	Scurfield (1967)
Lower Phycomycetes		15–25	Aronson & Preston (1960*a*)
Aspergillus nidulans protoplasts	Chitin	20	Gibson, Buckley & Peberdy (1976)
Mucor rouxii chitin synthase	Chitin	12–18	Ruiz-Herrera *et al.* (1975)

[a] At apex; [b] in distal regions.

The observation that the thickness of the primary wall remains approximately constant throughout the extension zone (Trinci & Collinge, 1975) suggests that the gradient in the rate of surface area expansion of the wall is paralleled by gradients in the rates of wall and membrane deposition.

Wessels & Sietsma (1979) suggest that in *Schizophyllum commune* there is an intimate relationship between the synthesis of chitin and R-glucan (a highly branched β-glucan with 1–3 and 1–6 linkages). In *S. commune* chitin is embedded in the R-glucan and may be covalently bound to it. Since these polymers form the inner layer of the mature wall they may, as in *Neurospora crassa*, also form the major part of the primary wall. The branched nature of R-glucan and its widespread occurrence in fungal walls has led to the suggestion that it may fulfil a special function in the wall and in particular that it may be involved in the formation of crosslinks between other wall polymers (Rosenberger, 1976).

The primary wall formed at the tip eventually forms the inner layer of the mature wall (Fig. 4). Observations made using fluorescent conjugated

wheat germ agglutinin (Galun *et al.*, 1976) and the fluorescent brightener, calcofluor (Gull & Trinci, 1974) show that chitin in the primary
wall is more accessible to these compounds than is the chitin in the
mature wall, perhaps because secondary wall material is deposited onto
the outer surface of the primary wall as 'maturation' proceeds below the
tip (Fig. 4).

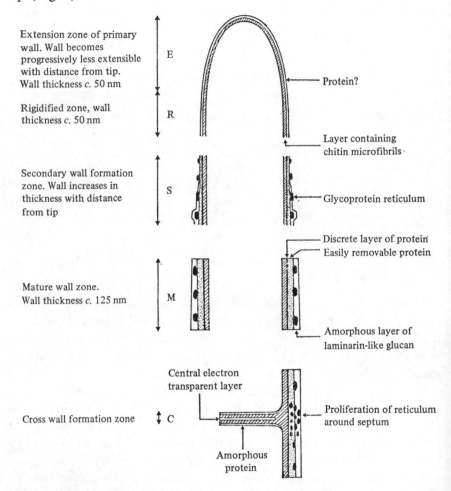

Fig. 4. Diagrammatic representation of the wall and septal structure of a hypha of *Neuro-
spora crassa*; based upon the work of Hunsley & Kay (1976), Hunsley & Gooday (1974) and
Trinci & Collinge (1975). Reproduced with permission from Trinci, A. P. J. (1978). *Science
Progress, Oxford*, **65**, 79–99. © W. Blackwell Scientific Publications Ltd. 1978.

Microfibrial orientation

The orientation of microfibrils in the walls of a number of tip growing
filamentous systems has been studied by Roelofsen and his colleagues

(Roelofsen, 1965). He found that microfibrils deposited on the inner surface of the wall of Stage I sporangiophores of *Phycomyces blakeslee-anus* exhibited a predominantly transverse orientation whilst microfibrils on the outside of these walls had an isotropic (random) orientation. Scurfield (1967) also found that the microfibrils of the inner surface of the hyphal wall of *Polyporus myllitae* showed a strong tendency towards transverse orientation, whilst those on the outer surface showed no preferred orientation (isotropic). Roelofsen & Houwink (1953) proposed the multi-net growth theory to explain the type of wall architecture found at the tips of sporangiophores of *Phycomyces blakesleeanus* and other tip growing filaments. They suggested that new microfibrils are deposited on the inner face of the wall and that these gradually shift outwards through the wall as a result of the deposition of further microfibrils. They envisaged that wall extension caused the orientation of the microfibrils to change with time from being transverse to being more nearly isotropic and eventually to being more nearly axial.

However, Aronson & Preston (1960*b*) found that microfibrils in the inner surface of the wall of *Allomyces macrogynus* were orientated parallel with the long axis of the rhizoid, whilst those in the outer layer were arranged randomly. Similarly, Tokunaga & Bartnicki-Garcia (1971) found a longitudinal orientation of microfibrils in the tip wall of *Phytophthora palmivora*. Burnett (1979) also failed to find evidence for multi-net growth in *Neurospora crassa*. It is possible that multi-net growth only occurs in moulds such as *Phycomyces blakesleeanus* which have very long extension zones and thick primary walls (Plate 6*a*; Peat & Banbury, 1967) composed of several lamellae.

Dickinson (1977) has shown that adhesion of germ tubes of *Puccinia coronata* to a supporting nitrocellulose membrane altered microfibril arrangement and orientation in the wall and caused the hyphae to grow in a zig-zag manner.

SECONDARY WALL

In *Neurospora crassa* the primary wall is about 50 nm thick and this increases below the extension zone until eventually, some 250 μm below the tip, the mature wall attains a thickness of about 125 nm (Fig. 4; Trinci & Collinge, 1975; Hunsley & Kay, 1976). In this organism approximately an hour elapses between deposition of the primary wall in the extension zone and maturation of the wall below the tip.

The primary wall of *Neurospora crassa* has been described above and is shown in Fig. 4, whilst the structure of the mature wall is shown in

Fig. 5. The mature wall of *N. crassa* consists of three co-axially arranged layers. There is an inner layer of chitin, whose thickness remains more or less constant at *c* 20 nm, overlaid by a layer of principally protein-aceous material. This protein layer is in turn overlaid by a glycoprotein

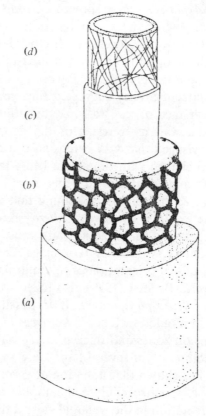

Fig. 5. Diagram of the principal regions of the wall of *Neurospora crassa*. The relative thickness of each layer is to scale but the wall/lumen ratio is not. Layers are (*a*) outer mixed glucans, (*b*) the glycoprotein reticulum with glucans merging into protein, (*c*) princi-pally protein and (*d*) chitin microfibrils embedded in protein. Reproduced with permission from Burnett, J. H. (1976). *Fundamentals of Mycology*, 2nd edn. © Edward Arnold (Publishers) Ltd. 1976.

reticulum and on the outside of the wall there is a layer of mixed α-and β-glucans. Burnett (1979) has emphasised that these regions of the mature wall are not discrete but instead grade into one another. The carbohydrate moiety of the glycoprotein includes glucose, galactose and glucuronate and is probably linked covalently by *O*-glycosyl serine bonds to an uncertain number of peptides. The glycoprotein forms a reticulum in the wall but the network is only recognisable below the tip

(Hunsley & Burnett, 1970); the strands of the reticulum are initially about 11 × 10 nm (tangential:radial dimensions) but increase up to 43 × 35 nm in the mature wall. The outermost layer of the wall is largely composed of a predominantly β1–3 linked glucan of unknown structure with some β1–6 linkages in it. Thus secondary wall growth involves the synthesis of the glycoprotein reticulum and the laminarin-like glucan and these polymers are deposited on the outside of the primary wall. It is not known how precursors are transported through the primary wall to the site of polymer synthesis or how this synthesis is regulated.

That secondary wall formation may not be essential for normal hyphal growth and morphogenesis is suggested by the observation that cultures of *Neurospora crassa* grown on media containing snail juice enzymes formed more or less normal hyphae which had thin walls (Rizvi & Robertson, 1965). Further, Polacheck & Rosenberger (1977) showed that hyphae of an *Aspergillus nidulans* mutant grew normally although their thin walls appeared to lack two outer layers (secondary wall layers?) found on the walls of wild-type hyphae. The mutant failed to produce α 1–3 glucan and melanin and was unable to produce clcistothecia. Melanin is deposited as a protective sheath on the outside of the mature wall of *Aspergillus nidulans* (Rowley & Pirt, 1972). Zonneveld (1972) has shown that *A. nidulans* uses the α1–3 glucan present in the outer layer of its wall (Bull, 1970) as a substrate for endogenous metabolism during carbon starvation and during cleisto-thecial formation. Similarly the R-glucan component of the secondary wall of *Schizophyllum commune* may also be used as a substrate for endogenous metabolism during carbon starvation (Wessels & Sietsma, 1979).

UNBALANCED WALL GROWTH

Under optimal conditions, growth of an asynchronous population of micro-organisms is 'balanced' (Campbell, 1957). During balanced growth, all extensive properties of the culture, such as the amounts of DNA, RNA, protein and wall polymers, increase at the same rate. It is important to distinguish between hyphal walls formed during balanced growth, such as in cultures which are growing exponentially, and hyphal walls formed during unbalanced growth, such as in nutrient-deprived cultures or in cultures treated with an inhibitor. Unbalanced wall growth occurs at the centre of old colonies where very thick walled hyphae may be formed. For example, the mature wall of *Neurospora crassa* is normally about 125 nm thick but walls harvested from the centre of a colony had a mean thickness of about 275 nm (Trinci & Collinge, 1975). Unbalanced wall growth will also occur in nitrogen limited cultures, i.e. in cultures where protein synthesis is restricted by the lack of a nitrogen source, although carbohydrate synthesis can proceed unchecked. Unfortunately the distinction between balanced and unbalanced wall growth has often been ignored by investigators analysing the gross composition of fungal walls. In particular it has often not been appreciated that hyphae required for analyses of wall composition should be harvested from the exponential and not the stationary phase of growth.

When protein synthesis in *Aspergillus nidulans* was inhibited by cycloheximide, Sternlicht, Katz & Rosenberger (1973) found that the culture continued to synthesise all the major components of the wall; five hours after the addition of cycloheximide mean wall thickness had increased from 84 ± 21 nm to 165 ± 61 nm and the wall was thickened evenly along the length of the hypha. Hyphal extension, and hence normal primary wall synthesis, is inhibited by cycloheximide (Katz & Rosenberger, 1971*b*). In such cultures chitin is formed more or less uniformly along the length of the hypha, i.e. its synthesis is depolarised.

L-Sorbose also changes the normal pattern of wall synthesis. Walls from cultures of *Neurospora crassa* grown in the presence of L-sorbose were three times thicker than normal walls (Burnett, 1979). Almost all components of such walls were affected by the treatment and in particular a much thicker layer of chitin was formed in treated than in normal walls (Fig. 6). In contrast the glycoprotein reticulum and microfibril orientation of the wall appeared little affected by the L-sorbose treatment. L-sorbose reduced the rate of hyphal extension in *N. crassa* but not the maximum rate of biomass production (Trinci & Collinge, 1973*a*). Mishra & Tatum (1972) showed that L-sorbose inhibited the activity of $\beta1$–3 glucan synthetase and caused a decrease in the $\beta1$–3 glucan content of the wall.

The primary effect of the anti-fungal antibiotic griseofulvin is that it inhibits mitosis (Gull & Trinci, 1973). However, secondary effects include hyphal 'curling', wall thickening and an inhibition of hyphal extension (Brian, 1960). Griseofulvin treatment appears to have little effect on the chitin layer of *Neurospora crassa* walls (Burnett, 1979). However, only a very thin layer of the laminarin-like glucan was formed under such conditions and the glycoprotein reticulum was abnormal; bundles of several strands varying from 90–225 nm were observed in treated hyphae. Inside the reticulum there was a wide amorphous layer not observed in the wall of normal hyphae (Fig. 6).

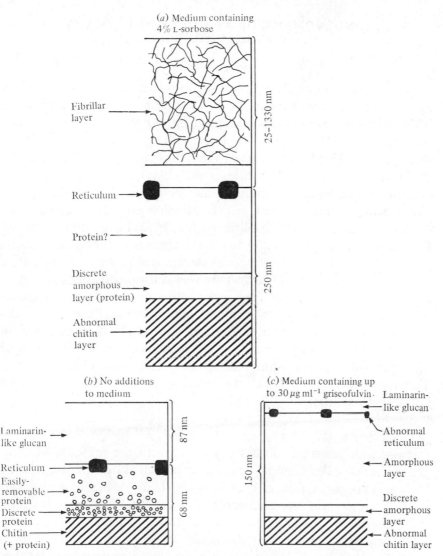

Fig. 6. To illustrate the effects of (*a*) 4% (w/v) L-sorbose and (*c*) up to 30 μg ml⁻¹ griseofulvin on the ultrastructural architecture of the walls of *Neurospora crassa*. (*b*) Composition of the normal wall. Redrawn from Burnett (1979).

In L-sorbose- or griseofulvin-treated hyphae, the chitin microfibrils show no increase, on average, in their dimensions from the tip of the hypha backwards. Thus a variety of treatments which reduce hyphal extension result in abnormal or increased secondary wall growth. It is not known if under conditions of balanced growth there is a mechanism which ultimately limits the extent of wall thickening.

INVOLVEMENT OF VESICLES IN PRIMARY WALL GROWTH

Spitzenkörper

Brunswick (1924) observed an apical iron-haematoxylin positive area at the apical dome of hyphae; he called the structure a Spitzenkörper. This phase-dark structure was subsequently detected in some septate moulds (Girbardt, 1955; 1957) but not all fungi (e.g. *Pythium ultimum*, Grove *et al.*, 1970; McClure *et al.*, 1968). Girbardt (1955; 1957), Grove & Bracker (1970) and McClure *et al.* (1968) found that extension growth stopped when the Spitzenkörper disappeared from the tip and the structure reappeared just before extension growth resumed. However, Howard & Aist (1977) observed that hyphal extension continued for a time in *Fusarium acuminatum* in the absence of the Spitzenkörper, presumably at the expense of vesicles transported to the tip from remote regions of the hypha (see below). It was subsequently shown that in septate fungi the Spitzenkörper represents a specialised region within the apical zone which either contains microvesicles (Plate 2) or lacks any type of vesicle (Grove & Bracker, 1970). The Spitzenkörper in *Neurospora crassa* and *Monilia fructigena* reduces redox dyes, such as methylene blue and neutral red, to their leucobase (Turian, 1978). Turian suggests ways in which this reducing power may be maintained in the Spitzenkörper.

Apical vesicles and primary wall growth

Primary wall growth is always associated with the presence of cytoplasmic vesicles. For example, vesicles are present at the tips of extending hyphae (Plate 2*a*; Grove *et al.*, 1970) and buds (McCully & Bracker, 1972), prior to germ tube emergence (Bracker, 1971; Grove & Bracker, 1978) and during the initiation and growth of branches (Nolan & Bal, 1974) and clamp connections (Girbardt, 1969; Raudaskoski, 1970). It is significant that similar vesicular systems have also been implicated in tip extension of root hairs (Bonnett & Newcomb, 1966), pollen tubes (Rosen, 1964) and filamentous algae (Sievers, 1965; Otto & Brown, 1974). The vesicular mode of growth may be a mechanism ensuring that membrane and wall growth are closely integrated during tip extension.

Vesicles are bounded by a single membrane and sometimes have granular contents of medium electron opacity. Two size classes of vesicles are usually present at the tips of growing hyphae. The large class has a diameter which varies from 100–400 nm whilst the small class varies from 30–100 nm (Table 2). The smaller vesicles are sometimes called microvesicles and are of a size similar to chitosomes (see below); as mentioned above, clusters of microvesicles (the Spitzenkörper) are present at the tips of septate fungi (Plate 2). There may be a positive correlation between mean vesicle diameter and the rate of hyphal extension (Table 2). Vesicles may fuse to form larger vesicles (Grove et al., 1970).

Table 2. *Mean diameter of vesicles at the tips of vegetative hyphae*

Organism	Growth rate (μm h^{-1})	Vesicle diameter (nm)	Reference
Penicillium chrysogenum	68	29 ± 6; 68 ± 10	Collinge *et al.* (1978)
Neurospora crassa spco 9 grown on media with 2% (w/v) L-sorbose	100	59 ± 17	Trinci & Collinge (1973a)
Aspergillus nidulans	c. 300	80 ± 13	Trinci & Morris (1979)
Ascodesmis sphaerospora	500	50–100	Grove (1978)
Pythium ultimum	1000	40–120; 200–300	Grove *et al.* (1970)
Neurospora crassa spco 9	1000	38 ± 3; 123 ± 19	Trinci & Collinge (1973a); Collinge & Trinci (1974)
Phycomyces blakesleeanus	2000–3000	300	Marchant *et al.* (1967)

In *Aspergillus parasiticus* germ tube formation is preceded by the accumulation of cytoplasmic vesicles and as the germ tube increases in length there is an increase in the number of vesicles at its apex (Grove, 1972). Further, there is a positive relationship between the size of the apical cluster of vesicles and the hypha's rate of extension (Grove et al., 1970). Like the Spitzenkörper, apical vesicles disappear when extension growth ceases. As in algae (Sievers, 1967), changes in the direction of hyphal growth may be associated with changes in the location of the apical cluster of vesicles from the central position it normally occupies during linear growth.

In hyphae which are growing linearly, vesicles are present at the apex to the exclusion of most other types of organelles (Fig. 3; Plate 2). Fig. 3 shows the distribution of vesicles and other organelles at the tip of a hypha of *Neurospora crassa cot* 3; the hypha had a diameter of 12.5 μm and was extending at a rate of about 18 μm min^{-1} prior to fixation. About 80% of the volume of the apical 1 μm of the hypha was occupied by vesicles. This value fell sharply to about 10%, 10 μm from the tip, and to about 5% at the base of the extension zone. A similar sharp decrease in vesicle concentration was observed by Girbardt (1969) in *Polystictus versicolor* and by Collinge et al. (1978) in *Penicillium chrysogenum*. In *N. crassa* the shape of the curve, concave upwards, is almost identical with that describing the rate of chitin synthesis in tips of the same organism (Fig. 3). Thus the decline in the rate of wall synthesis from tip to base of the extension zone (Figs. 2 and 3) is reflected by a comparable gradient in vesicle concentration in the tip. This in turn may reflect a gradient in the rate at which vesicles fuse with the protoplasmic membrane. However, as observed by Green (1974), all tip activities would be expected to show a similar steep descending gradient. The *spco* 9 mutant of *N. crassa* has a quite different distribution of vesicles to that shown in Fig. 3 (Collinge & Trinci, 1974).

It is generally assumed that the rapid expansion of the primary wall and protoplasmic membrane which occurs during tip growth is associated with vesicles which fuse with the protoplasmic membrane (Plate 2a). These vesicles liberate their contents into the wall and thus increase the surface area of the protoplasmic membrane and wall to give hyphal elongation. Estimates have been made of the number of vesicles which must fuse with the tip to maintain the observed rate of expansion of the protoplasmic membrane (it is assumed that the protoplasmic membrane only increases in surface area in this manner). Growth of a hypha of *Pythium ultimum* would need the fusion of 1000 large and 9000 small vesicles with the tip every minute (Grove *et al.*, 1970), whilst about 40 000 vesicles would be needed to maintain one minute of extension growth in *Neurospora crassa* (Collinge & Trinci, 1974).

Origin of apical vesicles

Individual particles (vesicles?) were seen at the apex of extending hyphae of *Pythium aphanidermatum*. These particles, which disappeared from view as the hyphae were elongated, were constantly replaced by vesicles that migrated from the sub-apical zone (Grove *et al.*, 1970). A small proportion of the particles did not fuse with the tip but instead returned to the sub-apical zone (Grove & Bracker, 1970; Grove, 1978).

Vesicles are formed either by dictyosomes (Grove *et al.*, 1970) or smooth endoplasmic reticulum (Girbardt, 1969). In *Pythium ultimum*, vesicles budded from the smooth endoplasmic reticulum fuse to form new dictyosome cisternae and as a cisterna traverses the dictyosome its membrane is gradually transformed until it assumes the characteristics of the protoplasmic membrane (Grove *et al.*, 1970). Collinge & Trinci (1974) suggest that most of the vesicles observed at the tip are formed in remote sub-apical regions of the hypha, i.e. they assume that the apical cluster of vesicles is continuously maintained by a supply of vesicles generated in the peripheral growth zone of the hypha (Trinci, 1971). Their hypothesis predicts a very gradual increase in vesicle concentration from the rear of the peripheral growth zone to the base of the extension zone, followed by a very steep increase within the extension zone as the cross sectional area of the tip declines rapidly. They found that they could predict the concentration of vesicles in the extension zone using the equation $V_p = R^2/r^2$, where V_p is the relative vesicle concentration, R is the radius of the hypha below the extension zone and r is the radius at the point in question.

The hypothesis of Collinge & Trinci (1974) has been successfully modelled (Prosser & Trinci, 1979). However, it is as yet only supported by the observation (see above) that particles (vesicles?) migrate from sub-apical to apical regions of hyphae and by indirect evidence. For example, vesicles have been observed in parts of hyphae of *Pythium ultimum* 4–5 nm from the edge of the colony (Grove *et al.*, 1970) and linear arrays of vesicles have been observed in remote, sub-apical regions of hyphae of *Neurospora crassa* (Trinci & Collinge, 1973*b*) and *Volvariella volvacea* (Tanaka & Chang, 1972). These linear arrays of vesicles may reflect active protoplasmic streaming (Tanaka & Chang, 1972).

Vesicle transport

It is assumed that vesicle transport is polarised in a hypha in the direction of the tip. The highly polarised nature of hyphal growth is indicated by the observation that when the sub-apical compartment of a *Coprinus* hypha was excised and cultured separately, a branch was formed just behind the clamp connection which had been closest to the tip in the intact hypha (Yanagita, 1977). A possible basis for this polarity is suggested by the observation that the membrane potential of a *Neurospora crassa* hypha rises from *c.* -25 mV at the tip to -127 ± 10 mV some 7–8 mm further back (Slayman, C. L. & Slayman, C. W., 1962). Roomans & Boekstein (1978) suggest that the high negative potential observed in *N. crassa* away from the tip cannot be explained solely in terms of a diffusion potential based upon local K^+ concentration but must in part be due to an electrogenetic potential resulting from H^+ pumps. Thus either the molecular mechanism underlying the H^+ pump is not present at the tip, or it is present in an inactive state. The presence of this electropotential gradient has been used to advance two hypotheses to account for vesicle transport (Jennings *et al.*, 1974). One hypothesis suggests that the electropotential gradient may generate suffiicent current to cause vesicles to move by eletrophoresis. The alternative hypothesis is that the potential gradient is caused by a decline in the number of K^+ pumps towards the tip, with the result that a standing gradient osmotic flow is established in the hypha and consequently there is bulk flow of liquids and vesicles to the tip. However, Jennings (1979) has raised objections to the latter hypothesis and instead suggests that a flow of water can be directed to the tip by other means, e.g. by the generation of a hydrostatic pressure within the mycelium.

Barstow & Lovett (1974) and Howard & Aist (1977) have suggested that cytoplasmic microtubules (diameter 25 nm) may be involved in guiding vesicles to the tips of hyphae. Howard & Aist based their hypothesis upon the observation that anti-tubulin agents such as methyl-benzimidazole-2-cycarbamate and griseofulvin (Gull & Trinci, 1973; Roobol, Gull & Pogson, 1977) cause hyphae to meander. And Dall'Olio & Vannini (1979) showed that coumarin, a supposed antitubulin agent, caused abnormal deposition of chitin in *Trichophyton metagrophytes*. It has been suggested that endoplasmic reticulum and microtubules in plants may guide the movement of Golgi-derived vesicles to sites in the cell where cellulose microfibrils are deposited (Gunning & Steer, 1975; Palevitz & Helper, 1976). A further suggestion is that microfilaments may be involved in intracellular movement in cells (Stewart & Rogers, 1978). Microfilaments have been observed in animal, plant and fungal cells (Gull, 1975). They occur in bundles in the *spco* mutant of *Neurospora crassa* (A. P. J. Trinci & A. J. Collinge, unpublished). Microfilaments have a diameter of 5–8 nm and may be associated in bundles. They are thought to be composed of polymerised actin. Cytochalasin B affects the synthesis of microfilaments and inhibits cytoplasmic streaming and growth of plant cells.

Apical vesicles other than chitosomes

The properties of apical vesicles have been investigated by histochemical and fractionation techniques. Histochemical methods have revealed that in several fungi the apical vesicles, the maturing face of dictyosomes, and the apical wall contain polysaccharides (Heath, Gay & Greenwood, 1971; Grove *et al.*, 1970; Dargent, 1975). Also, vesicles and the maturing face of dictyosomes of *Achlya ambisexualis* contain cellulase (Nolan & Bal, 1974) and alkaline phosphatase (Dargent, 1975).

Cell fractionation studies have revealed that vesicles (60–150 nm in diameter) from *Saccharomyces cerevisiae* contain exo- and endo-β1–3 glucanase, mannanase and mannan synthase (Cortat, Matile & Wiemken, 1972) whilst a similar study by Meyer, Parish & Hohl (1976) showed that two enzymes (β-glucanase and UDP-transferase) involved in wall growth in *Phytophthora parasitica* were probably transported to the tip in distinct vesicles. Fèvre (1979) found that in *Saprolegnia monoica*, cellulase, β1–3-glucanase, glucan synthase and a β1–4-linked polysaccharide were located in a dictyosome-rich fraction and a fraction which contained apical vesicles associated with endoplasmic reticulum and protoplasmic membrane.

Chitosomes

Chitosomes are intracellular microvesicles (Plate 3) that have been found in homogenates of *Allomyces macrogynus*, *Mucor rouxii*, *Saccharomyces cerevisiae*, *Neurospora crassa*, and *Agaricus bisporus*, a taxonomically diverse range of fungi (Bracker *et al.*, 1976; Bartnicki-Garcia *et al.*, 1979). They contain chitin synthase zymogen, and in the words of Bartnicki-Garcia *et al.* (1978), 'in-vivo chitosomes probably serve as conveyors of chitin synthetase zymogen from its point of synthesis to its final destination at the cell surface'. Chitosome-like microvesicles can be seen in the cytoplasm in electron micrographs of growing fungi, and Bartnicki-Garcia *et al.* (1978) contend that they are true cellular organelles in all chitinous fungi.

Chitosomes are mostly in the range 40–70 nm in diameter. When activated by proteolytic enzymes and incubated in the presence of substrate, chitosomes form chitin microfibrils (Plates 4 and 5). Negative staining reveals two types of chitosome particle, the proctoid type being the most abundant. Stain fails to penetrate the proctoid chitosomes (Plate 3) and reveals little of their internal structure; proctoid chitosomes are highly polymorphic and are the most prevalent type. Negative staining and thin section electron microscopy show that the cycloid type of chitosome has an outer, membrane-like shell which is 6.5–7 nm thick and an interior with a grainy texture (Plate 3). Thus the chitosome 'membrane' is appreciably thinner than the protoplasmic membrane (8–9 nm). Structures resembling chitosomes have also been observed in intact cells (Plates 2 and 6, Collinge & Trinci, 1974; Bracker *et al.*, 1976). These structures, the microvesicles, are found either free in the cytoplasm (Plate 2) or within multivesicular bodies (Plate 6). Multivesicular bodies were implicated in chitin synthesis before the isolation of chitosomes (Marchant, Peat & Banbury, 1967).

Bracker *et al.* (1976) suggest that when chitosomes are incubated in an activation and substrate mixture appropriate for chitin synthesis, the proctoid type of chitosome is transformed into the cycloid type. The latter then undergoes a gradual internal conversion which leads to the formation of a fibroid structure (Plates 4 and 5) containing fibrils 1–2 nm in diameter. The formation of this fibroid structure within the chitosome precedes microfibril formation. The 'membrane' of the chitosome is opened or shed and a slender microfibril emerges and is continuous with the fibroid particle. Fibroid particles are probably formed because chitin synthesis is initially confined to the lumen of the chitosome. Most chitosome shells disintegrate during fibrillogenesis leaving behind naked fibroids. Further incubation leads to the formation of thick chitin microfibrils (Plates 4 and 5) which are made up of slender fibrils in lateral association. The observation that a chitin microfibril can be formed by a single chitosome indicates that a chitosome must contain sufficient chitin synthase to synthesise the many chains which make up a microfibril. Each chitin chain is presumably made by a separate chitin synthase unit and the nascent chains collectively crystallise into a microfibril as they are synthesised. Chitin microfibrils formed *in vitro* are up to 12–18 nm wide and up to 2 μm long (Ruiz-Herrera *et al.*, 1975), i.e. they are the same size as microfibrils formed *in vivo* (Table 1). Chitosomes may arise by self-assembly from subunits located in the cytoplasm or within the lumen of macrovesicles (to form a multivesicular body (Plate 2*b*)) or may be budded from the smooth endoplasmic reticulum (Bartnicki-Garcia *et al.*, 1979).

ENZYMIC CONTROL OF WALL GROWTH

Two aspects must be considered for an understanding of the enzymology of fungal walls – their synthesis and their autolysis. As discussed by Bartnicki-Garcia (1973) and Gooday (1977*a*) these two processes must be co-ordinately regulated for orderly growth to proceed, but our current knowledge is imbalanced as we know much more of the former process than the latter.

Chitin synthesis

Chitin synthase preparations from fungi are by far the best characterised enzyme systems catalysing the formation of an insoluble polysaccharide (Table 3). The ease with which they can be made, and with which they 'rain down' a visible precipitate of chitin microfibrils in a test-tube must make biochemists struggling with cellulose, peptidoglycan and mucopolysaccharides biosyntheses very envious. Our initial microsomal preparations from *Coprinus cinereus* can have activities of over 100 nmol *N*-acetylglucosamine incorporated per min per mg protein.

There is general agreement that the product of the enzyme preparations in Table 3 is macromolecular chitin that can be related directly to the chitin of the cell wall of the fungus. In all cases examined, the product has all of the necessary chemical and physical criteria (for example, see Ruiz-Herrera & Bartnicki-Garcia, 1974).

Table 3. Properties of some chitin synthase preparations from fungi

Organism	Source	'K$_m$' for UDP-GlcNAc (mM)	UDP inhibition 'K$_i$' (mM)	Polyoxin inhibition 'K$_i$' (μM)	Optimum [Mg^{2+}] (mM)	Optimum pH	Reference
Chytridiomycotina							
Allomyces macrogynus	Microsomes	1.2	—	—	—	7.8	Porter & Jaworski (1966)
Blastocladiella emersonii	γ-Particles	1.8–4.1	—	—	20	8	Camargo et al. (1967)
Zygomycotina							
Mucor rouxii	Chitosomes	0.5	0.4	0.6	20	6.5	Ruiz-Herrera et al. (1977)
Phycomyces blakesleeanus	Microsomes	0.6	—	—	20	6.5	Jan (1974)
Mortierella vinacea	Microsomes	1.8	—	—	20	6	Peberdy & Moore (1975)
Ascomycotina and Deuteromycotina							
Saccharomyces cerevisiae	Solubilised from membranes	1.5	1.5	1.0	10	6.2	Duran & Cabib (1978); Keller & Cabib (1971)
Candida albicans	Solubilised from microsomes	0.7–4.5	2.0	2.0	20	8.5	Hardy & Gooday (1978), J. C. Hardy (unpublished)
Piricularia oryzae	Microsomes	3.3	1.4	1.2	—	—	Hori et al. (1971)
Aspergillus nidulans	Microsomes	3.1	—	—	2.5	7.5	Ryder & Peberdy (1977a)
Neurospora crassa	Microsomes	1.4	—	—	—	7.5	Endo, Kakiki & Misato (1970)
Basidiomycotina							
Coprinus cinereus	Solubilised from microsomes	0.9	0.5	3.0	30	8	Gooday & Rousset-Hall (1975)

The $\beta1$–4 linkages between the monomers of N-acetylglucosamine in their chain forms result in the single chitin chain being linear, a helix with two residues per turn stabilised by hydrogen bonds between the hydroxyl group at C-3 of one residue and the ring-oxygen of the adjacent residue. Such chains then stack on top of each other, stabilised by hydrogen bonds between the acetyl groups and the secondary amine groups. Current interpretations of X-ray data suggest that the α-chitin (the form that occurs in preparations of fungal walls) has an antiparallel structure, i.e. adjacent stacks of chains run in opposite directions (in contrast to β-chitin, in which adjacent chains run in the same direction) (Rudall & Kenchington, 1973).

Consideration of the biophysics of the α-chitin system leads to two factors that must be taken into account in order to understand chitin biosynthesis *in vivo*: (*a*) is the crystalline structure of the chitin microfibril (i.e. its dehydrated unit cell) formed directly, or does it 'crystallise' from a nascent, relatively disorganised macromolecular precursor; and (*b*) if the α-chitin unit cell is unequivocally of antiparallel units, how and when does this state occur? E. Cabib (personal communication) has found that chitin, newly-formed by the chitin synthase of *Saccharomyces cerevisiae*, was much more susceptible to attack by an endochitinase than older chitin, which suggests that the newly-formed material undergoes some change to make it more resistant to enzymic lysis. Colvin & Leppard (1977) suggest that the formation of bacterial cellulose microfibrils involves a dehydration of nascent polysaccharide chains, and such a process with chitin would be expected to increase its resistance to chitinase. W. H. Leith & G. W. Gooday (unpublished), working with a chitin-synthesising system from *Mucor mucedo*, found that the chitin formed by a membrane fraction of the fungus was rapidly degraded by an endogenous chitinase (see below). This process was not inhibited by exogenous α-chitin, which might be expected to compete with the newly-formed chitin as a substrate, and so again it seems that the immediate enzyme product exists for a short time in a form readily accessible to these chitinases. The chitin microfibrils, seen in electron micrographs to be directly attached to particles in incubations of enzyme preparations from a range of fungi (Bartnicki-Garcia *et al.*, 1978; Mills & Cantino, 1978), could be a result of crystallisation during preparation, or could indicate that the process of crystallisation of the nascent chains is very rapid. If such a process is taking place, it could directly involve the concomitant formation of the antiparallel arrangement of α-chitin by the folding-back of the chains on themselves.

The enzyme preparations listed in Table 3 come from a wide taxonomic range of fungi. Their properties in general are remarkably similar, suggesting a high degree of genetic conservation during evolution. Perhaps the constraint of having to make a rigid insoluble product may mean that very little variation in enzymic structure and activity is possible. Thus it is to the control of the enzyme activity that we must look for an understanding of the different arrangements of chitin synthesis that we find in different fungi.

The K_m values for UDP-N-acetylglucosamine are all of the order of 1 mM (Table 3). Enzyme activity with respect to substrate concentration is often sigmoidal, exhibiting allosteric kinetics (de Rousset-Hall & Gooday, 1975; Gooday, 1977b). Some enzyme preparations, however, yield linear Lineweaver–Burk plots (Duran & Cabib, 1978). Estimates of the concentrations of UDP-N-acetylglucosamine in fungi can be made from analyses of the total content and from estimates of the accompanying water content *in vivo* (Table 4). These estimates are minimum values as they take no account of any compartmentation of the metabolites that might occur. The values, from 0.4–3.8 mM, are similar to the K_m values for chitin synthase (Table 3), indicating that substrate availability cannot be a limiting factor for enzyme activity, unless substrate and enzyme are physically separated (as they seem to be in the case of the naked zoospore of *Blastocladiella emersonii* (Selitrennikoff, Allin & Sonneborn, 1976)).

All preparations are strongly inhibited by the antibiotic polyoxin, which is a competitive inhibitor of UDP-N-acetylglucosamine (Table 3).

The enzyme reaction is:

$$\text{UDP-GlcNAc} + (\text{GlcNAc})_n \longrightarrow (\text{GlcNAc})_{n+1} + \text{UDP}$$

There is no knowledge, however, of how the polymerisation is initiated. If, as is likely (Gooday & de Rousset-Hall, 1975), the chain elongates from its non-reducing end, then from somewhere such a primary acceptor must be provided. Impure enzyme preparations might carry along fragments of chitin from the walls to act as primers, but these have not been detected in purified solubilised preparations (G. W. Gooday, unpublished). The monomer of chitin, N-acetylglucosamine, activates chitin synthase but is not required for activity. It is not incorporated into the product (Gooday & de Rousset-Hall, 1975; Ruiz-Herrera, Lopez-Romero & Bartnicki-Garcia, 1977), except by the enzyme preparation from the naked zoospore of *Blastocladiella emersonii* (Camargo *et al.*, 1967). It may be that, as suggested for glycogen and dextran biosyntheses (Whelan, 1976), the enzyme itself could act as a

Table 4. *Minimum concentrations (mM) of uridine nucleotides in fungal cells*[a]

Organism	Source	UDP-GlcNAc	UDP-Glc	UTP	UDP	UMP	Reference
Chytridiomycotina							
Blastocladiella emersonii	Zoospores	0.5	—	0.1	0.1	0.2	D. R. Sonneborn &
	Cells, 6 h growth	1.8	—	0.8	0.4	1.5	C. P. Selitrennikoff
	Cells 1 h later, in sporulation medium	3.8	—	0.4	0.2	1.5	(personal communication)
Ascomycotina							
Neurospora crassa	Mycelium, 15 h growth	1.5	2.6	—	0.2	3.3	Slayman (1973)
	Mycelium, 15 h growth with 1% 2-deoxyglucose	0.2	0.1	—	2.0	0.2	
	Mycelium, 24 h growth	0.4	—	—	—	—	Schmit, Edson & Brody (1975)
	Conidia	0.4	—	—	—	—	
Basidiomycotina							
Agaricus bisporus	Mushroom	0.5	—	—	—	—	Mouri, Hashida & Shiga (1971)

[a]Calculated where necessary from authors' values by estimating total water content of fungus assuming dry weight as 5% of fresh weight. Quoted to nearest 0.1 mM.

protein primer to initiate the biosynthesis of the polysaccharide chain via a covalent enzyme–substrate complex.

Chitin-synthesising preparations of γ-particles from the zoospores of *Blastocladiella emersonii* also provide the only evidence of a lipid intermediate in chitin synthesis. This compound, a glycosyl diglyceride, was produced following incubation of the γ-particles with radioactively labelled UDP-*N*-acetylglucosamine. When extracted from the incubation and added to γ-particles with unlabelled substrate, radioactive chitin was formed (Mills & Cantino, 1978; Cantino & Mills, 1979). In no case, however, has a glycosylated polyprenol been found as intermediate in chitin biosynthesis, as has been described for example for mannan-glycoprotein biosynthesis in yeast (Lehle & Tanner, 1975).

As each successive unit in chitin is inverted compared to its neighbour, it may be that the active site of chitin synthase contains two substrate-binding sites so that two sugar units are added simultaneously. This would avoid the necessity of a mechanism to rotate either enzyme or product through 180° after each reaction. Reports of diacetylchitobiose being an enzyme product can probably be ascribed to chitinase activities in the preparations (Ruiz-Herrera *et al.*, 1977; W. H. Leith & G. W. Gooday, unpublished).

Chitin synthase requires Mg^{2+} (or a similar divalent cation, but Mg^{2+} is probably active *in vivo*) for activity, the optimum being at about 10–20 mM (Table 3). As discussed by Gooday (1979), availability of this cation could be a controlling factor for enzyme activity *in vivo*. UDP is inhibitory to chitin synthase (Table 3). Enzyme preparations from *Coprinus cinereus* have a uridine diphosphatase activity, which could regulate the rate of chitin synthesis by breaking down the UDP as it is formed to the much less inhibitory UMP (de Rousset-Hall & Gooday, 1975; Gooday, 1979). Values in Table 4 show that for *Blastocladiella emersonii* and *Neurospora crassa* that are actively synthesising chitin the ratio of UDP to UMP is low, so perhaps in these cases there is also an active uridine diphosphatase. The resulting UMP could then be recycled utilising two molecules of ATP to give UTP, the substrate for the synthesis of UDP-*N*-acetylglucosamine. In *N. crassa*, in the presence of the growth inhibitor, 2-deoxyglucose, there is a marked accumulation of UDP (Table 4).

Lopez-Romero, Ruiz-Herrera & Bartnicki-Garcia (1978) described a protein, mol. wt. about 17 500, that strongly inhibited chitin synthase from *Mucor rouxii*. It acted as a competitive inhibitor, with a K_i value of 0.63 μM.

It is now clear that a very important aspect of the regulation of chitin synthase in growing yeast cells and hyphae is the phenomenon of zymogenicity. This was described first by Cabib and coworkers for *Saccharomyces* species, and has now been observed for preparations from a wide variety of fungi (Cabib, Duran & Bowers, 1979). In nearly all cases, preparations of chitin synthase consist partly of active enzyme, and partly of inactive enzyme (zymogen) that can be activated by addition of proteolytic enzymes or incubation to allow endogenous proteolytic enzymes to work. The ratio of active enzyme to zymogen varies greatly according to source and type of preparation, and the activities of different proteases differ for different preparations. For the yeast/mould dimorphic fungi *Mucor rouxii* and *Candida albicans*, preparations from mycelium contained a much higher proportion of active enzyme than preparations from yeast cells (Ruiz-Herrera & Bartnicki-Garcia, 1976; Braun & Calderone, 1978; Hardy & Gooday, 1978). For *M. rouxii*, Ruiz-Herrera & Bartnicki-Garcia interpret this as reflecting a higher activity of endogenous protease in the mycelium; for *Candida albicans*, Braun & Calderone implicate this in the observed higher chitin content of the mycelial walls. Archer (1977) found that a chitin synthase preparation from protoplasts from hyphal tips of *Aspergillus fumigatus* was not activated by trypsin treatment, whereas a preparation from homogenised whole mycelium showed a four-fold stimulation in activity; Issac, Ryder & Peberdy (1978) found that 'early' protoplasts, from the hyphal tips of *Aspergillus nidulans*, had a higher proportion of active enzyme to zymogen than 'late' protoplasts, from sub-apical regions of the hyphae. For both fungi, the authors suggest that the active form is predominantly at the hyphal tips, while the zymogenic form is present along the length of the hypha.

The exact molecular mechanism of proteolytic activation of the zymogen remains unclear. It may be a specific cleavage of a precursor protein at one site, or it may be a less specific 'unmasking'. The identities of the endogenous proteases that presumably activate the zymogen *in vivo* remain to be clarified. In *Saccharomyces cerevisiae*, Hasilik & Holzer (1973) and Ulane & Cabib (1976) suggest that the chitin synthase activator is proteinase B but Wolf & Ehmann (1978) describe a proteinase B-deficient mutant that shows a normal pattern of growth and budding. J. M. Campbell & J. F. Peberdy (personal communication) have characterised a neutral protease that strongly activates chitin synthase zymogen from *Aspergillus nidulans*. The activity of this protease was highest during active growth, when synthesis would be highest. In *Mucor rouxii*, Sentandreu & Ruiz-Herrera (1978) suggest that the activator is an acid protease. The yeast proteinase was shown to be concentrated in vacuoles in the cell (Cabib, Ulane & Bowers, 1973), i.e. sequestered away from the membrane-bound chitin synthase zymogen, and Cabib *et al.* (1979) discuss how the migration of the vacuole to the site destined for synthesis and release of the proteinase would result in localised activation of the enzyme.

Ulane & Cabib (1974) have characterised a specific inhibitor of the activating proteinase. This inhibitor, a small protein mol. wt. about 8500, was present in the cytoplasm, but formed a tight one-to-one complex with the proteinase. Cabib *et al.* (1979) describe current views on the sites and interactions between zymogen, activator, and inhibitor of activator.

The proteolytic activating factors have a two-fold action on the zymogens: they rapidly activate them, and then slowly inactivate them (for example, Hasilik, 1974). Both processes appear to be irreversible, the latter presumably leading to degradation of the enzyme. In *Mucor rouxii*, Ruiz-Herrera & Bartnicki-Garcia (1976) found the chitin synthase preparations from mycelium to be largely as active enzyme, which was unstable, rapidly losing activity on storage, whereas the preparations from yeast cells were largely as zymogen but were much more stable. They attribute these differences to a high activity of activating protease in the mycelium, and a low activity in the yeast.

Chitin synthase activity can be obtained in a variety of different forms, according to preparation and source. Dating from the first report of Glaser & Brown (1957), the earlier preparations were particulate, usually as microsomal fractions, but often with activity associated with larger cell debris, such as fractions enriched in cell walls (McMurrough, Flores-Carreon & Bartnicki-Garcia, 1971). Since then, preparations have been more precise in three different ways: firstly, by solubilisation the enzyme has been substantially purified by standard techniques; secondly, techniques for producing plasma membrane preparations have been used to locate enzyme activity; thirdly, the intracellular particles containing enzyme activity, chitosomes, have been prepared (see above).

Gooday and de Rousset-Hall (1975) and de Rousset-Hall & Gooday (1975) described the solubilisation of chitin synthase from microsomal preparations from the stalks of fruiting bodies of *Coprinus cinereus* by treatment with a cold suspension (1 %, w/v) of digitonin. The resultant enzyme preparation has been characterised as to molecular size (D. J. Adams, A. de Rousset-Hall & G. W. Gooday, unpublished). On gel exclusion chromatography columns the enzyme appeared as a band of mol. wt. $> 10^6$ in low salt conditions, but predominantly as about 1.5×10^5 in high salt conditions. On a polyacrylamide gradient gel

(2–16%), enzyme activity coincided with two major protein bands corresponding to mol. wts. of 1.1×10^6 and 6.5×10^5, while on a poly-acrylamide disc electrophoresis gel it coincided with the major protein band, also staining for lipid, at 1.2×10^5. On a continuous 5–20% sucrose gradient, enzyme activity occurred as three distinct fractions. These results are consistent with the enzyme being a lipoprotein that can readily aggregate. These partially purified enzyme preparations from C. cinereus showed the same enzymic activities and properties as those of the original microsomal preparations, and showed no requirement for any extra factors. The digitonin-solubilised enzyme from Saccharomyces cerevisiae, after preliminary purification by gel exclusion chromatography, showed a requirement for phospholipids, especially phosphatidylserine or lysophosphatidylserine (Duran & Cabib, 1978). Conversely, treatment with phospholipase or with unsaturated fatty acids inhibited activity, as they did for the particulate preparation from Mucor rouxii (Lyr & Seyd, 1978). Duran & Cabib suggest that this points to the importance of specific hydrophobic interactions for activity of the enzyme.

Reports for a number of fungi, including Phycomyces blakesleeanus (Jan 1974), Aspergillus fumigatus (Archer, 1977), Candida albicans (Braun & Calderone, 1978) and Schizophyllum commune (Wessels & Sietsma, 1979) show that active chitin synthase is predominantly located in the plasma membrane fraction of cell homogenates or protoplast lysates. Here the enzyme would be accessible to substrates and effectors from cytoplasm, and as an integral protein spanning the membrane, could 'feed out' the growing chitin chain to the wall. An organised aggregate of enzyme monomers could in this way give rise directly to the microfibrils in the wall. For Saccharomyces cerevisiae, Duran, Bowers & Cabib (1975) showed that most of the chitin synthase zymogen was also associated with the plasma membrane fraction. By using autoradiography and fluorescent brightener to locate the chitinous product of enzyme activity, Cabib & Bowers (1979) confirmed this finding, and demonstrated that the zymogen was not localised in any one part of the cell membrane, but was generally dispersed over it.

Chitosomes were originally found by Ruiz-Herrera et al. (1975) in the supernatants of enzyme preparations that had been treated with substrate UDP-N-acetylglucosamine and activator N-acetyl-glucosamine in the cold (Ruiz-Herrera & Bartnicki-Garcia, 1974). This treatment presumably somehow released the chitosome from association with other components in the cell homogenate. The chitosomes may also be prepared directly by differential centrifugation to remove large components, followed by molecular exclusion chromatography, treatment of resultant 'mini-organelle' components with ribonuclease to remove ribosomes, and density gradient centrifugation. The zymogen of the chitosome can then be activated by a proteolytic enzyme, and chitin microfibrils will then be made, emanating from the chitosome, on incubation with substrate (Bartnicki-Garcia et al., 1978). Treatment of chitosomes with digitonin (0.5%, w/v) led to dissociation of the chitosomes to 16 S subunits (mol. wt. about 5×10^5) which then reassociated to give structures very similar to the original chitosomes when the digitonin was removed (Bartnicki-Garcia et al. (1979). The relationship between these 16 S subunits and the digitonin-solubilised preparations from Coprinus cinereus and Saccharomyces cerevisiae (discussed above) remains to be investigated.

Thus chitin synthase occurs in at least three states in the fungal cell: as zymogen in chitosomes; as zymogen in the plasma membrane; and as active enzyme in the plasma membrane. The relative proportions in these three states will vary with fungus and state of growth. Thus Bartnicki-Garcia et al. (1978) showed a much greater recovery of chitosomes compared with total activity from yeast cells than from mycelium of Mucor rouxii; and as reported above, protoplasts from hyphal tips of Aspergillus species had a much higher ratio of active enzyme to zymogen than those from sub-apical regions of hyphae (Archer, 1977; Isaac et al., 1978). Technically it is difficult to get good estimates of such ratios, because of the likely presence of the endogenous protease systems during extraction.

Cabib et al. (1979) present the model of localised conversion of zymogen to active enzyme, regulated by the protease and its inhibitor, as a means of precisely controlling chitin deposition in space and time to give rise to the septa formed at budding of Saccharomyces species. Such a model, incorporating the chitosomes, can be proposed to control the deposition of chitin in the growing apex of a hypha: chitosomes are produced sub-apically, perhaps on the endoplasmic reticulum; they migrate to the apex; they fuse with the plasma membrane; some of them are activated by a specific protease; apical chitin is produced, allowing the hypha to extend; and the active enzyme in the membrane is degraded by the protease as the membrane becomes sub-apical. This is consistent with the very marked apical deposition of chitin (Gooday, 1977a).

Sub-apical regions of hyphae are capable of chitin deposition, at side branch formation, at septum formation (Hunsley & Gooday, 1974), and uniformly along the wall, as shown by an increase in the size of chitin

microfibrils (Hunsley & Burnett, 1968) and by the action of cycloheximide on *Aspergillus nidulans* (Sternlicht *et al.*, 1973). As protoplasts
from lateral walls contain predominantly zymogen (see above), these
phenomena can be explained by localised (in the case of septa and
branches) or generalised (in the case of cycloheximide action) activation
of this zymogen, which is present uniformly through the lateral walls.
It could have arisen by apical deposition and escaped activation at the
apex, or it could have arisen by sub-apical insertion of chitosomes.

Ryder & Peberdy (1977*b*) found chitin synthase activity in conidiospores of *Aspergillus nidulans*. The specific activity however was much
less than that of mycelium, and the total activity of spores declined on
storage at 4 °C or ageing of the parent culture. Van Laere & Carlier
(1978) found the enzyme to be nearly absent in dormant sporangiophores of *Phycomyces blakesleeanus*. From experiments with inhibitors they concluded that the subsequent appearance of chitin synthase
on germination was probably dependent on protein synthesis on newly
formed mRNA.

The exception that proves the rule, as far as site and regulation of
chitin synthase are involved, is chitin synthesis in the naked zoospore
of *Blastocladiella emersonii*. As well as its action apparently involving a
lipid intermediate (see above) the enzyme itself is located within the
special intracellular organelles of this cell, the γ-particles (which are
quite distinct from chitosomes, being much larger and containing many
more components) (Myers & Cantino, 1974). When triggered, the
zoospore very rapidly settles and forms a chitin-rich wall. This process,
involving an increase in total hexosamine of at least an order of magnitude, is not inhibited by cycloheximide (Selitrennikoff *et al.*, 1976).
Cantino & Mills (1979) suggest a model whereby the chitin synthase
and the lipid intermediate are stored in the γ-particle, away from the
UDP-*N*-acetylglucosamine in the cytoplasm, but on triggering, the
γ-particle releases enzyme and lipid in vesicles to the plasma membrane,
so that chitin synthesis commences. Selitrennikoff *et al.* (1976) suggest
that in the zoospore, the UDP-*N*-acetylglucosamine concentration (see
Table 4) is kept constant (at about one tenth of the substrate needed for
total chitin biosynthesis) by its end-product inhibition of hexosamine
biosynthesis, which would then rapidly be relieved by the initiation of
chitin synthesis.

Synthesis of other wall polysaccharides

Many fungal walls contain glucans, often in large amounts, such as
$\beta1$–3 and $\beta1$–4 glucans in Oomycotina, $\beta1$–3, $\beta1$–6 and $\alpha1$–3 glucans in

Ascomycotina and Basidiomycotina. Amongst the Oomycotina, Wang & Bartnicki-Garcia (1976) characterised a cell-free preparation from *Phytophthora cinnamomi* that gave fibrillar β1–3 glucan as sole product, when incubated with UDP-glucose and Mg^{2+}. Other nucleotide sugars were not substrates. Fèvre & Dumas (1977) described a cell-free preparation from *Saprolegnia monoica* which produced an alkali-insoluble glucan when incubated with UDP-glucose. They identified the product as a β1–4 glucan, because it was hydrolysed by cellulase to cellobiose. Density gradient centrifugation of cytoplasmic particles showed that the major part of the enzyme activity with the highest specific activity occurred in the fraction containing Golgi dictyosomes and vesicles, and Fèvre & Dumas (1977) suggest that this represents enzyme being transported to the cell surface.

Amongst the Ascomycotina, Mishra & Tatum (1972) characterised a β1–3-glucan synthase activity and an α1–4-glucan synthase activity from *Neurospora crassa*. The former was associated with a cell-wall preparation, and the latter with a membrane preparation, and they suggested that both activities were associated with cell-wall biosynthesis. Both used UDP-glucose as substrate. Cells of *Saccharomyces cerevisiae*, made permeable to substrates by treatment with organic solvents, were able to utilise UDP-glucose to synthesise wall glucans, chiefly β1–3-glucan (Sentandreu, Elorza & Villanueva, 1975). Lopez-Romero & Ruiz-Herrera (1977, 1978) have made membrane preparations from this yeast that synthesise a β1–3-glucan with a proportion of β1–6 glycosidic bonds from UDP-glucose. The K_m value for UDP-glucose was 0.12 mM, the enzyme activity was stimulated by Mg^{2+}, and was competitively inhibited by UDP, with a K_i of 0.45 mM.

Thus, from these different fungi, preparations have been obtained that synthesise β1–3, β1–4, β1–6 and α1–4 linkages, all from UDP-glucose. As with UDP-N-acetylglucosamine, fungi contain a high concentration of UDP-glucose (Table 4), so that a problem to be solved is how synthesis is directed towards any one particular glucan in space and time.

Mannans, cross-linked to proteins, are major components of the *Saccharomyces* species and of related fungi such as *Candida albicans*. Considerable attention has been given to their biosynthesis (Ballou, 1976), particularly from the point of view of the involvement of poly-prenol lipid intermediates (Lehle & Tanner, 1975). Marriott (1977) showed that purified plasma membranes from *C. albicans* incorporated mannose from GDP-mannose to give a mannan-protein. The mannan-proteins of the plasma membranes were very similar to those found in purified wall preparations, suggesting that after formation in the membrane they passed out into the wall.

Role of lytic enzymes

The fungal wall, once formed, shows little evidence of turnover (Polacheck & Rosenberger, 1977; Sentandreu *et al.*, 1975). The fungal autolysins instead may be more concerned with the process of wall formation (Rosenberger, 1979). As the wall components such as cellulose and chitin microfibrils occur as mechanically intact layers over the hyphal apex, insertion of new material must involve some loosening of these. Various lines of evidence now implicate lytic enzymes in this process.

Kritzman, Chet & Henis (1978) detected localisations of β1–3 glucanase in hyphal preparations of *Sclerotium rolfsii* by indirect fluorescent antibody staining. The enzyme was just where wall synthesis was occurring, at the hyphal tips, clamp connections, side branch initiation sites, and new septa.

Preparations of chitin synthase from cultures of *Mucor mucedo* grown in Petri dishes on agar contain a strong membrane-bound chitinase activity (W. H. Leith & G. W. Gooday, unpublished). The product on incubating a microsomal preparation with UDP-*N*-acetylglucosamine is not chitin, but diacetylchitobiose. This membrane-bound lytic activity is intimately integrated with the synthase activity, as it does not effect chitin formation by microsomes of *Coprinus cinereus* when preparations from the two fungi are mixed. This chitinase activity may thus represent an autolysin active precisely at the site of wall synthesis, nicking chitin chains to allow insertion of new microfibrils into the wall.

Polacheck & Rosenberger (1975, 1978) have used pulse-chase labelling of walls of *Aspergillus nidulans* to detect the activity of wall-bound autolysins. Apical walls were better than lateral walls as substrate. Six classes of autolysin were detected as wall-bound enzymes, chitinase, β1–3-glucanase, β-glucosidase, α-glucosidase, β-*N*-acetylglucoaminidase, and protease. These bound enzymes could be released by incubation of the wall preparations (self-digestion), or by treatment with a cationic detergent, but once released they did not reassociate with the walls. Polachek & Rosenberger (1978) suggest that the enzymes are trapped in the wall in an inactive state, so that their localised activation could give rise to localised wall growth, such as branch initiation, by concomitant loosening of the wall and insertion of new material. The activities of all six classes of enzyme were strikingly derepressed in a starvation medium, however, as were those of autolytic β1–3-glucanases in *Penicillium italicum* (Santos *et al.*, 1979). The latter authors suggest that these enzymes could be involved *in vivo* in the mobilisation of wall glucans as an energy source, i.e. in wall turnover, rather than in the process of wall growth.

In the oomycetes, Fèvre (1977, 1979) and Mullins & Ellis (1974) describe localisations of vesicles containing celullase and glucanases in hyphae of *Saprolegnia monoica* and *Achlya ambisexualis*, respectively, at sites of wall formation, hyphal apices and branches. Fèvre identified these vesicles as products of the Golgi apparatus by sub-cellular fractionation. The role of autolysins in fungal wall growth, long suspected (see Bartnicki-Garcia, 1973; Gooday, 1977*a*), thus still remains unclear.

We thank the many authors that have sent us reprints and unpublished information, Drs B. W. Bainbridge and C. F. Thurston for helpful discussions, and Annette J. Collinge for help in preparing electron micrographs.

REFERENCES

AHLQUIST, C. N. & GAMOW, R. I. (1973). Phycomyces: mechanical behaviour of stage I and IV. *Plant Physiology*, **51**, 586–7.

AHLQUIST, C. N., IVERSON, S. L. & JHASMAN, N. E. (1975). Cell wall structure and mechanical properties of Phycomyces. *Journal of Biomechanics*, **8**, 357–62.

ARCHER, D. B. (1977). Chitin biosynthesis in protoplasts and subcellular fractions of *Aspergillus fumigatus*. *Biochemical Journal*, **164**, 653–8.

ARONSON, J. M. & PRESTON, R. D. (1960*a*). An electron microscopic and X-ray analysis of the walls of selected lower Phycomycetes. *Proceedings of the Royal Society of London, Series B*, **152**, 346–52.

ARONSON, J. M. & PRESTON, R. D. (1960*b*). The microfibrillar structure of the cell walls of the filamentous fungus, *Allomyces*. *Journal of Biophysical and Biochemical Cytology*, **8**, 247–56.

BALLOU, C. E. (1976). Structure and biosynthesis of the mannan component of the yeast envelope. *Advances in Microbial Physiology*, **14**, 93–158.

BARSTOW, W. E. & LOVETT, J. S. (1974). Apical vesicles and microtubules in rhizoids of *Blastocladiella emersonii*. Effects of actinomycin D and cycloheximide on development during germination. *Protoplasma*, **82**, 103–17.

BARTNICKI-GARCIA, S. (1968). Cell wall chemistry, morphogenesis and taxonomy of fungi. *Annual Review of Microbiology*, **22**, 87–105.

BARTNICKI-GARCIA, S. (1973). Fundamental aspects of hyphal morphogenesis. In *Microbial Differentiation, Symposium of the Society for General Microbiology*, **23**, ed. J. O. Ashworth and J. E. Smith, pp. 245–67. Cambridge University Press.

BARTNICKI-GARCIA, S., BRACKER, C. E., REYES, E. & RUIZ-HERRERA, J. (1978). Isolation of chitosomes from taxonomically diverse fungi and synthesis of chitin microfibrils *in vitro*. *Experimental Mycology*, **2**, 173–92.

BARTNICKI-GARCIA, S. & LIPPMAN, E. (1972). The bursting tendency of hyphal tips of fungi: presumptive evidence for a delicate balance between wall synthesis and wall lysis in apical growth. *Journal of General Microbiology*, **73**, 487–500.

BARTNICKI-GARCIA, S., RUIZ-HERRERA, J. & BRACKER, C. E. (1979). Chitosomes and chitin synthesis. In *Fungal Walls and Hyphal Growth*, ed. J. H. Burnett & A. P. J. Trinci, pp. 149–68. Cambridge University Press.

BERGMAN, K., BURKE, P. V., CERDÁ-OLMEDO, E., DAVID, C. N., DELBRÜCK, M., FOSTER, W. W., GOODALL, E. W., HEISENBERG, M., MEISSNER, G., ZALOKAR, M., DENNISON, D. S. & SHROPSHIRE, W. (1969). Phycomyces. *Bacteriological Reviews*, **33**, 99–157.

BONNETT, H. T. & NEWCOMB, E. H. (1966). Coated vesicles and other cytoplasmic components of growing root hairs of radish. *Protoplasma*, **1**, 59–75.

BRACKER, C. E. (1971). Cytoplasmic vesicles in germinating spores of *Gilbertella persicaria*. *Protoplasma*, **72**, 381–97.

BRACKER, C. E., RUIZ-HERRERA, J. & BARTNICKI-GARCIA, S. (1976). Structure and transformation of chitin synthetase particles (chitosomes) during microfibril synthesis *in vitro*. *Proceedings of the National Academy of Sciences, USA*, **73**, 4570–4.

BRAUN, P. C. & CALDERONE, R. A. (1978). Chitin synthesis in *Candida albicans*. Comparison of yeast and hyphal forms. *Journal of Bacteriology*, **135**, 1472–7.

BRIAN, P. W. (1960). Griseofulvin. *Transactions of the British Mycological Society*, **43**, 1–13.

BRUNSWICK, H. (1924). Untersuchungen über Geschlechts und Kernverhaltnisse bei der Hymenomyzetengattung Coprinus. In *Botanische Abhandlungen*, ed. K. Goebel, vol. 5. Jena: Gustav Fischer.

BULL, A. T. (1970). Chemical composition of wild-type and mutant *Aspergillus nidulans* cell walls. The nature of polysaccharide and melanin constituents. *Journal of General Microbiology*, **63**, 75–94.

BULL, A. T. & TRINCI, A. P. J. (1977). The physiology and metabolic control of fungal growth. *Advances in Microbial Physiology*, **15**, 1–84.

BURNETT, J. H. (1976). *Fundamentals in Mycology*. London: Edward Arnold.

BURNETT, J. H. (1979). Aspects of the structure and growth of hyphal walls. In *Fungal Walls and Hyphal Growth*, ed. J. H. Burnett & A. P. J. Trinci, pp. 1–25. Cambridge University Press.

CABIB, E. & BOWERS, B. (1979). Chitin synthetase distribution on the yeast plasma membrane. *Science*, **203**, 363–5.

CABIB, E., DURAN, A. & BOWERS, B. (1979). Localized action of chitin synthase in the initiation of yeast septum formation. In *Fungal Walls and Hyphal Growth*, ed. J. H. Burnett & A. P. J. Trinci, pp. 189–201. Cambridge University Press.

CABIB, E., ULANE, R. E. & BOWERS, B. (1973) Yeast chitin synthetase. Separation of the zymogen from its activating factor and recovery of the latter in the vacuole fraction. *Journal of Biological Chemistry*, **248**, 1451–8.

CAMARGO, E. P., DIETRICH, C. P., SONNERBORN, D. & STROMINGER, J. L. (1967). Biosynthesis of chitin in spores and growing cells of *Blastocladiella emersonii*. *Journal of Biological Chemistry*, **242**, 3121–8.

CAMPBELL, A. (1957). Synchronization of cell division. *Bacteriological Reviews*, **12**, 253–72.

CANTINO, E. C. & MILLS, G. L. (1979). The gamma particle in *Blastocladiella emersonii*: what is it? In *Viruses and Plasmids in Fungi*, ed. P. Lemke. New York: Marcel Dekker, in press.

CASTLE, E. S. (1942). Spiral growth and reversal by spiralling in Phycomyces, and their bearing on primary wall structure. *American Journal of Botany*, **29**, 664–72.

CASTLE, E. S. (1953). Problems of orientated growth and structure in Phycomyces. *Quarterly Review of Biology*, **28**, 364–72.

CASTLE, E. S. (1958). The topography of tip growth in a plant cell. *Journal of General Physiology*, **41**, 913–26.

COLLINGE, A. J., MILES, E. A. & TRINCI, A. P. J. (1978). Ultrastructure of hyphae of *Penicillium chrysogenum* Thom from colonies and chemostat cultures. *Transactions of the British Mycological Society*, **70**, 401–8.

COLLINGE, A. J. & TRINCI, A. P. J. (1974). Hyphal tips of wild-type and spreading colonial mutants of *Neurospora crassa*. *Archiv für Mikrobiologie*, **99**, 353–68.

COLVIN, J. R. & LEPPARD, G. G. (1977). The biosynthesis of cellulose by *Acetobacter xylinum* and *Acetobacter acetigenus*. *Canadian Journal of Microbiology*, **23**, 701–9.

CORTAT, M., MATILE, P. & WIEMKEN, A. (1972). Isolation of glucanase-containing vesicles from budding yeast. *Archiv für Mikrobiologie*, **82**, 189–205.

DALL'OLIO, G. & VANNINI, G. L. (1979). Coumarin-induced disturbances of morphological development and cell wall formation in *Trichophyton mentagrophytes*. *Cytobiologie*, **18**, 290–7.

DARGENT, R. (1975). Sur l'ultrastructure des hyphes en croissance de l'*Achlya bisexualis* Coker. Mise en'evidence d'une secretion polysaccharidique et d'une activite, phosphatasique alkaline dans l'appareil de Golgi et au niveau des vesicules cytoplasmiques apicales. *Comptes Rendus hebdomadaire des séances de l'Académie des Sciences*, **280**, 1445–8.

DAVIES, P. J. (1973). Current theories on the mode of action of auxin. *Botanical Review*, **39**, 139–71.

DICKINSON, S. (1977). Studies in the physiology of obligate parasitism. X. Induction of responses to a thigmotropic stimulus. *Phytopathologische Zeitschrift*, **89**, 97–115.

DOW, J. M. & RUBERY, P. M. (1975). Hyphal tips bursting in *Mucor rouxii*: antagonistic effects of calcium ions and acid. *Journal of General Microbiology*, **91**, 425–8.

DURAN, A., BOWERS, B. & CABIB, E. (1975). Chitin synthetase zymogen is attached to the yeast plasma membrane. *Proceedings of the National Academy of Sciences, USA*, **72**, 3952–5.

DURAN, A. & CABIB, E. (1978). Solubilization and partial purification of yeast chitin synthetase. Confirmation of the zymogenic nature of the enzyme. *Journal of Biological Chemistry*, **253**, 4419–25.

ENDO, A., KAKIKI, K. & MISATO, T. (1970). Mechanism of action of the antifungal agent polyoxin D. *Journal of Bacteriology*, **103**, 588–94.

FÈVRE, M. (1977). Subcellular localization of glucanase and cellulase in *Saprolegnia monoica* Pringsheim. *Journal of General Microbiology*, **103**, 287–95.

FÈVRE, M. (1979). Glucanases, glucan synthases and wall growth in *Saprolegnia monoica*. In *Fungal Walls and Hyphal Growth*, ed. J. H. Burnett & A. P. J. Trinci, pp. 225–63. Cambridge University Press.

FÈVRE, M. & DUMAS, C. (1977). β-Glucan synthetases from *Saprolegnia monoica*. *Journal of General Microbiology*, **103**, 297–306.

GALUN, M., BRAUN, A., FRENSDORFF, A. & GALUN, E. (1976). Hyphal walls of isolated lichen fungi. Autoradiographic localization of precursor incorporation and binding of fluoroscein-conjugated lectins. *Archives of Microbiology*, **108**, 9–16.

GIBSON, R. K., BUCKLEY, C. E. & PEBERDY, J. F. (1976). Wall ultrastructure in regenerating protoplasts of *Aspergillus nidulans*. *Protoplasma*, **89**, 381–7.

GIRBARDT, M. (1955). Lebendbeobachdungen an *Polystictus versicolor* (L.). *Flora*, **142**, 540–63.

GIRBARDT, M. (1957). Der Spitzenkörper von *Polystictus versicolor* (L.). *Planta, Berlin*, **50**, 47–59.

GIRBARDT, M. (1969). Die Ultrastruktur der Apikalregion von Pilzhyphen. *Protoplasma*, **67**, 413–41.

GLASER, L. & BROWN, D. H. (1957). The synthesis of chitin in cell-free extracts of *Neurospora crassa. Journal of Biological Chemistry*, **228**, 729–42.

GOODAY, G. W. (1971). An autoradiographic study of hyphal growth of some fungi. *Journal of General Microbiology*, **67**, 125–33.

GOODAY, G. W. (1977a). The enzymology of hyphal growth. In *The Filamentous Fungi*, ed. J. E. Smith & D. R. Berry, vol. 3, pp. 51–77. London: Edward Arnold.

GOODAY, G. W. (1977b). The biosynthesis of the fungal wall: mechanisms and implications. *Journal of General Microbiology*, **99**, 1–11.

GOODAY, G. W. (1979). Chitin synthesis and differentiation in *Coprinus cinereus*. In *Fungal Walls and Hyphal Growth*, ed. J. H. Burnett & A. P. J. Trinci, pp. 203–23. Cambridge University Press.

GOODAY, G. W. & DE ROUSSET-HALL, A. (1975). Properties of chitin synthetase from *Coprinus cinereus. Journal of General Microbiology*, **89**, 137–45.

GREEN, P. B. (1974). Morphogenesis of the cell and organ axis – biophysical models. *Brookhaven Symposium in Biology*, **25**, 166–90.

GROVE, S. N. (1972). Apical vesicles in germinating conidia of *Aspergillus parasiticus*. *Mycologia*, **64**, 638–41.

GROVE, S. N. (1978). The cytology of hyphal tip growth. In *The Filamentous Fungi*, ed. J. E. Smith & D. R. Berry, 3, pp. 28–50. London: Edward Arnold.

GROVE, S. N. & BRACKER, C. E. (1970). Protoplasmic organization of hyphal tips among fungi: vesicles and Spitzenkörper. *Journal of Bacteriology*, **104**, 989–1009.

GROVE, S. N. & BRACKER, C. E. (1978). Protoplasmic changes during zoospore encystment and cyst germination in *Pythium aphanidermatum. Experimental Mycology*, **2**, 51–98.

GROVE, S. N., BRACKER, C. E. & MORRÉ, D. J. (1970). An ultrastructural basis for hyphal tip growth in *Pythium ultimum. American Journal of Botany*, **59**, 245–66.

GULL, K. (1975). Cytoplasmic microfilament organization in two basidiomycete fungi. *Journal of Ulstrastructure Research*, **51**, 226–32.

GULL, K. & TRINCI, A. P. J. (1973). Griseofulvin inhibits fungal mitosis. *Nature, London*, **244**, 292–4.

GULL, K. & TRINCI, A. P. J. (1974). Detection of areas of wall differentiation in fungi using fluorescent staining. *Archiv für Mikrobiologie*, **96**, 53–7.

GUNNING, B. E. S. & STEER, M. W. (1975). *Ultrastructure and Biology of Plant Cells*. London: Edward Arnold.

HARDY, J. C. & GOODAY, G. W. (1978). Chitin biosynthesis by *Candida albicans*. *Proceedings of the Society for General Microbiology*, **5**, 106–7.

HASILIK, A. (1974). Inactivation of chitin synthase in *Saccharomyces cerevisiae*. *Archives of Microbiology*, **101**, 295–301.

HASILIK, A. & HOLZER, H. (1973). Participation of the tryptophan synthase inactivating system from yeast in the activation of chitin synthase. *Biochemical and Biophysical Research Communications*, **53**, 552–9.

HEATH, I. B., GAY, J. L. & GREENWOOD, A. D. (1971). Cell wall formation in the Saprolegniales: cytoplasmic vesicles underlying developing walls. *Journal of General Microbiology*, **65**, 225–32.

HEGNAUER, H. & HOHL, H. R. (1973). A structural comparison of cyst and germ tube wall in *Phytophthora palmirora*. *Protoplasma*, **77**, 151–63.

HOLLIDAY, L. (1966). *Composite Materials*. Amsterdam: Elsevier.

HORI, M., KAKIKI, K., SUZUKI, S. & MISATO, T. (1971). Studies on the mode of action of polyoxins. III. Relation of polyoxin structure to chitin synthetase inhibition. *Agricultural and Biological Chemistry, Japan*, **35**, 1280–91.

HOWARD, R. J. & AIST, J. R. (1977). Effect of MBC on hyphal tip organization, growth and mitosis of *Fusarium acuminatum*. *Protoplasma*, **92**, 195–210.

HUNSLEY, D. & BURNETT, J. H. (1968). Dimensions of microfibrillar elements in fungal walls. *Nature, London*, **218**, 462–3.

HUNSLEY, D. & BURNETT, J. H. (1970). The ultrastructural architecture of the walls of some hyphal fungi. *Journal of General Microbiology*, **62**, 203–18.

HUNSLEY, D. & GOODAY, G. W. (1974). The structure and development of septa in *Neurospora crassa*. *Protoplasma*, **82**, 125–46.

HUNSLEY, D. & KAY, D. (1976). Wall structure of the Neurospora hyphal apex: immunofluorescent localization of wall surface antigens. *Journal of General Microbiology*, **95**, 233–48.

ISAAC, S., RYDER, N. S. & PEBERDY, J. F. (1978). Distribution and activation of chitin synthase in protoplast fractions released during the lytic digestion of *Aspergillus nidulans* hyphae. *Journal of General Microbiology*, **105**, 45–50.

JAN, Y. N. (1974). Properties and cellular localization of chitin synthetase in *Phycomyces blakesleeanus*. *Journal of Biological Chemistry*, **249**, 1973–9.

JENNINGS, D. H. (1979). Membrane transport and hyphal growth. In *Fungal Walls and Hyphal Growth*, ed. J. H. Burnett & A. P. J. Trinci, pp. 279–94. Cambridge University Press.

JENNINGS, D. H., THORNTON, J. D., GALPIN, M. F. J. & COGGINS, C. R. (1974). Translocation in fungi. In *Transport at the Cellular Level. Symposia of the Society for Experimental Biology*, **28**, ed. M. A. Sleigh & D. H. Jennings, pp. 139–56. Cambridge University Press.

KATZ, D. & ROSENBERGER, R. F. (1971a). Lysis of *Aspergillus nidulans* mutant blocked in chitin synthesis and its relation to wall assembly and wall metabolism. *Archiv für Mikrobiologie*, **30**, 284–92.

KATZ, D. & ROSENBERGER, R. F. (1971b). Hyphal wall synthesis in *Aspergillus nidulans*: effect of protein synthesis inhibition and osmotic shock by chitin insertion and morphogenesis. *Journal of Bacteriology*, **108**, 184–90.

KELLER, F. A. & CABIB, E. (1971). Chitin and yeast budding: properties of chitin synthetase from *Saccharomyces carlsbergensis*. *Journal of Biological Chemistry*, **246**, 160–6.

KELLY, A. (1966). *Strong Solids*. Oxford: Clarendon Press.

KRITZMAN, G., CHET, I. & HENIS, Y. (1978). Localization of β-(1,3)-glucanase in the mycelium of *Sclerotium rolfsii*. *Journal of Bacteriology*, **135**, 470–5.

LEHLE, L. & TANNER, W. (1975). Formation of lipid-bound oligosaccharides in yeast. *Biochimica et Biophysica Acta*, **399**, 364–74.

LOPEZ-ROMERO, E. & RUIZ-HERRERA, J. (1977). Biosynthesis of β-glucans by cell-free extracts from *Saccharomyces cerevisiae*. *Biochimica et Biophysica Acta*, **500**, 372–84.

LOPEZ-ROMERO, E. & RUIZ-HERRERA, J. (1978). Properties of β-glucan synthetase from *Saccharomyces cerevisiae*. *Antonie van Leeuwenhoek, Journal of Microbiology and Serology*, **44**, 329–39.

LOPEZ-ROMERO, E., RUIZ-HERRERA, J. & BARTNICKI-GARCIA, S. (1978). Purification and properties of an inhibitory protein of chitin synthetase from *Mucor rouxii*. *Biochimica et Biophysica Acta*, **525**, 338–45.

LYR, H. & SEYD, W. (1978). Hemmung der Chitin-Synthetase von *Mucor rouxii in vitro* durch Fungicide und andere Wirkstoffe. *Zeitschrift für Allgemeine Mikrobiologie*, **18**, 721–9.

MCCLURE, W. K., PARK, D. & ROBINSON, P. M. (1968). Apical organisation in the somatic hyphae of fungi. *Journal of General Microbiology*, **50**, 177–82.

MCCULLY, E. K. & BRACKER, C. E. (1972). Apical vesicles in growing bud cells of heterobasidiomycetous yeasts. *Journal of Bacteriology*, **109**, 922–61.

MCMURROUGH, I., FLORES-CARREON, A. & BARTNICKI-GARCIA, S. (1971). Pathway of chitin synthesis and cellular localization of chitin synthetase in *Mucor rouxii*. *Journal of Biological Chemistry*, **246**, 3999–4007.

MADELIN, M. F., TOOMER, D. K. & RYAN, J. (1978). Spiral growth of fungus colonies. *Journal of General Microbiology*, **106**, 73–80.

MARCHANT, R., PEAT, A. & BANBURY, G. H. (1967). The ultrastructural basis of hyphal growth. *New Phytologist*, **66**, 623–9.

MARRIOTT, M. S. (1977). Mannan-protein location and biosynthesis in plasma membranes for the yeast form of *Candida albicans*. *Journal of General Microbiology*, **103**, 51–9.

METRAUX, J. P. & TAIZ, L. (1977). Cell wall extension in Nitella as influenced by acids and ions. *Proceedings of the National Academy of Sciences, USA*, **74**, 1565–9.

MEYER, R., PARISH, R. W. & HOHL, H. R. (1976). Hyphal tip growth in *Phytophthora*. Gradient distribution and ultrahistochemistry of enzymes. *Archives of Microbiology*, **110**, 215–24.

MILLS, G. L. & CANTINO, E. C. (1978). The lipid composition of the *Blastocladiella emersonii* γ-particle and the function of γ-particle lipid in chitin formation. *Experimental Mycology*, **2**, 99–109.

MISHRA, N. C. & TATUM, E. L. (1972). Effect of L-sorbose on polysaccharide synthetases of *Neurospora crassa*. *Proceedings of the National Academy of Sciences, USA*, **69**, 313–17.

MOURI, T., HASHIDA, W. & SHIGA, I. (1971). UDP-Acetylglucosamine from mushrooms. *Journal of Fermentation Technology*, **49**, 699–705.

MULLINS, J. T. & ELLIS, E. A. (1974). Sexual morphogenesis in *Achlya*: ultrastructural basis for the hormonal induction of antheridial hyphae. *Proceedings of the National Academy of Sciences, USA*, **71**, 1347–50.

MYERS, R. B. & CANTINO, E. C. (1974). The gamma particle. A study of cell–organelle interactions in the development of the water mold *Blastocladiella emersonii*. *Monographs in Developmental Biology*, ed. A. Wolsky, vol. 8. Basel: S. Karger.

NOLAN, R. A. & BAL, A. K. (1974). Cellulase localization in hyphae of *Achlya ambisexualis*. *Journal of Bacteriology*, **117**, 840–3.

OTTO, D. W. & BROWN, R. M. (1974). Developmental cytology of the genus Vaucheria. I. Organisation of the vegetative filament. *British Phycological Journal*, **9**, 111–26.

ORTEGA, J. K. E. (1976). Mechanical and Structural Dynamics of Cell Wall Growth. PhD thesis, University of Colorado, Boulder, USA.

ORTEGA, J. K. E., HARRIS, J. F. & GAMOW, K. I. (1974). The analysis of spiral growth in Phycomyces using a novel optical method. *Plant Physiology*, **53**, 485–90.

PALEVITZ, B. A. & HELPER, P. K. (1976). Cellulose microfibril orientation and cell shaping in developing guard cells of *Allium*: the role of microtubules and ion accumulation. *Planta, Berlin*, **132**, 71–93.

PEAT, A. & BANBURY, G. H. (1967). Ultrastructural protoplasmic streaming, growth and tropisms of Phycomyces sporangiophores. I. General Introduction. II. The ultrastructure of the growing zone. *New Phytologist*, **66**, 475–84.

PEBERDY, J. F. (1979). Wall biogenesis by protoplasts. In *Fungal Walls and Hyphal Growth*, ed. J. H. Burnett & A. P. J. Trinci, pp. 49–70. Cambridge University Press.

PEBERDY, J. F. & MOORE, P. M. (1975). Chitin synthase in *Mortierella vinacea*; properties, cellular location and synthesis in growing cultures. *Journal of General Microbiology*, **90**, 228–36.

POLACHEK, Y. & ROSENBERGER, R. F. (1975). Autolytic enzymes in hyphae of *Aspergillus nidulans*: their action on old and newly formed walls. *Journal of Bacteriology*, **121**, 332–7.

POLACHECK, Y. & ROSENBERGER, R. F. (1977). *Aspergillus nidulans* mutant lacking α(1–3)-glucan, melanin, and cleistothecia. *Journal of Bacteriology*, **132**, 650–6.

POLACHEK, I. & ROSENBERGER, R. F. (1978). Distribution of autolysins in hyphae of *Aspergillus nidulans*. Evidence for a lipid-mediated attachment to hyphal walls. *Journal of Bacteriology*, **135**, 741–7.

PORTER, C. A. & JAWORSKI, E. G. (1966). The synthesis of chitin by particulate preparations of *Allomyces macrogynus*. *Biochemistry*, **5**, 1149–54.

PROSSER, J. I. & TRINCI, A. P. J. (1979). A model for hyphal growth and branching. *Journal of General Microbiology*, **111**, 153–64.

RAUDASKOSKI, M. (1970). Occurrence of microtubules and microfilaments, and origin of septa in dikaryotic hyphae of *Schizophyllum commune*. *Protoplasma*, **70**, 415–22.

RAY, P. M. & BAKER, D. B. (1965). The effect of auxin on synthesis of oat coleoptile cell wall constituents. *Plant Physiology*, **40**, 353–68.

REINHARDT, M. O. (1892). Das Wachstum der Pilzhyphen. Ein Beitrag zur Kenntnis des Flächenwachstung vegetabilischer Zellmembranen. *Jahrbüch für Wissenschaftliche Botanik*, **23**, 479–566.

RIZVI, S. R. H. & ROBERTSON, N. F. (1965). Apical disintegration of hyphae of *Neurospora crassa* as a response to L-sorbose. *Transactions of the British Mycological Society*, **48**, 469–77.

ROBERTSON, N. F. (1958). Observations on the effect of water on the hyphal apices of *Fusarium oxysporum*. *Annals of Botany*, **22**, 159–73.

ROBERTSON, N. F. (1959). Experimental control of hyphal branching and branch form in hyphomycetous fungi. *Journal of the Linnean Society (Botany)*, **56**, 207–11.

ROBERTSON, N. F. (1965). The fungal hypha. *Transactions of the British Mycological Society*, **48**, 1–8.

ROELOFSEN, P. A. (1965). Ultrastructure of the wall in growing cells and its relation to the direction of growth. *Advances in Botanical Research*, ed. R. D. Preston, vol. 2, pp. 67–149. New York & London: Academic Press.

ROELOFSEN, P. A. & HOUWINK, A. L. (1953). Architecture and growth of the primary cell wall in some plant hairs and in the *Phycomyces* sporangiophore. *Acta Botanica Neerlandica*, **2**, 218–25.

ROOBOL, A., GULL, K. & POGSON, C. I. (1977). Griseofulvin-induced aggregation of microtubule protein. *Biochemical Journal*, **167**, 39–43.

ROOMANS, G. M. & BOEKSTEIN, A. (1978). Distribution of ions in *Neurospora crassa* determined by quantitative electron microprobe analysis. *Protoplasma*, **95**, 385–92.

ROSEN, W. G. (1964). Chemotropism and fine structure of pollen tubes. In *Pollen Physiology and Fertilization*, ed. H. F. Linskens, pp. 159–66. Amsterdam: North-Holland.

ROSENBERGER, R. F. (1976). The cell wall. In *The Filamentous Fungi*, ed. J. E. Smith & D. R. Berry, vol. 2, pp. 328–44. London: Edward Arnold.

ROSENBERGER, R. F. (1979). Endogenous lytic enzymes and wall metabolism. In *Fungal Walls and Hyphal Growth*, ed. J. H. Burnett & A. P. J. Trinci, pp. 265–77. Cambridge University Press.

DE ROUSSET-HALL, A. & GOODAY, G. W. (1975). A kinetic study of a solubilized chitin synthetase preparation from *Coprinus cinereus*. *Journal of General Microbiology*, **89**, 146–54.

ROWLEY, B. I. & PIRT, S. J. (1972). Melanin production by *Aspergillus nidulans* in batch and chemostat cultures. *Journal of General Microbiology*, **72**, 553–63.

RUDALL, K. M., & KENCHINGTON, W. (1973). The chitin system. *Biological Reviews*, **48**, 597–636.

RUIZ-HERRERA, J. & BARTNICKI-GARCIA, S. (1974). Synthesis of wall microfibrils *in vitro* by a 'soluble' chitin synthetase from *Mucor rouxii*. *Science*, **186**, 357–9.

RUIZ-HERRERA, J. & BARTNICKI-GARCIA, S. (1976). Proteolytic activation and inactivation of chitin synthase from *Mucor rouxii*. *Journal of General Microbiology*, **97**, 241–9.

RUIZ-HERRERA, J., LOPEZ-ROMERO, E. & BARTNICKI-GARCIA, S. (1977). Properties of chitin synthetase in isolated chitosomes from yeast cells of *Mucor rouxii*. *Journal of Biological Chemistry*, **252**, 3338–43.

RUIZ-HERRERA, J., SING, V. O., VAN DER WOUDE, W. J. & BARTNICKI-GARCIA (1975). Microfibril assembly by granules of chitin synthetase. *Proceedings of the National Academy of Sciences, USA*, **72**, 2706–10.

RYDER, N. S. & PEBERDY, J. F. (1977a). Chitin synthase in *Aspergillus nidulans*: Properties and proteolytic activation. *Journal of General Microbiology*, **99**, 69–76.

RYDER, N. S. & PEBERDY, J. F. (1977b). Chitin synthase activity in dormant conidia of *Aspergillus nidulans*. *FEMS Microbiology Letters*, **2**, 199–201.

SANTOS, T., SANCHEZ, M., VILLANUEVA, J. R. & NOMBELA, C. (1979). Derepression of β-1,3-glucaneses in *Penicillium italicum*: localization of the various enzymes and correlation with cell wall glucan mobilization and autolysis. *Journal of Bacteriology*, **137**, 6–12.

SAUNDERS, P. T. & TRINCI, A. P. J. (1979). Determination of tip shape in fungal hyphae. *Journal of General Microbiology*, **110**, 469–73.

SCHMIT, J. C., EDSON, C. M. & BRODY, S. (1975). Changes in glucosamine and galactosamine levels during conidial germination in *Neurospora crassa*. *Journal of Bacteriology*, **122**, 1062–70.

SCURFIELD, G. (1967). Fine structure of the cell walls of *Polyporus myllitae* Cke. et Mass. *Journal of the Linnean Society (Botany)*, **60**, 159–66.

SELITRENNIKOFF, C. P., ALLIN, D. & SONNEBORN, D. R. (1976). Chitin biosynthesis during *Blastocladiella* zoospore germination: evidence that the hexosamine biosynthetic pathway is post-transcriptionally activated during cell differentiation. *Proceedings of the National Academy of Sciences, USA*, **73**, 534–8.

SENTANDREU, R., ELORZA, M. V. & VILLANEUVA, J. R. (1975). Synthesis of yeastwall glucan. *Journal of General Microbiology*, **90**, 13–20.

SENTANDREU, R. & RUIZ-HERRERA, J. (1978). In-situ study of the localization and regulation of chitin synthetase in *Mucor rouxii*. *Current Microbiology*, **1**, 77–80.

SIETSMA, J. H., CHILD, J. J., NESBITT, L. R. & HASKINS, R. A. (1975). Ultrastructural aspects of wall regeneration of Pythium protoplasts. *Antonie van Leeuwenhoek, Journal of Microbiology and Serology*, **47**, 17–23.

SIEVERS, A. (1965). Elektronenmikroksopische Untersuchungen zur geotropischen Reaktion. I. Über Pesonderheiten im Feinbau der Rhizoide von *Chara foetida*. *Zeitschrift für Pflanzenphysiologie*, **53**, 193–213.

WIEVERS, A. (1967). Elektronenmikroskopische Untersuchungen zur geotropischen Reaktion. III. Die transversale Polarisierung der Rhizoid Spitze von *Chara foetida* nach 5 bis minuten Horizontallage. *Zeitschrift für Planzenphysiologie*, **57**, 462–73.

SLAYMAN, C. L. (1973). Adenine nucleotide levels in *Neurospora*, as influenced by conditions of growth and by metabolic inhibitors. *Journal of Bacteriology*, **114**, 752–66.

SLAYMAN, C. L. & SLAYMAN, C. W. (1962). Measurement of membrane potentials in Neurospora. *Science*, **136**, 876–7.

SMIDSRØD, O. & HANG, A. (1972). Properties of poly (1,4-hexuronates) in the gel state. II. Comparison of gells of different chemical composition. *Acta Chimica Scandinavica*, **26**, 79–88.

STEELE, G. C. & TRINCI, A. P. J. (1975). The extension zone of mycelial hyphae. *New Phytologist*, **25**, 583–7.

STERNLICHT, E., KATZ, D. & ROSENBERGER, R. F. (1973). Subapical wall synthesis and wall thickening induced by cycloheximide in hyphae of *Aspergillus nidulans*. *Journal of Bacteriology*, **114**, 819–23.

STEWART, P. R. & ROGERS, P. J. (1978). Fungal dimorphism: a particular expression of cell wall morphogenesis. In *The Filamentsous Fungi*, ed. J. E. Smith & D. R. Berry, Vol. 3, pp. 164–6. London: Edward Arnold.

TANAKA, K. & CHANG, SHU-TING (1972). Cytoplasmic vesicles in the growing hyphae of the basidiomycete, *Volvariella volvacea*. *Journal of General and Applied Microbiology*, **18**, 165–79.

THOMPSON, D'ARCY, W. (1917). *On Growth and Form*. Cambridge University Press.

TILBY, M. J. (1977). Helical shape and wall synthesis in bacteria. *Nature, London*, **266**, 450–1.

TOAZE-SOULET, J. M., RAMI, J., DARGENT, R. & NONTANT, C. (1978). Aspect particulier des hyphes de *Boletus edulis* Fr. ex Bull. et de leur paro, après action des filtrats de culture du mycoparasite *Hypomyces chlorinus* Tul. *Comptes Rendus hebdomadaire des séances de l'Académie des Sciences*, **289**, 33–5.

TOKUNAGA, J. & BARTNICKI-GARCIA, S. (1971). Structure and differentiation of the cell wall of *Phytophthora palmivora*: cysts, hyphae and sporangia. *Archiv für Mikrobiologie*, **79**, 293–310.

TRINCI, A. P. J. (1971). Influence of the peripheral growth zone on the radial growth rate of fungal colonies. *Journal of General Microbiology*, **67**, 325–44.

TRINCI, A. P. J. & BANBURY, G. H. (1967). A study of the growth of tall conidiophores of *Aspergillus giganteus*. *Transactions of the British Mycological Society*, **40**, 525–38.

TRINCI, A. P. J. & COLLINGE, A. J. (1973*a*). Influence of L-sorbose on the growth and morphology of *Neurospora crassa*. *Journal of General Microbiology*, **78**, 179–92.

TRINCI, A. P. J. & COLLINGE, A. J. (1973*b*). Structure and plugging of septa of wild-type and spreading colonial mutants of *Neurospora crassa*. *Archiv für Mikrobiologie*, **91**, 355–64.

TRINCI, A. P. J. & COLLINGE, A. J. (1975). Hyphal wall growth in *Neurospora crassa* and *Geotrichum sandidum*. *Journal of General Microbiology*, **91**, 355–61.

TRINCI, A. P. J. & HALFORD, E. A. (1975). The extension zone of stage I sporangiophores of *Phycomyces blakesleeanus*. *New Phytologist*, **74**, 81–3.

TRINCI, A. P. J. & MORRIS, N. R. (1979). Morphology and growth of a temperature sensitive mutant of *Aspergillus nidulans* which forms a septate mycelia at non-permissive temperatures. *Journal of General Microbiology*, **114**, 53–9.

TRINCI, A. P. J. & SAUNDERS, P. T. (1977). Tip growth of fungal hyphae. *Journal of General Microbiology*, **103**, 243–8.

TRINCI, A. P. J., SAUNDERS, P. T., GOSRANI, K. & CAMPBELL, K. A. S. (1979). Spiral growth of mycelial and reproductive hyphae. *Transactions of the British Mycological Society*, in press.

TURIAN, G. (1978). Spitzenkörper, centre of reducing power in growing hyphae apices of two septomycetous fungi. *Experientia*, **34**, 1277–9.

ULANE, R. E. & CABIB, E. (1974). The activating system of chitin synthetase from *Saccharomyces cerevisiae*. Purification and properties of an inhibitor of the activating factor. *Journal of Biological Chemistry*, **249**, 3418–22.

ULANE, R. E. & CABIB, E. (1976). The activating system of chitin synthetase from *Saccharomyces cerevisiae*. Purification and properties of the activating factor. *Journal of Biological Chemistry*, **251**, 3367–74.

VAN LAERE, A. J. & CARLIER, A. R. (1978). Synthesis and proteolytic activation of chitin synthetase in *Phycomyces blakesleeanus* Burgeff. *Archives of Microbiology*, **116**, 181–4.

WANG, M. L. & BARTNICKI-GARCIA, S. (1976). Synthesis of β-1,3-glucan microfibrils by a cell-free extract of *Phytophthora cinnamomi*. *Archives of Biochemistry and Biophysics*, **175**, 351–4.

WESSELS, J. G. H. & SIETSMA, J. H. (1979). Wall structure and growth in *Schizophyllum commune*. In *Fungal Wall and Hyphal Growth*, ed. J. H. Burnett & A. P. J. Trinci, pp. 27–48. Cambridge University Press.

WHELAN, W. J. (1976). On the origin of primer for glycogen synthesis. *Trends in Biochemical Sciences*, **1**, 13–15.

WOLF, D. H. & EHMANN, E. (1978). Isolation of yeast mutants lacking proteinase B activity. *FEBS Letters*, **92**, 121–4.

YANAGITA, T. (1977). Cellular age in microorganisms. In *Growth and Differentiation in Microorganisms*, ed. T. Ishikawa, Y. Marayama & H. Matsumiya, pp. 1–26. Tokyo University Press.

ZONNEVELD, B. J. M. (1972). Morphogenesis in *Aspergillus nidulans*. The significance of α(1–3)glucan of the cell wall and α(1–3) glucanase for cleistothecium development. *Biochimica et Biophysica Acta*, **273**, 174–87.

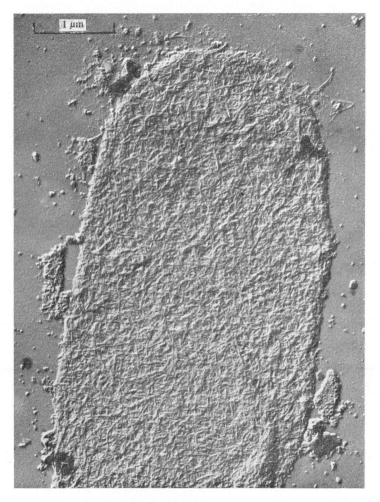

PLATE 1

PLATE 2

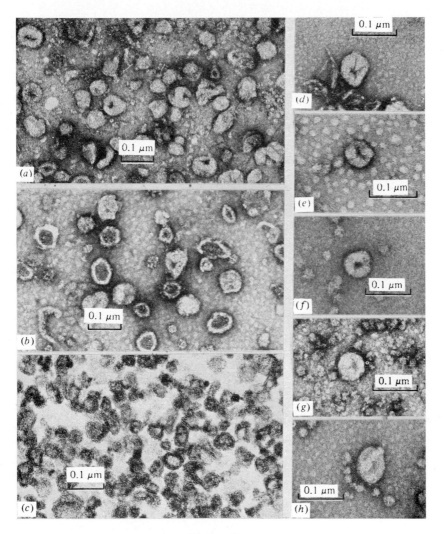

PLATE 3

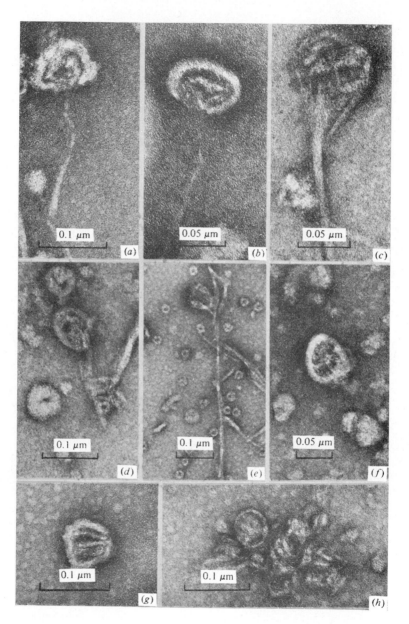

PLATE 4

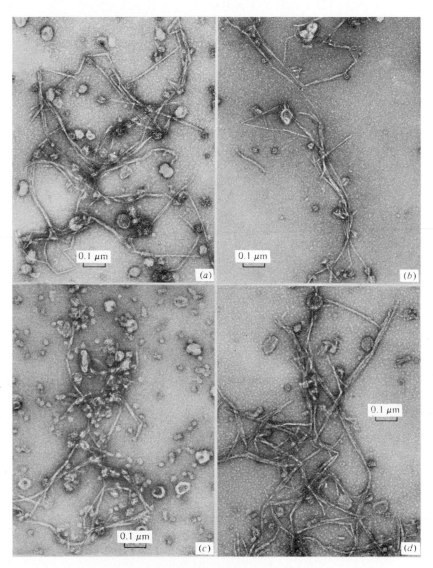

PLATE 5

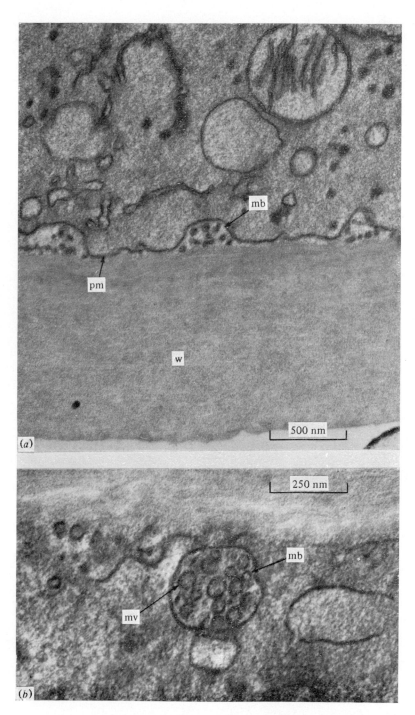

PLATE 6

EXPLANATION OF PLATES

PLATE 1

Microfibrils in the tip of a hyphae of *Syncephalastrum racemosum* revealed by chemical treatment. Micrograph kindly provided by Mr J. A. Hobot and K. Gull.

PLATE 2

(*a*) Longitudinal section of a hyphal tip of *Penicillium chrysogenum* showing a Spitzenkörper containing microvesicles (mv) surrounded by apical vesicles (v). From Collinge *et al.* (1978). (*b*) Transverse section through the tip of a hypha of *Neurospora crassa spco* 9 showing the central region in which microvesicles are surrounded by apical vesicles. From Collinge & Trinc (1974).

PLATE 3

Chitosomes from various fungi. Negatively stained and thin-sectioned preparations of peak samples from sucrose density gradients. Fields of chitosomes from *Mucor rouxii*: (*a*) most with a proctoid appearance, negatively stained; (*b*) most with a cycloid appearance, negatively stained; (*c*) in thin section. Examples of proctoid chitosomes from: (*d*) mycelium of *M. rouxii*; (*e*) mycelium of *Allomyces macrogynus*; (*f*) budding cells of *Saccharomyces cerevisiae*; (*g*) mycelium of *Neurospora crassa*; (*h*) basidiocarp of *Agaricus bisporus*. Parts (*a*), (*b*), (*d–h*) kindly supplied by S. Bartnicki-Garcia, J. Ruiz-Herrera & C. E. Bracker (1979).

PLATE 4

Chitin microfibrils synthesised *in vitro* by chitosome preparations. Negatively stained samples from: (*a*) yeast cells of *Mucor rouxii*; (*b*) mycelium of *M. rouxii*; (*c*) budding cells of *Saccharomyces cerevisiae*; (*d*) mycelium of *Allomyces macrogynus*. Plate kindly supplied by S. Bartnicki-Garcia, J. Ruiz-Herrera & C. L. Bracker (1979).

PLATE 5

Physical connection between chitosomes and microfibrils. (*a, b*) Extended microfibrils arising from partly opened chitosomes: note that the straight microfibrils are continuous with a microfibrillar coil – 'fibroid' – inside the chitosomes. (*c*) Naked 'fibroid' with its associated microfibril; the chitosome shell or membrane is usually lost during chitin synthesis. Parts (*a–c*) are of chitosome samples from yeast cells of *M. rouxii* incubated with substrate and activators. Other examples of fibroids in (*d*) *Saccharomyces cerevisiae*; (*e*) *Allomyces macrogynus*; (*f*) *Neurospora crassa*; (*g*) mycelium of *Mucor rouxii*; (*h*) *Agaricus bisporus*. Plate kindly supplied by S. Bartnicki-Garcia, J. Ruiz-Herrera & C. E. Bracker (1979).

PLATE 6

(*a*) Transverse section through the lower part of the extension zone of a Stage IVb sporangiophore of *Phycomyces blakesleeanus* showing the thick wall (w) and multivesicular (mb) bodies apparently fusing with the protoplasmic membrane (pm). (*b*) Longitudinal section through the lower part of the extension zone of a Stage IVb sporangiophore of *Phycomyces blakesleeanus* showing a multivesicular body (mb). Compare the microvesicles (mv) in the multivesicular body with the chitosome shown in Plate 3(*c*). Plate kindly supplied by A. Peat & G. H. Banbury (1967).

BIOCHEMISTRY OF ACTOMYOSIN-DEPENDENT CELL MOTILITY IN *ACANTHAMOEBA CASTELLANII*

EDWARD D. KORN

Laboratory of Cell Biology, National Heart, Lung, and Blood Institute, National Institutes of Health, Bethesda, Maryland 20014, USA

INTRODUCTION

Approximately 14 years ago, actin was purified from *Physarum polycephalum* by Hatano & Oosawa (1966) and 11 years ago Adelman & Taylor (1969) purified myosin from the same source. These papers were probably the first definitive demonstration of the presence of the two major 'muscle' proteins, not only in micro-organisms but in any non-muscle cell. We had previously initiated studies on the biochemistry and ultrastructure of phagocytosis in *Acanthamoeba castellanii* (Weisman & Korn, 1967; Korn & Weisman, 1967), an amoeba that can be grown on a soluble medium in the kilogramme quantities useful for biochemical studies. It soon became obvious from general morphological studies (Bowers & Korn, 1968) that the amoeba contained numerous microfilaments, especially underlying the plasma membrane (Pollard *et al.* 1970), and that the microfilaments were even more enriched in regions of active phagocytosis (Korn *et al.*, 1974). Moreover, highly purified plasma membranes, but not the membranes of the phagocytic vesicles, contained numerous associated microfilaments (Pollard & Korn, 1973*a*) and the major protein of the isolated plasma membranes had the same electrophoretic mobility as muscle actin (Korn & Wright, 1973). We decided, therefore, to digress from our studies of membrane biochemistry and phagocytosis *per se* in order to investigate the role of actomyosin in phagocytosis.

Since then it has become clear that in all eukaryotic cells actomyosin is involved in many motile processes including amoeboid motility, phagocytosis, exocytosis, cytokinesis, possibly chromosome movement, and probably maintenance of cell shape and membrane organisation and transmembrane signalling (for references see: Korn, 1978; Edelman, 1976; Nicolson, 1976; Clarke & Spudich, 1977; Silverstein, Steinman & Cohn, 1977; Pollard & Weihing, 1974; Weihing, 1976*a*). Complete comprehension of the mechanisms of these actomyosin-dependent motile processes will require integration of the physical chemistry and enzymology of macromolecules and supramolecular structures with membrane, organelle and cellular ultrastructure. The ultrastructural approach to the problem is the localisation by electron microscopy, light microscopy and fluorescence microscopy of the several proteins known to be involved in the motile activities. The biochemical approach is the identification and purification to homogeneity of the proteins followed by a complete characterisation of the physical, chemical and enzymatic properties of each protein alone and in combination with the others. As the biochemist studies progressively more integrated structures, he approaches the microscopist. In practice, of course, the two approaches proceed simultaneously, often in the same laboratory.

At the same time that we diverted our major efforts from studies of membrane biochemistry and phagocytosis to the biochemistry of *Acanthamoeba castellanii* actin and myosin, many other investigators made similar decisions. As a result, there is now appreciable, although still incomplete, information about the properties of the contractile proteins from many non-muscle sources including, most recently, prokaryotes (Niemark, 1977; Nakamura, Takahashi & Watanabe, 1978). In fact, development in our understanding of non-muscle actomyosin systems has been so extensive in recent years that, where previously our research was highly dependent on principles derived from muscle biochemistry, the reverse is now often the case. It is useful to remember, in this context, that muscle is but the most recently evolved, and most highly specialised example of a biochemical system that is extremely widespread, if not universal, in cellular organisms.

In this chapter, I will develop in some detail our present understanding of the biochemistry of the actomyosin systems in *Acanthamoeba castellanii*. It has been our general philosophy that, for a complete understanding of the processes of cell motility in all cells, it is necessary to investigate the many aspects of this problem as they are represented in one cell. With this as the central theme, occasional reference will be made to the many important discoveries in other systems.

ACTIN

Actin can comprise as much as 20% of the total protein of actively motile cells such as *Acanthamoeba castellanii* (Gordon, Eisenberg & Korn, 1976). It can be purified readily in high yield and, therefore, more is known about *A. castellanii* actin than about many of the proteins with which it interacts in the cell. *A. castellanii* actin has an amino acid composition strikingly similar to that of rabbit skeletal muscle actin (Gordon *et al.*, 1976) including the presence of 1 mol of N$^\tau$-methyl-histidine but differs from any other known actin in containing about 1 mol of methyllysine. Almost complete amino acid sequence analysis shows approximately 94% identity of sequence between actins from *A. castellanii* and rabbit skeletal muscle (Table 1) making actin probably

Table 1. *Positions of amino acid residues that differ in actins from* Acanthamoeba castellanii *and muscle*

	Position													
Actin	16	106	129	176	228	266	286	294	296	298	305	316	357	364
Muscle[a]	Leu	Thr	Val	Met	Ala	Ile	Ile	Ala	Asn	Met	Tyr	Ile	Thr	Ala
Acanthamoeba[b]	Met	Val	Thr	Leu	His	Leu	Val	Gly	Val	Leu	Phe	Leu	Ser	Ser

[a]Rabbit skeletal muscle (Elzinga *et al.*, 1973); total residues, 374.
[b]Elzinga and Lu, 1976; Elzinga, personal communication. Number of residues sequenced, 228.

the most highly conserved protein in nature. Actins from amoebae and non-muscle mammalian sources are even more similar; in fact, more similar than the latter are to muscle actins of the same species (Vandekerckhove & Weber, 1978*a*, *b*). *A. castellanii* actin is a single isoelectric species with a significantly more alkaline isoelectric point than skeletal muscle actin (Gordon, Boyer & Korn, 1977). *Dictyostelium discoideum* actin and *Physarum polycephalum* actin also are single isoelectric species but with different isoelectric points (Uyemura, Brown & Spudich, 1978; Vandekerckhove & Weber, 1978*a*). In contrast, vertebrate non-muscle and smooth-muscle cells contain two major species of actins with isoelectric points lying between those of skeletal muscle and *A. castellanii* actins (for references see Korn, 1978). These differences in isoelectric points have been shown to be due to different ratios of aspartic and glutamic acid residues in the first four amino acids at the N-terminal end of the molecules (Vandekerckhove & Weber, 1978*b*,*c*). Despite their striking similarities in amino acid sequences, it is nonetheless possible to identify certain residues that seem to be characteristic of muscle actins and of non-muscle actins (Elzinga & Lu, 1976; Vandekerckhove & Weber, 1978*a*, *b*, *c*). The functional consequences of these differences in amino acid sequences, which occur mostly between residues 2 to 17 and 259 to 298, are not known at this time.

Regulation of the physical state of actin

Actin is a globular protein of mol. wt. 42 000. In the presence of relatively low ionic strength or divalent cations, monomeric actin polymerises to form very long double-stranded helical filaments. Filamentous F-actin is the functional form whether as the thin filaments in muscle or as the microfilaments in non-muscle cells. In contrast to muscle, however, regulation of the state of polymerisation of actin and of the organisation of actin filaments seems to be a major biochemical mechanism for the control of motile processes in non-muscle cells. In *Acanthamoeba castellanii* (Gordon *et al.*, 1977), as in other non-muscle cells (Bray & Thomas, 1976; Markey, Lindberg & Eriksson, 1978; Harris & Weeds, 1978; Blikstad *et al.*, 1978; Carlsson *et al.*, 1977), approximately 50% of the actin seems to be in the non-polymerised form despite the fact that the ionic conditions of the cell are such that essentially all of the actin would be expected to be polymerised (Table 2). In quantita-

Table 2. *Comparison of the properties of actins from* Acanthamoeba castellanii *and muscle*

Actin	Critical concentration (mg ml^{-1})		K_{app} (μM)[a]
Muscle	0.03[b]	0.03[c]	7.1
Acanthamoeba	0.06[b]	0.09[c]	21.7

[a]Concentrations of F-actin required for half-maximal activation of rabbit skeletal muscle heavy meromyosin.
[b]Concentrations required for polymerisation of G- to F-actin in 2 mM-MgCl$_2$, 25 °C.
[c]Concentrations required for polymerisation of G- to F-actin in 0.1 M-KCl, 25 °C.
Data are from Gordon *et al.* (1976, 1977).

tive terms the concentration of non-polymerised actin in *A. castellanii* at any time is probably about 100 times greater than would be expected from measurements of the concentration of non-polymerised G-actin that is in equilibrium with polymerised F-actin when the pure protein is studied under equivalent conditions (Gordon *et al.*, 1977). The identical conclusion has been reached for actins from human platelets (Gordon *et al.*, 1977), embryonic chick brain (Gordon *et al.*, 1977) and *Dictyostelium discoideum* (Uyemura *et al.*, 1978).

Maintenance of non-polymerised actin under polymerising conditions seems to result from the formation of a 1:1 equimolar complex between actin and a protein, profilin, of mol. wt. 14 000–16 000 originally isolated from bovine spleen and human platelets (Markey *et al.*, 1978; Harris & Weeds, 1978; Blikstad *et al.*, 1978; Carlsson *et al.*, 1977) but also purified in high yield from *Acanthamoeba castellanii* (Reichstein & Korn, 1979). Studies of this protein (Table 3) and of its interaction with

Table 3. *Physical properties of profilin from*
Acanthamoeba castellanii

Molecular weight	
Sephadex chromatography	14 000
SDS gel electrophoresis	12 000
Isoelectric point	6.4
α-Helix	28%
β-Structure	29%

actin are not yet very far advanced but it seems to function in reconstituted systems primarily by inhibiting the first step of actin polymerisation (Fig. 1), the formation of 'nuclei' by the interaction of three to four actin monomers, and to have little if any effect on the rate of the second step of polymerisation, elongation of the actin nuclei (Reichstein & Korn, 1979). There is enough profilin in *A. castellanii* (Reichstein &

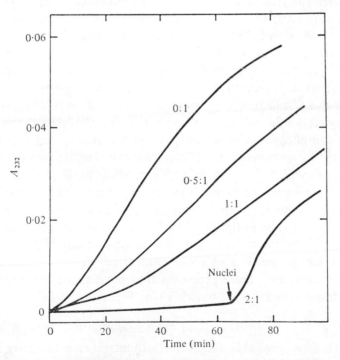

Fig. 1. Inhibition of actin polymerisation by profilin. The rate of polymerisation of G-actin (0.5 mg ml^{-1}) was followed by the increase in absorbance at 232 nm in the presence of profilin in a molar ratio of 0, 0.5, 1 and 2. Sonicated F-actin nuclei were added at the arrow. Data are from Reichstein & Korn (1979).

Korn, 1979) and in platelets (Harris & Weeds, 1978; Markey *et al.*, 1978) to form a 1:1 complex with all the non-polymerised actin.

Once formed, actin filaments probably associate with each other to form bundles or networks and probably also associate with proteins in the plasma membrane of the cell. Initiation of polymerisation may occur by interactions of the non-polymerised actin with specific sites on the plasma membrane. We have no experimental data on this latter possibility in *Acanthamoeba castellanii* but there may be an analogy in the ability of spectrin–actin complexes, derived from the membrane-associated cytoskeleton of erythrocytes, to induce polymerisation of actin (Pinder, Bray & Gratzer, 1977; Brenner & Korn, 1979).

We know somewhat more about the biochemical mechanisms for the association of actin filaments with each other. Proteins that crosslink F-actin filaments are of widespread occurrence (for references, see Korn, 1978) and include macrophage actin-binding protein (dimer of subunits of mol. wts. about 250 000) filamin (dimer of subunits of mol. wts. about 250 000) of non-muscle and smooth muscle cells, and erythrocyte spectrin (a heterodimer with subunits of mol. wts. about 240 000 and 220 000). Four relatively low mol. wt. proteins (subunit mol. wts. of about 23 000, 28 000, 32 000 and 38 000) have been purified from *Acanthamoeba castellanii* each of which alone is capable of inducing the crosslinking of F-actin filaments to form highly viscous gels (Maruta & Korn, 1977a). Recent evidence suggests that the most active of these 'gelactins' may be a heterodimer of two different subunits of mol wt about 38 000 that can exist in a phosphorylated, active state, and a non-phosphorylated, inactive state. Active gelactin can be inactivated by treatment with *Escherichia coli* alkaline phosphatase and reactivated by phosphorylation with a partially purified kinase isolated from *A. castellanii*. The kinase is inhibited by Ca^{2+} and this possibly may be related to the Ca^{2+} inhibition of gelation of crude *A. castellanii* extracts described by Pollard (1976). Continued study of the mechanisms for regulation of the polymerisation and associations of actin filaments should teach us much about the control of vital cellular functions. Parenthetically, it is these processes that seem to be inhibited by the cytochalasins (Hartwig & Stossel, 1976; Weihing, 1976b).

Interaction of actin with myosin

Although actin probably functions without myosin in many essential cytoskeletal phenomena, it is almost certainly true that active motile processes will necessarily involve both actin and myosin. In general, F-actin from all cells has been shown to interact with skeletal muscle myosin, as well as its homologous myosin, in the same stoichiometry of one myosin molecule for each actin subunit in the F-actin filament (Korn, 1978). Frequently, and most usefully, the proteolytic fragments of muscle myosin, heavy meromyosin or subfragment-1, are used in the biochemical studies because of their greater solubility under physiologically relevant conditions. This interaction can be visualised with the electron microscope as decorated filaments as the heavy meromyosin forms 'arrowhead' projections on the F-actin. Reaction with heavy meromyosin is frequently used to characterise cytoplasmic filaments as actin *in situ* (Ishikawa, Bischof & Holzer, 1969).

Biochemically, the interaction of F-actin and heavy meromyosin is assayed by the actin activation of the Mg^{2+}-ATPase activity of heavy meromyosin. At infinite actin concentration, the ATPase activity (V_{max}) of actin–heavy meromyosin is the same for *Acanthamoeba castellanii* actin (Gordon *et al.*, 1976) and other non-muscle actins (Gordon *et al.*, 1977; Uyemura *et al.*, 1978) as for the homologous skeletal muscle actin. However, the concentrations of actins required to reach half-maximal ATPase activity (K_{app}) are different. The K_{app} for *A. castellanii* actin is about three times greater than for muscle actin (Table 2) with vertebrate non-muscle actins having values in between (Gordon *et al.*, 1977). *Dictyostelium discoideum* actin is similar to *A. castellanii* actin (Uyemura *et al.*, 1978). It is reasonable to assume, although definitive data are not available, that the various actins will, correspondingly, have greater affinities for their homologous myosins than will muscle actin. These differences in the affinities of different actins for the same myosin must be one consequence of the slight differences in amino acid compositions among the actins.

The ability of *Acanthamoeba castellanii* actin to activate the ATPase activity of muscle heavy meromyosin has allowed an interesting study that illustrates some of the potential of comparative biochemistry. Skeletal muscle F-actin is, in part, regulated through its interaction with tropomyosin and troponin. One molecule of tropomyosin binds to seven actin subunits in the filament and, under appropriate experimental conditions, inhibits the actin–heavy meromyosin ATPase activity. We have found that muscle tropomyosin (*Acanthamoeba castellanii* probably does not contain tropomyosin) will also bind to *A.castellanii* F-actin in the same stoichiometry of one tropomyosin to seven actins but that a higher ionic strength is required for the association to occur (Yang *et al.*, 1977; Korn & Eisenberg, 1979). Interestingly, however, muscle tropomyosin either has no effect (Yang *et al.*, 1979), or actually enhances (Yang *et al.*, 1977), the *A. castellanii* actin–muscle heavy meromyosin ATPase activity, depending on the condition of assay. Therefore although the tropomyosin-binding sites occur in *A. castellanii* actin, the structure of the tropomyosin–*A. castellanii* actin complex must be significantly different from that of the tropomyosin–muscle actin complex.

More interestingly, we have found that *Acanthamoeba castellanii* actin and muscle actin form random copolymers that bind muscle tropomyosin with binding characteristics dependent on the molar ratios of the two actins in the copolymers (Yang *et al.*, 1979). Unexpectedly, the effect of bound tropomyosin on the two actins in the copolymers is the same as it is in the homopolymers, i.e. tropomyosin inhibits the muscle actin–heavy meromyosin ATPase activity and activates the *A. castellanii* actin–heavy meromyosin ATPase activity even when the two actins are adjacent subunits in the same copolymer interacting with the same tropomyosin molecule (Yang *et al.*, 1979). These data have major implications for the possible mechanism of action of tropomyosin in muscle but it is inappropriate to discuss them further in this article. However, the implications of these data for the actins of vertebrate non-muscle systems are relevant. The fact that *A. castellanii* and muscle actin readily copolymerise strongly suggests that where isoactins occur in the same cell (see Korn, 1978) they will also form copolymers unless there are specific biochemical mechanisms to prohibit their interaction. Moreover, our observations on the effects of tropomyosin on the copolymers of *A. castellanii* and muscle actin suggest that any differential properties of cellular isoactins will be expressed even if they are in copolymers.

MYOSINS

Myosins show much more tissue and species variation than do actins but the typical myosin is a large protein of mol. wt. about 500 000 consisting of six polypeptide chains: a pair of heavy chains of mol. wt. about 200 000 and two pairs of light chains of mol. wt. about 20 000 and 16 000. The heavy chains have a rod and a head portion, the rods associating to form a relatively rigid double-stranded helix. The two pairs of light chains are associated with the globular head portions of these two-headed molecules. The ATPase, actin-binding and regulatory sites are all thought to be located in the globular heads. At low ionic strength, myosin molecules associate, through interaction of their rod portions, to form bipolar filaments that form the thick filaments of muscle and which are interact with the thin filaments of F-actin to produce contraction.

Myosin itself is an ATPase with maximal activity generally expressed in the presence of K^+ and EDTA (to chelate divalent cations), with less activity as a Ca^{2+}-ATPase, and with very low Mg^{2+}-ATPase activity until activated by F-actin. It is, however, the actin-activated Mg^{2+}-ATPase that is the physiologically relevant activity in muscle and presumably in all non-muscle cells.

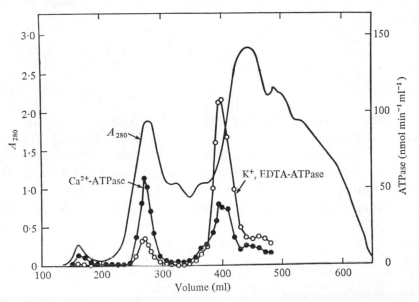

Fig. 2. Separation of *Acanthamoeba castellanii* myosins by chromatography on Biogel A-15M (Maruta & Korn, 1977*a*).

When we fractionated an appropriate extract of *Acanthamoeba castellanii* on an agarose column (Fig. 2), monitoring for K^+, EDTA- and Ca^{2+}-ATPase activities, peaks were observed at positions corresponding to mol. wts. of about 400 000 and 180 000 (Pollard & Korn, 1973*b*; Maruta & Korn, 1977*a*). From such extracts, four different myosin-like enzymes have been purified: a two-headed myosin II (Maruta & Korn, 1977*b*) and single-headed myosins IA, IB and IC (Maruta, *et al.*, 1979).

Acanthamoeba *Myosin II*

The native mol. wt. of 400 000 deduced from gel chromatography of *Acanthamoeba castellanii* myosin II (Table 4) has been more firmly established by sedimentation equilibrium ultracentrifugation (Pollard, Stafford & Porter, 1978). This molecular weight is compatible with a subunit composition of two heavy chains of about 170 000, one pair of light chains of 17 500 and one pair of light chains of 17 000 mol. wt. derived from polyacrylamide gel electrophoresis in dodecyl sulphate

Table 4. *Subunit compositions of myosins from*
Acanthamoeba castellanii

Myosin	Native mol. wt.	Heavy chain		Light chains		
IA	180 000	130 000(1)[a]	—		17 000(1)	14 000 (< 0.5)
IB	180 000	125 000(1)[a]	27 000(1)		—	14 000 (< 0.5)
IC	180 000	130 000(1)[a]	20 000(1)	17 000(1)	14 000 (< 0.5)
II	400 000	170 000(2)[a]	17 500(1.5)[a]	17 000(2)	—

[a]Phosphorylatable.
Data are from Maruta *et al.* (1979).

(Maruta & Korn, 1977*b*). The molecule has been confirmed to be a two-headed structure by electron microscopy (Pollard *et al.*, 1978), the length of its rod portion being fully compatible with the lower molecular weight of the heavy chain relative to myosins from other sources. The enzyme forms very short bipolar filaments *in vitro* (Pollard *et al.*, 1978). As an enzyme, *A. castellanii* myosin II has a higher Ca^{2+}-ATPase than K^+, EDTA-ATPase activity and very low Mg^{2+}-ATPase activity (Table 5). Myosin II binds to F-actin to form typical arrowhead decorations

Table 5. *ATPase activities of myosins from*
Acanthamoeba castellanii

	ATPase activity (μmol $min^{-1}mg^{-1}$)				
	K^+, EDTA	Ca^{2+}		Mg^{2+}	
Myosin			− Actin	+ Actin	Phosphorylated + actin
IA	3.00	1.09	0.11	0.15	1.67
IB	4.00	1.24	0.10	0.22	2.36
IC	1.06	0.02	0.01	0.01	0.46
II	0.16	0.60	0.02	0.03	0.10

Data are from Maruta *et al.* (1979).

(Pollard *et al.*, 1978) but its Mg^{2+}-ATPase activity is activated only 1.5 to 5-fold by F-actin (Table 5), which is less than one might have expected (see below). In all other ways, however, *A. castellanii* myosin II is a reasonably typical myosin, albeit of somewhat low molecular weight mostly because of the reduced length of the rod portion of its heavy chains.

Acanthamoeba *myosins IA, IB and IC*

In addition to myosin II, three different myosins of native mol. wts. of approximately 180 000 have been purified from extracts of *Acanthamoeba castellanii* (Tables 4 & 5). They co-chromatograph on DEAE ion exchange columns but are separable by chromatography on ADP-agarose (Maruta *et al.*, 1979). Although of similar native molecular weights, the three enzymes differ in subunit composition (Table 4). *A. castellanii* myosin IA has a single heavy chain of mol. wt. 130 000 and single light chains of mol. wts. 17 000 and 14 000; *A. castellanii* myosin IB has a single heavy chain of 125 000 and single light chains of 27 000 and 14 000; *A. castellanii* myosin IC has a single heavy chain of 130 000 and single light chains of 20 000, 17 000 and 14 000. The mol. wt. of *A. castellanii* myosin IA is 150 000 by sedimentation equilibrium ultracentrifugation confirming that it, and by inference also IB and IC, is a single-headed enzyme. These are the only single-headed myosins yet to be isolated from any source. None of these enzymes has yet been shown to be capable of forming bipolar filaments.

Despite their unusual physical properties, *Acanthamoeba castellanii* myosins IA, IB and IC have enzymatic activities (Table 5) more similar to those of skeletal muscle myosin than does any other non-muscle myosin characterised so far. The enzymes typically have a higher K^+, EDTA-ATPase than Ca^{2+}-ATPase activity and very low Mg^{2+}-ATPase activity, and the absolute specific activities of these enzymes are higher than have been described for any other non-muscle myosin. The enzymatic activities of myosin IC are significantly lower, however, than those of IA and IB. But we have been able to convert myosin IC into a molecule that closely resembles IA: treatment of myosin IC with LiCl removes the light chain of mol. wt. 20 000 to produce a molecule with subunit composition indistinguishable from that of myosin IA (Maruta *et al.*, 1979). This derived enzyme also now has the higher enzymatic activities characteristic of myosin IA. Myosin IA is unaffected by LiCl but treatment of myosin IB with 2 M-LiCl removes both of its light chains leaving a heavy chain of mol. wt. 125 000 that has all of the enzymatic activity of the original molecule (Maruta, 1978). This latter observation is the clearest demonstration for any myosin that the fundamental properties of the enzyme reside solely in its heavy chain and that the light chains must, therefore, serve regulatory functions only.

Regulation of Acanthamoeba *myosins*

In skeletal muscle, the Mg^{2+}-ATPase activity of actomyosin is converted to a Ca^{2+}-dependent, Mg^{2+}-ATPase activity by the binding of tropomyosin and troponins to the actin filament. In the absence of Ca^{2+}, the native or reconstituted complex is inactive but when the troponin C subunit binds Ca^{2+} the inhibition of actomyosin Mg^{2+}-ATPase activity is derepressed. Recent developments have shown a different mechanism for Ca^{2+} regulation of mammalian non-muscle and smooth-muscle myosins. In these cases, a light chain kinase specifically phosphorylates the larger of the two pairs of light chains (mol. wt. approx. 20 000) and the phosphorylated, but not the non-phosphorylated, myosin is an actin-activated Mg^{2+}-ATPase (for references see Korn, 1978). The kinase, but apparently not the actomyosin ATPase, is a Ca^{2+}-dependent enzyme so that Ca^{2+} regulation of mammalian non-muscle and smooth muscle actomyosins seems to be indirect through the myosin light chain kinase.

We have found that *Acanthamoeba castellanii* myosins IA, IB and IC are similarly converted into actin-activated Mg^{2+}-ATPases through phosphorylation (Maruta & Korn, 1977c) but with a significant difference; it is the heavy chain of all three enzymes, and not the light chains, that is phosphorylated by a specific Ca^{2+}-independent kinase. The same enzyme acts on all three myosins I but has no activity towards *A. castellanii* myosin II. This kinase, which we have now purified about 800-fold, may have a molecular weight of about 100 000. As mentioned, it does not require Ca^{2+} nor do we have any evidence for Ca^{2+} regulation of this system.

The results of phosphorylation of the heavy chains of the myosins I led us to seek evidence for a similar enzyme that would act on *Acanthamoeba castellanii* myosin II. Partially purified preparations of a specific *A. castellanii* myosin II were, in fact, obtained that were capable of incorporating almost one mole of phosphate into each of the two heavy chains of myosin II. However, in contrast to the phosphorylated myosins I, phosphorylated myosin II had the same low actin-activated Mg^{2+}-ATPase as the non-phosphorylated myosin II (Table 5). The possibility that heavy chain phosphorylation of myosin II might be a non-physiological artifact was tested by isolating myosin II from amoebae grown in the presence of radioactive phosphate. The purified heavy chain contained about 0.83 moles of phosphate to which only about 0.1 mole of additional phosphate could be added by the partially purified myosin II heavy-chain kinase. Therefore, we feel reasonably confident

that the enzyme that we have partially purified does act *in situ* to phos-
phorylate the heavy chain of myosin II although we do not know the
role of this process.

There is precedent for this situation, however, in that skeletal muscle
myosin is known to undergo phosphorylation *in vivo*, and the kinases
have been purified, although the role of that phosphorylation is also
unknown. An even closer analogy can be made between *Acanthamoeba
castellanii* myosin II and *Dictyostelium discoideum* myosin. In the latter
case, only a conventional two-headed enzyme has been isolated in which
both the heavy chains (mol. wt. 215 000) and the larger of the two pairs
(18 000 and 16 000) of light chains are phosphorylated (Kuczmarski &
Spudich, 1979). This myosin is highly activated by actin. Removal of the
phosphate groups from the heavy chains with alkaline phosphatase
seems to have no effect on the enzymatic activity but removal of the
phosphate from the heavy and light chains with acid phosphatase
results in the loss of its actin-activated Mg^{2+}-ATPase activity. It is
possible, therefore, that *A. castellanii* myosin II requires phosphorylation of one of its two pairs of light chains in order to express its actomyosin in ATPase activity. Although preliminary experiments suggest
that just such a myosin II light chain kinase exists in *A. castellanii* it has
not yet been purified nor has its effect on the actomyosin ATPase been
tested.

Acanthamoeba *myosins II, IA and IB are isoenzymes*

The fact that myosins II, IA, IB and IC can all be isolated from cultures
derived from a single amoeba establishes that the multiple myosins are
not constituents of different cells in a mixed population. The existence of
myosin isoenzymes in a single cell is not, itself, unexpected since several
examples of myosin isoenzymes are known in muscle (Starr & Offer,
1973; Pope, Wagner & Weeds, 1977; Holt & Lowey, 1977; Wagner &
Weeds, 1977; Hoh, 1978; Schachat, Harris & Epstein, 1977; Burridge
& Bray, 1975) and, as discussed above, isoactins are common at least in
vertebrate non-muscle cells. However, myosin II and the myosins I seem
to have very different structures and enzymatic properties and there is
no precedent for single-headed myosins such as the myosins I. It
became necessary, therefore, to establish that the myosins were indeed
isoenzymes and that the myosins I were not artifacts derived from a
two-headed myosin.

From their subunit compositions alone (Table 4), it is not possible for
any two of the three myosins I to be derived from the other, but one
could imagine a possible common precursor molecule with the heavy

chains of myosin IA and IC and the light chains of myosin IB. Similarly, myosin II cannot be the precursor of the three myosins I but a common precursor might be imagined with the heavy chains of myosin II and with light chains of molecular weight sufficient to be the precursors of all of the light chains of all four myosins. Such an hypothesis requires that this putative common precursor would have to undergo marked changes in its enzymatic properties in being converted to the isolated myosin II and myosins I and marked change in its substrate properties for the several myosin kinases that have been identified.

One way, however, to test directly this very unlikely possibility is to see if myosin II can be converted to an enzyme that resembles any of the myosins I by controlled proteolytic digestion similar to that which converts muscle myosin to heavy meromyosin and subfragment-1. It has been possible to degrade myosin II with trypsin and papain into two-headed and single-headed forms with molecular weights even smaller than those of the myosins I. However, these degradation products retain all of the original enzymatic properties of myosin II and also do not become substrates for myosin I heavy chain kinase (Gadasi et al., 1979).

An alternative approach to test for a possible common origin for the several Acanthamoeba castellanii myosins is to compare peptide maps derived from their heavy chains. If myosin IB heavy chain (mol.wt. 125 000) were derived from myosin IA heavy chain (mol. wt. 130 000) the amino acid sequences of the two polypeptides would be the same and, therefore, their peptide maps would be essentially identical. Similarly, if myosin IA and IB heavy chains were derived from myosin II heavy chain (mol. wt. 170 000) essentially all of the peptides in the maps of myosin IA and IB heavy chains should be present in the map of myosin II heavy chain and about 75% of the peptides in the map of myosin II heavy chain should also occur in the maps of the heavy chains of myosin IA and IB. Even in the unlikely event that myosin II and the myosins I were derived from a common precursor with a heavy chain mol. wt. of 215 000 (the largest known for any myosin), the heavy chains of myosin II and IA and IB would share a minimum sequence of mol. wt. about 85 000.

To test these possibilities we have compared single-dimensional poly-acrylamide gel electrophoretic peptide maps of the heavy chains of myosin II, IA, IB and IC after degradation by Staphylococcus aureus protease (specific for aspartic and glutamic residues), chymotrypsin (specific for aromatic residues), cyanogen bromide (cleavage at methio-nine residues) and cyanylation (cleavage at cysteine residues) (Gadasi *et al.*, 1979). The results of these analyses were apparently unequivocal. As anticipated from the earlier data, the heavy chains of myosin IA and IC were indistinguishable (Plate 1*a*). From this evidence and the previous observations on the effects of LiCl on it, myosin IC seems to be myosin IA plus a regulatory chain of mol. wt. 20 000. Whether one or both forms of myosin IA/IC exist in the cell is not known. On the other hand, the peptide maps of the heavy chains myosin IA, IB and II showed essentially no peptide in common (Plate 1(*b*) for example). Therefore, these three enzymes seem to be products of three different structural genes in the same cell.

REVIEW AND PREVIEW

We can summarise our present understanding of the biochemistry of actomyosin-dependent cell motility in *Acanthamoeba castellanii* by the scheme shown in Fig. 3. It appears highly likely that the polymerisation of actin is regulated, at least in part, by the interaction of monomeric G-actin with profilin which inhibits the first step in the polymerisation

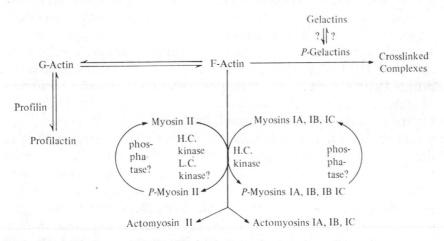

Fig. 3. Summary of biochemistry of actomyosin-dependent cell motility.

process. Polymerised F-actin, on the other hand, can be crosslinked into greater networks by any one of several relatively low molecular weight gelactins the activity of which may be modulated by phosphorylation and dephosphorylation by specific kinases and phosphatases (for whose existence there is now only encouraging evidence but no proof).

Acanthamoeba castellanii contains three myosin isoenzymes. The most typical of these, *A. castellanii* myosin II, is a two-headed enzyme, the heavy chains of which can be phosphorylated by a specific kinase but the effect of that phosphorylation is unknown. There must be also a phosphatase that removes the phosphate group from the heavy chain but it has not yet been isolated. By analogy with other systems, we assume there will also be a myosin II light chain kinase (and phosphatase) and it is possible, but not known, that phosphorylation of the myosin II light chain will enhance the actin-activation of its Mg^{2+}-ATPase activity. In addition, two single-headed myosins, IA and IB, exist in the cell: *A. castellanii* myosin IA can exist with or without an associated chain of molecular weight 20 000 that modulates its ATPase activities. We do not know if both forms, or only the regulated form, exist in the cell. The same heavy chain kinase phosphorylates the heavy chains of all three forms of myosin I resulting in marked enhancement of their actin-activated Mg^{2+}-ATPase activities. There must also be a phosphatase, not yet discovered, that removes the phosphate.

Although the enzymology of the *Acanthamoeba castellanii* motility systems is becoming clear, its physiology is still quite obscure. With little or no direct evidence, it is generally assumed that non-muscle actomyosin motility systems function, as does the skeletal muscle contractile system, through the relative sliding of thin (actin) and thick (myosin) filaments. The circumstantial evidence in support of this assumption is the fact that at least some actin filaments in *Acanthamoeba castellanii* (Pollard & Korn, 1973a) as in other non-muscle cells are oriented with respect to their attachment sites on the plasma membrane as are the thin filaments in muscle towards the Z-line, and the further fact that non-muscle myosins in general, including *A. castellanii* myosin II, are capable *in vitro* of forming bipolar filaments. The general inability to detect myosin filaments in cells (actin filaments are prolific) can be quite reasonably rationalised by the low concentration of myosin relative to actin (and to myosin in muscle cells) and to the small size of the anticipated filaments (Niederman & Pollard, 1975). Nonetheless, although it seems likely to occur, the sliding filament model is unproven in non-muscle cells. In any case, it requires an imaginative leap to explain all modes of motile activity on the basis of sliding filaments.

The probable existence of single-headed myosins IA, IB and IC *in situ* presents yet a different problem. A sliding filament model is possible for a single-headed myosin as long as it is able to form bipolar filaments and just such models have been produced and studied *in vitro* (Cooke & Franks, 1978). But if, as now seems the case, the *Acanthamoeba castellanii* myosins I cannot form bipolar filaments, a different mechanism for their function must be envisaged. Perhaps the simplest alternative can be derived from the proposed mechanism for flagellar movement in which microtubules are thought to slide relative to one another through the action of dynein ATPase, individual molecules of which apparently link one microtubule with the adjacent one (Sale & Satir, 1976). In a similar manner, one could imagine that a myosin I molecule could be firmly attached to one actin filament and 'walk' along an adjacent actin filament with the result that the actin filaments would slide relative to each other. Alternative models can be imagined in which the myosin is attached, not to an actin filament, but to an organelle for example, such that walking of the organelle-attached myosin along an actin filament would produce movement of the organelle.

These alternative models pre-suppose – indeed the existence of such distinct isomyosins in *Acanthamoeba castellanii* predicts – that the several myosins will have very different functional roles in the amoeba. This may be best tested by the use of antibodies directed against the myosins. With fluorescent antibodies it may be possible to localise the myosins to different regions of the motile cell and, if the antibodies inactivate the actomysin ATPase activities of their respective antigens, it may be possible to inhibit selectively one or more motile activities of the cell. We have now produced the antibodies and, at the time of this writing are beginning to characterise them. We hope we will be able to use them productively.

REFERENCES

ADELMAN, M. R. & TAYLOR, E. W. (1969). Further purification and characterization of slime mold myosin and slime mold actin. *Biochemistry*, **8**, 4976–88.

BLIKSTAD, I., MARKEY, F., CARLSSON, L., PERSSON, T. & LINDBERG, U. (1978). Selective assay of monomeric and filamentous actin in cell extracts, using inhibition of deoxyribonuclease I. *Cell*, **15**, 935–43.

BOWERS, B. & KORN, E. D. (1968). The fine structure of *Acanthamoeba castellanii* I. The trophozoite. *Journal of Cell Biology*, **39**, 95–111.

BRAY, D. & THOMAS, C. (1976). Unpolymerized actin in fibroblasts and brain. *Journal of Molecular Biology*, **105**, 527–44.

BRENNER, S. L. & KORN, E. D. (1979). Spectrin–actin interaction. Phosphorylated and dephosphorylated spectrin crosslink F-actin. *Journal of Biological Chemistry*, **254**, 8620–7.

BURRIDGE, K. & BRAY, D. (1975). Purification and structural analysis of myosins from brain and other non-muscle tissues. *Journal of Molecular Biology*, **99**, 1–14.

CARLSSON, L., NYSTRON, L.-E., SUNDKVIST, I., MARKEY, F. & LINDBERG, U. (1977). Actin polymerizibility is influenced by profilin, a low molecular weight protein in non-muscle cells. *Journal of Molecular Biology*, **115**, 465–83.

CLARKE, M. & SPUDICH, J. A. (1977). Non-muscle contractile proteins: The role of actin and myosin in cell motility and shape determination. *Annual Review of Biochemistry*, **46**, 797–820.

COOKE, R. & FRANKS, K. E. (1978). Generation of force by single headed myosin. *Journal of Molecular Biology*, **120**, 361–73.

EDELMAN, G. M. (1976). Surface modulation in cell recognition and cell growth. *Science*, **192**, 218–26.

ELZINGA, M., COLLINS, J. H., KUEHL, W. M. & ADELSTEIN, R. S. (1973). Complete amino acid sequence of actin of rabbit skeletal muscle. *Proceedings of the National Academy of Sciences, USA*, **70**, 2687–91.

ELZINGA, M. & LU, R. C. (1976). Comparative amino acid sequence studies on actins. In *Contractile Systems in Non-muscle Tissues*, ed. J. V. Perry, A. Margreth & R. S. Adelstein, pp. 29–37. Amsterdam: North-Holland.

GADASI, H., MARUTA, H., COLLINS, J. H. & KORN, E. D. (1979). Peptide maps of the myosin isoenzymes of *Acanthamoeba castellanii*. *Journal of Biological Chemistry*, **254**, 3631–6.

GORDON, D. J., BOYER, J. L. & KORN, E. D. (1977). Comparative biochemistry of non-muscle actins. *Journal of Biological Chemistry*, **252**, 8300–9.

GORDON, D., EISENBERG, E. & KORN, E. D. (1976). Characterization of cytoplasmic actin isolated from *Acanthamoeba castellanii* by a new method. *Journal of Biological Chemistry*, **251**, 4778–86.

HARRIS, H. E. & WEEDS, A. G. (1978). Platelet actin: sub-cellular distribution and association with profilin. *FEBS Letters*, **90**, 84–8.

HARTWIG, J. H. & STOSSEL, T. P. (1976). Interactions of actin, myosin, and an actin-binding protein of rabbit pulmonary macrophages III. Effects of cytochalasin B. *Journal of Cell Biology*, **71**, 295–303.

HATANO, S. & OOSAWA, F. (1966). Isolation and characterization of plasmodium actin. *Biochimica et Biophysica Acta*, **127**, 488–98.

HOH, J. D. Y. (1978). Light chain distribution of chicken skeletal muscle myosin. *FEBS Letters*, **90**, 297–300.

HOLT, J. C. & LOWEY, S. (1977). Distribution of alkali light chains in myosin: isolation of isoenzymes. *Biochemistry*, **16**, 4398–402.

ISHIKAWA, H., BISCHOFF, R. & HOLZER, H. (1969). Formation of arrowhead complexes with heavy meromyosin in a variety of cell types. *Journal of Cell Biology*, **43**, 312–28.

KORN, E. D. (1978). Biochemistry of actomyosin-dependent cell motility (a review). *Proceedings of the National Academy of Sciences, USA*, **75**, 588–99.

KORN, E. D., BOWERS, B., BATZRI, S., SIMMONS, S. R. & VICTORIA, E. J. (1974). Endocytosis and exocytosis: Role of microfilaments and involvement of phospholipids in membrane fusion. *Journal of Supramolecular Structure*, **2**, 517–28.

KORN, E. D. & WEISMAN, R. A. (1967). Phagocytosis of latex beads by *Acanthamoeba* II. Electron microscopic study of the initial events. *Journal of Cell Biology*, **34**, 219–27.

KORN, E. D. & WRIGHT, P. C. (1973). Macromolecular composition of an amoeba plasma membrane. *Journal of Biological Chemistry*, **248**, 439–47.

KUCZMARSKI, E. R. & SPUDICH, J. A. (1979). Phosphorylation of *Dictyostelium* myosin. *Federation Proceedings*, **38**, 629 (abstract).

MARKEY, F., LINDBERG, U. & ERIKSSON, L. (1978). Human platelets contain profilin, a potential regulator of actin polymerisability. *FEBS Letters*, **88**, 75–9.

MARUTA, H., GADASI, H., COLLINS, J. H. & KORN, E. D. (1978). The isolated heavy chain of an *Acanthamoeba* myosin contains full enzymatic activity. *Journal of Biological Chemistry*, **253**, 6297–300.

MARUTA, H., GADASI, H., COLLINS, J. H. & KORN, E. D. (1979). Multiple forms of *Acanthamoeba* myosin I. *Journal of Biological Chemistry*, **254**, 3624–30.

MARUTA, H. & KORN, E. D. (1977a). Purification from *Acanthamoeba castellanii* of proteins that induce gelation and syneresis of actin. *Journal of Biological Chemistry*, **252**, 399–402.

MARUTA, H. & KORN, E. D. (1977b). *Acanthamoeba* myosin II. *Journal of Biological Chemistry*, **252**, 6501–9.

MARUTA, H. & KORN, E. D. (1977c). *Acanthamoeba* cofactor protein is a heavy chain kinase required for actin-activation of the Mg^{2+}-ATPase activity of *Acanthamoeba* myosin I. *Journal of Biological Chemistry*, **252**, 8329–32.

NAKAMURA, K., TAKAHASHI, K. & WATANABE, S. (1978). Myosin and actin from *Escherichia coli* K12 C600. *Journal of Biochemistry, Tokyo*, **84**, 1453–8.

NICOLSON, G. L. (1976). Transmembrane control of the receptors on normal and tumor cells I. Cytoplasmic influence over cell surface components. *Biochimica et Biophysica Acta*, **457**, 57–108.

NIEDERMAN, R. & POLLARD, T. D. (1975). Human platelet myosin II. In-vitro assembly and structure of myosin filaments. *Journal of Cell Biology*, **67**, 79–92.

NIEMARK, H. C. (1977). Extraction of an actin-like protein from the prokaryote *Mycoplasma pneumoniae*. *Proceedings of the National Academy of Sciences, USA*, **74**, 4041–5.

PINDER, J. C., BRAY, D. & GRATZER, W. B. (1977). Control of interaction of spectrin and actin by phosphorylation. *Nature, London*, **270**, 752–4.

POLLARD, T. D. (1976). The role of actin in the temperature-dependent gelation and contraction of extracts of *Acanthamoeba*. *Journal of Cell Biology*, **68**, 579–601.

POLLARD, T. D. & KORN, E. D. (1973a). Electron microscopic identification of actin associated with isolated amoeba plasma membranes. *Journal of Biological Chemistry*, **248**, 448–50.

POLLARD, T. D. & KORN, E. D. (1973b). *Acanthamoeba* myosin I. Isolation from *Acanthamoeba castellanii* of an enzyme similar to muscle myosin. *Journal of Biological Chemistry*, **248**, 4682–90.

POLLARD, T. D., SHELTON, E., WEIHING, R. R. & KORN, E. D. (1970). Ultrastructural characterization of F-actin isolated from *Acanthamoeba castellanii* and identification of cytoplasmic filaments as F-actin by reaction with rabbit heavy meromycin. *Journal of Molecular Biology*, **50**, 91–7.

POLLARD, T. D., STAFFORD, W. F., III & PORTER, M. E. (1978). Characterization of a second myosin from *Acanthamoeba castellanii*. *Journal of Biological Chemistry*, **253**, 4798–808.

POLLARD, T. D. & WEIHING, R. R. (1974). Actin and myosin and cell movement. *CRC Critical Reviews of Biochemistry*, **2**, 1–65.

POPE, B. J., WAGNER, P. D. & WEEDS, A. G. (1977). Heterogeneity of myosin heavy chains in subfragment-1 isoenzymes from rabbit skeletal muscle. *Journal of Molecular Biology*, **109**, 470–3.

REICHSTEIN, E. & KORN, E. D. (1979). *Acanthamoeba* profilin. A protein of low molecular weight from *Acanthamoeba castellanii* that inhibits actin nucleation. *Journal of Biological Chemistry*, **254**, 6174–80.

SALE, W. S. & SATIR, P. (1976). Splayed *Tetrahymena* cilia. A system for analyzing sliding and axonemal spoke arrangements. *Journal of Cell Biology*, **71**, 589–605.

SCHACHAT, F. H., HARRIS, H. E. & EPSTEIN, H. F. (1977). Two homogeneous myosins in body-wall muscle of *Caenorhabditis elegans*. *Cell*, **10**, 721–8.

SILVERSTEIN, S. C., STEINMAN, R. M. & COHN, Z. A. (1977). Endocytosis. *Annual Review of Biochemistry*, **46**, 669–722.

STARR, R. & OFFER, G. (1973). Polarity of the myosin molecule. *Journal of Molecular Biology*, **81**, 17–31.

UYEMURA, D. G., BROWN, S. S. & SPUDICH, J. A. (1978). Biochemical and structural characterization of actin from *Dictyostelium discoideum*. *Journal of Biological Chemistry*, **253**, 9088–96.

VANDEKERCKHOVE, J. & WEBER, K. (1978a). The amino acid sequence of *Physarum* actin. *Nature, London*, **276**, 720–1.

VANDEKERCKHOVE, J. & WEBER, K. (1978b). Actin amino acid sequences. *European Journal of Biochemistry*, **90**, 451–62.

VANDEKERCKHOVE, J. & WEBER, K. (1978c). Mammalian cytoplasmic actins are the products of at least two genes and differ in primary structure in at least 25 identified positions from skeletal muscle actins. *Proceedings of the National Academy of Sciences, USA*, **75**, 1106–10.

WAGNER, P. D. & WEEDS, A. G. (1977). Studies on the role of myosin alkali light chains. Recombination and hybridization of light and heavy chains in subfragment-1 preparations. *Journal of Molecular Biology*, **109**, 455–70.

WEIHING, R. R. (1976a). Occurrence of microfilaments in non-muscle cells and tissues. Physical and chemical properties of microfilaments in non-muscle cells and tissues. Biochemistry of microfilaments in cells and tissues. In *Cell Biology*, ed. P. C. Altman & D. S. Dittmer, pp. 341–56. Bethesda: Federation of American Societies of Experimental Biology.

WEIHING, R. R. (1976b). Cytochalasin B inhibits actin-related gelation of HeLa cell extracts. *Journal of Cell Biology*, **71**, 303–7.

WEISMAN, R. A. & KORN, E. D. (1967). Phagocytosis of latex beads by *Acanthamoeba* I. Biochemical properties. *Biochemistry*, **6**, 485–97.

YANG, Y.-Z., GORDON, D. J., KORN, E. D. & EISENBERG, E. (1977). Interaction between *Acanthamoeba* actin and rabbit skeletal muscle tropomyosin. *Journal of Biological Chemistry*, **252**, 3374–8.

YANG, Y.-Z., KORN, E. D. & EISENBERG, E. (1979). Binding of tropomyosin to copolymers of *Acanthamoeba* and muscle actin. *Journal of Biological Chemistry* **254**, 2084–8.

EXPLANATION OF PLATE

PLATE 1

(a) Peptide maps of the heavy chains of *Acanthamoeba castellanii* myosins IA, IB, IC produced by limited proteolysis by *Staphylococcus aureus* protease (Gadasi *et al.*, 1979). Samples were run on 13% acrylamide gels with 10% sodium dodecyl sulphate. (b) Peptide maps of the heavy chain of *Acanthamoeba castellanii* myosin IA, IB and II produced by limited proteolysis by chymotrypsin followed by dodecyl sulphate acrylamide (9–13%) gel electrophoresis (Gadasi *et al.*, 1979).

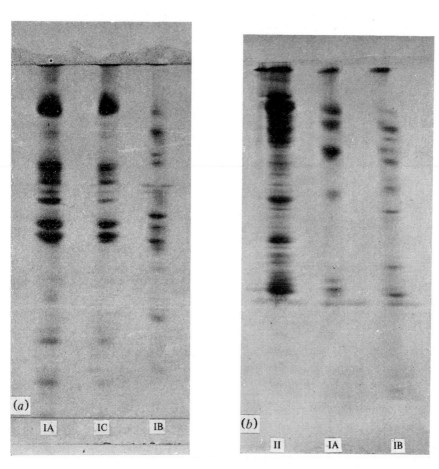

(a)

IA   IC   IB

(b)

II   IA   IB

PLATE 1

MOVEMENT OF CILIA

M. E. J. HOLWILL

Physics Department, Queen Elizabeth College, London, UK

INTRODUCTION

The number of laboratories concerned with studies of the structure and function of cilia has increased significantly in the last 15 years, and there is now a substantial body of knowledge relating to these organelles. It is not my purpose here to review this information in detail, for many excellent articles have been written on the subject (e.g. Sleigh, 1974a), but rather to discuss recent developments, concepts and ideas which have led to a greater understanding of ciliary behaviour. Throughout this chapter the word 'cilia' will be used to include eukaryotic cilia and flagella, since the two types of organelle have an apparently identical structural organisation and display common patterns of movement.

The observed motion of a cilium is a dynamic consequence, at a macroscopic level, of molecular interactions taking place within the organelle and, possibly, the lumen of the cell of which it is a part. Primary aims of those involved in the study of ciliary movement are to understand the mechano-chemical processes which deform the organelle, and the way in which the processes are co-ordinated and controlled by the cell. The cilium is particularly suited to a study of general mechano-chemical processes in cells, and of microtubular interactions in particular, since it is a well-defined organelle and can be observed in a motile condition even when isolated from the remainder of the cell.

A complete description of the working of a cilium will involve an explanation of how internal structural changes lead to the observed patterns of movement. It is therefore important to know the forms which can be adopted by a motile organelle so that a functional interpretation of the structure can be achieved. I propose, therefore, first to describe the movement of cilia as seen through the light microscope, second to discuss the internal structure as determined by electron microscopy and then to examine experiments based on a variety of techniques which have been performed to provide a functional interpretation of the structural features.

FORMS OF MOVEMENT

The primary function of the majority of motile cilia is to create relative motion between their cell surfaces and the liquid environment. The purpose of generating the relative movement is varied: free swimming ciliated protozoa do so to feed and, in some cases, to reproduce sexually; sessile organisms such as *Vorticella* and certain chrysomonads use their cilia to produce feeding currents; the latero-frontal cilia in the gills of molluscs maintain a flow of water over the gill surface so that the animal can obtain oxygen while the much larger abfrontal cilia are used to keep the gill clean; fluid movements are induced by cilia in the mammalian lung and female reproductive tract to aid the functions of these organs.

Externally, a cilium is roughly cylindrical, with the length considerably greater than the diameter. The length varies from species to species and usually falls within the range 5–100 μm, but the diameter remains essentially constant at about 0.2 μm for all cilia. The constant external diameter reflects a specific internal ultrastructure to be considered in a later section.

A single isolated cilium attached to a cell surface generally executes an undulatory motion, which may be two- or three-dimensional, with a frequency between 5 and 100 Hz. In practice the majority of ciliary waves are three-dimensional, though in certain cases the departure from a planar form is relatively small. Examples of planar waveforms are found on the sperm tails of various marine invertebrates (e.g. Brokaw, 1974b) and on certain trypanosomid flagellates (e.g. Holwill, 1965, Plate 1). Three-dimensional flagellar wave patterns are observed on most free-swimming protozoa bearing a single organelle, with the euglenoid flagellates providing typical examples. It is rare for a flagellum to contain significantly more than one wavelength, although the number of wavelengths on the *Ochromonas* flagellum has been observed to change abruptly from one to several for short periods of time (c. 1 s) before reverting to the single wavelength. The amplitude in the multiple wave-mode is significantly smaller than when the flagellum carries a single wave-length. For those flagella carrying roughly planar waves, the amplitude is approximately one-sixth of the wavelength, which indicates that the flagellum is operating close to its optimum hydrodynamic efficiency (Holwill, 1974).

The other form of cyclic motion characteristic of cilia is a tonsate (oar-like) movement (e.g. Sleigh 1974b), in which the ciliary shaft is straight and a bend formed at the base of the cilium causes the tip to trace a circular arc (Fig. 1). This part of the cycle is often known as the effective or power stroke. In the recovery phase of the cycle the bend propagates along the cilium, thereby returning it to a position from which the effective stroke can be generated (Fig. 1). Although it is convenient to divide the tonsate motion into two components for descriptive purposes, the actual motion is not discontinuous and the two parts of the cycle blend smoothly into each other. Tonsate movement is generally observed when a cilium is one of a field of similar organelles, as on a ciliated epithelium or the surface of some protozoa (e.g. *Paramecium* or *Tetrahymena*).

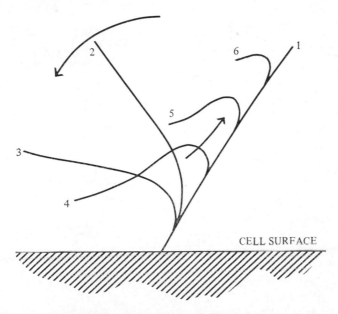

Fig. 1. Diagrammatic form of the tonsate pattern of ciliary movement. Positions 1, 2 and 3 constitute the effective or power stroke while 4, 5 and 6 are the recovery stroke. Arrows indicate the direction in which the tip of the cilium moves.

Within the field, neighbouring cilia are separated by distances of, at most, a few times the ciliary diameter, so that unless some form of co-ordination exists the cilia will interfere with each other during motion and little useful work will be achieved. The co-ordination which exists in practice is such that along a line drawn in a particular direction on the cell surface, all cilia beat synchronously; along a line normal to this direction, neighbouring cilia are slightly out of phase. The net result of this co-ordination is that waves, the so-called metachronal waves, travel across the ciliated surface giving an appearance such as that shown in Plate 2. In some systems, such as the digestive epithelium of *Cirriforma tentaculata* (Mellor & Hyams, 1978) both the recovery and effective phases of the cycle are planar. In other cases, e.g. cilia of *Paramecium* (Machemer, 1974) and of *Mytilus edulis* (Aiello & Sleigh, 1972), the effective stroke occurs in a plane while the bend propagated in the recovery stroke is three-dimensional so that the cilium returns to its 'starting' position by a route which is not in the beat plane of its immediate neighbour along the line of maximum asynchrony (see also Barlow & Sleigh, 1979).

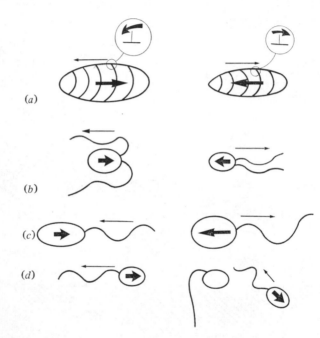

Fig. 2. Avoiding responses in various organisms. In each case normal motion is shown on the left with the avoiding reaction on the right. Wave direction is indicated by the long narrow arrow beside each organism while the direction of motion of the organism is shown by the arrow within each cell. (*a*) *Paramecium*, effective stroke is shown inset and crests of metachronal waves are shown by curved lines; (*b*) *Chlamydomonas*; (*c*) trypanosomes; (*d*) zoospores of *Allomyces*.

Many ciliated cells exhibit characteristic behaviour on encountering unfavourable conditions. The behaviour frequently takes the form of an avoiding reaction, in which the cell alters its direction of motion to reach a more hospitable environment. To change its direction, a cell modifies the movement of its cilium and is thus able to exert control over the organelle. The avoiding reaction in *Paramecium* involves a changed orientation of the effective stroke (e.g. Naitoh & Eckert, 1974; Fig. 24*a*), while for *Chlamydomonas* the two undulating cilia are changed in position from being directed posteriorly to pointing anteriorly (e.g.

Hyams & Borisy, 1978; Fig. 2b). In the trypanosomes and trypanosomid protozoa, changes in direction are effected by a reversal of the direction of wave propagation; normally a wave moves from the tip of the flagellum towards the base, but during the avoiding reaction waves are initiated at the base and travel distally (Holwill, 1965; Fig. 2c).

The tactic responses of certain cells also involve modification of the ciliary beat pattern. This is especially clear for some zoospores in which wave propagation ceases while the cell body and basal portion of the flagellum are re-oriented in the appropriate direction. Wave propagation is then resumed with the cell swimming towards the origin of the stimulus (Miles & Holwill, 1969; Fig. 2d). Further examples indicating that cells have the ability to control movement of their cilia are found in organisms such as *Peranema trichophorum*, the flagellum of which is able to execute a variety of movements (e.g. Holwill, 1966). In the usual form of movement the distal third of the cilium executes a tonsate movement. On other occasions, two- or three-dimensional waves are observed, with the direction of propagation apparently variable.

Bend shapes

So far in this section, I have described the wide variety of bend patterns observed on cilia, patterns which ultimately require explanation in terms of the basic molecular mechanism responsible for movement. Important clues to the mechanism may be provided by a detailed analysis of the wave-shape present on cilia. It has been recognised for some time that cilia do not adopt a sinusoidal configuration, although it has proved to be convenient for hydrodynamic analysis to assume that undulating cilia are sine waves. In early studies of bend-shape, certain cilia were shown to be matched closely by waves consisting of circular arcs linked by straight lines (Brokaw, 1965; Brokaw & Wright, 1963). This arc-line wave is similar in form to meander and sine-generated waves with the same ratio of amplitude to wavelength, and it is of importance to establish which, if any, of these three waveforms is actually present on cilia. Silvester & Holwill (1972) provided a theoretical method, based on Fourier analysis, to distinguish between the shapes, but in practice the technique requires considerable amounts of data for its application. To acquire sufficient information for analysis, an automatic data collection system has been developed (Silvester & Johnston, 1976; Johnston & Silvester, 1979) and used to record co-ordinates from projected images of the cilium of *Crithidia oncopelti*. Fourier analysis of the data reveals that the wave-shape is arc-line in character with the arc occupying about 84% of the flagellar length. Hiramoto & Baba (1978) and

Rikmenspoel (1978) have carried out a careful examination of the wave shapes of echinoderm spermatozoa, and found that, over small regions, the angular direction of the sperm tail changes sinusoidally with distance along the organelle, although both the amplitude and wavelength of the sinusoid changed with distance. They were able to represent the curvature of the wave by a sinusoidal function of time with a phase term which depended on distance along the flagellum. It would be of interest to perform a comparative study of the wave-shapes on various flagella by applying the Fourier technique of Johnston, Silvester & Holwill (1979), so that any common features would be revealed.

ULTRASTRUCTURE

The eukaryotic cilium is enclosed by an extension of the cell membrane and contains the *axoneme*, which consists essentially of two single microtubules surrounded by nine doublet microtubules, together with a number of inter-microtubular linkages (Fig. 3). All microtubules, which consist of the protein tubulin, run continuously from one end of a cilium to the other, with the peripheral microtubules extending (as triplets) into a specialised basal region at which the central microtubules

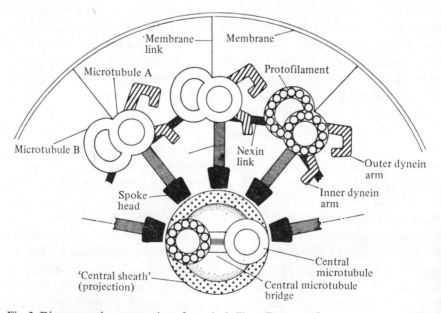

Fig. 3. Diagrammatic cross-section of a typical cilium. For ease of representation only three of the nine peripheral doublet microtubules are shown. (Based on diagrams by many authors, particularly Warner, 1974.)

terminate on a disk. The diameter of an individual central microtubule and each component of a peripheral doublet is about 24 nm. Viewed by electron microscopy, the microtubules appear hollow, with a wall thickness (including the dividing wall of a peripheral doublet) of about 4 nm. The apparent dimensions of the components of a particular organelle depend somewhat on the procedures used to prepare the specimen for electron microscopy, and the measurements quoted here are for negatively stained specimens (see Warner, 1974). In cross section, a doublet microtubule is oriented so that its axis of symmetry makes an angle of 10° with the tangent (at the doublet) to a circle drawn through the outer microtubules. By convention the innermost microtubule of the doublet is designated 'A' and the other 'B'.

Attached to the microtubule A at longitudinal intervals of 16–22 nm (depending on the organisms) are pairs of projections known as the dynein arms which project towards microtubule B of the neighbouring doublet. Early electron micrographs invariably showed the dynein arms separated from the B microtubule, but in certain preparations the dynein arms form a complete bridge between adjacent microtubules (e.g. Gibbons, 1975). Corresponding arms of the inner and outer rows may not be in precise lateral array, but in certain organisms appear to be displaced longitudinally by 3–4 nm (Warner, 1976).

The arrangement of the microtubules in a cross section of the axoneme is such that a plane perpendicular to a line joining the central microtubules intersects a particular peripheral microtubule. For identification purposes, this doublet is numbered '1' and the other doublets are numbered sequentially in such a way that the arms on doublet n point towards doublet $(n + 1)$ (e.g. Warner, 1974). Looking along the axoneme from the base towards the tip, doublet $(n + 1)$ is reached by moving clockwise from doublet n (Fig. 3).

The peripheral doublets are joined together by another linkage, somewhat thinner and more difficult to characterise than the dynein arms, but which seems to have the same longitudinal periodicity as the arms along the axoneme. Known as the 'interdoublet' or 'nexin' link, its precise attachment points to the microtubules have not yet been unequivocally established, and it is not clear whether the link is intimately associated with the dynein arms or is structurally unrelated to them.

Another important structure attached to the A component of a peripheral doublet is the spoke, which is directed towards the central region of the axoneme. The spoke has a shaft some 38 nm in length and 5 nm in diameter, and terminates in an electron-dense head which may be attached to the central structures of the axoneme (e.g. Warner, 1974). Depending on the species, the spokes occur in repeating groups of two or three along the A microtubule, with an average spacing between groups of about 90 nm. Where three spokes occur in a group, the spoke separations within a group are unequal and the larger spacing occurs towards the ciliary base. In a straight axoneme, the spokes attached to the nine doublets are arranged helically, although there appears to be no inter-connecting helical structure (Sale & Satir, 1976).

Each of the central microtubules bears one or two rows of projections (e.g. Warner, 1974). The projections on the two microtubules are parallel to each other, and each projection makes an acute angle with the microtubule axis. The projections may form a continuous link between the microtubules and their true nature may be a closed loop, inclined to the axis of the axoneme and joining the central microtubules. The separation of adjacent projections in *Elliptio* cilia is 14.3 nm, a figure which is just one-sixth of the spoke group repeat. This relationship may be significant in a consideration of the mechanism of ciliary motion.

While some cilia have smooth external surfaces, others bear surface appendages, some of which have a significant effect on the propulsive effect of the organelle. The appendages may take the form of scales (e.g. Manton & Parke, 1960), of fine, apparently flexible, hairs (flimmer) as in *Distigma proteus* (Plate 3*a*) or of thicker, rigid projections (mastigonemes) seen on the flagellum of *Ochromonas danica* (Plate 3*b*). The flimmer are at least 3 μm long and 10 nm in diameter in *D. proteus* and are longer and thinner than the mastigonemes (1 μm long, 25 nm in diameter) of *O. danica*. The ciliary membrane in some species has been examined using freeze-fracture techniques. A variety of particle patterns has been observed, including the ciliary necklace, which consists of between two and six rows of membrane particles located just distal to the basal plate on which the central microtubules terminate (e.g. Gilula & Satir, 1972). These authors also describe particles arranged in rows which correspond in position to the underlying microtubules. These two types of particle may be associated with links between the membrane and the peripheral microtubules. Freeze-fracture studies of the membrane of *Crithidia oncopelti* reveal particle patches at sites where the flagellum emerges from an invagination in the cell and where the flagellar membrane and cell membrane closely appose (M. E. J. Holwill & P. Satir, unpublished). These patches probably correspond to the desmosome-like structures observed in transmission electron micrographs of the entrance to the invagination and which have also been reported for *Crithidia fasciculata* (Kusel, Moore & Weber, 1967).

MECHANISMS OF MOVEMENT

Sliding microtubules

Critical examinations of the motile behaviour of cilia under a variety of conditions and detailed electron microscope studies of cilia in known configurations have provided considerable information about the internal mechanism by which an organelle is bent. Of fundamental importance was the realisation by Gray (1955) that bends on cilia can be maintained without loss in observed amplitude only if a source of energy for bending is available along the whole length of an organelle. It was later shown analytically by Machin (1958) that the waveforms observed on cilia could not be achieved by a passive elastic rod with an energy source at one end. More recently, flagella from *Crithidia oncopelti* have been amputated at various points by laser irradiation (Goldstein, Holwill & Silvester, 1970) or by micromanipulation (Holwill & McGregor, 1974) and the severed section was found to be capable of active bending. These experiments demonstrate that for *C. oncopelti*, at least, active bending elements are distributed throughout the ciliary length.

It is now generally accepted that the forces required to bend a cilium are derived internally from active sliding between the outer doublet microtubules. The dynein arms which are attached to a doublet microtubule are thought to interact with the neighbouring doublet tubule in a manner similar to that in which myosin cross bridges are believed to walk along actin filaments in muscle. Dynein has long been intimately associated with the fundamental motile process since it is known to be an ATPase. Evidence for the sliding microtubule mechanism was first presented by Satir (1965), who made an electron microscope study of the tips of cilia sectioned when the organelles were in known bent states. He found that, for a cilium containing a single bend, the ends of microtubules on the inside of the bend projected further than those on the outside, as would be expected if the microtubules were inextensible and sliding relative to each other. To show that the sliding is active, Summers & Gibbons (1971) performed experiments, using optical microscopy, on cilia which had been treated with detergent to remove the membrane and with trypsin to digest the nexin links. When solutions containing ATP were added to this preparation, individual organelles became longer and thinner, as predicted by a sliding microtubule hypothesis.

That sliding is taking place in the type of preparation described by Summers & Gibbons has been confirmed using electron microscopy by Sale & Satir (1977). They were able to demonstrate that all nine of the microtubules from *Tetrahymena* axonemes were able to slide with respect to each other. In these preparations, the spokes remained attached to the microtubules, so that, because of the asymmetry in spoke group configuration referred to earlier, the distal and proximal ends of individual microtubules can be identified unequivocally. Furthermore, by examining the positions of the dynein arms the numerical ordering of the microtubules can be established; that is, if one microtubule be assigned number n, microtubule number $(n + 1)$ can be identified. The results obtained by Sale & Satir indicate that relative sliding occurs in one direction only, with microtubule n pushing microtubule $(n + 1)$ towards the cilium tip. The direction of sliding in microtubules from *Tetrahymena* is unaltered by the addition of calcium, which is thought to produce a reversal of the direction of beating in this cell, thus suggesting that the direction of microtubular sliding is polarised, and can occur in one sense only.

In a cilium containing a single planar bend, with the plane of bending normal to a line joining the central microtubules, sliding in the two halves of the axoneme defined by the bend plane will be symmetric, as indicated in Fig. 4. To achieve this configuration from an initially

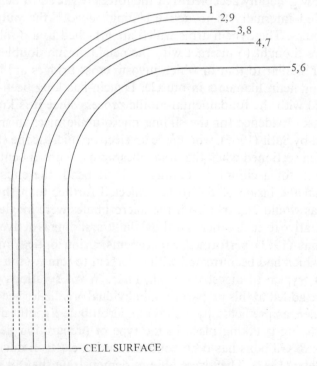

Fig. 4. Showing relative sliding of the microtubules in a cilium containing a single bend in a plane containing microtubule 1.

straight position, those microtubules which are effective in producing the required bending moment, given polarised sliding, will be numbers 1 to 5. Interactions between microtubules 6 to 9 and 1 during the formation of this bend would produce a moment opposing the bending. Satir (1979) has suggested that the two sets of microtubules above are activated or inactivated according to the direction of bending. Thus while one set is actively sliding to produce a bend, the other set is passive and slides in response to the bending. Some form of control is needed to activate or inactivate the two halves of the axoneme and Satir speculates that the central microtubules and associated projections may be involved in this process since they have the appropriate symmetry.

An alternative possibility, which may not be as efficient as Satir's speculation but which requires less control, is that the microtubules are continuously active but that the force generated between them depends on the relative direction of sliding. If the system were completely uniform, continuous activation of the microtubules would not lead to bending. However, a perturbation of the system, caused perhaps by spoke interactions, could favour bending in one sense and cause its initiation. By assuming that relative sliding in the direction opposite to that which would be induced by the dynein arms inhibits the dynein–tubulin interaction, an effect similar to that of Satir's suggested switch can be achieved (Brokaw, 1977).

Conversion of sliding into bending

To produce the bends observed on cilia, the microtubules must be restrained from sliding freely so that a couple is exerted on the system. Additionally, the coherent propagation of bends requires that the couple be applied in a regulated manner. It is believed that the microtubules cannot slide relative to each other in the basal portion of the cilium, so that forces between the microtubules tending to generate sliding would, in practice, produce a bending moment in the structure. The site and extent of bending along the cilium depends on the resistance offered by the organelle to the applied couples, i.e. to the local stiffness of the system. A uniform couple applied to the entire length of a cilium by virtue of microtubule sliding would produce even bending along the cilium if the stiffness were constant. However, if the stiffness varies along the cilium, those regions of lower stiffness would bend to a greater extent than others and local bending could be achieved. Thus, one method of achieving bend propagation is to control the local stiffness of an organelle. The microtubules have their own inherent resistance to bending, which probably varies little along their lengths, but the stiffness of an assembly of tubules can be affected by varying the number or arrangement of linkages formed between them. As noted earlier, there are several linking structures in a cilium, but as yet there is no definitive evidence which associates any of them with stiffness control or the local regulation of sliding. It is, however, believed that the radial spokes are involved in this function. Warner & Satir (1974) have presented evidence which strongly suggests that the spokes undergo a cycle of activity involving transient attachments to the central complex. In *Chlamydomonas* mutants that lack the spokes or central complex, bending is not observed, although the de-membranated axoneme slides apart when treated with ATP (Witman, Fay & Plummer, 1975; Witman, Plummer & Sander, 1978). There are, however, other observations which suggest that the radial spoke and central complex are not essential features of the bending mechanism. The flagellum of the male gamete of the gregarine *Lecundina tuzetae* has a structure which contains no central apparatus and only six outer microtubules, and yet it propagates bending waves (Goldstein, Besse & Schrevel, 1979). In this case, and perhaps more generally, other components of the axoneme than the spokes must be effective in controlling the co-ordination of bends. It is also interesting to note that in split *Chlamydomonas* flagella, Nakamura & Kamiya (1978) suggest that bundles containing fewer than four doublet microtubules are capable of oscillatory motion.

Extent of active sliding

The forces which must be overcome by the active elements in a flagellum are the viscous resistance of the organelle as it moves through its liquid environment and forces due to the natural elasticity of the structures in the axoneme. Little quantitative information about the elasticity is available, although estimates based on particular models have been made (e.g. Holwill, 1965; Brokaw, 1975*b, c*; Rikmenspoel, 1976), but Dryl (1975) has proposed a mechanism of ciliary movement which relies heavily on the elastic properties of the system. The model assumes that the cilium has a natural resting state to which it will return when there is no internal activity and that active sliding occurs only in the bending portion of a cilium. Starting from the resting position, a bend is developed at the base of the cilium by active sliding in a limited region of the organelle. During bend development, sliding occurs in the distal region of the cilium but it is unaccompanied by the dynein–tubulin interaction, i.e. it is passive. Once the bend is formed it propagates distally as the active sliding progresses along the organelle, mediated by radial spoke–central complex interactions; in the proximal region the dynein–tubulin interaction continues, so this section of the cilium remains stiff. In bending the organelle, elastic energy is stored in deformed structures such as the nexin links and radial spokes, and Dryl suggests that this energy is used to restore the flagellum to its original resting position.

In his original model, Dryl (1975) suggested that all nine microtubules were active simultaneously, and thus assumed active relative sliding between adjacent microtubules to be occurring in opposite directions in the two halves of the axoneme (see p. 282 and Fig. 4). The finding that sliding is probably polarised indicates that the model should be modified, but the concept of elastic forces acting to produce a bend rather than to oppose it is attractive from the viewpoint of efficiency. In their analysis of wave-shapes, Johnston et al. (1979) have considered the possibility that the act of straightening a bend is essentially passive, and brought about by the elastic properties of the system. If a constant elasticity is assumed, the time-course of such an unbending process can be predicted; the experimental data for Crithidia oncopelti were at variance with the theoretical prediction, suggesting that unbending requires the development of active forces within the cilium. However, the results do not preclude the possibility that elastic forces assist the unbending process and this might reasonably be expected if the microtubules are straight when not under stress (see also Brokaw, 1975a; Rikmenspoel, 1976). In some experiments involving sliding and splaying of detergent-treated cilia, there is a tendency for the microtubules to bend when under no apparent stress, so the phase of the beat cycle during which elastic forces are acting in favour of a bend could be the bending rather than the unbending phase (e.g. Costello, 1973a, b). A mechanism which relies on changes in the stiffness of the system for bend propagation is also suggested to explain the movement of bends along the microtubular axostyle in Pyrsonympha vertens (Langford & Inoué, 1979). The patterns of bending are similar in many respects to those observed on cilia, and, since the structure is based on microtubule interactions, common features for the basis of movement are to be expected.

Although the microtubules extend for the entire length of a cilium, active sliding may occur in a limited region of the organelle only, as suggested by Dryl (1975). Other authors (e.g. Rikmenspoel & Rudd, 1973) consider that the microtubules undergo active interactions with their neighbours along the whole length of the cilium. For a cilium executing a tonsate motion, where a single bend formed at the base causes movement of an otherwise predominantly straight appendage, a mechanism in which activity occurs along the length of the microtubules is energetically advantageous, since the maximum possible portion of the energy producing system would be contributing to the local bending. The same is, of course, true for a flagellum which is undulating but the details of the mechanism for this case are difficult to visualise if the sliding forces are polarised. It is of interest to note that if a bend configuration (be it a single bend or a wave) propagates unchanged along an organelle, no microtubule sliding will occur in regions distal or proximal to the bend (assuming these regions to remain 'inactive'); having established the configuration, it can be propagated by simply altering the stiffness of the organelle in the appropriate way. The direction of bend propagation is determined by the direction of the stiffness change which, if reversed, will cause wave reversal. In real cilia, it is rare for a bend configuration to propagate without variation, but the basic mechanism of propagation by stiffness change in a 'primed' system could still operate, but with additional sliding superimposed (see also Holwill, 1974; Rikmenspoel, 1976).

Hiramoto & Baba (1978) found that the curvature of a particular point on a sea urchin or starfish sperm flagellum changed sinusoidally with time and hence argued that the force generated by the sliding microtubules is also a sine function of time. In common with others (e.g. Rikmenspoel, 1971; Machemer, 1977) these authors suggest that active sliding is localised and that the active site propagates around the axoneme from one pair of microtubules to the next in a helical path, as well as along the axoneme. This type of mechanism would lead to a three-dimensional waveform, as is seen on the echinoderm sperm tails, with the three-dimensional character determined by structural constraints such as the central microtubules and the bridge which exists between peripheral microtubules 5 and 6 in these organelles. Because of the geometry of the axoneme, sliding forces between a pair of microtubules will exert a maximum bending moment in a plane containing that pair; there is hence a tendency for a cilium to adopt a three-dimensional pattern. This is particularly true if the direction of active sliding is polarised, since couples generated between all pairs of doublets will be acting to cause helical bending of the same sense, and, incidentally, twisting of the doublet microtubules about the cilium axis.

In his examination of the tips of *Elliptio complanatus* cilia, Satir (1963) noted asymmetry (skew) in the displacements of microtubules which could not be matched by a model which assumed bending in a plane perpendicular to that containing the central microtubules. By assuming that the microtubules may twist about the ciliary axis, quantitative agreement between prediction and experiment can be obtained (Holwill, Cohen & Satir, 1979). For *E. complanatus* cilia the results are consistent with a planar effective stroke and a three-dimensional recovery stroke, similar to that reported for *Mytilus edulis* cilia by Aiello & Sleigh (1972). The plane of the effective stroke contains microtubule 2, rather than 1, to produce the observed skew. Gibbons (1975) has reported twisting the microtubules in the transitional regions between the bend and straight sections of sea urchin sperm tails, while many cilia and flagella execute three-dimensional bending (e.g. Woolley, 1977; Machemer, 1974) which may be accompanied by axonemal twisting. The amount of twisting observed depends on the interactions between the sliding microtubules and other axonemal structures, such as the radial spokes and nexin links. In their analysis of axonemal twisting Holwill *et al.* (1979) consider rotation of the microtubule complex as a unit. Omoto & Kung (1979) report that the central microtubules of *Paramecium tetraurelia* rotate relative to the outer doublets during the beat cycle, and they suggest that the sliding of the peripheral microtubules could be regulated by this behaviour. A detailed study of twisting in cilia from a variety of micro-organisms may therefore provide information which will lead to a more complete understanding of the mechanism underlying ciliary motility.

Bend initiation

In discussing the mechanism of wave propagation, no account has been taken of the way in which bends are initiated on a cilium and the consequent implications for the sliding microtubule interactions. Information about the amount of sliding within an organelle can be obtained from the angle between the tangents at two points on the cilium. Careful angular measurements have been made of developing bends on echinoderm spermatozoa (Goldstein, 1976, 1977) and the results indicate that, if sliding is prevented by basal structures, established propagating bends can tolerate substantial extrinsic sliding with no effect on bend shape or propagation. Brokaw, Josselin & Bobrow (1974) report a similar effect in sea urchin sperm induced to beat asymmetrically by high calcium concentrations. Their observations reveal that a bend in one direction (the principal bend) is often the only developing bend, so that sliding of all microtubules distal to the bend must be occurring synchronously, and yet there is no effect on bends present on the distal region of the sperm tail. It is apparent that, although it is generally accepted that the basis of ciliary movement is an active sliding microtubule mechanism, the way in which sliding forces are converted into bending moments and the manner of bend propagation are still subject to speculation and experiment.

Dynein–tubulin interaction

In the discussion so far we have considered the mechanics and geometry of sliding microtubules but not the detailed interaction between dynein and tubulin which almost certainly provides the motive force for sliding. Little is known about the details of the interaction but structural information has recently become available which may be relevant to the mechano-chemical cycle of dynein. For many years, electron micrographs of ciliary cross sections showed a gap between the 'free' end of a dynein arm and the neighbouring doublet towards which the arm is directed. By removing ATP rapidly from a detergent-treated axoneme, Gibbons (1975) obtained a preparation in which the dynein arms linked the A microtubule of one doublet to the B microtubule of its neighbour. More recently Sale & Satir (1977) and Warner & Mitchell (1978) have obtained images of microtubules arrested while sliding and found two additional dynein arm configurations, in which the arm is tilted with respect to the long axis of the microtubule. In one, called the extended form, the arm makes an angle of 50° with the microtubule axis and is sufficiently long (*c.* 24 nm) to bridge the interdoublet gap; in the other, the flattened form, the angle is reduced to about 20° and the arm is shortened to a length less than the interdoublet gap. (It is believed that this latter image corresponds to that observed in the early work.) Although it is possible that the images are to some degree artefacts, or may represent differences between the inner and outer arms, as suggested by Warner & Mitchell (1978), Satir (1979) is sufficiently confident that the images represent real configurational changes to suggest a cross-bridge cycle based on them, and to associate each stage in the cycle with a corresponding reaction in the enzymic activity of dynein.

By suitable extraction techniques, the dynein arms can be removed from the microtubules and the chemical properties of the dynein studied (e.g. Gibbons, B. H. & Gibbons, I. R., 1976). Using gel electrophoresis, more than six bands, identified either numerically or alphabetically according to the author, occur in the dynein region and several of these have ATPase activity. The major band observed is known as dynein 1 and this enzyme, in the isolated form, has the same requirements for activity as does an *in vitro* axoneme prepared for bending or sliding. The basic requirements are ATP and Mg^{2+}, with the latter present in at least as high a concentration as the ATP. It appears that dynein 1 is present in both arms (Kincaid, Gibbons & Gibbons, 1973) and is probably the component directly involved in the mechano-chemical event responsible for sliding in the axoneme. Baccetti *et al.* (1979) have studied the dynein from axonemes which naturally lack the inner or outer arms. Their results indicate that the characteristics of dynein 1 vary between different phyla and, as noted by Mabuchi and his coworkers (1974, 1976), between different species. They also conclude that dynein from the inner and outer arms migrates on gels at different rates, and that there is dynein associated with the radial spokes and central microtubule complex, an observation which assumes significance in relation to active spoke movements.

While the characterisation of dynein *in vitro* is essential, it is also necessary to obtain information about its behaviour *in vivo*, as it is difficult to obtain in a test tube conditions that mimic those in the axoneme. Some basic information about its chemical interactions has been obtained by experimentation with axonemes prepared for reactivation, but even this situation does not provide precisely the same environment as experienced by dynein within the axoneme. To investigate the relationships between the in-vivo and in-vitro conditions, some experimental method is needed which exposes both types of system to the same constraints and allows a comparison to be made. One approach to this problem is to examine the effects of physical variables, such as temperature and pressure on the movement of cilia, since these variables are known to affect the rate at which a chemical reaction proceeds and will therefore affect the activity of dynein.

In practice, the most sensitive parameter to changes in temperature and pressure is the beat frequency of an organelle (Holwill & Silvester, 1967; Coakley & Holwill, 1974; Holwill & Wais, 1979) and it has been possible to interpret the results of experiments by assuming that the beat frequency reflects a rate constant of a reaction within the organelle. By plotting suitable graphs the general theory of reaction rates (e.g. Johnson, Eyring & Stover, 1974) provides a method for evaluating changes in entropy, enthalpy and volume associated with a reaction characterised by the beat frequency. The experimental results obtained over a range of temperatures (0–40 °C) and pressures (0.1–60 MPa, 0.1 MPa = 1 atmosphere) suggest that the beat frequency does reflect the rate of an enzymic reaction, presumably that involving dynein, but the situation is complicated by denaturation effects. It appears that the rate-limiting reaction with flagella, i.e. that which is characterised by the beat frequency, is the breakdown of an ATP–ATPase complex. The results obtained in cells *in vivo* and *in vitro* indicate that the reactions affected are the same, with modifications in character due to the changed environmental conditions.

In a recent study, Holwill & Wais (1979) have examined the combined effects on the flagellar movement of *Crithidia oncopelti* of changing temperature, pressure and viscosity. They find that, up to a certain viscosity, which depends on the temperature, the beat frequency decreases with increasing pressure, while at the 'critical' viscosity, an increase in pressure causes an increase in beat frequency. This result can be explained if it is assumed that the effect of pressure is directly on a chemical reaction whereas the effect of viscosity on the reaction is exerted by mechanical constraints, for example by limiting the sliding rate. The sliding rate would affect the time for which a dynein arm and the active site on the appropriate doublet were opposed and could therefore, influence the reaction rate. The full significance of this result is still under investigation, but it is possible that the application of pressure with changed viscosity allows the identification of separate characteristics of the mechano-chemical process.

CONTROL OF CILIARY MOVEMENT

As noted in an earlier section some cilia and flagella are able to exert control over their movements as is clearly demonstrated in the avoiding response of certain cells and by the reversal or arrest of ciliary movement in certain metazoan epithelia cells (e.g. Aiello, 1974). By modifying the ionic environment of intact (using ionophores) and de-membranated cells it has been demonstrated in all the cells studied so far that the ciliary movement which characterises the behaviour described is sensitive to the level of calcium in the cell (e.g. Naitoh & Eckert, 1974 – *Paramecium*; Brokaw, 1974*a* – bracken spermatozoids; Holwill & McGregor, 1976 – *Crithidia oncopelti*; Murakami & Takahashi, 1975; Tsuchia, 1976 – *Mytilus*; Miller, 1975 – *Tubularia crocea* sperm; Satir, 1975; Walter & Satir, 1978 – *Elliptio complanatus*; Hyams & Borisy, 1978 – *Chlamydomonas*). Below a critical concentration (about 10^{-6} mol dm^{-3}) of calcium in de-membranated preparations normal movement is observed while above it the avoiding response develops. In most cells studied, the calcium required for the avoiding reaction by the cell *in vivo* is derived from the surrounding medium. When a hostile environment is encountered, channels in the membrane open to admit Ca^{2+} to the cell interior and later, once the avoiding response is complete, the ions are removed from the cell by a calcium 'pump'. In *C. oncopelti* calcium required for wave reversal appears to be supplied from a pool within the cell (Holwill & McGregor, 1976) since the avoiding response is observed when the cell is swimming in a medium devoid of calcium. Furthermore, *C. oncopelti* is insensitive to changes in the concentration of calcium in the medium, in contrast to the behaviour of *Paramecium* where ciliary reversal occurs with a periodicity which depends on the ratio of the numbers of Ca^{2+} to K^+ in the solution (see e.g. Dryl, 1970; Naitoh & Eckert, 1974).

Since the response to calcium can be evoked in organelles which have been separated from the cell (e.g. Hyams & Borisy, 1978; Walter & Satir, 1978) it appears that calcium acts directly on the axoneme or the basal structures to produce the observed reaction. It is of importance to identify the site of action of Ca^{2+} in the axoneme, and histochemical experiments on *Paramecium aurelia* have shown calcium-sensitive electron-dense deposits in the basal region of the cilia, on the cytoplasmic surface of the ciliary membrane (Plattner, 1975; Fisher, Kaneshiro & Peters, 1976; Tsuchia & Takahashi, 1976). Since the concentration of calcium needed to obtain these deposits is several orders of magnitude greater than that to which the cilia are sensitive, caution should be exercised in associating the deposits with active sites directly implicated in the behavioural response. There are, however, specialised membrane structures in the basal region of the cilia from many species, including the ciliary necklace (Gilula & Satir, 1972) and a ring of granule plaques, in which membrane particles are regularly arranged (Plattner, 1975). It is believed that these membrane particles are the channels through which calcium is admitted to the cilium.

The calcium-specific ATPase present in *Chlamydomonas reinhardii* flagella (Watanabe & Flavin, 1976) could be associated with the avoiding reaction in this cell, but its location is unknown. Since the enzyme is present in mutant flagella which lack the central microtubules, it may not be associated with these structures, although the absence of these microtubules as an integrated structure may indicate that the cell is unable to polymerise them; the components of the microtubules, including the enzyme may be present but would remain undetected by electron microscopy. Walter & Satir (1977) have shown that Ca^{2+} does not inhibit sliding of microtubules from mussel gill cilia, so that it is presumably not affecting the dynein–tubulin interaction. If the sliding process is not influenced by Ca^{2+}, the observed behavioural response could be produced if the mechanism which converts sliding into bending were affected by it.

A report by Salisbury & Floyd (1978) concerning contractile root structures may be of relevance to a consideration of control mechanisms in algal flagellates. These authors show that the rhizoplasts which join pairs of basal bodies in the quadriflagellate *Platymonas subcordiformis* undergo rhythmic contraction in the presence of calcium and ATP. It is suggested that this activity could contribute to bend initiation and co-ordination and to the orientation of the flagella relative to the cell. The smaller fibrous structures which link the basal bodies of *Chlamydomonas* flagella may perform a similar function, since it is observed that during the avoiding response the angle between the basal bodies decreases (Hyams & Borisy, 1978; Sleigh, 1979). The concentration of calcium to which the root structures of *P. subcordiformis* respond is considerably higher (10^{-3} mol dm^{-3}) than that (10^{-7} mol dm^{-3}) which induces the avoiding reaction in *Chlamydomonas*, but since little is known about the flagellar movements of *P. subcordiformis* constructive comparisons are not possible. It is possible, as suggested by Hyams & Borisy (1978), that in the avoiding response of *Chlamydomonas*, Ca^{2+} trigger two separate processes, one being the re-orientation of the flagella and the other being a change in the flagellar waveform.

Divalent cations, including Ca^{2+}, have been shown to affect the amount of dynein cross-bridging and of microtubule sliding observed in cilia isolated from *Tetrahymena pyriformis* (Warner, 1978; Warner & Mitchell, 1978; Zanetti, Mitchell & Warner 1979). The number of cross-bridges formed in the cilium is dependent on the concentration of Mg^{2+} or of Ca^{2+} and appears to be in simple equilibrium with the cation concentration, with 100% bridging occurring at 3 mmol dm^{-3} for the two ions respectively. Bridging is observed in the absence of ATP when Mg^{2+} is present, suggesting that this ion is involved in the attachment phase of the dynein–tubulin cross-bridge cycle. Addition of ATP to the preparation induces the arms to detach from the B micro-tubule. Microtubule sliding is observed under conditions of partial dyneim arm bridging, although a quantitative interpretation of the sliding results in terms of dynein bridges and ATPase activity is not yet possible. It is interesting to note that significant amounts of sliding can be obtained under conditions where only between 10% and 15% of the dynein arms are seen to make contact with the B microtubule. Many other divalent cations are found to affect bridging and sliding, but in several cases no correlation is found between the two processes. For example, Co^{3+}, Ba^{2+} and Ni^{3+} are able to induce a high percentage of cross-bridging, but do not support sliding. Mn^{2+} produces sliding equivalent to the corresponding magnesium concentration, but is the least efficient of all the ions in the production of bridging. Whether any of these cations has a physiologically important role in the regulation of sliding is not yet clear, and an answer to this question awaits further experimentation.

Considerable attention has been given to the role of divalent ions in the control of ciliary motion, but other biologically important ions affect the behaviour of these organelles and may also be involved in regulatory processes. For example, Gibbons *et al.* (1978) have shown that the vanadate ion at concentrations in the region of 1 μmol dm^{-3} inhibits the ATPase activity of dynein and also the motion of reactivated sea urchin sperm tails and cilia of sea urchin embryos. For these organelles, concentrations of vanadate below that causing complete inhibition produce a reduction in frequency but little change in waveform. Wais & Satir (1979) also report inhibition of mussel gill cilia by vanadate and note that the cilia stop at a different stage of the beat cycle from that found when the cilia are arrested by Ca^{2+}. The possibility of two switch points, each sensitive to a different ion, is suggested by Wais & Satir, and tentatively associated with sliding in the two halves of the axoneme separated by the diameter passing through microtubule 1. Since vanadium is present in many biological systems, Gibbons *et al.* (1978) suggest that it might have a regulatory role in living cilia.

The beat frequency of cilia can be varied without changing the bend shape by altering the ATP concentration (e.g. Brokaw, 1975a). Exposure of de-membranated sea-urchin sperm to CO_2 causes the bend angle to change but does not affect the frequency (Brokaw, 1977; Brokaw & Simonik, 1976). These results may support the suggestion that cilia have control mechanisms which independently regulate bend angle and beat frequency (e.g. Gibbons, 1974) but the results do not preclude the possibility that CO_2 has a more general effect on the mechanism which generates bending. As an alternative to a direct effect on the mechanism which regulates the bend angle, Brokaw (1977) suggests that CO_2 may influence a control mechanism which selectively inactivates cross-bridges during the normal bending cycle. As noted earlier, bending requires that microtubules on one side of the cilium produce a larger bending moment than those on the other side, a situation presumably achieved in practice by inhibition of the dynein–tubulin interaction. If this inhibition is partially destroyed, the bending moments produced in opposite senses by the two halves of the axoneme would become more equal and a lower bend amplitude (and angle) would result. The precise mode of action of CO_2 is, however, not clear, and further experiments are required to determine the real inter-ciliary mechanism responsible for this important effect.

The effect of ions on the behaviour of cilia suggests that electrical activity at the cellular level is involved in the responses described above. The association of a membrane action potential with the avoiding reactions of *Paramecium* and *Stylonychia mytilus* has been reported by several authors (e.g. Naitoh & Eckert, 1974; Eckert, Naitoh & Machemer, 1976; de Peyer & Machemer, 1978*b*; Nelson & Kung, 1978). One demonstration of the close relationship between the action potential and the avoiding response in *Paramecium* is seen when the behaviour of wild-type cells is compared with that of a mutant known as 'pawn', so-called because it does not exhibit the avoiding response and can only move forwards (e.g. Oertel, Schein & Kung, 1977). Electrophysiological experiments reveal that the mutant pawn does not produce an action potential. However, if the membrane of pawn is removed using detergents, the avoiding response can be elicited in the de-membranated cell by altering the Ca^{2+} concentration as described earlier (Kung & Naitoh, 1973). The membrane signal in the living cell is thus an essential feature of the control process.

The avoiding reaction of many cells occurs in response to a variety of stimuli, such as a changed chemical environment, increased light intensity and touch. For experimental purposes it is convenient to have a technique by which the response can be elicited in a controlled manner. This has been achieved for *Paramecium* and *Stylonychia mytilus* which are sensitive to mechanical stimuli and can be induced to reverse the direction of ciliary beating by application of a microneedle. Electrophysiological studies of mechano-sensitivity in these two cells have shown that electrical signals are produced in the membrane by mechanical stimulation of the cell, but that the type of signal depends on the site of stimulation. If the anterior portion of the cell is stimulated, depolarising receptor potentials, followed by action potentials, driven by calcium batteries are produced. Stimulation of the cell posterior activates a potassium battery which gives rise to a hyperpolarising receptor potential (e.g. Takahashi & Naitoh, 1978; de Peyer & Machemer, 1978*a*; Ogura & Machemer, personal communication). Mechanical stimulation of de-ciliated cells causes the same depolarisation and hyperpolarisation as in the living organism, thus indicating that the mechano-receptors are present in the somatic membrane only. Voltage-clamp experiments showed that the early inward current carried by Ca^{2+} as a prelude to ciliary reversal and seen in the living organism, is not present in the de-ciliated cell. The early inward current is restored during ciliary regeneration, thus indicating that the calcium channels associated with this current are located on the ciliary membrane. It is, however, not yet possible to use the results to test for a uniform or localised distribution of channels on the membrane and hence to correlate the results of these studies with known membrane structural features.

Although the detailed response varies from species to species, the sequence of activities which leads to modified ciliary movement is probably similar in form in all cases. The stimulus, be it physical or chemical, induces electrical activity of the membrane which allows calcium to enter the cilium from either outside the membrane or from binding sites within the cell. The calcium acts directly on the axoneme to alter the beat pattern, and must be actively removed before normal beating can resume.

Since the beat parameters such as frequency, bend angle and propagation rate alter according to the environmental conditions, they must, to some extent, be under cellular control. There is some evidence that bioelectric activity of the membrane is associated with changes in beat frequency (Naitoh & Eckert, 1974) and it is possible that changes in the other variables are also mediated in this way. There is, however, no direct evidence for the way in which bend propagation is controlled, although Brokaw (1975b, 1976) has developed theoretical models based on control of microtubular sliding by either local curvature or local shear which closely mimic the waveforms of real flagella.

EPILOGUE

In this brief and incomplete survey I have endeavoured to indicate major recent advances in our knowledge of ciliary behaviour. Certain areas of research, such as those using hydrodynamics, computer modelling and immunology, have not been treated here although they have made substantial contributions to the field. From the discussion, it is apparent that although much is known about events associated with microtubular sliding and the control of ciliary movement, a detailed description of the mechano-chemical processes responsible for bend propagation and its control await further experimentation.

REFERENCES

AIELLO, E. (1974). Control of ciliary activity in Metazoa. In *Cilia and Flagella*, ed. M. A. Sleigh, pp. 353–76. London: Academic Press.

AIELLO, E. & SLEIGH, M. A. (1972). The metachronal wave of lateral cilia of *Mytilus edulis*. *Journal of Cell Biology*, **54**, 493–506.

BACCETTI, B., BURRINI, A. G., DALLAI, R. & PALLINI, V. (1979). The dynein electrophoretic bands in axonemes naturally lacking the inner or the outer arm. *Journal of Cell Biology*, **80**, 334–40.

BARLOW, D. I. & SLEIGH, M. A. (1979). Freeze substitution for preservation of ciliated surfaces for scanning electron microscopy. *Journal of Miscroscopy*, **115**, 81–95.

BROKAW, C. J. (1965). Non-sinusoidal bending waves of sperm flagella. *Journal of Experimental Biology*, **43**, 155–69.

BROKAW, C. J. (1974a). Calcium and flagellar response during the chemotaxis of bracken spermatozoids. *Journal of Cell Physiology*, **83**, 151–8.

BROKAW, C. J. (1974b). Movement of the flagellum of some marine invertebrate spermatozoa. In *Cilia and Flagella*, ed. M. A. Sleigh, pp. 93–109. London: Academic Press.

BROKAW, C. J. (1975a). Effects of viscosity and ATP concentration on the movement of reactivated sea-urchin sperm flagella. *Journal of Experimental Biology*, **62**, 701–19.

BROKAW, C. J. (1975b). Molecular mechanism for oscillation in flagella and muscle. *Proceedings of the National Academy of Sciences, USA*, **72**, 3102–6.

BROKAW, C. J. (1975c). Spermatozoan motility: a biophysical survey. In *The Biology of the Male Gamete*, ed. J. G. Duckett & P. A. Racey, *Biological Journal of the Linnean Society*, 7, Suppl. 1., pp. 423–39.

BROKAW, C. J. (1976). Computer simulation of flagellar movement. IV. Properties of an oscillatory two-state cross-bridge model. *Biophysical Journal*, **16**, 1029–41.

BROKAW, C. J. (1977). CO_2-inhibition of the amplitude of bending of triton-demembranated sea-urchin sperm flagella. *Journal of Experimental Biology*, **71**, 229–40.

BROKAW, C. J., JOSSELIN, R. & BOBROW, L. (1974). Calcium ion regulation of flagellar beat symmetry in reactivated sea-urchin spermatozoa. *Biochemical and Biophysical Research Communications*, **58**, 795–800.

BROKAW, C. J. & SIMONICK, T. F. (1976). CO_2 regulation of the amplitude of flagellar bending. In *Cell Motility*, ed. R. Goldman, T. Pollard & J. Rosenbaum, pp. 933–40. New York: Cold Spring Harbor Laboratory.

BROKAW, C. J. & WRIGHT, L. (1963). Bending waves on the posterior flagellum of *Ceratium*. *Science*, **142**, 1169–70.

COAKLEY, C. J. & HOLWILL, M. E. J. (1974). Effects of pressure and temperature changes on the flagellar movement of *Crithidia oncopelti*. *Journal of Experimental Biology*, **60**, 605–29.

COSTELLO, D. P. (1973a). A new theory on the mechanics of ciliary and flagellar bending. I. Supporting observations. *Biological Bulletin*, **145**, 279–91.

COSTELLO, D. P. (1973b). A new theory on the mechanics of ciliary and flagellar motility. II. Theoretical considerations. *Biological Bulletin*, **145**, 292–309.

DE PEYER, J. & MACHEMER, H. (1978a). Hyperpolarizing and depolarizing mechanoreceptor potentials in *Stylonychia*. *Journal of Comparative Physiology*, **127**, 255–66.

DE PEYER, J. & MACHEMER, H. (1978b). Are receptor-activated ciliary motor responses mediated through voltage or current? *Nature, London*, **276**, 285–7.

DRYL, S. (1970). Response of ciliate protozoa to external stimuli. *Acta Protozoologica*, **7**, 325–33.

DRYL, S. (1975). Local microtubules interaction theory of ciliary and flagellar motion. *Bulletin de l'Académie Polonaise des Sciences*, **23**, 339–46.

ECKERT, R., NAITOH, Y. & MACHEMER, H. (1976). Calcium in the bioelectric and motor functions of *Paramecium*. In *Symposia of the Society for Experimental Biology*, **30**, ed. C. J. Duncan, 233–55.

FISHER, G., KANESHIRO, E. S. & PETERS, P. D. (1976). Divalent cation affinity sites in *Paramecium aurelia*. *Journal of Cell Biology*, **69**, 429–42.

GIBBONS, B. H. & GIBBONS, I. R. (1976). Functional recombination of dynein-1 with demembranated sea-urchin sperm partially extracted with KCl. *Biochemical and Biophysical Research Communications*, **73**, 1–6.

GIBBONS, I. R. (1974). Mechanisms of flagellar motility. In *The Functional Anatomy of the Spermatozoon*, ed. B. A. Afzelius, pp. 127–40. Oxford: Pergamon.

GIBBONS, I. R. (1975). The molecular basis of flagellar motility. In *Molecules and Cell Movement*, ed. S. Inoué & R. E. Stephens, pp. 207–32. New York: Raven Press.

GIBBONS, I. R., COSSON, M. P., EVANS, J. A., GIBBONS, B. H., HOUCK, B., MARTINSON, K. H., SALE, W. S. & TANG, W-J. Y. (1978). Potent inhibition of dynein adenosinetriphosphatase and of the motility of cilia and sperm flagella by vanadate. *Proceedings of the National Academy of Sciences, USA*, **75**, 2220–4.

GILULA, N. B. & SATIR, P. (1972). The ciliary necklace. *Journal of Cell Biology*, **53**, 494–509.

GOLDSTEIN, S. F. (1976). Form of developing bends in reactivated sperm flagella. *Journal of Experimental Biology*, **64**, 173–84.

GOLDSTEIN, S. F. (1977). Asymmetric waveforms in echinoderm sperm flagella. *Journal of Experimental Biology*, **71**, 157–70.

GOLDSTEIN, S. F., BESSE, C. & SCHREVEL, J. (1979). Axoneme fonctionnel de base 6 + 0. *Journal of Protozoology*, in press.

GOLDSTEIN, S. F., HOLWILL, M. E. J. & SILVESTER, N. R. (1970). The effects of laser microbeam irradiation on the flagellum of *Crithidia (Strigomonas) oncopelti*. *Journal of Experimental Biology*, **53**, 401–9.

GRAY, J. (1955). The movement of sea-urchin spermatozoa. *Journal of Experimental Biology*, **32**, 775–801.

HIRAMOTO, Y. & BABA, S. A. (1978). A quantitative analysis of flagellar movement in echinoderm spermatozoa. *Journal of Experimental Biology*, **76**, 85–104.

HOLWILL, M. E. J. (1965). The motion of *Strigomonas oncopelti*. *Journal of Experimental Biology*, **42**, 125–37.

HOLWILL, M. E. J. (1966). Physical aspects of flagellar movement. *Physiological Reviews*, **46**, 696–785.

HOLWILL, M. E. J. (1974). Hydrodynamic aspects of ciliary and flagellar movement. In *Cilia and Flagella*, ed. M. A. Sleigh, pp. 143–75. London: Academic Press.

HOLWILL, M. E. J., COHEN, H. J. & SATIR, P. (1979). A sliding microtubule model incorporating axonemal twist and compatible with three-dimensional sliding. *Journal of Experimental Biology*, **78**, 265–80.

HOLWILL, M. E. J. & MCGREGOR, J. L. (1974). Micromanipulation of the flagellum of *Crithidia oncopelti*. I. Mechanical effects. *Journal of Experimental Biology*, **60**, 437–44.

HOLWILL, M. E. J. & MCGREGOR, J. L. (1976). Effects of calcium on flagellar movement in the trypanosome *Crithidia oncopelti*. *Journal of Experimental Biology*, **65**, 229–42.

HOLWILL, M. E. J. & SILVESTER, N. R. (1967). Thermodynamic aspects of flagellar activity. *Journal of Experimental Biology*, **47**, 249–65.

HOLWILL, M. E. J. & WAIS, J. (1979). Thermodynamic and hydrodynamic studies relating to the mechanochemical cycle in the flagellum of *Crithidia oncopelti*. *Journal of Experimental Biology*, in press.

HYAMS, J. S. & BORISY, G. G. (1978). Isolated flagellar apparatus of *Chlamydomonas*: characterization of forward swimming and alteration of waveform and reversal of motion by calcium ions *in vitro*. *Journal of Cell Science*, **33**, 235–53.

JOHNSON, F. H., EYRING, H. & STOVER, B. J. (1974). *The Theory of Rate Processes in Biology and Medicine*. New York: Wiley.

JOHNSTON, D. N. & SILVESTER, N. R. (1979). A digitally controlled curve follower. *Journal of Physics E: Scientific Instruments*, **12**, 235–6.

JOHNSTON, D. N., SILVESTER, N. R. & HOLWILL, M. E. J. (1979). An analysis of the shape and propagation of waves on the flagellum of *Crithidia oncopelti*. *Journal of Experimental Biology*, **80**, 299–315.

KINCAID, H. L. JR., GIBBONS, B. H. & GIBBONS, I. R. (1973). The salt extractable fraction of dynein from sea-urchin flagella: an analysis by gel electrophoresis and adenosine triphosphatase activity. *Journal of Supramolecular Structure*, **1**, 461–70.

KUNG, C. & NAITOH, Y. (1973). Calcium-induced ciliary reversal in the extracted models of 'Pawn', a behavioral mutant of *Paramecium*. *Science*, **179**, 195–6.

KUSEL, J. P., MOORE, K. E. & WEBER, M. M. (1967). The ultrastructure of *Crithidia fasciculata* and morphological changes induced by growth in acriflavin. *Journal of Protozoology*, **14**, 283–96.

LANGFORD, G. M. & INOUÉ, G. M. (1979). Motility of the microtubular axostyle in *Pyrsonympha*. *Journal of Cell Biology*, **80**, 521–38.

MABUSHI, I. & SHIMIZU, T. (1974). Electrophoretic studies on dyneins from *Tetrahymena* cilia. *Journal of Biochemistry* (*Tokyo*), **76**, 991–9.

MABUCHI, I., SHIMIZU, T. & MABUSHI, Y. (1976). A biochemical study of flagellar dynein from starfish spermatozoa: protein components of the arm structure. *Archives of Biochemistry and Biophysics*, **176**, 561–76.

MACHEMER, H. (1974). Ciliary activity and metachronism in Protozoa. In *Cilia and Flagella*, ed. M. A. Sleigh, pp. 199–286. London: Academic Press.

MACHEMER, H. (1977). Motor activity and bioelectric control of cilia. *Fortschritte der Zoologie*, **24**, 195–210.

MACHIN, K. E. (1958). Wave propagation along flagella. *Journal of Experimental Biology*, **25**, 796–806.

MANTON, I. & PARKE, M. (1960). Further observations on small green flagellates with special reference to possible relatives of *Chromulina pusilla Butcher*. *Journal of the Marine Biological Association, UK*, **39**, 275–98.

MELLOR, J. S. & HYAMS, J. S. (1978). Metachronism of cilia of the digestive epithelium of *Cirriformia tentaculata*. *Micron*, **9**, 91–4.

MILES, C. A. & HOLWILL, M. E. J. (1969). Asymmetric flagellar movement in relation fo the orientation of the spore of *Blastocladiella emersonii*. *Journal of Experimental Biology*, **50**, 683–7.

MILLER, R. L. (1975). Effect of calcium on *Tubularia* sperm chemotaxis. *Journal of Cell Biology*, **67**, 285A.

MURAKAMI, M. & TAKAHASHI, K. (1975). The role of calcium in the control of ciliary motility in *Mytilus*. II. The effects of calcium ionophores X537A and A23187 on the lateral gill cilia. *Journal of the Faculty of Science of the Imperial University of Tokyo. Section IV Zoology*, **13**, 251–6.

NAITOH, Y. & ECKERT, R. (1974). The control of ciliary activity in Protozoa. In *Cilia and Flagella*, ed. M. A. Sleigh, pp. 305–52. London: Academic Press.

NAKAMURA, S. & KAMIYA, R. (1978). Bending motion in split flagella of *Chlamydomonas*. *Cell Structure and Function*, **3**, 141–4.

NELSON, D. L. & KUNG, C. (1978). Behavior of *Paramecium* – chemical physiology and genetic studies. In *Taxis and Behavior*, ed. G. L. Hazelbauer, vol. 5, pp. 77–100. New York: Chapman & Hall.

OERTEL, D., SCHEIN, S. J. & KUNG, C. (1977). Separation of membrane currents using a mutant. *Nature, London*, **268**, 120–4.

OMOTO, C. K. & KUNG, C. (1979). The pair of central tubules rotates during ciliary beat in *Paramecium*. *Nature, London*, **279**, 532–4.

PLATTNER, H. (1975). Ciliary granule plaques: membrane-intercalated particle aggregates associated with Ca^{2+}-binding sites in *Paramecium*. *Journal of Cell Science*, **18**, 257–69.

RIKMENSPOEL, R. (1971). Contractile mechanisms in flagella. *Biophysical Journal*, **11**, 446–63.

RIKMENSPOEL, R. (1976). Contractile events in the cilia of *Paramecium, Opalina, Mytilus* and *Phragmatopoma*. *Biophysical Journal*, **16**, 445–69.

RIKMENSPOEL, R. (1978). Movement of sea urchin sperm flagella. *Journal of Cell Biology*, **76**, 310–22.

RIKMENSPOEL, E. & RUDD, W. G. (1973). The contractile mechanism in cilia. *Biophysical Journal*, **13**, 955–93.

SALE, W. S. & SATIR, P. (1976). Splayed *Tetrahymena* cilia. A system for analysing sliding and axonemal spoke arrangements. *Journal of Cell Biology*, **71**, 589–605.

SALE, W. S. & SATIR, P. (1977). Direction of active sliding of microtubules in *Tetrahymena* cilia. *Proceedings of the National Academy of Sciences, USA*, **74**, 2045–9.

SALISBURY, J. L. & FLOYD, G. L. (1978). Calcium-induced contraction of the rhizoplast of a quadriflagellate green alga. *Science*, **202**, 975–7.

SATIR, P. (1963). Studies on cilia. The fixation of the metachronal wave. *Journal of Cell Biology*, **18**, 345–65.

SATIR, P. (1965). Studies on cilia. II. Examination of the distal region of the ciliary shaft and the role of the filaments in motility. *Journal of Cell Biology*, **26**, 805–34.

SATIR, P. (1975). Ionophore-mediated calcium entry induces mussel gill ciliary arrest. *Science*, **190**, 586–8.

SATIR, P. (1979). Basis of flagellar motility in spermatozoa: current status. In *Spermatozoan Symposium*, ed. D. W. Fawcett, in press.

SILVESTER, N. R. & HOLWILL, M. E. J. (1972). An analysis of hypothetical flagellar waveforms. *Journal of Theoretical Biology*, **35**, 505–23.

SILVESTER, N. R. & JOHNSTON, D. (1976). An electro-optical curve follower with analogue control. *Journal of Physics E: Scientific Instruments*, **9**, 990–5.

SLEIGH, M. A. (1974a). *Cilia and Flagella*. London: Academic Press.

SLEIGH, M. A. (1974b). Patterns of movement of cilia and flagella. In *Cilia and Flagella*, ed. M. A. Sleigh, pp. 79–92. London: Academic Press.

SLEIGH, M. A. (1979). Contractility of the roots of flagella and cilia. *Nature, London*, **277**, 263–4.

SUMMERS, K. E. & GIBBONS, I. R. (1971). Adenosine triphosphate-induced sliding of microtubules in trypsin treated flagella of sea-urchin sperm. *Proceedings of the National Academy of Sciences, USA*, **68**, 3092–6.

TAKAHASHI, M. & NAITOH, Y. (1978). Behavioral mutants of *Paramecium caudatum* with defective membrane electrogenesis. *Nature, London*, **271**, 656–9.

Tsuchia, T. (1976). Ca-induced arrest response in Triton-extracted lateral cilia of *Mytilus* gill. *Experientia*, **32**, 1439–40.

Tsuchia, T. & Takahashi, K. (1976). Localization of possible calcium-binding sites in the cilia of *Paramecium caudatum*. *Journal of Protozoology*, **23**, 523–6.

Wais, J. & Satir, P. (1979). Effect of vanadate on gill cilia: switching mechanisms in ciliary beat. *Biophysical Journal*, in press.

Walter, M. F. & Satir, P. (1977). Calcium does not inhibit sliding of microtubules from mussel gill cilia. *Journal of Cell Biology*, **75**, 287a.

Walter, M. F. & Satir, P. (1978). Calcium control of ciliary arrest in mussel gill cells. *Journal of Cell Biology*, **79**, 110–20.

Warner, F. D. (1974). The fine structure of the ciliary and flagellar axoneme. In *Cilia and Flagella*, ed. M. A. Sleigh, pp. 11–37. London: Academic Press.

Warner, F. D. (1976). Ciliary intermicrotubule bridges. *Journal of Cell Science*, **20**, 101–14.

Warner, F. D. (1978). Cation-induced attachment of ciliary dynein crossbridges. *Journal of Cell Biology*, **77**, R19–26.

Warner, F. D. & Mitchell, D. R. (1978). Structural conformation of ciliary dynein arms and the generation of sliding forces in *Tetrahymena* cilia. *Journal of Cell Biology*, **76**, 261–77.

Warner, F. D. & Satir, P. (1974). The structural basis of ciliary bend formation. Radial spoke positional changes accompanying microtubule sliding. *Journal of Cell Biology*, **63**, 35–63.

Watanabe, T. & Flavin, M. (1976). Nucleotide-metabolizing enzymes in *Chlamydomonas* flagella. *Journal of Biological Chemistry*, **251**, 182–92.

Witman, G. B., Fay, R. & Plummer, J. (1975). *Chlamydomonas* mutants: evidence for the roles of specific axonemal components in flagellar movement. In *Cell Motility*, eds. R. Goldman, J. Rosenbaum & T. Pollard, pp. 969–86. New York: Cold Spring Harbor Laboratory.

Witman, G. B., Plummer, J. & Sander, G. (1978). *Chlamydomonas* flagellar mutants lacking radial spokes and central tubules. Structure, composition and function of specific axonemal components. *Journal of Cell Biology*, **76**, 729–47.

Woolley, D. M. (1977). Evidence for 'Twisted Plane' undulations in golden hamster sperm tails. *Journal of Cell Biology*, **75**, 851–65.

Zanetti, N. C., Mitchell, D. R. & Warner, F. D. (1979). Effects of divalent cations on dynein cross-bridging and ciliary microtubule sliding. *Journal of Cell Biology*, **80**, 573–88.

EXPLANATION OF PLATES

PLATE 1

The planar wave observed on the flagellum of *Crithidia oncopelti*. (Micrograph kindly provided by Dr D. N. Johnston.)

PLATE 2

Scanning electron micrograph showing metachronal waves on the ciliated surface of *Paramecium candatum*. The organism was prepared using the rapid-freezing technique described by D. I. Barlow and M. A. Sleigh (1979), who also kindly supplied this micrograph.

PLATE 3

(a) Flimmer on the flagellum of *Distigma proteus*. (b) Mastigonemes on the flagellum of *Ochromonas danica*. (Micrographs kindly provided by Professor J. Schrevel.)

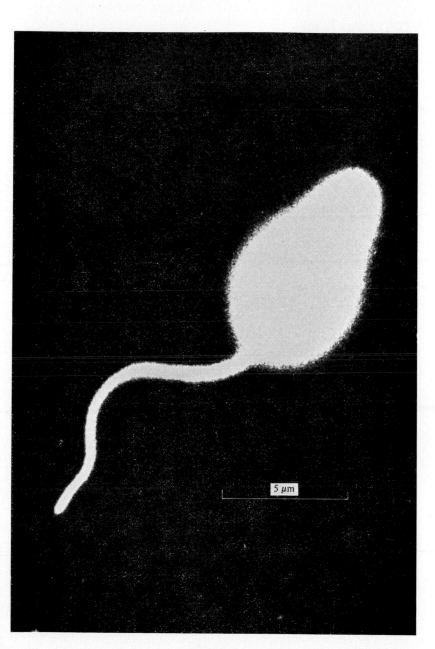

5 μm

PLATE 1

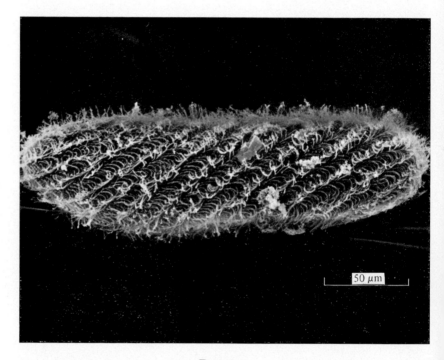

PLATE 2

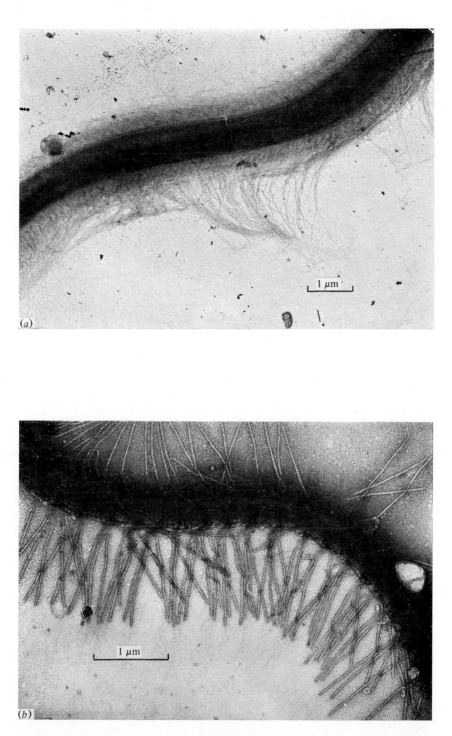

(a)

1 μm

(b)

1 μm

PLATE 3

SEXUAL MICROBIOLOGY: MATING REACTIONS OF *CHLAMYDOMONAS REINHARDII*, *TETRAHYMENA THERMOPHILA* AND *SACCHAROMYCES CEREVISIAE*

URSULA W. GOODENOUGH

*Department of Biology, Washington University,
St Louis, Missouri 63130, USA*

INTRODUCTION

This review will consider sexual differentiation and the mating strategies employed by three unicellular eukaryotes: the flagellate *Chlamydomonas reinhardii*, the ciliate *Tetrahymena thermophila* and the yeast *Saccharomyces cerevisiae*. Reviews of all three mating systems have recently appeared (Crandall, 1977; Goodenough, 1977; Goodenough & Forest, 1978; Goodenough *et al.*, 1979; MacKay, 1978; Manney & Meade, 1977; Miyake, 1978; Nanney, 1977). Emphasis will therefore be given here to literature that has escaped earlier review, to very recent publications, and to areas where research opportunities appear particularly ripe. In limiting consideration to only three unicellular organisms, the enormous array of sexual strategies employed by eukaryotic microbes will be depicted most inadequately. On the other hand, most of the individual strategies utilised by these three species have been widely adopted by other micro-organisms. It is also appropriate to note at the outset that many of the cellular and genetic phenomena described for these three organisms have conspicuous analogues in the Metazoa. Thus the elaboration of a microfilament-filled fertilisation tubule by gametes of *Chlamydomonas reinhardii* is strikingly similar to the acrosome reaction of echinoderm sperm (Tilney, 1977). The ubiquitous requirement for phase G_1 arrest prior to sexual fusion is found as well for fusing myoblasts (Konigsberg, Sollmann & Mixter, 1978). And the movement of 'cassettes' of mating-type genetic information from one yeast chromosomal location to another is highly reminiscent of the movement of genetic information for immunoglobulin synthesis during lymphocyte differentiation (Sakano *et al.*, 1979). The devoted microbiologist does not, of course, need these analogues to sustain interest in his or her research: an understanding of the micro-organisms themselves is a fully valid and stimulating goal.

CHLAMYDOMONAS REINHARDII

Stimulation and course of sexual differentiation

The biflagellate Chlorophyte *Chlamydomonas reinhardii* is ordinarily capable of three states of differentiation: the mitotic haploid vegetative cell, the non-dividing gametic cell, and the diploid zygote destined to undergo meiosis. The gametes engage in copulatory interactions that are very similar to those employed by 'higher' multicellular relatives of *C. reinhardii*, such as species of *Pandorina*, *Volvox*, and *Ulva* (Coleman, 1979; Løvlie & Bryhni, 1976). Indeed, the hierarchy shown in the Volvocales provides a classical example of a virtually unchallenged biological law: no matter how complex the eukaryotic soma becomes, sexual interactions are in the end the province of single cells.

Gametes of *Chlamydomonas reinhardii* are of two mating types (*mt*) known as *plus* and *minus*. Gametogenesis of mt^+ cells can take place in the complete absence of mt^- cells and vice versa: there is no evidence that any physical or hormonal input from the opposite type contributes to the sexual differentiation process, and we can thus designate the differentiation as totally self-stimulatory.

For both mating types, sexual differentiation is triggered in the laboratory by nitrogen starvation (Sager & Granick, 1954), whereupon the cells shift from their vegetative mode of mitotic growth and embark upon gametogenesis. Presumably, nitrogen starvation most commonly occurs as a gradual process in the organism's natural soil habitat. Such a gradual starvation can be simulated in the laboratory by plating cells to an agar-containing medium and allowing them to deplete the growth medium of NH_4^+ (Martin & Goodenough, 1975). Such 'plate gametogenesis' is to be contrasted with the abrupt withdrawal of nitrogen effected when synchronous vegetative cultures are centrifuged and resuspended in NH_4^+-free medium (Kates & Jones, 1964; Chiang *et al.*, 1970; Martin & Goodenough, 1975). 'Synchronous gametogenesis' is, of course, more amenable to experimental study, and has been shown to have the following properties.

(1) One or two rounds of DNA replication, followed by one or two mitotic cell divisions, almost inevitably occur during synchronous gametogenesis (Kates, Chiang & Jones, 1968), the nitrogen for the biosynthesis of deoxyribonucleotides and histones apparently deriving in large part from the catabolism of pre-existing rRNA, ribosomal proteins, and chloroplast membranes (Siersma & Chiang, 1971; Martin, Chiang, & Goodenough, 1976).

(2) Although small numbers of cells will differentiate into gametes if nitrogen starvation is imposed at any stage in the 24-hour synchronous vegetative cycle, optimal (100%) gametogenesis is achieved when vegetative cells are deprived of NH_4^+ in the mid-G_1 phase of the cell cycle (Kates, 1966; Schmeisser, Beaumgartel & Howell, 1973). Under these conditions, differentiation requires about 12 hours and culminates in the gametic mitosis noted above.

(3) Unmated cells can remain for months in a non-dividing G_0 state under appropriate laboratory conditions (Martin & Goodenough, 1975). If returned at any time to an NH_4^+-containing medium, such gametes will de-differentiate in about ten hours, and undergo their first vegetative mitosis some 12 hours later (Kates & Jones, 1964).

Several investigators have documented that major cellular and metabolic changes accompany a synchronous *Chlamydomonas reinhardii* gametogenesis (Kates & Jones, 1966, 1967; Martin & Goodenough, 1975; Minami, 1977): the cells quite literally dismantle themselves. The vast majority of these changes, however, probably represent adjustments to the stress of sudden nitrogen starvation and do not reflect sexual differentiation *per se*. The two phenotypic changes that bear obvious relevance to mating ability are changes in the flagellar surface and the assembly of mating structures; these are described below.

At all mitotic divisions, *Chlamydomonas reinhardii* cells dismantle their flagella at preprophase and grow out new flagella after cytokinesis (Buffaloe, 1958; Johnson & Porter, 1968). The flagella that grow out after the gametic mitosis, however, differ in at least two respects from their vegetative counterparts. First, both *plus* and *minus* gametic flagella respond to the presence of such bound surface ligands as concanavalin A (Wiese & Shoemaker, 1970; McLean & Brown, 1974) or antibody (Goodenough & Jurivich, 1978) by moving them to their flagellar tips, a response that has been designated 'tipping' (Goodenough & Jurivich, 1978). The development of the capacity for the tipping response suggests that gametic surface components are differently organised in the flagellar membrane; the molecular basis for this distinctive organisation is being sought in several laboratories. The second difference between vegetative and gametic flagella is the presence on the latter of *mt*-specific agglutinins and agglutinin receptors. Thus, *mt*+ gametes bear *plus* agglutinins that recognise and adhere to receptors on *minus* gametic flagella, and vice-versa; more information on these molecules is given in a later section.

The mating structures of *Chlamydomonas reinhardii* gametes are first detected on the cleavage-furrow membrane of the gametic mitosis (Martin & Goodenough, 1975), suggesting that they, too, are assembled in the final minutes of gametogenesis (unknown, however, is the time during the 12-hour differentiation process that the genetic information for these gametic traits is first expressed). Both *plus* and *minus* mating structures are intimately associated with a microtubule-rootlet system of the basal apparatus (Goodenough & Weiss, 1978), and both consist of a small differentiated region of the plasma membrane underlain by a plaque of dense cytoplasmic material (Friedmann, Colwin & Colwin, 1968; Goodenough & Weiss, 1975; Weiss, Goodenough & Goodenough, 1977). The mt^+ structure, however, possesses a more complex cytoplasmic component known as the doublet zone, and can be readily distinguished from its mt^- counterpart in thin section (Friedmann *et al.*, 1968; Goodenough & Weiss, 1975; Cavalier-Smith, 1975; Triemer & Brown, 1975).

Genetics of sexual differentiation

Nothing is known about the gene(s) that respond to nitrogen starvation in *Chlamydomonas reinhardii*, nor about the mechanism that prevents the expression of gamete-specific genes during vegetative growth. A search for constitutive mutants that differentiate in the presence of NH_4^+ should be undertaken, as should a search for pleiotropic gametogenesis mutants, defective in all gametic functions, which might carry mutations in key 'early genes'.

By contrast, limited information is accumulating on the genes that determine the gametic traits. Ultimate control is exerted by the mating-type (*mt*) locus itself, located in linkage-group VI (Ebersold *et al.*, 1962). The mt^+ and mt^- versions of this locus segregate in a Mendelian fashion in genetic crosses (Smith & Regnery, 1950; Sager, 1955) and have never been reported to interconvert. In addition to controlling gametic phenotype, the *mt* loci influence the pattern of inheritance of chloroplast DNA in the zygote (Sager, 1954; Gillham, 1969). It is not known whether *mt* is expressed during the vegetative phase of the organism's life cycle.

To date, one mutation (*imp-1*) affecting gametic fusion (Goodenough & Weiss, 1975; Goodenough, Hwang & Martin, 1976) and two (*mat-1* and *mat-2*) affecting chloroplast DNA transmission (Sager & Ramanis, 1974) have been found to be closely linked to the *mt* locus; whether they actually lie within the locus cannot be assessed due to an apparent suppression of recombination in this sector of the genome (Gillham, 1969). Of the remaining genes controlling gametic traits, the most interesting are marked by 'sex-limited' mutations, which are unlinked to *mt* but affect sexuality in one mating type only. Thus the conditional *gam-1* mutation (Forest & Togasaki, 1975) blocks sexual signalling (Forest, Goodenough & Goodenough, 1978) only when carried by *mt⁻* cells; it has no effect on *mt⁺* 'carriers'. Similarly, the five closely-linked mutations marking the *sag-1* (*sexual agglutination*) locus and the single mutation affecting *sag-2* are expressed only in *mt⁺* cells (Good-enough, Hwang & Warren, 1978); they are without effect on the flagellar agglutinability of *mt⁻* (*sag*) 'carriers'. The existence of sex-limited loci reveals that *mt* must act as a 'master locus', affecting the expression of selected genes lying elsewhere in the genome. Whether it acts at the transcriptional or post-transcriptional level is not yet known.

Sexual recognition

When *mt⁺* and *mt⁻* gametes are mixed together, they instantly recognise one another by the protease-sensitive agglutinins associated with their flagellar membranes (Wiese, 1965; Wiese & Metz, 1969; Bergman *et al.*, 1975; Snell, 1976), and these are mobilised to the tip where a tight adhesion is established (Mesland, 1976; Goodenough & Jurivich, 1978). The unsuccessful attempts to identify these agglutinins and their recep-tors biochemically have been reviewed (Goodenough, 1977). A promis-ing approach has more recently been discovered by W. S. Adair, C. Hwang, and U. W. Goodenough. They find that both the *mt⁺* and *mt⁻* agglutinins remain active after flagellar-membrane extraction by the detergent octylglucoside (Whittenberger *et al.*, 1978). Moreover, the *mt⁺* agglutinin is selectively retained on an affinity column bearing derivatised IgG antibody raised against gametic *mt⁺* flagella. In contrast a parallel column carrying antibody raised against vegetative *mt⁻* flagella does not retain the *mt⁺* agglutinin but does retain many other membrane glycopolypeptides present in the detergent extract, including the prominent 'major membrane glycoprotein' (Witman *et al.*, 1972; Bergman *et al.*, 1975; Snell, 1976) which has the most antigenic activity (Adair, Jurivich & Goodenough, 1978). Thus it would appear that a tentative identification of the *mt⁺* agglutinin is close at hand. This identification will be considerably strengthened if it can then be shown that at least one of the six *sag* mutant strains, all of which lack *mt⁺* agglutination activity (Bergman *et al.*, 1975), also exhibits an absence or electrophoretic modification of the (glyco)polypeptide suspected to serve as the agglutinin.

Once the *mt⁺* and *mt⁻* agglutinins are identified biochemically, it will be important to determine whether they interact with one another or whether each interacts with an *mt*-specific receptor displayed on opposite-type flagella. It will also be possible to compare the carbohydrate and peptide composition of the two molecular species, determine with which polypeptides they associate in the membrane, and ascertain at what stage during synchronous gametogenesis they are synthesised. Finally, it may be possible to probe the mechanism by which zygotic flagella lose their agglutinability at the time of cell fusion (Lewin, 1954; Goodenough, 1977) and to ask why it is that living gametes of one *mt* can agglutinate with membranes, isolated flagella, or glutaraldehyde-fixed gametes of opposite *mt*, whereas mixtures of the non-living material exhibit no agglutination (Goodenough, 1977).

Mesland and Goodenough have recently discovered (Goodenough *et al.*, 1979; D. A. M. Mesland & U. W. Goodenough, unpublished) that sexual adhesion in *Chlamydomonas reinhardii* is normally followed by the appearance of a dense material in the flagellar tips which fills the space between the nine singlet axonemal microtubules and the flagellar membrane. The material accumulates during the one to two minutes that agglutination takes place and disappears once cell fusion has occurred. Its reversible appearance is accompanied by a reversible elongation in the length of the tip microtubules. This transformation in tip morphology, termed *flagellar tip activation*, is blocked by colchicine and by an antiserum directed against tubulin, both of which block zygotic fusion but not adhesion, demonstrating that the activation process can be experimentally uncoupled from adhesion. Similarly, limited chymotrypsin digestion and the *imp-1* mutation block fusion but do not affect either adhesion nor tip activation; therefore, the activation process can be experimentally uncoupled from fusion. These and other experimental approaches have led D. A. M. Mesland and colleagues to conclude that sexual adhesion normally elicits tip activation and that activation appears to represent an intermediate event in the mating reaction.

Sexual signal generation

The most attractive function for tip activation is to postulate that it plays a role in sexual signalling, informing the cells that a 'meaningful' recognition event has taken place and that it is 'time to fuse'. It is not known whether the gametes detect the presence of the dense material and/or the elongated tip region, and the mechanisms by which such sensations are collected and conveyed to the cell are as yet only subjects for speculation.

Sexual signalling is rendered defective by the *gam-1* mutation: when *mt⁻ gam-1* mutants are differentiated into gametes at non-permissive temperature, their flagella agglutinate but fail to signal normally (Forest, Goodenough & Goodenough, 1978). D. A. M. Mesland has found (unpublished) that many of the flagella of *gam-1* cells grown at 35 °C bear activated tips even when *mt*+ gametes are not present, revealing a defective control over the activation process. Moreover, isolated *gam-1* flagella are able to stimulate wild-type *mt*+ tip activation only when presented at six times the concentration necessary for normal *mt*− flagella to elicit activation, revealing that a quantitative defect is also incurred by the mutation. Since tip activation normally occurs in both mating types, it is intriguing to ask why the *gam-1* mutation should be limited in its expression to *mt*− gametes. One speculation holds that the normal *gam-1* allele is responsible for the proper association between the *mt*− agglutinin and the signal apparatus. Indeed, available information on the properties of the mutant is compatible with the possibility that the normal *gam-1* allele is in fact the structural gene for the *mt*− agglutinin and that the *gam-1* mutation affects its correct positioning in the membrane rather than its agglutination properties *per se*.

Response to sexual signalling

Gametes of *Chlamydomonas reinhardii* give at least two responses to sexual signalling: they release an autolysin which digests their glyco-proteinaceous walls (Claes, 1971; Schlösser, Sachs & Robinson, 1976; Solter & Gibor, 1977a) and they activate their mating structures. The specific site of autolysin release is unknown but it clearly resides at the cell's anterior end (Goodenough & Weiss, 1975); release can be stimulated by non-specific adhesion events (Claes, 1977), and it is reported to be more active in *mt*+ than in *mt*− gametes (Matsuda, Tamaki & Tsubo, 1978). Mating-structure activation entails the extension of a microfilament-filled fertilisation tubule from the doublet zone of the *mt*+ mating structure (Friedmann *et al.*, 1968; Goodenough & Weiss, 1975; Cavalier-Smith, 1975; Triemer & Brown, 1975) and a rearrange-ment of the specialised region of the plasma membrane in *mt*− gametes (Weiss *et al.*, 1977). Since the mating structures are also localised at the cell's anterior end, and are connected to flagellar basal bodies via a microtubule system (Goodenough & Weiss, 1978), it is reasonable to propose that just as signals appear to be generated at the flagellar tips, signals are transmitted to the cells at the flagellar bases. Nothing is yet known, however, about the signal–transmission mechanism(s), although speculations have been offered (Solter & Gibor, 1977a; Goodenough & Weiss, 1978; Goodenough *et al.*, 1979).

A number of experimental manipulations have been reported to inhibit cell-wall digestion and/or mating structure activation: these include fixation of one mating partner (Goodenough & Weiss, 1975; Solter & Gibor, 1977a; Mesland & van den Ende, 1979), cold temperature (Goodenough et al., 1979), proteolysis (Solter & Gibor, 1978; Goodenough et al., 1979), metabolic inhibitors (Solter & Gibor, 1978), and short flagellar length (Solter & Gibor, 1977b; Ray, Solter & Gibor, 1978). Most of these agents have not yet been tested for their effects on flagellar tip activation, and it is therefore not yet known whether they inhibit signal generation or signal response. As noted earlier, however, chymotrypsin digestion clearly affects some feature of signal response (D. A. M. Mesland, E. Caligor, J. L. Hoffman & U. W. Goodenough, unpublished) as does the imp-1 mutation (Goodenough & Weiss, 1975).

Cell fusion

The final event in the Chlamydomonas reinhardii mating response is the fusion of cell pairs at the site where the mt^+ fertilisation tubule makes contact with the mt^- mating structure. The fused cells are initially joined only by a narrow cytoplasmic bridge, but this quickly opens so that the two cells become confluent. The fusion event is mt-specific: gametes of one mt, induced by antibody agglutination to activate their mating structures, fail to fuse with one another (Goodenough & Jurivich, 1978). Mechanisms of membrane fusion are presently unknown for any system; speculations on C. reinhardii fusion mechanisms have been presented by Weiss et al. (1977).

TETRAHYMENA THERMOPHILA

Stimulation and course of sexual differentiation

As with Chlamydomonas reinhardii, sexual differentiation of Tetrahymena thermophila (formerly syngen 1 of Tetrahymena pyriformis) and other ciliates is triggered in the laboratory by starvation, e.g. by suspending the cells in 10mM-Tris buffer. Cells that have freshly emerged from a conjugation event fail to differentiate when so shifted to non-nutrient medium, and are said to be sexually immature. Such cells must undergo some 50–80 mitotic fissions before sexual maturity is attained; the molecular basis for attaining maturity is unknown.

Sexually mature Tetrahymena thermophila can undergo the first stage of differentiation, known as sensitisation or initiation, under one of two laboratory conditions. In the first, mitotic cells of complementary mating type (e.g. types BIII and BVII) are starved separately for about 18 hours; during this time, organisms already in the G_1 phase of the cell cycle remain in G_1, while cells in later stages complete mitosis and then arrest in G_1 (Wolfe, 1973). There is no reported period in the cell cycle at which differentiation following starvation is optimal, as with *Chlamydomonas reinhardii*.

The second laboratory protocol for inducing *Tetrahymena thermophila* sensitisation is to mix mitotic cells of *both* mating types in starvation medium. Under these conditions, the G_2 cells induce one another to traverse the cell cycle, a synchronous wave of mitosis takes place, and a population of sensitised G_1 cells of complementary type is generated in only two hours (Wolfe, 1974). The mechanism of the induction process is unknown except that direct cell contacts appear to be necessary: if a Millipore filter separates the two types during the two-hour starvation, no mitotic wave takes place (Wolfe, 1974).

Sensitisation is inhibited by treatment with cycloheximide, actinomycin D, and high concentrations of Tris buffer (Bruns & Brussard, 1974; Allewell, Oles & Wolfe, 1976; Wellnitz & Bruns, 1979). The one physiological parameter that has been studied during the sensitisation period is uridine transport. Adair & Wolfe (1977) have demonstrated that the rate of uridine uptake drops by an order of magnitude within two hours after starvation, and they argue that the effect is caused by a reduction in the number of functional transport molecules at the cell surface. Whether this change in the cell surface is causally related to the acquisition of mating competence could not be ascertained.

Following sensitisation, *Tetrahymena thermophila* cells enter the second stage of sexual differentiation, known as *co-stimulation* (Wolfe, 1973; Bruns & Brussard, 1974). Co-stimulation is initiated by mixing sensitised cells of opposite type; if non-sensitised cells are mixed, co-stimulation will not begin until the two hours needed for sensitisation have elapsed.

Two phases have been experimentally detected during the one-hour co-stimulation period (Allewell *et al.*, 1976; Allewell & Wolfe, 1977): events occurring during the initial *activation* phase (about 20 min) can be inhibited by subjecting cells either to mechanical agitation or to actinomycin D, while events occurring during the subsequent *maturation* phase (about 40 minutes) are not disrupted by either treatment. The sensitivity to agitation during the initial activation phase is believed to reflect a need for 'meaningful' contacts between sensitised cells of opposite type. The nature of this co-stimulation interaction is as yet undefined, since no obvious cell–cell interactions (e.g. agglutination) characterise the co-stimulation period. It has, however, been shown (Adair *et al.*, 1978) that a cell-free factor is necessary for co-stimulation.

The factor is produced by cells of both mating types during the preceding sensitisation period, and is therefore present in the starvation medium in which mating mixtures are usually made. If sensitised cells are washed and suspended in fresh starvation medium before mixing, the co-stimulation period is delayed while the cells provide the new medium with appropriate levels of the factor. The factor is non-dialysable and heat-stable, and a more detailed biochemical characterisation is underway. Meanwhile, available data allow two models for factor action (Adair *et al.*, 1978): the factor components may act passively, holding cells of opposite type together long enough for the elusive cell-surface interactions; alternatively, the factor macromolecules may play an active role in co-stimulation, perhaps serving as transiently shared effectors that are active only when cells of opposite type are physically contiguous.

It is during the maturation phase of the co-stimulation period that a morphological change, designated 'tip transformation', heralds the onset of mating receptivity. The anterior end of an unmated *Tetrahymena thermophila* cell ordinarily tapers to a point and exhibits a marked ridge along the midline. During tip transformation, which affects both mating types, the ridged appearance disappears and the ventral surface between the oral apparatus and the cell tip becomes smooth; simultaneously, the pointed tip is transformed to a stub, as if sawn off (Wolfe & Grimes, 1979). Since cells pair at their transformed tips (see below), it is probable that these morphological changes represent final events in the sexual differentiation of *T. thermophila*.

Comparing what is known about sexual differentiation in *Tetrahymena thermophila* and in *Chlamydomonas reinhardii*, then, the two are similar in that starvation serves as the environmental trigger and sexually competent cells are uniformly in the G_1 phase of the cell cycle. Strikingly different are the prominence of mating-type-specific interactions during the ciliate differentiation process: thus the opportunity for the mutual stimulation of mitosis is present during the sensitisation period, and the requirement for meaningful contacts characterises the co-stimulation period. None of these interactions has as yet been shown to be mediated by mating-type-specific hormones in *T. thermophila*, in contrast to the hormone-mediated sexual receptivity described for the ciliate *Blepharisma intermedium* (Miyake, 1978), but a factor secreted by both types is required for co-stimulation. Since gametic differentiation by *C. reinhardii* cells exhibits no detectable requirement for either soluble factors or cell–cell interactions, the two organisms have clearly evolved different strategies for sexual differentiation.

A final point of comparison between the two kinds of differentiation relates to their relative permanence. Both *Chlamydomonas reinhardii* gametes and sensitised *Tetrahymena thermophila* cells will dedifferentiate into vegetative cells and resume mitotic growth if the nutrient supply in their medium is replenished. Once *C. reinhardii* mating competency has been acquired, however, some six to ten hours must elapse before de-differentiation occurs (Kates & Jones, 1964); by contrast, if nutrients are provided to *T. thermophila* cells, even an hour after cell–cell pairing has begun, the pairs at once disperse and all further mating is blocked (Nanney, 1977).

Genetics of sexual differentiation

Of the three organisms being considered in this review, knowledge of the genetics of sexuality is at present least developed for *Tetrahymena thermophila*. The determination of mating type in the Ciliata is in general a complex subject (Sonneborn, 1978). For *T. thermophila*, seven mating types are known, the rule being that a cell can mate with all but its own type. The available data are consistent with a model wherein two alleles, mt^A and mt^B, occupy the *mt* locus, each allele permitting the selection of a restricted range of the seven possible phenotypic mating types. Specifically, it is proposed that the mt^A allele restricts development to mating types I, II, III, V, or VI, while the mt^B allele restricts development to types II, III, IV, V, and VI. An mt^A/mt^B heterozygote, by this model, has a choice of all seven mating types.

Given these restricted ranges, the actual choice of mating type selected for expression is made during the period of macronuclear differentiation which follows conjugation. The sexually immature daughter exconjugants, in producing their new macronuclei which will control their subsequent phenotypes, select a single mating type for expression from among the allowed options. The frequency with which particular mating types are selected can be influenced by the temperature at which cells are maintained during macronuclear development; otherwise, little is known about the choice mechanism, or about its regulation.

Expression of a single mating type is not rigorously upheld throughout the life history of a clone: selfing often takes place within a clone, and 'mutation' to other mating types occurs with relatively high frequency (Orias, 1963). Such 'mutations' usually appear to affect macronuclear phenotypic expression rather than sexually heritable micronuclear genes, however, and *bona fide* mutations affecting sexual differentiation or conjugation are as yet unreported. Wolfe (1979) has recently developed methods for screening for such mutations, so that this situation should soon change.

Sexual recognition, signalling, and fusion

In *Chlamydomonas reinhardii*, sexual differentiation can be distinguished readily from the cell–cell interactions of the mating process whereas, for *Tetrahymena thermophila*, as yet ill-defined interactions with cells of unlike mating type are required if the differentiation process is to be completed. If we arbitrarily choose the onset of cell pairing as the first event in the mating process, then the events of conjugation can be listed as the development of stable pairs, the establishment of cytoplasmic continuity, the elongation and migration of the micronuclei (Wolfe, Hunter & Adair, 1976), and the occurrence of meiosis and nuclear fusion. The later events are described by Elliott (1973); here we can consider the establishment of stable junctions between paired cells, a phenomenon described by Elliott and Tremor (1958) and by Wolfe & Loyter (1975).

The 'sawn-off' ends of differentiating *Tetrahymena thermophila* cells are the initial sites of conjugal pairing. As yet unknown is the mechanism by which these specialised sites effect recognition of heterologous mating type, but an extracellular factor has been demonstrated to be necessary for this interaction (Wolfe *et al.*, 1979), possibly the same factor as that used during co-stimulation (Adair *et al.*, 1978). The initial contact region is relatively small, but it gradually enlarges to occupy over one quarter the length of the cell. A major structural reorganisation accompanies this expansion: such organelles as cilia, basal bodies, alveolar membranes, and mitochondria all disappear, and the cell membranes of the paired areas each become underlain by a thick electron-dense plaque called the *fusion plate*. Subsequently, this plate develops perforations which allow limited cytoplasmic continuity and, eventually, nuclear exchange.

Intriguing questions can be asked about the establishment and development of this cell–cell junction in *Tetrahymena thermophila*. How is recognition effected? What signals following recognition trigger the elaboration of the fusion plates? How are the pores created within these plates, and what limits the diameter of these pores – why do they not spread in size so as to effect a complete fusion of the two cells into one, as in *Chlamydomonas reinhardii*? These phenomena merit far more extensive analysis.

SACCHAROMYCES CEREVISIAE

Stimulation and course of sexual differentiation

Saccharomyces cerevisiae has a sexual life cycle that is to a first approximation the reciprocal of *Chlamydomonas reinhardii* and most other eukaryotic microbes (Crandall, 1977): *S. cerevisiae* cells mate in a nutrient medium and are induced to sporulate (enter meiosis) in a depleted medium, whereas *C. reinhardii* cells require starvation for gametogenesis and nutritive conditions for zygote germination. Intuitively the strategy of *C. reinhardii* appears the more adaptive to adverse environmental conditions, and it is the mode found in most other yeasts (e.g. *Schizosaccharomyces pombe* and *Hansenula wingei*). Presumably, therefore, the *S. cerevisiae* life cycle evolved in response to a different set of environmental constraints.

If the *Chlamydomonas reinhardii* differentiation process, where no cell–cell interactions are required, is labelled self-stimulatory, and the *Tetrahymena thermophila* process partially co-stimulatory, then yeast sexual differentiation can be termed highly co-stimulatory: cells of one mating type carry out most if not all of their sexual differentiation only when they are presented with cells or sexual factors of opposite mating type. When the two cell types (called *a* and α) are mixed, an obligatory two-hour 'courtship' period (Hartwell, 1973) must precede the formation of stable diploid zygotes. That this courtship involves a cellular differentiation and not simply a conditioning of the medium can be shown by allowing genetically marked *a* and α cells to court and then presenting the mixture with differently marked 'naive' cells. The diploids that form in the ensuing two hours are all of the 'courted' genotype; only after an additional lag do the newly introduced cells begin to participate in zygote formation (Hartwell, 1973).

Courtship is marked by three events. First, the yeast cells agglutinate to one another by their cell walls, and a concomitant modification of cell-wall structure is observed. Second, the cells induce one another to arrest their mitotic cycle in G_1 (Hartwell, 1973; Reid & Hartwell, 1977); thus, for all three micro-organisms considered in this review, mating systems have evolved in which only G_1 cells can participate. Finally, agglutinated cell walls break down sufficiently to allow plasma-membrane contact and, subsequently, membrane fusion, cytoplasmic confluence, and the fusion of synchronised nuclei (Byers & Goetsch, 1975).

For *Chlamydomonas reinhardii* and *Tetrahymena thermophila* we considered sexual recognition, signalling, and fusion as topics separate from sexual differentiation. The stringent co-stimulation pattern in yeast makes it much more difficult to sort out differentiation from mating interactions; they may, in fact, prove to be totally interdependent and thereby inseparable. Consideration will therefore first be given to studies of courtship/mating events that can be induced in one mating type by sex factors (pheromones) from the opposite type. Courtship/mating events wherein both cell types are participating will then be considered.

Pheromone-elicited responses

Yeast a-cells secrete constitutively into the medium a pheromone known as a-factor (Wilkinson & Pringle, 1974) while α-cells produce α-factor (Levi, 1956; Duntze, MacKay & Manney, 1970). Each factor is capable of inducing both G_1 arrest and cell-wall changes in cells of opposite mating type but not in cells of homologous mating type. Therefore, each factor is suspected (but not yet proved) to participate in the co-stimulation of sexual differentiation.

The better characterised of the two substances is α-factor. Duntze and collaborators (Duntze et al., 1973; Stötzler, Kiltz & Duntze, 1976) have identified four related oligopeptides in highly purified α-factor preparations, and Ciejek, Thorner & Geier (1977) have synthesised two active forms. All have the basic sequence N-(Trp)-His-Trp-Leu-Gln-Leu-Lys-Pro-Gly-Gln-Pro-Met (or MetSO)-Tyr-C. A radioimmuno-assay has been developed for α-factor, using antisera raised against either the natural or the chemically synthesised oligopeptides, which can detect as little as 0.5–1.0 pmol of pheromone (J. Thorner, personal communication). To date, detection of α-factor has been performed by bioassay (see below).

The a-factor has proved far more intractable to analysis. It associates in a high-molecular weight, carbohydrate-containing complex from which, with prolonged dialysis, a smaller dialysable activity is eventually released (Betz, MacKay & Duntze, 1977). W. Duntze (personal communication) reports success in purifying an oligopeptide a-factor from the carbohydrate material, and the oligopeptide is now being sequenced. In the future, therefore, studies with both a- and α-factor may become equally plentiful; at the present time, however, α-factor studies predominate.

The very earliest response of yeast cells to α-factor may be a suppression in activity of a membrane-associated adenylate cyclase, as α-factor addition inhibits the enzyme by about 40% within a few minutes (J. Thorner, personal communication); moreover, exogenous cyclic AMP will interfere with the ability of α-factor to stimulate G_1 arrest and cause morphological changes in a cells (see below).

The next stage in responsiveness to pheromone is an enhancement in a/α adhesiveness. By using very sensitive assay systems Fehrenbacher, Petty & Thorner (1978) find that homologous cell mixtures (a with a and α with α) adhere only 3–5% and that pheromone treatment of such mixtures does not enhance their agglutinability. Heterologous mixtures of 'naive' a and α cells exhibit a basal level of weak adhesive interactions involving 5–20% of the cells, depending on ionic conditions. After pheromone incubation, on the other hand, 40–50% of the cells in a heterologous mixture are strongly agglutinated, forming clumps that do not dissociate at low ionic strength. The same degree of stimulation is observed whether both cell types are pre-treated with hormone or whether only the a cells are pre-conditioned with α-factor, and the stimulation is blocked by inhibitors of RNA- and protein-, but not polysaccharide synthesis. The stimulation of adhesiveness requires only 30 to 60 minutes, and is therefore the earliest response to α-factor for which there exists a published bioassay. Whether, and how, this increase relates to the reported cyclic AMP effects is not yet known.

The next detectable response to pheromone presentation, and the one most extensively studied, is the induction of G_1 arrest. Within a two-hour period after α-factor presentation, a cells accumulate as unbudded, mononucleate cells that do not participate in DNA replication (Bücking-Throm et al., 1973). The arrest is transient (Throm & Duntze, 1970): the higher the concentration of α-factor, the longer the G_1 arrest, but recovery and resumption of DNA synthesis eventually occurs (Chan, 1977). Several investigators have presented evidence that a cells actively destroy α-factor when it is presented to them (Hicks & Herskowitz, 1976a; Chan, 1977), and E. Ciejak and J. Thorner (personal communication) have recently demonstrated that a cells have a surface-bound endopeptidase which degrades α-factor, the first scission preferentially occurring between Leu-6 and Lys-7. They further show that this peptidase activity serves to remove pheromone from the environment rather than to potentiate the pheromone's biological activity, as had been proposed by Manness and Edelman (1978). Finkelstein & Strausberg (1979) have also reported proteolytic degradation of α-factor.

The final effect of the yeast sex factors, one that requires several hours to develop, is the induction of morphological changes: treated cells transform into asymmetric, pear-shaped cells aptly described by H. Roman as 'shmoos'. The response is dose-related: the more factor present, the more extreme the morphological transformation (Lipke, Taylor & Ballou, 1976). Since shmoos are considerably larger than untreated cells, the effect is presumably the result of de-novo cell wall synthesis and/or alterations in cell-wall structure to allow expansion.

Three recent studies of the 'shmooing' response have focused on different features of wall structure. Lipke *et al.* (1976) provide evidence that the walls of α-factor-treated *a* cells contain more glucan and less mannan than controls, and that the mannan of stimulated cells is less highly branched. They also find that the shmoos are more sensitive to glucanases (but not proteases), and that their walls appear thinner in the extended portions of the cell. Finally, they demonstrate that these changes, and the shmoo transformation, are blocked by cycloheximide and by inhibitors of cell-wall biosynthesis. Schekman and Brawley (1979) report a striking accumulation of chitin along the tapered, elongated surfaces of shmooing cells, and demonstrate that the α-factor stimulates rates of chitin synthesis, rates of synthesis of the chitin synthase zymogen, and activates a surface-membrane fraction of the synthase. And finally, Tkacz and MacKay (1979) show that fluorescein-labelled concanavalin A associates preferentially with the elongating projections of shmoos whereas it stains the perimeters of vegetative cells uniformly.

Cell–cell interactions

Available studies on the kinetics of the yeast mating reaction (Fehrenbacher *et al.*, 1978) indicate that about 50% of the cells in a mating mixture participate in tight aggregate formation during the first hour; two to four hours later, growth stops and perhaps 30–50% of the cells (the same cells that agglutinated?) participate in zygote formation. Both the new diploids and the unmated haploids then resume vegetative growth and no appreciable further mating takes place. Such relatively low mating efficiencies contrast sharply with *Tetrahymena thermophila* and *Chlamydomonas reinhardii* where mating efficiencies of 90–100% are routinely achieved in the laboratory (Orias & Bruns, 1976; Chiang *et al.*, 1970). Possibly only a certain proportion of the yeast cells in a mating mixture become appropriately positioned for adhesion and/or fusion and, being non-motile, the unfavourably positioned cells are unable to correct the situation. While unique sites for cell interactions in yeast, analogous to the flagellar tips and fusing organelles in *C. reinhardii* and *T. thermophila*, have not yet been identified, the fact that they fuse only in pairs suggests that some specificity must characterise the adhesion/fusion process.

The studies on pheromone induction of yeast sexual responsiveness can be reconstructed into a mechanistic account of the mating reaction *in vivo.* When cultures of vegetatively growing *a* and α cells are mixed, the culture media contain, and the cells are actively producing, *a*-factor and α-factor. The low-level agglutinability manifested by such 'naive' cells (Fehrenbacher *et al.*, 1978) may serve to bring opposite-type cells into proximity, perhaps to allow pheromone presentation at optimal concentration. The initial response to the opposite-type factors, which may be cyclic-AMP-mediated, stimulates the conversion of the loose associations into tight cell–cell adhesions. Continued exposure to opposite-type cells (and therefore, by definition, to opposite-type factor) prevents the mitotically-cycling cells from passing beyond G_1, creating a synchronous population of nuclei. Concurrently, a number of changes in cell-wall structure are elicited, possibly at the sites of adhesive contacts and again, presumably, in response to pheromone. Once walls and membranes fuse and the nuclei become diploid (Byers & Goetsch, 1975; Conde & Fink, 1976), the resultant *a*/α cells are immune to the effects of both *a*- and α-factor and proceed into the S phase of the cell cycle. Meanwhile, any unmated *a* or α cells also recover from G_1 arrest, perhaps by surface-endopeptidase destruction of ambient sex factors.

Genetics of sexual differentiation

Genetic control of sexual differentiation in yeast is, in a number of respects, remarkably parallel in organisation to that of *Chlamydomonas reinhardii*. The 'master' yeast mating-type locus, designated **MAT**, resides to one side of the centromere on linkage group III and exists in two stable allelic forms, *a* and α, which segregate 2:2 at meiosis. The **MAT** locus controls sexual type and sporulation [the participation of the *C. reinhardii mt* locus during the zygote phase is inferred from its influence over the meiotic inheritance of chloroplast genes (Sager, 1954)].

Mutations that map to **MAT** (MacKay & Manney, 1974*a, b*; Kassir & Simchen, 1976) have provided clues as to its functions. Of mutations within the α locus, at least two complementation groups have been defined (Strathern, 1977), one marked by mutations we can denote as α1⁻ and the other by α2⁻ mutations. Haploid strains carrying α1⁻ mutations are unable to produce α-factor or to mate with any efficiency; but the rare α1⁻/*a* diploids that can be forced to mate show normal sporulation. Strains carrying α2⁻ mutations are also unable to produce α-factor, and do not mate with *a* cells. They do, however, exhibit at least two properties of *a* cells: they mate as *a* cells with very low efficiency (about 1% of *bona fide a* cells) and they degrade α-factor. The rare

$\alpha2^-/a$ diploids, moreover, are grossly defective in their ability to sporulate (Strathern, 1977). These observations have led to the proposal that the $\alpha1$ cistron in the **MAT** locus is involved with turning on the expression of α-specific mating genes, while the $\alpha2$ cistron turns off a-specific mating genes. An a cell is viewed as being constitutive for a-specific functions. Finally, in an a/α diploid, the $\alpha2$ and $a1$ functions are thought to interact in turning off mating functions and turning on sporulation genes.

That the **MAT** locus acts to turn on the expression of mating genes is indicated by the existence of 'sex-limited mutations' which, as in *Chlamydomonas reinhardii*, affect mating ability when carried in one mating type but have no effect on mating ability when carried in the other mating type. Thus mutations at the *ste2* locus, which is unlinked to **MAT**, block the ability to respond to α-factor and are expressed only in a cells, while mutations at *ste3* block the ability to respond to a-factor and are expressed only in α cells (MacKay & Manney, 1974b).

The map location of the structural gene(s) for the pheromones is not yet known, for of the many non-mating mutant strains that have been isolated to date, none is solely defective in pheromone production. Specifically, none of the α mutants can be 'cured' by supplying them with α-factor, nor can a mutants be corrected with a-factor; all exhibit additional mating defects. One possible but unproved explanation for the absence of mutations of this class is that defective factor production generates additional mating defects in a cell that cannot be corrected by exogenous factor. Alternatively, functions needed for α-factor production may be required for other mating-related cell processes, in which case mutations in such genes would generate pleiotropic mating-defective mutants. Meanwhile, putative structural genes for the pheromones have not as yet been identified, nor are the gene products of such loci as *ste2* and *ste3*, which are involved with factor responsiveness, yet known.

Mating-type switches

Some of the most active and exciting research on mating-type determination in eukaryotic microbes is concerned with mechanisms by which *Saccharomyces cerevisiae* is able to change its mating type. It is hoped that the abbreviated account presented here will encourage readers to tackle the complex but fascinating original literature on the subject (Takano & Oshima, 1970; Naumov & Tolstorukov, 1973; Harashima, Nogi & Oshima, 1974; Harashima & Oshima, 1976; Hicks & Herskowitz, 1976b, 1977; Hicks, Strathern & Herskowitz, 1977; Klar & Fogel, 1977; Klar, Fogel & MacLeod, 1979; Klar, Fogel &

Radin, 1979; Rine *et al.*, 1979; Strathern, Blair & Herskowitz, 1979; Strathern & Herskowitz, 1979).

As noted earlier, genetic crosses have established that the **MAT** locus occupies a medial position on chromosome III, and tetrad analysis reveals that the *a* and α alleles at this locus segregate in a stable Mendelian fashion. It turns out, however, that in addition to this 'expressed' **MAT** locus, chromosome III carries two additional loci that bear 'silent' copies of **MAT** genetic information. On the distal left arm of chromosome III lies a silent copy of the α mating-type locus which can be designated [*MAT*α]; on the distal right arm of chromosome III is located the corresponding silent [*MATa*]. Although all three copies of **MAT** lie on the same chromosome, they are separated by long map distances and in no way constitute a gene 'cluster'.

The [*MAT*] copies are ordinarily not expressed and, in fact, at least four unlinked genes in yeast have to date been identified that act as negative regulators to suppress [*MAT*] expression. Mutations at these loci allow the silent copies to be expressed; one mutant class permits the expression of [*MAT*α] but not [*MATa*]. The existence of these regulatory genes argues that the silent **MAT** copies do not represent aberrant duplications that have accumulated in chromosomes of a laboratory strain, but are instead carefully controlled components of the genome. When a haploid strain carries a mutation in one of these negative-regulatory genes so that the silent copies are expressed, the strain behaves as an *a*/α diploid, being sterile and unresponsive to either *a*- or α-factor.

A second condition under which the silent copies are expressed involves the action of a dominant homothallism gene known as *HO*, which has been crossed into certain *Saccharomyces cerevisiae* stocks from another yeast strain. Under the influence of *HO*, and as often as once every mitotic cell division, the information at the expressed **MAT** locus is physically removed from the chromosome and a copy of one of the distal [*MAT*] sequences is inserted into the medial position. In its new location, the [*MAT*] sequence is expressed and is able to determine the mating phenotype of the cell. Thus, to describe a specific homothallic conversion (Hicks & Herskowitz, 1976*b*), a meiotic spore was micromanipulated onto an agar surface that contained α-factor. The spore proceeded to divide in the presence of the pheromone, revealing its α phenotype. The spore and its daughter were separated and each proceeded to bud again, demonstrating that the α phenotype had been retained. When the four resultant cells were separated, however, the original spore and its most recent daughter were found to arrest in G_1

and to shmoo; these cells had been converted to **MAT**a. The other two cells were unresponsive to α factor and continued to bud; these had retained an expressed **MAT**α. At the next division, however, these cells might undergo a similar switching event; moreover, the new *a* cells might, at a later division, re-assume an α phenotype.

That the *HO* gene acts by moving copies of [*MAT*] information to the medial locus has been demonstrated genetically. Thus homothallic cells carrying the α1⁻ mutation at the expressed **MAT** locus can be observed to convert to *a* phenotypes in one generation; at the next switch, these acquire a normal α⁺ **MAT** locus, presumably recruited from the normal information housed in the [*MAT*α] 'cassette'. The reciprocal experiment has also been performed; a strain constructed to carry wild-type information at the expressed **MAT** locus and mutant information in one of the cassettes was found to acquire, under the influence of *HO*, mutant information in the expressed locus.

Two additional features of the *HO*-directed interconversion are of interest. First, it has been shown that the *HO* gene is not required to maintain a genotype created under its influence: converted chromosomes that are separated from *HO* by meiotic assortment retain their newly acquired genotype in a fully stable fashion. Second, it is found that if, after conversion, a *HO*α cell mates with its *HO*a sibling to form a *HO*/*HO* *a*/α diploid, no further mating-type interconversions take place. In other words, a feedback mechanism exists such that *HO* is turned off when an *a*/α diploid is formed. It is therefore tenable to argue that the 'purpose' of *HO* is to make diploids from haploids, that is, that it serves as a 'diploidisation system'. Whether the many homothallic species of *Chlamydomonas reinhardii* are governed by similar genetic systems is an intriguing and unanswered question.

Returning now to 'standard' (*ho*) heterothallic laboratory strains of yeast, we can consider conditions under which their usually stable mating type is reversed. The first, already noted above, is the occurrence of rare mutations in regulatory genes which normally function to suppress [*MAT*] expression. The second is the occurrence of rare switches of the sort that occur under the aegis of *HO*; in other words, while *HO* greatly stimulates the transposition of [*MAT*] information, the process can, very infrequently, occur in the absence of *HO*. And finally, mating-type switches have been found to be caused by chromosomal deletions. In one case (Hawthorne, 1963), an α strain was changed to an *a*, while in another (R. K. Mortimer and D. C. Hawthorne, unpublished) an *a* strain changed to an α. Both phenotypes have been shown to be associated with long lethal deletions in chromosome III

(Hawthorne, 1963; J. N. Strathern, C. S. Newlon, I. Herskowitz & J. B. Hicks, personal communication). The 'Hawthorne deletion' appears to have involved the loss of MATα and the DNA lying between MATα and [MATa]; as a result, the [MATa] cassette was brought into proximity with MAT elements (promoters?) that permit its expression. The 'Mortimer–Hawthorne rearrangement' involved a similar fusion between MAT controlling elements and the leftward [MATα] cassette, and, in the process, generated a 63μ circular fragment of chromosome III which has been isolated and is being characterised biochemically.

J. Hicks and his colleagues (personal communication) have cloned a yeast DNA fragment containing [MATα]. When a non-mating a mutant is transformed by this plasmid its progeny express a weak α phenotype. If the recipient also carries a mutation in one of the negative-regulator genes that normally repress cassette expression, the plasmid α information is strongly expressed. It would appear, therefore, that we can soon expect considerable information on such interesting questions as the degree of sequence homology between MATα and MATa loci and on the mechanism of MAT transposition.

Ira Herskowitz and Jeremy Thorner made many helpful suggestions on the yeast sections; they plus Jason Wolfe, Vivian Mackay, and Jim Hicks kindly provided unpublished information and reprints invaluable to preparing this review.

REFERENCES

ADAIR, W. S., BARKER, R., TURNER, R. S. & WOLFE, J. (1978). Demonstration of a cell-free factor involved in cell interactions during mating in *Tetrahymena*. *Nature, London*, **274**, 54–5.

ADAIR, W. S., JURIVICH, D. & GOODENOUGH, U. W. (1978). Localization of cellular antigens in sodium dodecyl sulfate-polyacrylamide gels. *Journal of Cell Biology*, **79**, 281–5.

ADAIR, W. S. & WOLFE, J. (1977). Induced alteration in uptake properties of *Tetrahymena* and its association with the development of mating competency. *Journal of Cellular Physiology*, **92**, 77–90.

ALLEWELL, N., OLES, J. & WOLFE, J. (1976). A physiochemical analysis of conjugation in *Tetrahymena pyriformis*. *Experimental Cell Research*, **97**, 394–405.

ALLEWELL, N. & WOLFE, J. (1977). A kinetic analysis of the induction and memory of a developmental interaction. Mating interactions in *Tetrahymena*. *Experimental Cell Research*, **109**, 15–24.

BERGMAN, K., GOODENOUGH, U. W., GOODENOUGH, D. A., JAWITZ, J. & MARTIN, H. (1975). Gametic differentiation in *Chlamydomonas reinhardtii*. II. Flagellar membranes and the agglutination reaction. *Journal of Cell Biology*, **67**, 606–22.

BETZ, R. MACKAY, V. L. & DUNTZE, W. (1977). a-Factor from *Saccharomyces cerevisiae*: Partial characterization of a mating hormone produced by cells of mating type a. *Journal of Bacteriology*, **132**, 462–72.

BRUNS, P. & BRUSSARD, T. (1974). Pair formation in *Tetrahymena pyriformis*, an inducible developmental system. *Journal of Experimental Zoology*, **188**, 337–44.

BÜCKING-THROM, E., DUNTZE, W., HARTWELL, L. H. & MANNEY, T. R. (1973). Reversible arrest of haploid yeast cells at the initiation of DNA synthesis by a diffusible sex factor. *Experimental Cell Research*, **76**, 99–110.

BUFFALOE, N. D. (1958). A comparative cytological study of four species of *Chlamydomonas*. *Bulletin of the Torrey Botanical Club*, **85**, 157–78.

BYERS, B. & GOETSCH, L. (1975). Behaviour of spindles and spindle plaques in the cell cycle and conjugation of *Saccharomyces cerevisiae*. *Journal of Bacteriology*, **124**, 511–23.

CAVALIER-SMITH, T. (1975). Electron and light microscopy of gametogenesis and gamete fusion in *Chlamydomonas reinhardtii*. *Protoplasma*, **86**, 1–18.

CHAN, R. K. (1977). Recovery of *Saccharomyces cerevisiae* mating-type *a* cells from G1 arrest by α factor. *Journal of Bacteriology*, **130**, 766–74.

CHAING, K. S., KATES, J. R., JONES, R. F. & SUEOKA, N. (1970). On the formation of homogeneous zygotic populations in *Chlamydomonas reinhardi*. *Developmental Biology*, **22**, 655–69.

CIEJEK, E., THORNER, J. & GEIER, M. (1977). Solid phase peptide synthesis of α-factor, a yeast mating pheromone. *Biochemical and Biophysical Research Communications*, **78**, 952–61.

CLAES, H. (1971). Autolyse der Zellwand bei Gameten von *Chlamydomonas reinhardii*. *Archives für Mikrobiologie*, **78**, 180–8.

CLAES, H. (1977). Non-specific stimulation of the autolytic system in gametes from *Chlamydomonas reinhardii*. *Experimental Cell Research*, **108**, 221–9.

COLEMAN, A. W. (1979). Sexuality in colonial green flagellates. In *Biochemistry and Physiology of Protozoa*, 2nd edn, vol. 1, pp. 307–40. New York: Academic Press.

CONDE, J. & FINK, G. R. (1976). A mutant of *Saccharomyces cerevisiae* defective for nuclear fusion. *Proceedings of the National Academy of Sciences, USA*, **73**, 3651–5.

CRANDALL, M. (1977). Mating-type interactions in micro-organisms. In *Receptors and Recognition*, Series A, ed. P. Cuatrecasas & M. F. Greaves, vol. 3, pp. 47–100. London: Chapman and Hall.

DUNTZE, W., MACKAY, V. L. & MANNEY, T. R. (1970). *Saccharomyces cerevisiae*: a diffusible sex factor. *Science*, **168**, 1472–3.

DUNTZE, W., STÖTZLER, D., BÜCKING-THROM, E. & KALBITZER, S. (1973). Purification and partial characterization of α-factor, a mating-type specific inhibitor of cell reproduction from *Saccharomyces cerevisiae*. *European Journal of Biochemistry*, **35**, 357–65.

EBERSOLD, W. T., LEVINE, R. P., LEVINE, E. E. & OLMSTED, M. A. (1962). Linkage maps in *Chlamydomonas reinhardi*. *Genetics*, **47**, 531–43.

ELLIOTT, A. M. (1973). Life cycle and distribution of *Tetrahymena*. In *Biology of Tetrahymena*, ed. A. M. Elliott, pp. 259–86. Stroudsburg, Pennsylvania: Dowden, Hutchinson & Ross.

ELLIOTT, A. M. & TREMOR, J. W. (1958). The fine structure of the pellicle in the contact area of conjugating *Tetrahymena pyriformis*. *Journal of Biophysical and Biochemical Cytology*, **4**, 839–40.

FEHRENBACHER, G., PERRY, K. & THORNER, J. (1978). Cell–cell recognition in *Saccharomyces cerevisiae*: Regulation of mating-specific adhesion. *Journal of Bacteriology*, **134**, 893–901.

FINKELSTEIN, D. B. & STRAUSBERG, S. (1979). Metabolism of α-factor by α mating type cells of *Saccharomyces cerevisiae*. *Journal of Biological Chemistry*, **254**, 796–803.

FOREST, C. L., GOODENOUGH, D. A. & GOODENOUGH, U. W. (1978). Flagellar membrane agglutination and sexual signaling in the conditional *gam*-1 mutant of *Chlamydomonas*. *Journal of Cell Biology*, **79**, 74–84.

FOREST, C. L. & TOGASAKI, R. K. (1975). Selection for conditional gametogenesis in *Chlamydomonas reinhardi*. *Proceedings of the National Academy of Sciences, USA*, **72**, 3652–5.

FRIEDMANN, I., COLWIN, A. L. & COLWIN, L. H. (1968). Fine structural aspects of fertilization in *Chlamydomonas reinhardi*. *Journal of Cell Science*, **3**, 115–28.

GILLHAM, N. W. (1969). Uniparental inheritance in *Chlamydomonas reinhardi*. *American Naturalist*, **103**, 355–88.

GOODENOUGH, U. W. (1977). Mating interactions in *Chlamydomonas*. In *Microbial Interactions. Receptors and Recognition*, Series B, ed. J. L. Reissig, vol. 3, pp. 323–50. London: Chapman and Hall.

GOODENOUGH, U. W., ADAIR, W. S., CALIGOR, E., FOREST, C. L., HOFFMAN, J. L., MESLAND, D. A. M. & SPATH, S. (1979). Membrane–membrane and membrane–ligand interactions in *Chlamydomonas* mating. In *Membrane–Membrane Interactions*, ed. N. B. Gilula, pp. 131–52. New York: Raven Press.

GOODENOUGH, U. W. & FOREST, C. L. (1978). Genetics of cell–cell interactions in *Chlamydomonas reinhardi*. *Birth Defects: Original Articles Series*, **14**, 429–38.

GOODENOUGH, U. W., HWANG, C. & MARTIN, H. (1976). Isolation and genetic analysis of mutant strains of *Chlamydomonas reinhardi* defective in gametic differentiation. *Genetics*, **82**, 169–86.

GOODENOUGH, U. W., HWANG, C. & WARREN, A. J. (1978). Sex-limited expression of gene loci controlling flagellar membrane agglutination in the *Chlamydomonas* mating reaction. *Genetics*, **89**, 235–43.

GOODENOUGH, U. W. & JURIVICH, D. (1978). Tipping and mating structure activation induced in *Chlamydomonas* gametes by flagellar membrane antisera. *Journal of Cell Biology*, **79**, 680–93.

GOODENOUGH, U. W. & WEISS, R. L. (1975). Gametic differentiation in *Chlamydomonas reinhardi*. III. Cell wall lysis and microfilament-associated mating structure activation in wild-type and mutant strains. *Journal of Cell Biology*, **67**, 623–37.

GOODENOUGH, U. W. & WEISS, R. L. (1978). Interrelationships between microtubules, a striated fiber, and the gametic mating structures of *Chlamydomonas reinhardi*. *Journal of Cell Biology*, **76**, 430–8.

HARASHIMA, S., NOGI, Y. & OSHIMA, Y. (1974). The genetic system controlling homothallism in *Saccharomyces* yeasts. *Genetics*, **77**, 639–50.

HARASHIMA, S. & OSHIMA, Y. (1976). Mapping of the homothallic genes *HMα* and *HMa* in *Saccharomyces* yeasts. *Genetics*, **84**, 437–51.

HARTWELL, L. H. (1973). Synchronization of haploid yeast cell cycles, a prelude to conjugation. *Experimental Cell Research*, **76**, 111–17.

HAWTHORNE, D. C. (1963). A deletion in yeast and its bearing on the structure of the mating type locus. *Genetics*, **48**, 1727–9.

HICKS, J. B. & HERSKOWITZ, I. (1976a). Evidence for a new diffusible element of mating pheromones in yeast. *Nature, London*, **260**, 246–8.

HICKS, J. B. & HERSKOWITZ, I. (1976b). Interconversion of yeast mating types. I. Direct observations of the action of the homothallism (*HO*) gene. *Genetics*, **83**, 245–8.

HICKS, J. B. & HERSKOWITZ, I. (1977). Interconversion of yeast mating types. II. Restoration of mating ability to sterile mutants in homothallic and heterothallic strains. *Genetics*, **85**, 373–93.

HICKS, J. B., STRATHERN, J. & HERSKOWITZ, I. (1977). The cassette model of mating type interconversion. In *DNA Insertion Elements, Plasmids, and Episomes*, ed. A. Bukhari, J. Shapiro & S. Adhaya, pp. 457–62. New York: Cold Spring Harbor Laboratory.

JOHNSON, U. G. & PORTER, K. R. (1968). Fine structure of cell division in *Chlamydomonas reinhardi*: basal bodies and microtubules. *Journal of Cell Biology*, **38**, 403–25.

KASSIR, Y. & SIMCHEN, G. (1976). Regulation of mating and meiosis in yeast by the mating-type region. *Genetics*, **82**, 187–206.

KATES, J. R. (1966). Biochemical aspects of synchronized growth and differentiation in *Chlamydomonas reinhardtii*. PhD Thesis. Princeton University, Princeton.

KATES, J. R., CHIANG, K. S. & JONES, R. F. (1968). Studies on DNA replication during synchronized vegetative growth and gametic differentiation in *Chlamydomonas reinhardtii*. *Experimental Cell Research*, **49**, 121–35.

KATES, J. R. & JONES, R. F. (1964). The control of gametic differentiation in liquid cultures of *Chlamydomonas*. *Journal of Comparative Physiology*, **63**, 157–64.

KATES, J. R. & JONES, R. F. (1966). Pattern of CO_2 fixation during vegetative development and gametic differentiation in *Chlamydomonas reinhardtii*. *Journal of Cellular Physiology*, **67**, 101–6.

KATES, J. R. & JONES, R. F. (1967). Periodic increases in enzyme activity in synchronized cultures of *Chlamydomonas reinhardtii*. *Biochimica et Biophysica Acta*, **145**, 153–8.

KLAR, A. & FOGEL, S. (1977). The action of homothallism genes in *Saccharomyces* diploids during vegetative growth and the equivalence of *hma* and *HMα* loci functions. *Genetics*, **85**, 407–16.

KLAR, A., FOGEL, S. & MACLEOD, K. (1979). *MAR*1 – A regulator of the *HMa* and *HMα* loci in *Saccharomyces cerevisiae*. *Genetics*, in press.

KLAR, A., FOGEL, S. & RADIN, D. (1979). Switching of a mating type *a* mutant allele in budding yeast *Saccharomyces cerevisiae*. *Genetics*, in press.

KONIGSBERG, I. R., SOLLMAN, P. A. & MIXTER, L. O. (1978). The duration of the terminal G_1 of fusing myoblasts. *Developmental Biology*, **63**, 11–26.

LEVI, J. D. (1956). Mating reaction in yeast. *Nature, London*, **177**, 753–4.

LEWIN, R. A. (1954). Sex in unicellular organisms. In *Sex in Micro-organisms*, ed. D. H. Wenrich, pp. 100–33. Washington: American Association for the Advancement of Science.

LIPKE, P. N., TAYLOR, A. & BALLOU, C. E. (1976). Morphogenic effects of α-factor on *Saccharomyces cerevisiae a* cells. *Journal of Bacteriology*, **127**, 610–18.

LØVLIE, A. & BRYHNI, E. (1976). Signal for cell fusion. *Nature, London*, **263**, 779–81.

MACKAY, V. L. (1978). Mating-type specific pheromones as mediators of sexual conjugation in yeast. In *Molecular Control of Proliferation Differentiation*, pp. 243–59. New York: Academic Press.

MACKAY, V. & MANNEY, T. R. (1974a). Mutations affecting sexual conjugation and related processes in *Saccharomyces cerevisiae*. I. Isolation and phenotypic characterization of nonmating mutants. *Genetics*, **76**, 255–71.

MacKay, V. & Manney, T. R. (1974*b*). Mutations affecting sexual conjugation and related processes in *Saccharomyces cerevisiae*. II. Genetic analysis of nonmating mutants. *Genetics*, **76**, 273–88.

Manness, P. & Edelman, G. M. (1978). Inactivation and chemical alteration of mating factor α by cells and spheroplasts of yeast. *Proceedings of the National Academy of Sciences, USA*, **75**, 1304–8.

Manney, T. R. & Meade, J. H. (1977). Cell–cell interactions during mating in *Saccharomyces cerevisiae*. In *Microbial Interactions. Receptors and Recognition*, Series B, ed. J. L. Reissig, vol. 3, pp. 281–322. London: Chapman and Hall.

Martin, N. C., Chiang, K. S. & Goodenough, U. W. (1976). Turnover of chloroplast and cytoplasmic ribosomes during gametogenesis in *Chlamydomonas reinhardi*. *Developmental Biology*, **51**, 190–201.

Martin, N. C. & Goodenough, U. W. (1975). Gametic differentiation in *Chlamydomonas reinhardtii*. I. Production of gametes and their fine structure. *Journal of Cell Biology*, **67**, 587–605.

Matsuda, Y., Tamaki, S. & Tsubo, Y. (1978). Mating type specific induction of cell wall lytic factor by agglutination of gametes in *Chlamydomonas reinhardtii*. *Plant and Cell Physiology*, **19**, 1253–61.

McLean, R. J. & Brown, R. M. (1974). Cell surface differentiation of *Chlamydomonas* during gametogenesis. I. Mating and concanavalin A agglutinability. *Developmental Biology*, **36**, 279–85.

Mesland, D. A. M. (1976). Mating in *Chlamydomonas eugamatos*, a scanning electron microscopical study. *Archiv für Mikrobiologie*, **109**, 31–5.

Mesland, D. A. M. & van den Ende, H. (1979). The role of flagellar adhesion in sexual activation of *Chlamydomonas eugamatos*. *Protoplasma*, **98**, 115–29.

Minami, S. A. (1977). Protein synthesis during differentiation of *Chlamydomonas reinhardi*. PhD thesis, Harvard University, Cambridge, Massachusetts.

Miyake, A. (1978). Cell communication, cell union, and initiation of meiosis in ciliate conjugation. *Current Topics in Developmental Biology*, **12**, 37–82.

Nanney, D. L. (1977). Cell–cell interactions in ciliates: Evolutionary and genetic constraints. In *Microbial Interactions. Receptors and Recognition*, Series B, ed. J. L. Reissig, vol. 3, pp. 323–50. London: Chapman and Hall.

Naumov, G. I. & Tolstorukov, I. I. (1973). Comparative genetics of yeast. X. Reidentification of mutators of mating type in *Saccharomyces*. *Genetika*, **9**, 82–91.

Orias, E. (1963). Mating-type determination in variety 8, *Tetrahymena pyriformis*. *Genetics*, **48**, 1509–18.

Orias, E. & Bruns, P. J. (1976). Induction and isolation of mutants in *Tetrahymena*. In *Methods in Cell Biology*, ed. D. M. Prescott, vol. 13, pp. 274–82. New York: Academic Press.

Ray, D. A., Solter, K. M. & Gibor, A. (1978). Flagellar surface differentiation. Evidence for multiple sites involved in mating of *Chlamydomonas reinhardi*. *Experimental Cell Research*, **114**, 185–9.

Reid, B. J. & Hartwell, L. H. (1977). Regulation of mating in the cell cycle of *Saccharomyces cerevisiae*. *Journal of Cell Biology*, **75**, 355–65.

Rine, J., Strathern, J., Hicks, J. & Herskowitz, I. (1979). A suppressor of mating type locus mutations in *Saccharomyces cerevisiae*: Evidence for identification of cryptic mating type locus. *Genetics*, in press.

SAGER, R. (1954). Mendelian and non-Mendelian inheritance of streptomycin resistance in *Chlamydomonas*. *Proceedings of the National Academy of Sciences, USA*, **40**, 356–63.

SAGER, R. (1955). Inheritance in the green alga *Chlamydomonas reinhardi*. *Genetics*, **40**, 476–89.

SAGER, R. & GRANICK, S. (1954). Nutritional control of sexuality in *Chlamydomonas reinhardi*. *Journal of General Physiology*, **37**, 729–42.

SAGER, R. & RAMANIS, Z. (1974). Mutations that alter the transmission of chloroplast genes in *Chlamydomonas*. *Proceedings of the National Academy of Sciences, USA*, **71**, 4698–702.

SAKANO, H., ROGERS, J. H., HÜPPI, K., BRACK, C., TRAUNECKER, A., MAKI, R., WALL, R. & TONEGAWA, S. (1979). Domains and the hinge region of an immunoglobulin heavy chain are encoded in separate DNA segments. *Nature, London*, **277**, 627–33.

SCHEKMAN, R. & BRAWLEY, V. (1979). Localized deposition of chitin on the yeast cell surface in response to mating pheromone. *Proceedings of the National Academy of Sciences USA*, **76**, 645–9.

SCHLÖSSER, U. G., SACHS, H. & ROBINSON, D. G. (1976). Isolation of protoplasts by means of a 'species-specific' autolysine in *Chlamydomonas*. *Protoplasma*, **88**, 51–64.

SCHMEISSER, E. T., BAUMGARTEL, D. M. & HOWELL, S. H. (1973). Gametic differentiation in *Chlamydomonas reinhardi*: Cell cycle dependency and rates of attainment of mating competency. *Developmental Biology*, **31**, 31–7.

SIERSMA, P. W. & CHIANG, K. S. (1971). Conservation and degradation of cytoplasmic and chloroplast ribosomes in *Chlamydomonas reinhardtii*. *Journal of Molecular Biology*, **58**, 167–85.

SMITH, G. M. & REGNERY, D. C. (1950). Inheritance of sexuality in *Chlamydomonas reinhardi*. *Proceedings of the National Academy of Sciences, USA*, **36**, 246–8.

SNELL, J. W. (1976). Mating in *Chlamydomonas*: A system for the study of specific cell adhesion. I. Ultrastructural and electrophoretic analysis of flagellar surface components involved in adhesion. *Journal of Cell Biology*, **68**, 48–69.

SOLTER, K. M. & GIBOR, A. (1977a). The release of carbohydrates during mating in *Chlamydomonas reinhardi*. *Plant Science Letters*, **8**, 227–31.

SOLTER, K. M. & GIBOR, A. (1977b). Evidence for role of flagella as sensory transducers in mating of *Chlamydomonas reinhardi*. *Nature, London*, **265**, 444–5.

SOLTER, K. M. & GIBOR, A. (1978). Removal and recovery of mating receptors on flagella of *Chlamydomonas reinhardi*. *Experimental Cell Research*, **115**, 175–81.

SONNEBORN, T. M. (1978). Genetics of cell–cell interactions in ciliates. *Birth Defects: Original Article Series*, **14**, 417–27.

STÖTZLER, D., KILTZ, H. & DUNTZE, W. (1976). Primary structure of α-factor peptides from *Saccharomyces cerevisiae*. *European Journal of Biochemistry*, **69**, 397–400.

STRATHERN, J. (1977). Regulation of cell type in *Saccharomyces cerevisiae*. PhD thesis, University of Oregon, Eugene.

STRATHERN, J., BLAIR, L. & HERSKOWITZ, I. (1979). Healing of *mat* mutations and *cerevisiae. Proceedings of the National Academy of Sciences, USA*, **76**, 3425–9.

cerevisiae. Proceedings of the National Academy of Sciences, USA, in press.

STRATHERN, J. & HERSKOWITZ, I. (1979). Asymmetry and directionality in production of new cell types during clonal growth: The switching pattern of homothallic yeast. *Cell*, **17**, 371–81.

TAKANO, I. & OSHIMA, Y. (1970). Mutational nature of an allele-specific conversion of the mating type of the homothallic gene HOα in *Saccharomyces*. *Genetics*, **65**, 421–7.

THROM, E. & DUNTZE, W. (1970). Mating-type-dependent inhibition of deoxyribonucleic acid synthesis in *Saccharomyces cerevisiae*. *Journal of Bacteriology*, **104**, 1388–90.

TILNEY, L. G. (1977). Actin: Its association with membranes and the regulation of its polymerization. In *International Cell Biology*, ed. B. R. Brinkley & K. R. Porter, pp. 388–402. New York: Rockefeller University Press.

TKACZ, J. S. & MACKAY, V. L. (1979). Sexual conjugation in yeast. Cell surface changes in response to the action of mating hormones. *Journal of Cell Biology*, **80**, 326–33.

TRIEMER, R. E. & BROWN, R. M. (1975). Fertilization in *Chlamydomonas reinhardi*, with special reference to the structure, development, and fate of the choanoid body. *Protoplasma*, **85**, 99–107.

WEISS, R. L., GOODENOUGH, D. A. & GOODENOUGH, U. W. (1977). Membrane differentiations at sites specialized for cell fusion. *Journal of Cell Biology*, **72**, 144–60.

WELLNITZ, W. R. & BRUNS, P. J. (1979). The pre-pairing events in *Tetrahymena thermophila*. Analysis of blocks imposed by high concentrations of Tris-Cl. *Experimental Cell Research*, **119**, 175–80.

WHITTENBERGER, B., RABEN, D., LIEBERMAN, M. A. & GLASER, L. (1978). Inhibition of growth of 3T3 cells by extract of surface membranes. *Proceedings of the National Academy of Sciences, USA*, **75**, 5457–61.

WIESE, L. (1965). On sexual agglutination and mating-type substances (gamones) in isogamous heterothallic Chlamydomonads. I. Evidence of the identity of the gamones with surface components responsible for sexual flagellar contact. *Journal of Phycology*, **1**, 46–54.

WIESE, L. & METZ, C. B. (1969). On the trypsin sensitivity of fertilization as studied with living gametes in *Chlamydomonas*. *Biological Bulletin*, **136**, 483–93.

WIESE, L. & SHOEMAKER, D. W. (1970). On sexual agglutination and mating-type substances (gamones) in isogamous heterothallic Chlamydomonads. II. The effect of concanavalin A upon the mating-type reaction. *Biological Bulletin (Woods Hole)*, **138**, 88–95.

WILKINSON, L. E. & PRINGLE, J. R. (1974). Transient G1 arrest of *S. cerevisiae* cells of mating type α by a factor produced by cells of mating type *a*. *Experimental Cell Research*, **89**, 175–87.

WITMAN, G. B., CARLSON, K., BERLINER, J. & ROSENBAUM, J. L. (1972). *Chlamydomonas* flagella I. Isolation and electrophoretic analysis of microtubules, matrix, membranes, and mastigonemes. *Journal of Cell Biology*, **54**, 507–39.

WOLFE, J. (1973). Conjugation in *Tetrahymena*: The relationship between the division cycle and cell pairing. *Developmental Biology*, **35**, 221–31.

WOLFE, J. (1974). Reciprocal induction of cell division by cells of complementary mating types in *Tetrahymena*. *Experimental Cell Research*, **87**, 39–46.

WOLFE, J. (1979). Mass selection of conditional mating mutants of *Tetrahymena thermophila*. *Journal of General Microbiology*, **115**, 451–6.

WOLFE, J. & GRIMES, G. (1979). Tip transformation in *Tetrahymena*: a morphogenetic response to interactions between mating types. *Journal of Protozoology*, **26**, 82–9.

WOLFE, J., HUNTER, B. & ADAIR, W. S. (1976). A cytological study of micronuclear elongation during conjugation in *Tetrahymena*. *Chromosoma, Berlin*, **55**, 289–308.

WOLFE, J. & LOYTER, A. (1975). Plasma membrane fusion and pore formation in conjugating *Tetrahymena*: A freeze-etch study. *Journal of Cell Biology*, **67**, 459a.

WOLFE, J., TURNER, R., BARKER, R. & ADAIR, W. S. (1979). The need for an extracellular component for cell pairing in *Tetrahymena*. *Experimental Cell Research*, in press.

VIRUSES AND KILLER FACTORS OF FUNGI

K. W. BUCK

Department of Biochemistry, Imperial College of Science and Technology, London SW7 2AZ, UK

DISCOVERY AND OCCURRENCE

The study of fungal viruses (mycoviruses) is a comparatively new subject. It is less than 20 years since the first report of virus particles able to infect and multiply in a fungus, namely the cultivated mushroom *Agaricus bisporus* (Hollings, 1962) and most of the known mycoviruses have been discovered within the last decade. One of the main reasons for the belated discovery of mycoviruses is that many produce latent infections and cause no overt effects on their hosts. A major stimulus for searches for mycoviruses came from studies of the antiviral- and interferon-inducing activities in animals of cell-free extracts of two *Penicillium* species. Ellis & Kleinschmidt (1967) demonstrated that the interferon-inducing activity of statolon, an antiviral agent isolated from culture filtrates of *Penicillium stoloniferum*, was associated with the presence of isometric virus-like particles. In the same year Lampson *et al.* (1967) identified double-stranded RNA (dsRNA) of presumed viral origin as the active agent of helenine, an interferon inducer isolated from a strain of *Penicillium funiculosum*. Shortly afterwards, Banks *et al.* (1968) isolated and purified the virus-like particles from *P. stoloniferum*; they showed that the particles contained dsRNA and also isolated another dsRNA virus from a strain of *P. funiculosum*. Discovery of virus-like particles and of dsRNA in other species of *Penicillium* and of *Aspergillus* soon followed (Banks *et al.*, 1969*a*, *b*; 1970). A detailed review of the early work on fungal viruses and interferon induction, and of the current status of dsRNA and interferon as antiviral agents in animals, has been published (Buck, 1979*a*).

Subsequent screening programmes (Bozarth, 1972) have indicated that 10 to 15 per cent of randomly sampled fungal isolates contain virus-like particles, as detected by electron microscopy of subcellular fractions. This figure probably represents a lower limit since some fungi may contain particles below the level of detection by this method and, moreover, the procedure used for subcellular fractionation may have artificially selected certain classes of virus-like particles and rejected others

(Bozarth, 1979). To date, virus-like particles have been discovered in over 100 species from more than 60 genera, spread among all classes of fungi. Several different morphological types of particle have been described (Table 1). By far the most frequently occurring are the small

Table 1. *Morphological types of virus-like particles detected in fungi*[a]

Particle morphology	Genome[b]	Number of fungal species in which detected
Isometric (diameter 25–50 nm)	dsRNA	70
Club-shaped	dsRNA	2
Rigid rods	ssRNA?	7
Filamentous	ssRNA?	4
Bacilliform	ssRNA	2
Pleomorphic with membrane envelope	ssRNA	1
Head and tail, similar to some DNA bacteriophages	dsDNA	6
Isometric (diameter 100–200 nm)	dsDNA	4
Herpes-like	dsDNA	1
Isometric (diameter 20 nm), particles geminate	ssDNA?	1

[a]Based on electron microscopy. For lists of fungal species containing virus-like particles see Bozarth (1972; 1979), Moffit & Lister (1975), Yamashita, Doi & Yora (1975), Saksena & Lemke (1978). Bozarth (1972) lists a further 30 fungal species containing virus-like particles of unspecified morphology.
[b]ds, double stranded; ss, single-stranded.

isometric particles and so far about 30 of these have been isolated and shown to contain dsRNA. Interestingly, particles with club-shaped morphology, found so far in only two fungi, may also contain dsRNA (Day & Dodds, 1979). Moffitt & Lister (1975) detected dsRNA in 14 of 70 fungal isolates tested using a serological method. It is noteworthy that nearly all strains of some species e.g. *Saccharomyces cerevisiae* appear to contain dsRNA and virus particles, whereas no dsRNA or particles have been detected in any strains of other species e.g. *Podospora anserina*.

Because of their relatively recent discovery and because of their possible implications in many areas of fungal biology, mycoviruses have been the subject of many review articles (Hollings & Stone, 1969, 1971; Bozarth, 1972; Wood, 1973; Lemke & Nash, 1974; Lemke, 1976, 1977; Buck, 1977; Hollings, 1978; Saksena & Lemke, 1978; Rawlinson & Buck, 1980), two symposium volumes (Hasegawa, 1975; Molitoris, Wood & Hollings, 1979) and a multi-author book (Lemke, 1979). The present chapter will not therefore attempt to give a comprehensive coverage, but will focus attention on what the author believes to be the most interesting aspects of dsRNA mycovirus research at the present time. Because many dsRNA mycovirus infections are latent, because virus particles appear to be transmitted only by intracellular routes, and because infectivity of isolated particles has rarely been demonstrated, many investigators refer to dsRNA mycoviruses as virus-like particles. Justification for the use of the word 'virus', which will be used here, has been discussed in detail previously (Buck, 1979b).

BIOPHYSICAL PROPERTIES AND GENOME ORGANISATION

The isometric dsRNA mycoviruses do not form a homogenous group. Although in most cases, virus capsids are constructed from one major polypeptide, the molecular weights of capsid polypeptides range from 25 000 to 130 000 with different viruses. Moreover, numbers of dsRNA components associated with a given virus vary from one to eight with molecular weights of individual dsRNAs ranging from 0.27×10^6 to 6.3×10^6. It is clear that it will be necessary eventually to create several groups in order to accommodate all the known isometric dsRNA mycoviruses, but sufficient detailed knowledge of the properties of most of them has not yet been obtained. The problem of classification is compounded by the fact that where more than one dsRNA component is associated with a virus, with the exception of a few well-studied viruses to be discussed later, it is not known whether: (a) all the dsRNA components are required for virus replication; (b) some dsRNA components are derived from others e.g. by deletion mutation, resulting in defective interfering particles (Huang & Baltimore, 1977); (c) some dsRNA components are satellites i.e. they utilise the coat protein of a helper virus for encapsidation and helper virus functions for their replication, but they contain genetic information not present in the helper genome in an analogous way to RNA-5 of cucumber mosaic virus (Mossop & Franki, 1978).

The physical properties of some of the better characterised isometric dsRNA mycoviruses are given in Tables 2 and 3. Viruses containing a single dsRNA molecule have been grouped together, as have those with genomes consisting of two or more dsRNA molecules, but further subdivision of these two groups will doubtless be necessary as more information accumulates. Coding capacities of dsRNA molecules have been calculated by making the assumptions that: (a) only one strand codes, (b) three nucleotides encode one amino acid (average molecular weight of nucleotide and amino acid residues, 330 and 110, respectively), (c) the whole of an RNA strand codes and non-coding sequences

Table 2. *Physical properties of isometric dsRNA mycoviruses with undivided genomes*

Virus	Particle diameter (nm)	Mol. wt. of major capsid polypeptide ($\times 10^{-3}$)	Mol. wt. of virion dsRNA ($\times 10^{-6}$)	Calculated coding capacity of dsRNA (daltons $\times 10^{-3}$)	Reference
Helminthosporium maydis virus	48	121	6.3	350	Bozarth (1977)
Gaeumannomyces graminis virus F10 A	40	94	6	330	Almond (1979)
Saccharomyces cerevisiae virus[a]	40	88[b]	3.4	189	Hopper *et al.* (1977)
Ustilago maydis virus[c]	41	75	4.1	229	Holm *et al.* (1978)
Mycogne perniciosa virus	42	69	4.3	239	Barton (1978)
Helminthosporium victoriae virus (190 S particles)	35–40	88 (+ 83?)	3.0[d]	–	Sanderlin & Ghabrial (1978)

[a]Virions in sensitive yeast strains contain a single dsRNA component (L dsRNA).
[b]Oliver *et al.* (1977) reported 75 000 but this value is considered to be less reliable, because the determination involved an extrapolation from known standards; minor virion polypeptides of 53 000 and 37 000 daltons (Oliver *et al.* 1977) and 140 000, 82 000 and 78000 daltons (Hopper *et al.*, 1977) have been reported for this virus.
[c]Virions in certain non-killer *Ustilago maydis* strains, derived from killer strains by mutation, contain a single dsRNA component (Koltin, 1977).
[d]Almost certainly an underestimate, since the value was obtained by linear extrapolation from reovirus dsRNA standards (see Bozarth & Harley, 1976).

are neglected. The values obtained approximate the maximum size of single polypeptides which could be produced by individual dsRNA molecules. The effective total coding capacity could be considerably larger because of the possibilities of producing more than one polypeptide from the same genomic region as a result of (*a*) overlapping genes in different reading frames, e.g. as in phage ϕX 174 (Sanger *et al.*, 1977) or in simian virus 40 (Fiers *et al.*, 1978), (*b*) splicing of pre-messenger RNAs in different ways, e.g. as in adenovirus (Berk & Sharp, 1978*a*) or simian virus 40 (Berk & Sharp, 1978*b*; Paucha *et al.*, 1978), (*c*) by initiating translation at two different sites in the same reading frame and terminating at the same site, e.g. as in phage ϕX 174 (Pollack, Tessman & Tessman, 1978), (*d*) by initiating translation at the same site and terminating at two different sites, e.g. as in phage Qβ (Weiner & Weber, 1971) or tobacco mosaic virus (Pelham, 1978) or (*e*) by protein modification (Pollack *et al.*, 1978). The presence of non-coding sequences, or of duplicated genes, would, of course, reduce the calculated coding capacity.

Table 3. *Physical properties of isometric dsRNA mycoviruses with segmented genomes*

Virus	Particle diameter (nm)	Mol. wt. of virion polypeptides ($\times 10^{-3}$)	No. of copies of polypeptides per virion	Mol. wt. of virion dsRNA components ($\times 10^{-6}$)	Coding capacity of dsRNA components (daltons $\times 10^{-3}$)
Penicillium chrysogenum virus[a]	35–40			2.2	123
				2.1	117
		130	60	2.0	111
Aspergillus foetidus[b] virus F	35–40	125	1	2.7	150
		100	1	1.87	104
		87	120	1.70	94
				1.44	80
				1.24	69
Mushroom virus 4[c]	35	64	ND	1.5	83
				1.4	78
Penicillium stoloniferum virus F[d]	30	59	1	0.99	55
		47	120	0.89	49
				0.46	27
Penicillium stoloniferum virus S[d]	30	56	1	1.10	61
		42	120	0.94	52

[a]Wood & Bozarth (1972); Buck & Girvan (1977); [b]Ratti & Buck (1972); Buck & Ratti (1975); [c]Barton & Hollings (1979); [d]Bozarth *et al.* (1971), Buck & Kempson-Jones (1974), Bozarth & Harley (1976).
ND, not determined.

In order for a dsRNA virus to promote its own replication within a fungus it may be expected that as a minimum its genome will encode a polypeptide for construction of the virus capsid and an RNA-dependent RNA polymerase for transcription and replication of the genome. Of the viruses listed in Tables 2 and 3 it has been shown, using *in vitro* translation methods, that the genomes of *Saccharomyces cerevisiae* virus (Hopper *et al.* (1977), *Penicillium chrysogenum* virus (K. A. Bostian, J. E. Hopper & R. F. Bozarth, unpublished), and *Penicillium stoloniferum* virus S (Buck, 1979c) encode their own capsid polypeptide. Virions of *S. cerevisiae* virus (Herring & Bevan, 1977), *P. chrysogenum* virus (Nash *et al.*, 1973), *Aspergillus foetidus* virus F (Ratti & Buck, 1975) and *P. stoloniferum* viruses S and F (Chater & Morgan, 1974) have been shown to contain associated RNA polymerase activity, but in no case has formal proof been obtained that the polymerases are encoded by the virus dsRNA. However in all cases the genome appears to have ample capacity to code for both capsid polypeptide and an RNA polymerase. In the case of the viruses with undivided genomes (Table 2) the major capsid polypeptide accounts for only a third to a half of the calculated coding capacity of the dsRNA and these dsRNAs are therefore likely to be at least dicistronic. It is noteworthy that dsRNA viruses with undivided genomes have not been reported in any other host taxa.

Viruses have been assumed to have divided genomes (Table 3) when most of the calculated coding capacity of one dsRNA species is required for the major capsid polypeptide. In such cases these dsRNA molecules may be monocistronic, bearing in mind the provisos discussed above. For example, *Penicillium stoloniferum* virus S has two dsRNA components, one just large enough to encode the major capsid polypeptide and the second about the right size to code for a minor polypeptide, present in one copy per particle and believed to be the RNA polymerase. *P. stoloniferum* virus F is basically similar, but contains an additional smaller RNA of unknown status. In the case of *Aspergillus foetidus* virus F the three larger dsRNA components have sufficient capacity to encode the major capsid polypeptide and the two minor virion polypeptides, one or both of which may be part of the RNA polymerase, but the status of the two smaller dsRNA components is unknown. For *Penicillium chrysogenum* virus it has been found recently that all three species of dsRNA are translated (after denaturation) *in vitro* into distinct polypeptides. By comparative peptide analysis it was shown that the smallest dsRNA species codes for the major capsid polypeptide (K. A. Bostian, J. E. Hopper & R. F. Bozarth, unpublished).

In a number of other cases, it is still not possible to decide whether the viruses have divided or undivided genomes. For example, *Aspergillus foetidus* virus S has three dsRNA components, mol. wts. 4.1×10^6 (60% G + C), 2.6×10^6 (47% G + C) and 0.27×10^6, respectively, plus a major capsid polypeptide, mol. wt. 83 000 and a minor virion polypeptide, possibly the RNA polymerase, mol. wt. 85 000 (Buck & Ratti, 1975; G. Ratti & K. W. Buck, unpublished). The largest dsRNA component (calculated coding capacity 229×10^3 daltons) is of sufficient size to encode both the major and minor (RNA polymerase?) virion polypeptides. However, the base compositions of this dsRNA and that of the 2.6×10^6 component are sufficiently different as to exclude a precursor–product relationship between these RNAs. Whether the 2.6×10^6 and 0.27×10^6 components are required for virus replication is unknown. Another example is mushroom virus 1 (Barton & Hollings, 1979) which has a major capsid polypeptide of mol. wt. 24 500 and two dsRNA components, mol. wt. about 1.4×10^6 each (calculated coding capacity of each, 78 000 daltons). It is clear that much more knowledge of the molecular biology of most isometric dsRNA mycoviruses is required before the diversity in the organisation of their genomes can be fully appreciated.

With some dsRNA mycoviruses, e.g. viruses isolated from *Allomyces arbuscula* (Khandjian, Turian & Eisen, 1977) and from *Helmintho-sporium victoriae* (Sanderlin & Ghabrial, 1978), more than one major capsid polypeptide has been reported. However, it will need to be established that some of these polypeptides do not result from degrada-tion before it can be accepted that these viruses represent a divergence from the usual pattern of one major capsid polypeptide found in most isometric dsRNA mycoviruses. The major capsid polypeptide, $\sigma 1$, of *Aspergillus foetidus* virus S is susceptible to proteolysis *in vitro* to generate a second polypeptide, $\sigma 2$, mol. wt. 78 000. Virus preparations, isolated under conditions which allow partial proteolysis, may contain equal amounts of $\sigma 1$ and $\sigma 2$, giving the spurious effect of two major capsid polypeptides (Buck & Ratti, 1975).

The postulated requirement of a virus-coded RNA-dependent RNA polymerase for transcription and replication of mycovirus dsRNA genomes deserves comment. dsRNA will not act as a mRNA; indeed in some systems (Robertson & Matthews, 1973; Farrell *et al.*, 1977), but not in others (Jay, Abrams & Kaempfer, 1974), dsRNA acts as an inhibitor of initiation of protein synthesis. Hence an enzyme is required to transcribe the dsRNA genome to produce single-stranded mRNA. The question arises as to whether or not a cellular enzyme can do the job. DNA-dependent RNA polymerase of *Escherichia coli* has been shown to be able to transcribe a variety of dsRNAs, including those of *Penicillium chrysogenum* virus, with low efficiency *in vitro* (Sugiura & Miura, 1977). Moreover the replicative form dsRNA of poliovirus is infective (Baltimore, 1969), indicating that at least some transcription of this dsRNA *in vivo* by a cellular enzyme takes place (or else the two strands become separated, which seems less likely). On the other hand a wealth of genetic experiments with poliovirus mutants has shown that at least one virus-specific polypeptide is required for the replication of poliovirus RNA (Cooper, 1977). A single-chain RNA polymerase, capable of transcribing poliovirus RNA, has been isolated from polio-virus infected cells and identified as a cleavage product from non-structural virus protein, NCVP-2 (Flanegan & Baltimore, 1977; Baltimore *et al.*, 1978). A similar situation exists with a number of single-stranded (ss) RNA plant viruses. A cellular enzyme, isolated from uninfected tobacco plants, has been found to be capable of transcribing *in vitro* a number of plant virus RNAs, including tobacco mosaic virus

(Ikegami & Fraenkel-Conrat, 1978). However, experiments with temperature-sensitive mutants of tobacco mosaic virus have shown that virus functions are necessary for replication of the virus RNA *in vivo* (Dawson & White, 1978, 1979). An RNA polymerase has been isolated from plant cells infected with tobacco mosaic virus, which is capable of transcribing tobacco mosaic virus RNA *in vitro* and which has tentatively been equated with the 110 000 and 160 000 dalton polypeptides synthesised *in vitro* in cell-free systems programmed with tobacco mosaic virus RNA (White & Murakishi, 1977; Pelham, 1978). The requirement of virus-coded RNA polymerase is well-established for the replication of small ssRNA bacteriophages, such as Qβ, although the holoenzyme in this case is constituted of one virus and three host polypeptides; moreover an additional host factor is required for transcription of Qβ (+) strand (Kamen, 1975). Virus-coded RNA polymerases are also requirements for the replication of negative strand RNA viruses, such as vesicular stomatitis virus (Banerjee, Abraham & Colonno, 1977). Overall it appears that cellular enzymes are capable of transcribing ssRNA and dsRNA *in vitro*, and probably also dsRNA *in vivo* with low efficiency, but that these enzymes are unlikely to be solely responsible for the replication of virus RNA *in vivo*. Efficient transcription and replication of virus RNAs *in vivo* require virus-coded RNA polymerases, which are highly specific for virus RNA. However, additional host polypeptides may well be required either as part of the polymerase holoenzyme, or for other functions in virus RNA replication.

All dsRNA viruses that have been examined, have been found to carry virion-associated RNA polymerases. These include members of the Reoviridae, e.g. reovirus (Silverstein, Christman & Acs, 1976) and Cystovividae (phage φ6; van Etten *et al.*, 1973) and the recently described group of animal viruses with bisegmented dsRNA genomes (Dobos *et al.*, 1978; Barnard & Petitjean, 1978). In the case of reovirus (Cross & Fields, 1977) and phage φ6 (Rimon & Haselkorn, 1978), genetic evidence indicates that the polymerases are virus-coded. It is noteworthy that in the case of reovirus, dsRNA is never released from virus cores throughout the entire replicative cycle (Silverstein *et al.*, 1976). This, together with fact that replication is confined to the cytoplasm, means that dsRNA is never available as a template for a cellular polymerase; moreover enclosure of the dsRNA in cores prevents its possible inhibition of cellular protein synthesis. dsRNA is not released from nucleocapsids during the replicative cycle of phage φ6 (Mindich, Sinclair & Cohen, 1976) and it is possible that dsRNA is never released from isometric dsRNA mycovirus particles either. There is, in fact, no convincing evidence for the existence of free dsRNA during the replication of any of these viruses. In summary, although formal proof is still lacking, it seems certain that virus-coded RNA polymerases are required for the replication of dsRNA mycoviruses.

The isometric dsRNA mycoviruses, diverse though they may be, all differ from dsRNA viruses of other host taxa. Members of all other groups have divided genomes. Human type 3 reovirus, as a prototype of the Reoviridae, has icosahedral virions (diameter 75 nm), with complex double-shelled capsids, composed of ten polypeptides of which eight are primary gene products, with mol. wts. ranging from 43 000 to 153 000; it has ten monocistronic dsRNA components, with mol. wts. ranging from 0.6×10^6 to 2.7×10^6, total genome size 22 kilobase pairs (kb) (Joklik, 1974). Phage $\phi6$ has a polyhedral nucleocapsid (diameter 60 nm), enclosed in a lipoprotein envelope. Virions contain ten polypeptides with mol. wts. ranging from 6000 to 77 000 and three polycistronic dsRNA components with mol. wts. of 2.0×10^6, 2.9×10^6 and 4.9×10^6, and total genome size 15 kb (van Etten et al., 1976). Drosophila X virus, and other similar viruses, have bisegmented genomes, molecular weight of each dsRNA component being about 2.5×10^6, and total genome size 7.5 kb. Virions are isometric, diameter c. 60 nm, with capsids composed of several polypeptides, in three size classes. (Dobos et al., 1978). With all the above viruses, virions contain one copy of all the genome segments. Isometric dsRNA mycoviruses are smaller, have simple capsids, total genome sizes of 3–9 kb, and in the case of those viruses with divided genomes, each dsRNA component, with rare exceptions, is contained in a separate virion, giving rise to a multi-component virus system.

TRANSMISSION

The isometric dsRNA mycoviruses appear to be transmitted only by intracellular routes. There is no evidence, in any case, for release of virus particles, followed by infection from without. Transmission occurs in the flow of protoplasm towards the tip during hyphal growth, in a variety of sexual and asexual spores and via heterokaryosis (Buck, 1979b). In many cases host growth appears to be unaffected by the presence of virus and the virus particles and associated dsRNAs appear to be relatively stable cytoplasmic features of particular fungal strains.

However, it should be noted that spore transmission apparently does not occur invariably. In an investigation of transmission of viruses S and F during conidiogenesis of *Penicillium stoloniferum*, DeMarini *et al.* (1977) found that virus levels remained constant throughout a series of transfers when the inoculum consisted of a large population of conidia (10^6 to 10^7) or mycelial fragments. However, when 43 random single conidial isolates derived from the same strain were examined, it was found that, whereas most cultures contained virus levels similar to that of the original strain, a small number contained higher or lower levels; three isolates contained no detectable virus F and a further two isolates contained no detectable virus particles at all. In the case of virus-infected strains of *Gaeumannomyces graminis* (Rawlinson *et al.*, 1973), *Endothia parasitica* (Day *et al.*, 1977) and *Helminthosporium maydis* (Bozarth, 1977), all ascospore progeny appeared to be free from detectable dsRNA or virus particles.

One interesting question which has been raised in recent years is whether or nor dsRNA mycoviruses can be transmitted as DNA proviruses. There are a number of observations for which provirus induction would offer an attractive explanation. One of these is provided by the wheat 'take-all' fungus, *Gaeumannomyces graminis* var. *tritlci* (henceforth called *G. graminis*). Rawlinson *et al.* (1973) found that isolates of *G. graminis* from first year cereals are generally free from detectable virus particles. This is consistent with the possibility that initial colonisation of fields previously free from *G. graminis* could be by virus-free ascospores (Brooks, 1965). In a study of the natural introduction of *G. graminis* and its viruses into cereal crops grown in a field which had carried no cereals for over 100 years previously, Rawlinson & Muthyalu (1976) found, as expected, no virus particles in the first *G. graminis* isolates from spring crops grown in alternate years of a period of three years. However within a further period of three months, virus-like particles, 27 nm in diameter and apparently mainly empty as judged by electron microscopy, were found in 17 out of 38 isolates taken from widely separated plants in the same field now growing a consecutive crop of winter wheat. After a further period of nine months, larger (35 nm) particles, later shown to contain dsRNA, appeared in many isolates. A possible explanation is that *G. graminis* carries endogenous DNA proviruses, which can be induced to produce virus particles and dsRNA under certain conditions. Sequential induction of a virus with a divided genome might involve production of mRNA for capsid polypeptide, resulting in formation of empty capsids, followed later by production of mRNA for RNA polymerase, resulting in dsRNA formation. However, at the present time, alternative explanations, such as stimulation of undetected low levels of virus to multiply, cannot be ruled out. Such stimulations have been reported for certain insect (Smith, 1967) and plant (Smith, 1977) viruses. Barton (1978) reported that repeated sub-culture on synthetic media of virus-infected mycelium of *Mycogone perniciosa* led to a marked reduction of virus concentration. The concentration could be restored by growing the fungus on healthy mushrooms or on mushroom extract media. A number of fungal metabolites, such as mycophenolic acid, patulin and gliotoxin (Detroy & Worden, 1979), have been shown to inhibit the replication of certain dsRNA mycoviruses at concentrations where host growth is minimally affected. Hence the possibility must be considered that virus replication in *G. graminis* could be influenced by factors, possibly microbial metabolites, present in soil. Conditions in first cereal crops may not be conducive to virus replication, but alterations possibly induced by changing microbial flora in subsequent crops, could result in an environment favourable to virus multiplication.

Adler (1979) detected no dsRNA or virus particles in several isolates of *Histoplasma capsulatum* grown in the mycelial phase; however, after conversion to the yeast phase, dsRNA and particles were readily detected. In one case, dsRNA and particles were maintained after converting back to the mycelial phase, but with another isolate no dsRNA could be detected after such reconversion. Here again provirus induction offers one explanation for the phenomenon, but gross alterations in virus or dsRNA levels as a result of cellular differentiation are also possible.

The occurrence of empty isometric capsids (27 nm diameter), constructed of a single polypeptide of mol. wt. 18 300, in a strain of *Aspergillus flavus*, in which no dsRNA or other 'viral' nucleic acids could be detected (Wood *et al.*, 1974; Bozarth, 1979) could also be explained by a DNA provirus hypothesis. The production of empty capsids in this fungus is a stable characteristic, transmissible through conidia, and the genetic information for capsid production could reside in endogenous provirus DNA. In all of the cases discussed above, the provirus hypothesis is amenable to testing, and further experimentation using nucleic acid hybridisation techniques is essential in order to resolve the problems.

Vodkin (1977) reported sequence homology between *Saccharomyces cerevisiae* virus L dsRNA (Table 2), derived from several yeast strains, and *S. cerevisiae* DNA, prepared both from strains containing dsRNA and from those containing no dsRNA. The evidence was based mainly on the increased rates of conversion of denatured dsRNA from ribonuclease sensitivity to ribonuclease resistance under annealing conditions in the presence of excess denatured yeast DNA, but not in the presence of native yeast DNA or denatured calf thymus DNA. The increased rate of RNA association was assumed to result from the formation of DNA/RNA hybrids; indeed some evidence for formation of a DNA/RNA hybrid was obtained from the appearance of a band of density lower than that of dsRNA after analysis of hybridisation mixtures by density gradient centrifugation in isopycnic caesium sulphate. The results implied several copies of L dsRNA per yeast genome equivalent. However, Hastie, Brennan & Bruenn (1978) have been unable to duplicate these experiments and found no effect of denatured yeast DNA (prepared from one of strains used by Vodkin) on the kinetics of reassociation of yeast L dsRNA. Moreover these investigators also found that the rate of hybridisation of a ssRNA copy of yeast L dsRNA (prepared by transcription *in vitro* with the virion-associated RNA polymerase) to denatured L dsRNA was unaffected by addition of excess denatured yeast DNA. The kinetics of reassociation of single copy yeast DNA in the presence of excess denatured yeast DNA were measured under the same conditions as a positive control. On the basis of the results obtained it was concluded that there is at most one copy of L dsRNA per 13 yeast genome equivalents, assuming that the rate of DNA/RNA hybridisation is three times as slow as DNA/DNA reassociation (Davidson *et al.*, 1975), or possibly fewer than one copy of L dsRNA per 40 yeast genome equivalents, assuming that the two reactions occur at similar rates. Using similar methods, Wickner & Leibowitz (1977) and Hopper (unpublished results quoted in Hastie *et al.*, 1978) have also failed to detect sequence homology between yeast DNA and L dsRNA.

In view of the negative results from three independent laboratories the results of Vodkin (1977) require explanation. Contamination of dsRNA by DNA or vice versa was ruled out by carefully controlled nuclease tests and by the inclusion of one strain not containing dsRNA. Vodkin (1977) does not state the method used to shear his DNA or the size distribution of the driver DNA. If the yeast DNA contained a significant proportion of long molecules, reassociation might result in network formation (Britten, Graham & Neufeld, 1974). Such networks might entrap RNA molecules and sterically hinder the action of ribonuclease in the subsequent assay. Another possibility is that Vodkin's DNA contained sequences which were lost during preparation of DNA by the other investigators. The problem could be resolved by a more definitive experiment. L ssRNA (prepared free of dsRNA) could be hybridised to excess yeast DNA (prepared by Vodkin's method), and to separated restriction fragments of yeast DNA, under conditions where DNA/DNA reassociation does not occur (Casey & Davidson, 1977). Since dsRNA reassociation would also be eliminated by the use of ssRNA, hybridisation with excess RNA, as well as those with excess DNA, could be carried out.

HOST RANGE

The natural host range of dsRNA mycoviruses may be expected to be rather limited in view of their intracellular mode of transmission. Anastomosis of hyphae from unrelated fungal species seems unlikely and, even within a species, opportunities for virus spread by heterokaryosis may be limited by systems of vegetative incompatibility (Caten, 1972; Carlile, this volume). On the other hand if virus infection arose early in the phylogeny of a particular fungus the intracellular mode of replication could result in virus persistence during subsequent evolution and in its widespread occurrence in species between which transfer via heterokaryosis would be unlikely to occur. Such a theory would explain not only the occurrence of serologically related viruses in fungi fairly closely related phylogenetically, e.g. in *Penicillium brevi-compactum*, *Penicillium cyaneo-fulvum* and *Penicillium chysogenum* (Wood & Bozarth, 1972; Buck & Girvan, 1977), and in *Aspergillus niger* and *Aspergillus foetidus* (Buck, Girvan & Ratti, 1973), but also in fungi between which relationships are less obvious, e.g. virus particles serologically related to *Penicillium stoloniferum* virus S have been found in *Diplocarpon rosae* (Bozarth, Wood & Goenaga, 1972) and more recently in *Aspergillus ochraceous* (T. Kong & R. F. Bozarth, unpublished). By and large though, reports of serological relationships between dsRNA mycoviruses have been few. While this may be due to lack of appropriate tests having been carried out in some instances, it is likely that serological relatedness, implying relatedness between only a small fraction of a virus genome, is too specific a criterion to use in conducting a general search for related viruses in different fungal species. As a fungal species evolves, so may the viruses that it carries. Studies of sequence homology of virus dsRNAs of different fungal species, preferably carried out by direct comparison of sequences, may provide a greater insight into this interesting problem.

REPLICATION

There is ample evidence that dsRNA mycoviruses replicate in parallel with the growth of their hosts. Evidence from electron microscopy of thin sections of fungal hyphae, indicates that virus assembly, and probably also replication, takes place in the cytoplasm. Although virus particles appear to be absent from the growing hyphal tip, in many instances they have been observed free in the cytoplasm, in the peripheral growth zone (Trinci, 1971), including the apical hyphal compartment. Under suitable conditions virus replication may continue in non-proliferating cultures and highest virus yields are frequently obtained in shaken liquid cultures grown to stationary phase. Replication of mycovirus dsRNA has been compared to that of the 'relaxed' mode of replication of certain bacterial plasmids, which are produced at a controlled copy number during exponential bacterial growth, but which increase considerably in numbers in the stationary phase (Timmis, Cabello & Cohen, 1974). In older hyphal compartments virus particles often become enclosed in vesicles and, at a later stage, in large vacuoles. Aged hyphal compartments may contain as many as 100 000 particles (Yamashita, Doi & Yora, 1973), but numbers are usually very much smaller. For a detailed review of mycovirus replication in relation to fungal growth, see Buck (1979b).

Very little is known concerning the various stages of the replicative cycles of dsRNA mycoviruses. This is largely due to the lack of suitable systems for obtaining synchronous infection. Although fungal protoplasts have been infected with virus particles in a number of instances (Lhoas, 1971; Pallett, 1976), the systems have not been developed sufficiently in order to obtain synchronous infection. Indeed since mycelial colonies, derived from protoplasts after infection, have invariably been used to assay virus multiplication, there is no evidence as to whether or not virus replication takes place in the protoplasts, prior to their germination. The small amount of information that is available on mycovirus replication cycles comes from studies on virion-associated RNA polymerases in vitro. Because of the diversity of types of genome organisation it cannot be assumed that all dsRNA mycoviruses will employ the same strategy of replication. Here RNA polymerases of three viruses will be discussed: one, illustrating a virus with an un-divided genome, namely Saccharomyces cerevisiae virus; one with a divided genome, namely Penicillium stoloniferum virus S; and one with a genome of unknown status, namely Aspergillus foetidus virus S.

Saccharomyces cerevisiae virus (L virions)

Two types of RNA polymerase activity have been shown to be associated with virus particles isolated from Saccharomyces cerevisiae: (a) ss $\longrightarrow$ dsRNA polymerase activity was found to be associated with particles which sedimented more slowly than virions containing L dsRNA and was isolated from log phase cells (Bevan & Herring, 1976); (*b*) ds ⟶ ssRNA polymerase (transcriptase) activity was associated with virions containing L dsRNA isolated from stationary phase cultures (Herring & Bevan, 1977); the product of reaction was L ssRNA, i.e. a full length copy of one of the two strands of L dsRNA. On the basis of these two types of activity, Herring & Bevan (1977) postulated that L dsRNA may replicate asynchronously in a similar way to reovirus dsRNA (Silverstein *et al.*, 1976).

However, there are a number of differences from the reovirus system. Reovirions have latent methylase, as well as transcriptase activity (activated after removal of outer shell polypeptides with proteinase)

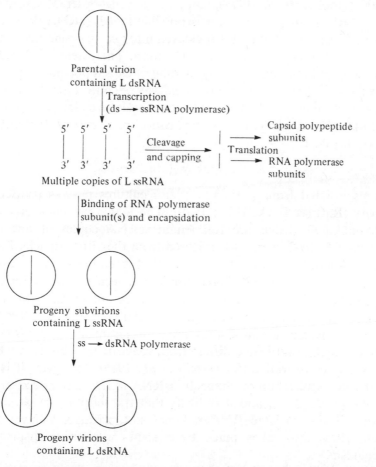

Fig. 1. A model for the replication of *Saccharomyces cerevisiae* L virions.

and their ssRNA transcripts (mRNAs) and the (+) strand of reovirus dsRNA have capped (7MeG ppp Gm . . .) 5' termini (Furuichi *et al.*, 1975; Furuichi, Muthukrishnan & Shatkin, 1975). On the other hand yeast L dsRNA has pppG at the 5' terminus of both strands (Bruenn & Keitz, 1976). If *Saccharomyces cerevisiae* virus mRNAs are capped, like most yeast mRNAs, capping must occur after transcription and a mechanism must exist to distinguish virus mRNA and RNA destined to be encapsidated in sub-virus particles. Since L ssRNA is at least dicistronic (unlike reovirus mRNAs which are all monocistronic), and since efficient translation in eukaryotic systems is usually confined to the cistron closest to the 5' terminus, it is likely that sub-genomic mRNAs are formed *in vivo*. *In-vitro* translation studies with denatured L dsRNA gave as the principal product, capsid polypeptide, mol. wt. 88 000 (Hopper *et al.*, 1977), implying a gene order 5'-capsid poly-peptide–RNA polymerase–3'. It is possible that L ssRNA does not act as a mRNA *in vivo*, but that it is cleaved into two sub-genomic mRNAs, which are then capped prior to translation. The extent of cleavage of L ssRNA could be controlled by binding of capsid polypeptide near the cleavage site, as suggested as a possible way of controlling the formation of a sub-genomic mRNA from tobacco mosaic virus RNA (Zimmern & Wilson, 1976). A model for the replication of *S. cerevisiae* L virions is shown in Fig. 1.

Aspergillus foetidus *virus S*

Virions isolated from stationary phase cultures possess transcriptase activity (Ratti & Buck, 1975; 1979). The major products of reaction with untreated virions are full length ssRNA copies of one of the strands of RNA-2, which are released from the virus particle. Efficient re-initiation of transcription occurs, so that several ssRNA-2 transcripts are produced from each dsRNA template molecule; since virions prob-ably contain only a single molecule of RNA polymerase, this implies that the dsRNA template is held in a circular conformation. Transcrip-tion takes place by a semi-conservative displacement mechanism (Ratti & Buck, 1978), and hence differs from reovirus transcription which is completely conservative (Silverstein *et al.*, 1976). However, it is note-worthy that replication of phage ϕ6 dsRNA *in vivo* is semi-conservative (Coplin *et al.*, 1975), and it is likely that transcription of ϕ6 dsRNA by the virion-associated RNA polymerase (Partridge, Vidaver & van Etten, 1978) also takes place by a semi-conservative displacement mechanism.

If virions of *Aspergillus foetidus* virus S are treated with proteinase K, full length ssRNA transcripts of all its three dsRNA components are produced with equal efficiency (G. Ratti & K. W. Buck, unpublished). Even though the two larger dsRNAs may be at least dicistronic, like yeast L dsRNA, there is no evidence for the synthesis of sub-genomic mRNAs *in vitro*.

Penicillium stoloniferum *virus S*

Virions isolated from stationary phase cultures have dsRNA replicase activity (Buck, 1975). Progeny dsRNA molecules remain encapsidated, so that diploid virions are formed. Replication of dsRNA takes place semi-conservatively (Buck, 1978) and the mechanism probably consists of synthesis of daughter (+) strand with concomitant displacement of parental (+) strand, followed by synthesis of daughter (−) strand on the displaced parental (+) strand as template, in a similar way to semi-conservative replication of adenovirus DNA (Lechner & Kelly, 1977). Whether this type of synchronous dsRNA replication takes place *in vivo* or whether it is an artefact of the *in-vitro* system, resulting from failure to release the displaced parental (+) strand, is not known. Particles containing one molecule each of dsRNA and a complete ssRNA transcript have been isolated from infected mycelium, but diploid virions have not (Buck, 1979c). Particles containing only ssRNA (virus mRNA) have also been isolated, and although in the *in-vitro* RNA polymerase assay system these particles showed no activity, they contain the putative RNA polymerase polypeptide, suggesting that replication of *Penicillium stoloniferum* virus S dsRNA may be asynchronous *in vivo* (cf. Fig. 1).

Since each dsRNA molecule is probably monocistronic, further cleavage of primary transcripts would not be expected to be a requirement for *Penicillium stoloniferum* virus S. Indeed it is possible that the dsRNA mycoviruses with divided genomes evolved from those with undivided genomes as a convenient method of producing monocistronic mRNAs. Separate encapsidation of individual segments of a multipartite genome would not be an impediment for viruses which are transmitted intracellularly, and could be an advantage in allowing variation in numbers of individual genome segments, and, in the case of viruses with virion-associated RNA polymerases, may facilitate control of production of individual mRNAs. It is noteworthy that viruses with multipartite monocistronic genome segments have been found only in eukaryotes. There would be no selection pressure for such a feature in viruses of prokaryotes in which polycistronic mRNAs and internal initiation of translation are commonplace. For a detailed review of the origin of small multicomponent RNA viruses, see Reijnders (1978).

BIOLOGICAL EFFECTS

It has already been mentioned that many dsRNA mycoviruses are latent in causing no overt effects on their hosts and in producing no measurable effects on host growth, or on host RNA, DNA or protein synthesis. There are probably two main reasons for this: firstly, no viral products potentially deleterious to the host are synthesised (or the host is resistant to such products), and secondly, the viruses replicate too slowly to effect host growth significantly by direct competition. It has been calculated (Buck, 1977) that even in the case of *Penicillium stoloniferum* ATCC 14586, which produces one of the highest known virus yields, the virus particles account for only 0.5% of the fungus dry weight when grown in shaken liquid batch culture and only about 20% of these particles are synthesised during exponential fungal growth, the remainder being produced during the deceleration or stationary phases of growth (Still, Detroy & Hesseltine, 1975). It is possible that virus replication during exponential fungal growth is controlled by competition for a host factor (or factors) present in limiting amounts and required for semi-conservative replication of both host dsDNA and virus dsRNA.

The latent nature of many dsRNA mycoviruses notwithstanding, the realisation of the widespread occurrence of viruses in fungi prompted many investigators to search for possible relationships between dsRNA or virus infection and various properties of fungi, e.g. production of secondary metabolites, phytopathogenicity and phenotypes known to be cytoplasmically inherited, such as ability to secrete 'killer' toxins or inheritance of the 'ragged' phenotype of *Aspergillus amstelodami* and senescence in *Podospora anserina*. In many cases negative results have been obtained, for example, production of secondary metabolites in fungi appears to be unaffected by virus infection (reviewed by Detroy & Worden, 1979), the 'ragged' phenotype of *A. amstelodami* is probably due to a mitochondrial defect (Caten & Handley, 1978) and senescence in *P. anserina* may be determined by a closed circular DNA plasmid (Stahl *et al.*, 1978). In many other cases results are of little value because isogenic fungal strains were not compared and proposed correlations could equally well be explained by host chromosomal differences (reviewed by Buck, 1979b). However, it has become increasingly clear in recent years that specific dsRNA molecules can influence aspects of the biology of some fungi. For example it is now known, as a result of some excellent genetical and biochemical studies, that the killer toxins produced by some strains of *Saccharomyces cerevisiae*, and of *Ustilago maydis*, are encoded by dsRNA molecules, and there is evidence that certain dsRNA molecules may be determinants of hypovirulence in *Endothia parasitica*. These studies will be summarised in the following sections.

DOUBLE-STRANDED RNA AND THE KILLER PHENOTYPE

Production of killer toxins by yeasts

Killer strains of *Saccharomyces cerevisiae* secrete a toxin, or killer factor (Woods & Bevan, 1968) to which they are themselves insensitive, but which is lethal to certain other strains of the same or related species, designated sensitives. Toxin from killer strain K12-1 has been purified and shown to be a protein of mol. wt. 11 500 (Palfree & Bussey, 1979). Genetic studies have shown that the killer character and immunity to killer are both inherited cytoplasmically (Somers & Bevan, 1968; Bevan & Somers, 1969), and several lines of evidence (reviewed by Wickner, 1976a; Bevan & Mitchell, 1979; Vodkin, 1979) have indicated that the killer determinant is a dsRNA plasmid. Most sensitive yeast strains are infected with virus particles which contain a single dsRNA molecule (L dsRNA, mol. wt. 3.4×10^6, Table 2) and which are able to replicate autonomously (see previous section). Killer strains are also infected with L virions, but contain, in addition, a dsRNA component, designated M dsRNA, of mol. wt. about 1.4×10^6 (Bevan, Herring & Mitchell, 1973). Under a variety of conditions where sensitive clones have been derived from killer parents loss of killer function was accompanied by loss of M dsRNA. This suggested that M dsRNA determines both killer toxin and immunity factor. Recently, unequivocal proof that M dsRNA encodes the killer toxin has been obtained by *in vitro* translation studies (Bostian *et al.*, 1979).

Very little is known concerning the biochemistry of M dsRNA replication. M dsRNA is apparently not transcribed from host DNA since no sequence homology between M dsRNA and yeast DNA could be detected (Wickner & Leibowitz, 1977). However, M dsRNA has been found only in strains also harbouring L dsRNA, suggesting that it is unable to replicate autonomously. M dsRNA is not a cleavage product of L dsRNA, since no sequence homology between these two RNAs could be detected by fingerprinting and hybridisation studies (Bruenn & Kane, 1978). M dsRNA is enclosed in virus-like particles of the same size as L virions and with a capsid constructed from a polypeptide with the same molecular weight as, and serologically related to, the capsid polypeptide of L virions (Herring & Bevan, 1975). It is therefore likely that L dsRNA encodes the polypeptide required for its own encapsidation and that of M dsRNA; it may also encode RNA polymerase for the replication of both these two RNAs. M dsRNA is therefore probably a satellite dsRNA, which replicates only in the presence of helper yeast virus (L virions).

Replication and expression of M dsRNA take place efficiently in the absence of mitochondrial (mt) DNA (Al-Aidroos, Somers & Bussey, 1973) and appear to be independent of other yeast plasmids, namely 2 μm closed circular DNA (Livingston, 1977), [URE 3] (Aigle & Lacroute, 1975) and [PSI] (Young & Cox, 1971). However, recessive mutations in any one of at least 29 nuclear genes (mak 1 through to mak 27, pet 18 and spe 2) have been described which result in loss of the killer phenotype and of M dsRNA (Bevan et al., 1973; Wickner, 1978 and unpublished results; Cohn et al., 1978). Many of these genes have been mapped and have been found to be distributed over 15 of the 18 yeast chromosomes (Wickner & Leibowitz, 1976a and unpublished results; Wickner, 1978 and unpublished results; Cohn, Tabor & Tabor, 1978). The spe 2 gene codes for S-adenosyl methionine decarboxylase, an enzyme in the biosynthetic pathway for the polyamines spermidine and spermine. Strains that are deficient in this enzyme, when grown in a medium containing these polyamines, are able to replicate M dsRNA and secrete toxin, but they lose M dsRNA when grown on polyamine-free medium (Cohn et al., 1978). None of the biochemical functions of any other of these so-called maintenance genes is known, but some of the genes exhibit pleiotropic effects. For example, mutations in the pet 18 gene result in inability to respire and absence of detectable mtDNA, and temperature sensitivity for growth (Leibowitz & Wickner, 1978); one mak-1 allele resulted in temperature sensitivity for growth (Wickner & Leibowitz, 1976a).

By analogy with the replication of small prokaryotic DNA viruses and plasmids (Kornberg, 1977) it would be anticipated that replication of M dsRNA would require a number of host functions. However, it is surprising that none of the products of the 29 nuclear genes that are required for M dsRNA replication, are needed for L dsRNA replication; in fact, no yeast mutants incapable of replicating L dsRNA have been described. Wickner (1976a) has estimated that killer strains contain about 100 molecules of L dsRNA and 12 molecules of M dsRNA per cell; this ratio is similar in different strains. Replication of L and M dsRNAs may have some requirements in common, e.g. (a) capsid polypeptide encoded by L dsRNA, (b) an RNA-dependent RNA polymerase, probably also encoded by L dsRNA and (c) possible common host factors. The M/L dsRNA ratio could be controlled by any shared component(s) of the replication complex, present in limiting concentration. If the affinity of a given component for L dsRNA were much greater than for M dsRNA and its concentration dropped below a critical level, M dsRNA could be eliminated with little effect on L dsRNA. Some support for this hypothesis comes from the observation that growth of killer yeast cells in the presence of low concentrations of cycloheximide leads to elimination of M dsRNA, but not of L dsRNA (Fink & Styles, 1972; Vodkin, Katterman & Fink, 1974). It is possible that the products of many of the mutant maintenance genes could cause lowering of the concentration of limiting factors below their critical levels, leading indirectly to loss of M dsRNA. If several maintenance genes affected the level of one particular component of the replication machinery, this could explain the effects of mutations in 29 maintenance genes without the necessity to postulate that 29 different host poly-peptides are required for the replication of M dsRNA, but not L dsRNA.

According to the above theory it may be expected that host mutations would be found which result in an increase in the level of a limiting replication factor and in some cases to an increase in the M/L dsRNA ratio; in fact *ski* (superkiller) mutations (recessive in diploids) with increased M/L ratios have been described (Toh-e, Guerry & Wickner, 1978); moreover, as may be expected, some of these are able to bypass several *mak* mutations (Toh-e & Wickner, 1979) with regard to M dsRNA replication. Bypass of *mak* 10-1 mutations by deletion of mtDNA (Wickner, 1977) could be explained in a similar way if replication of mtDNA and M dsRNA required a factor in common.

Two chromosomal genes (*kex* 1 and *kex* 2) are required for the secretion of toxin in strains carrying M dsRNA (Wickner & Leibowitz, 1976b) and a third host gene (*rex* 1) is required for expression of the resistance to toxin determined by M dsRNA (Wickner, 1974). The *kex* 2 gene product is required also for normal mating by α strains and meiotic sporulation in all strains (Leibowitz & Wickner, 1976). *Kex* 2 mutants of α mating type fail to secrete a pheromone α-factor or to respond to *a*-factor II pheromone. Diploids that are homozygous for the *kex* 2 mutation fail to undergo sporulation with the defect occurring in the fungal spore maturation stage. Although the nature of the *kex* 2 gene product is not known, it could well be a proteinase. Proteinases are known to be required for processing the N-terminal signal sequences and for activation of several mammalian secretory proteins (Shields & Blobel, 1978), and also for cleavage of polygenic primary translation products of several eukaryotic viruses (Villa-Komaroff *et al.*, 1975; Glanville *et al.*, 1978). A major primary translation product of denatured M dsRNA in reticulocyte lysate and wheat germ *in vitro* systems is a polypeptide of mol. wt. 32 000 which contains all but one of the tryptic peptides of the killer toxin (Bostian *et al.*, 1979). If this polypeptide is also a major primary translation product *in vivo* then it is clear that one or more host proteinases must play an essential role in killer expression. The mRNA species responsible for toxin production *in vivo* have not yet been identified. The 5′ termini of both strands of M dsRNA are pppG . . . (Bruenn & Keitz, 1976), and, as suggested for L dsRNA, capped sub-genomic mRNAs may be produced *in vivo*.

Killer mutants with altered M dsRNA

Several types of yeast killer mutants which owe their phenotype to mutations in M dsRNA have been described. Neutral mutants, some of which occur as wild-type strains (Somers & Bevan, 1968), have lost the ability to produce killer, but carry immunity to killing determined by an M dsRNA component of the same mol. wt. (1.4×10^6) as in wild-type killers (Bevan *et al.*, 1973; Vodkin *et al.*, 1974). *In vitro* translation of denatured M dsRNA from a neutral strain gave a polypeptide of 32 000 daltons which could be precipitated with anti-toxin antibody, suggesting that M dsRNA in this strain carries a point mutation (or very small deletion) either in the toxin gene itself or in a gene required for production of active toxin (Bostian *et al.*, 1979). Immunity-minus mutants may be derived similarly by a mutation in the immunity region of M dsRNA; some of these mutants, however, contain in addition to L and M dsRNAs, a small dsRNA (mol. wt. 0.25×10^6) of unknown status (Vodkin *et al.*, 1974; Sweeney, Tate & Fink, 1976). Superkiller plasmid mutants (Vodkin *et al.*, 1974; Sweeney *et al.*, 1976) have approximately 2.5 times more M dsRNA (1.4×10^6 daltons) and greater killing activity than standard killers. Toxin from one superkiller strain was found to be three to four times as stable as toxin from a standard killer strain. This could explain the higher killer activity in superkiller cultures, although the possibility that the superkiller produced more toxin on account of its higher content of M dsRNA could not be eliminated (Palfree & Bussey, 1979). The nature of the mutation in plasmid superkillers is not known. One possibility is that a mutation in the toxin gene (i.e. in the positive strand of M dsRNA) could give rise to a more stable killer. If the toxin gene maps near the 5′ terminus of the positive strand, the complementary mutation at the same site near the 3' terminus of the negative strand (possibly part of a binding site for RNA polymerase or other replication factor) could result in increased transcription and replication of M dsRNA. Superkiller-dependent mutants have been described (Toh-e & Wickner, 1979), in which replication of M dsRNA depends on the presence of a nuclear chromosomal *ski* mutation. M dsRNA in such mutants has the same molecular weight as in wild-type killers, and since its replication depends on an altered host protein, either directly or indirectly, it probably contains a mutation in a binding site for a replication factor, or in a site required for producing a polypeptide needed for replication in wild-type cells. Diploid-dependent mutants (Wickner, 1976b) contain an M dsRNA, which may carry mutations in similar, but not identical, sites to *ski*-dependent mutants. Such mutants may be dependent for M dsRNA replication on over-production in yeast diploids of certain host factors.

Suppressive non-killer mutants (Somers, 1973) lack M dsRNA, but contain a small (S) dsRNA of mol. wt. $< 1.0 \times 10^6$ as well as L dsRNA (Vodkin et al., 1974; Sweeney et al., 1976). When suppressive non-killers are mated with wild-type killers, the resulting diploids are frequently non-killers and all four meiotic products are sensitive and lack M dsRNA (but contain L and S dsRNAs). S dsRNAs have the ability to suppress the replication of M dsRNA. One suppressive dsRNA, S1 (mol. wt. 0.46×10^6), was shown by electron microscopic heteroduplex analysis (Fried & Fink, 1978) and fingerprinting studies (Bruenn & Kane, 1978) to be derived from M dsRNA by internal deletion. A second such dsRNA, S3, apparently arose by duplication of S1, and a third, S4, was derived from S3 by a small internal deletion. Translation of denatured S1, S3 or S4 dsRNA *in vitro* gave rise to a polypeptide, mol. wt. 8000, but in mixtures with denatured M or L dsRNAs, denatured S ds RNAs did not suppress the translation of these RNAs (Bostian *et al.*, 1979). It is likely that the suppressive properties of S dsRNAs rely on their ability to out-compete successfully M dsRNA for RNA polymerase or other replication factors. Since S dsRNAs lack killer and immunity functions it is probable that a greater proportion of S dsRNA transcripts will be available as templates for dsRNA synthesis than in the case of M dsRNA where a considerable proportion of transcripts will be required to be processed into mRNAs for toxin and immunity functions. It is noteworthy that all three S dsRNAs retained a 220 base terminal sequence and a 500 base terminal sequence found in M dsRNA, implying that these sequences are important for replication. S dsRNAs have been compared with defective interfering RNAs of animal viruses (Fried & Fink, 1978; Bruenn & Kane, 1978), which interfere with the replication of their standard helper virus (Huang & Baltimore, 1977). In fact, both S and M dsRNAs may interfere with the replication of their helper virions, since cycloheximide-cured strains, which lack M dsRNA, show an increase in the level of L dsRNA (Vodkin et al., 1974). Spontaneous non-killers, which segregate from killers with very low frequency, also lack M dsRNA and contain a lowered amount of L dsRNA. There is evidence, however, that these strains contain an altered L dsRNA (Vodkin et al., 1974), which may replicate inefficiently.

Yeast killer toxins of different specificities

All the above investigations of *Saccharomyces cerevisiae* killer character have used laboratory stock strains whose origins are uncertain, but which may have a common parentage (Rogers & Bevan, 1978). Killers have also been found in *S. cerevisiae* strains, isolated as brewery contaminants (Maule & Thomas, 1973) and from wine fermentations (Naumov & Naumova, 1973). These killers, designated K2, had a different pH optimum to the laboratory, K1 killers; moreover K2 killers could kill K1 killers and vice versa. Further work has revealed killer activities in strains of yeasts from seven different genera (Phillis-kirk & Young, 1975; Bussey & Skipper, 1975). Rogers & Bevan (1978) classified 14 killer yeasts into four groups (TOX1 to TOX4), based on their killing and immunity cross-reactions with one another, and their ability to kill resistant *S. cerevisiae* mutants, containing chromosomal mutations affecting the cell wall binding site for killer factor (Al-Aidroos & Bussey, 1978). Young & Yagiu (1978) classified 20 killer yeasts into ten groups (K1 to K10), based on cross reactivity. K1, K2 and K3 killers corresponded to the TOX1, TOX2 and TOX4 killers respectively of Rogers & Bevan (1978) and comprised *Saccharomyces* species and *Candida albicans*. These killers contained an L dsRNA species and an M dsRNA species (different molecular weight in each group), and could be cured by treatment with cycloheximide. Whether the L dsRNA species is related in all three killer groups is not clear, but it is possible that one helper virus (L virions), or its variants, could support the replication (in different strains) of different satellite M dsRNAs. It is noteworthy that when K2 killers are crossed with K1 killers, the diploids and meiotic segregants are all K1. A hypothesis to explain such incompatibility, based on differential affinities of RNA polymerase for two different satellite RNAs, has been proposed (Buck, 1979b).

The K4 to K10 toxins appear to be proteins, but none of these killer strains contained dsRNA or could be cured of their killer trait with cycloheximide (Young & Yagiu, 1978); nothing is known concerning the genetics of transmission of the killer character in these cases. Woods, Ross & Hendry (1974) obtained a killer strain, which arose as a spontaneous mutant from a K1-sensitive strain of *Saccharomyces cerevisiae* and which contained only L dsRNA. K1 killers were able to kill 'Woods' killers and vice versa. Crosses of K1 sensitive and 'Woods' killers suggested that in the latter strains killer toxin may be encoded by a nuclear gene and raised again the question of a possible M dsRNA 'provirus' (Mitchell, Herring & Bevan, 1976).

Ustilago maydis *killer strains*

Killer strains of *Ustilago maydis* (Puhalla, 1968) secrete toxins, which kill or inhibit the replication of strains of the same or related species (Koltin & Day, 1975). Ability to produce toxin is cytoplasmically transmitted and occurs only in strains carrying dsRNA virus particles (Wood & Bozarth, 1973; Koltin & Day, 1976a). Immunity to killer is also cytoplasmically transmitted on dsRNA determinants, but can also be determined by nuclear genes (when it is termed resistance). Nuclear resistance genes are recessive to their sensitive alleles (Koltin & Day, 1976a); they may be analogous to the killer resistant mutants of *Saccharomyces cerevisiae* lacking cell wall toxin binding sites (Al-Aidroos & Bussey, 1978). Three killer specificities (P1, P4 and P6) are known in *U. maydis*; each killer is insensitive to its own killer substance, but sensitive to the other two.

Purified toxins of the three specificities were all proteins with mol. wts. of *c.* 10 000. However they differed in their migration patterns in gel electrophoresis at pH 8.3, in thermolability, pH dependence and biological activity. P1 and P4 toxins did not cross-react with an anti-serum prepared to P6 toxin (Kandel & Koltin, 1978). Specific dsRNA components are associated with the three killer phenotypes (Koltin & Day, 1976a); their molecular weights are given in Table 4.

The capsids in all types of *Ustilago maydis* killer are probably related or identical and are constructed from polypeptides of the same molecular weight (Table 2). The functions of individual dsRNA molecules have been investigated using mutants with partial genomes, obtained from killer strains either by mutagenesis (Koltin, 1977; Koltin & Kandel, 1978), or by crossing with sensitive strains (Koltin, Mayer & Steinlauf, 1978). These studies have indicated that H1 dsRNA, found alone in the deletion mutant of P6, NK-10, contains sufficient information to encode both capsid polypeptide and other polypeptides required for its own replication. It is possible that H1 dsRNA can act as a helper for the replication of other satellite dsRNAs, as suggested for the role of L dsRNA in yeast killers. However, it is noteworthy that P4 deletion mutants, lacking H1 dsRNA and containing only H3 and H4 dsRNAs,

Table 4. *dsRNA components in* Ustilago maydis *killer strains of different specificities*

DsRNA component[a]	Mol. wt.[b] $\times 10^{-6}$	*U. maydis* killer strain		
		P1	P4	P6
H1	4.1	$+$[c]	$+$	$+$
H2	3.1	$+$	$+$	$+$
H3	2.2		$+$	
H4	1.8		$+$	
M1	1.0	$+$		
M2	0.8			$+$
M3	0.7	$+$	$+$	$+$
M4	0.6	$+$	$+$	
L1	0.3	$+$	$+$	$+$

[a]Nomenclature of Koltin *et al.*, (1978).
[b]Values determined by Lentz & Bozarth (Bozarth, 1979), except for those of H3, H4 and M2 which were interpolated from the data of Koltin & Day (1976a) and Lentz & Bozarth (Bozarth, 1979).
[c]$+$ Indicates presence of specific dsRNA component (Koltin & Day, 1976a); sequence identity of components of the same molecular weight in different strains has not been established.

have been obtained (Koltin *et al.*, 1978). H3 and H4 may be deletions of H1. In P6 killers, M2 and L1 dsRNAs are associated with killer expression, whereas in P4 killers, M3 and L1 ds RNAs are associated with these functions. No host nuclear genes required for killer genome replication or expression, similar to those described for yeast, have been found in *U. maydis*, possibly reflecting the greater complexity of the dsRNA genome. Perhaps some of the functions encoded by *mak* genes in yeast are encoded by dsRNA components in *U. maydis*.

Non-killer suppressive strains of *Ustilago maydis* have been described (Koltin & Day, 1976b); at least in some cases their killer suppressive properties appeared to be due to the presence of suppressive dsRNA and may have a basis similar to that suggested for yeast suppressive mutants. Crosses of two *U. maydis* killers of different specificities, as with yeast K1 and K2 killers, can lead to exclusion of one or both killer phenotypes (Koltin & Day, 1976a). For example, a P4 × P6 cross yielded only P4 progeny among the killers. Koltin et al. (1978) have shown that the exclusion phenomenon is abolished by deletion from P4 strains of M3 and L1 dsRNAs, i.e. the same dsRNAs as required for killer expression. It was suggested that the exclusion relations result from identification phenomena in which possibly the P4 toxin acts as a nuclease.

Koltin & Day (1975) suggested that the *Ustilago maydis* killer phenomenon could be used to control cereal smuts. If the genetic information for killer protein, either as dsRNA or a DNA copy, could be introduced into a cereal host, and if this information could be expressed in the plant cytoplasm, the plant could be rendered resistant to toxin-sensitive cereal smuts. The method would appear to have little potential for control of *Ustilago maydis* itself, since strains are known which carry nuclear resistance to P1, P4 and P6 killers. However, most other cereal smuts were found to be sensitive to at least one of the three killer toxins. For further details of the *Ustilago maydis* killer phenomenon, the reviews of Day & Dodds (1979) and Koltin (1979) may be consulted.

Comparison of fungal killer toxins and bacteriocins

Fungal killer toxins have a number of similarities to plasmid-coded bacteriocins (Hardy, 1975; Tagg, Dajani & Wannamaker, 1976). Killer toxins and bacteriocins are both proteins, which are lethal or inhibitory to strains of the same or related species as the producing strains. Killer toxins and bacteriocins are encoded by dsRNA and closed circular DNA plasmids, respectively. Immunity, which is also encoded by dsRNA or DNA plasmid, probably involves synthesis of a polypeptide which directly neutralises the toxin in both cases. Nuclear encoded resistance, probably as a result of mutations in receptor sites, is known for both killer toxins and bacteriocins. Moreover, bacteriocins of different specificities may sometimes share common host receptors; a similar situation has been found with yeast killer toxins of different specificities (Rogers & Bevan, 1978). Some fungal killer toxins and bacteriocins may have similarities in their modes of actions. For example, *Saccharomyces cerevisiae* K1 toxin causes permeability changes in the plasma membrane of sensitive strains, resulting in the efflux of K^+ and ATP (Bussey & Sherman, 1973; Skipper & Bussey, 1977), and may resemble the action of colicins E1 or K on sensitive cells of *Escherichia coli* (Holland, 1975). Nuclease activity is associated with *Ustilago maydis* killer proteins, purified by isoelectric focussing (Y.

Koltin, unpublished). Colicin E2 is a DNA endonuclease (Schaller & Nomura, 1976) and colicin E3 inactivates ribosomes in sensitive cells, by causing cleavage of 16 S rRNA near its 3′ terminus (Senior & Holland, 1971).

Possible killer toxins of other fungi

The advantage of killer toxin production by immune yeasts over sensitive organisms in a crowded environment seems obvious. Indeed Young & Philliskirk (1977) have shown that when killer and sensitive yeast strains are grown together in a chemostat at the pH optimum for killer activity, the sensitive strain is displaced from the culture. The significance of killer strains of *Ustilago maydis* is less obvious. Propagation of this fungus depends on mating of two compatible strains within the host. Apparently, the killer phenotype is not expressed, or killer toxin is rapidly inactivated, in the plant host, since matings between killer and sensitive strains *in vivo* occur readily.

Whether the killer phenomenon occurs in filamentous fungi is not known. However, the widespread occurrence of dsRNA in fungi, coupled with the similarities of fungal killer toxins and plasmid-coded bacteriocins, prompted Romanos *et al.* (1979) to search field isolates of the wheat take-all fungus, *Gaeumannomyces graminis* var. *tritici* for possible killer activity. Although three isolates out of 20 were found to produce inhibitors which resembled *Saccharomyces cerevisiae* killer proteins in the low pH associated with their activities, one of these inhibitors, studied in more detail than the others, was found to be active against a wide range of unrelated fungi and to be insensitive to two proteinases; hence it resembled broad spectrum antifungal antibiotics more closely than fungal killer proteins. However, the search for killer strains of fungi, other than yeasts or *Ustilago maydis*, has only just begun; too few isolates have as yet been investigated. Philliskirk & Young (1975) found 59 killers out of 964 yeasts examined from the National Collection of Yeast Cultures (Redhill, UK). However, Puhalla (1968) recovered only four *U. maydis* killer strains from 70 monosporidial strains obtained from 37 North American smut gall samples, and no killers were found from 37 Polish or 10 Mexican smut gall samples (Day & Dodds, 1979). The incidence of dsRNA in *U. maydis* strains may, however, be much higher. Day & Dodds (1979) report that dsRNA was recovered in six of 15 monosporidial isolates from 15 independent naturally produced infections. Some strains have been recovered which apparently contain partial genomes. Day & Dodds (1979) considered that new killers could arise *de novo* by mating or heterokaryon exchange between strains carrying the necessary complementary dsRNA genomes. In-vivo complementation of some non-killer mutants has, in fact, been observed (Koltin & Kandel, 1978). Almond, Buck & Rawlinson (1978) found that 20 isolates of *G. graminis* contained different numbers of dsRNA components, varying from one to 12; no two isolates contained identical patterns of dsRNA components. Whether some of these patterns could represent partial or complete killer genomes is not known. However it is clear that further search for the killer phenomenon among a wide range of fungi would be well worthwhile.

VIRUS DISEASES OF FUNGI

There are now several cases known where an association between the presence of dsRNA virus particles and a fungus disease has been reasonably well established. Transfer of disease and specific dsRNA components by heterokaryosis provides good evidence for involvement of a cytoplasmic determinant, which may be dsRNA or other plasmid transmitted with dsRNA, but proof of virus as causative agent of a disease requires infection with cell-free preparations. This has been achieved in only two of the four cases discussed below. Hypovirulence in *Endothia parasitica*, although debilitating in the sense of reduced pathogenicity, reduced capacity for sporulation and lack of pigmentation, does not affect saprophytic growth of the fungus and is probably caused by a dsRNA virus significantly different from the isometric dsRNA viruses described so far.

Formation of lytic plaques in Penicillium chrysogenum

When grown on surface culture, certain mutant strains of *Penicillium chrysogenum* produced erumpent patches of sterile white mycelia after a few days' growth and localised plaques after one month (Lemke, Nash & Pieper, 1973). Experiments involving heterokaryosis with genetically marked strains (Lemke, 1975) have shown that plaque formation depends on: (*a*) the presence of dsRNA virus particles; (*b*) host mutation, wild-type strains carry a nuclear gene (dominant in diploids) for resistance to lysis; (*c*) use of a medium containing a high lactose concentration; plaque formation on a similar medium has also been observed with a wild-type strain of *Penicillium citrinum*, which contains two dsRNA viruses (Benigni, Ignazzitto & Volterra, 1977) and a wild-type virus-containing strain of *Penicillium variabile* (Borre *et al.*, 1971).

Die-back disease of mushrooms

Die-back disease of *Agaricus bisporus* is caused by virus infection (Hollings, 1962) and several different viruses appear to be associated with the disease (for a detailed review see Dieleman-van Zaayen, 1979). The disease and the viruses are transmitted by heterokaryosis and in basidiospores. Infection of healthy mushrooms by inoculation with cell-free virus preparations, although difficult and inefficient, has been achieved in two independent laboratories (Hollings, 1962; Dieleman-van Zaayen & Temmink, 1968). At least two of the isometric mushroom viruses, MV1 (25 nm diameter) and MV4 (35 nm diameter), contain dsRNA. Particles of these two sizes have been detected in infected mushrooms in England, the Netherlands, France and the USA, although particles of the same size found in different countries are not necessarily identical. Barton & Hollings (1979) reported dsRNA components (1.5×10^6 and 1.4×10^6 daltons) from MV1, and two dsRNA components of closely similar size ($c.$ 1.4×10^6 daltons) from MV4 obtained from diseased mushrooms in England. In contrast nine dsRNA components, mol. wts. 2.17, 1.89, 1.76, 1.70, 1.60, 0.67, 0.4, 0.33 and 0.30×10^6 were obtained from diseased mushrooms in the USA and apparently derived from a mixture of particles of the same size as MV1 and MV4 (Marino *et al.*, 1976; Lemke, 1977). Low levels of particles of the same size as MV1 and MV4, and serologically related to similar particles from diseased mushrooms, have been detected in many apparently disease-free mushroom spawns (Del Vecchio, Dixon & Lemke, 1978); there is also a report of high levels of virus particles in symptomless mushrooms (Passmore & Frost, 1974).

Die-back disease of mushrooms arose apparently spontaneously in the USA in 1949, and appeared in Europe shortly afterwards, mushroom growers not having observed a similar disease in crops over a period of more than 20 years previously (Sinden & Hauser, 1950). This suggests that the disease may have arisen by a virus mutation. Alternatively, if the latent infections result from viruses lacking those parts of the genome essential for pathogenicity, complementation of two different partial genomes at heterokaryon formation could result in formation of a pathogenic virus. Unfortunately, no data on the genomes of the viruses in the latent infections are available at present.

Disease of Helminthosporium victoriae

Among cultures of *Helminthosporium victoriae* newly isolated from diseased oats, Lindberg (1959) encountered several abnormally stunted colonies. Closer examination revealed that some of the colonies had started normal growth, but that abnormal sectors began to develop at which time the growth of the entire colony rapidly stopped and the aerial mycelium, which had appeared normal, collapsed. Sub-cultures of the two colony types, normal and abnormal, always yielded colonies like the originals. The 'disease' could be transmitted to healthy cultures by hyphal anastomosis with abnormal cultures, or sometimes with apparently healthy 'carrier' cultures. Over 20 years ago Lindberg suggested that the disease of *H. victoriae* may be caused by a virus, but it is only recently that evidence supporting this hypothesis has been obtained.

Sanderlin & Ghabrial (1978) found that normal isolates of *Helminthosporium victoriae* were either virus-free or contained a 190 S virus, with a single dsRNA component (Table 2). On the other hand diseased isolates always contained, in addition to the 190 S virus, a serologically unrelated 145 S virus with four dsRNA components, mol. wts. 2.4, 2.2, 2.1 and 2.0×10^6. Colonies derived from protoplasts of diseased cultures were of three types: Type 1 or normal colonies; Type 2 colonies similar to the original diseased culture; and Type 3 colonies which were severely stunted and produced only sparse aerial mycelium. Type 1 colonies were found to contain the 190 S particles with little or none of the 145 S particles, whereas Type 2 and Type 3 colonies contained both types of particles with disease severity correlating well with the 145 S particle content of mycelia (Ghabrial, Sanderlin & Calvert, 1979). These results suggested that the 145 S virus was the causative agent of the disease. Whether the 145 S virus is able to replicate alone or requires the 190 S virus as a helper is not known, but no colonies containing only 145 S particles have been obtained. Recently successful transmission of the disease was achieved by fusing protoplasts from a normal virus-free isolate of *H. victoriae* in the presence of a partially purified virus preparation which contained both 190 S and 145 S particles. (S. A. Ghabrial, unpublished). Of approximately 200 colonies produced from regenerated protoplasts in each of two experiments, three and five colonies respectively showed disease symptoms. Thin sections of hyphae from the new diseased colonies, but not from control colonies, revealed the presence of aggregates of virus particles when examined by electron microscopy.

Rhizoctonia *decline*

Rhizoctonia decline is a degenerative disease of *Rhizoctonia solani*, characterised by cultures of irregular appearance, production of few or no sclerotia and an extremely reduced growth rate (Castanho & Butler, 1978a). Three diseased isolates of *R. solani* were found to contain dsRNA, although the dsRNA segments from each isolate were distinct. By contrast, healthy isolates of the fungus contained no dsRNA (Castanho, Butler & Shepherd, 1978). One diseased isolate, 189a, contained three dsRNA components of mol. wts. 2.2, 1.5 and 1.1×10^6; hyphal tip cultures (e.g. 189 HTS) were healthy and usually contained no dsRNA, although occasionally traces of the three dsRNA components were detected. The disease and the three dsRNA components could be transmitted to isolate 189 HT5 by anastomosis with isolate 189a. The possibility of the involvement of other plasmids, transmitted with dsRNA, has not been excluded. Some evidence for the existence of closed circular DNA plasmids in *R. solani* has been presented (Grimaldi, Guardiola & Martini, 1978). No virus particles could be detected in either healthy or diseased isolates of *R. solani*. Whether this indicates replication of unencapsidated dsRNA or failure to isolate the particles is not known. Failure to detect virus particles in some strains of *Ustilago maydis* containing dsRNA has also been reported (Koltin & Day, 1976 b).

Diseased isolates of *Rhizoctonia solani* have reduced virulence and have potential use for biological control of this plant pathogen. Encouraging results were obtained by addition of viable, 189a mycelium isolate to the seed furrow of soil previously infected with culture 189 HT5, which resulted in a five-fold reduction in post emergence damping-off of sugar beets (Castanho & Butler, 1978 b). Control was considered to result from transmission of the disease by anastomosis of hyphal from isolates 189a and 189 HT5. However, it should be noted that, so far, attempts to transmit the disease to four other field isolates of *R. solani* of the *same* anastomosis group have failed.

We need to know much more about the molecular biology of the virus-induced diseases in all four cases described above, in particular to identify the dsRNA components and gene products responsible for disease symptoms. Whether any of the diseases are connected with the killer phenomenon is unknown, but it is noteworthy that mutant killer strains of *Saccharomyces cerevisiae* which are sensitive to their own toxin as a result of mutation in a nuclear *rex* gene (Wickner, 1974) or in the immunity region of M dsRNA (Sweeney *et al.*, 1976), have been described and such 'suicide' strains grow poorly near the pH optimum of toxin activity. Such mutations are equivalent to changing a host from tolerance to sensitivity to virus infection, or to changing a virus infection from non-pathogenic to pathogenic.

DOUBLE-STRANDED RNA AND HYPOVIRULENCE

Hypovirulence in *Endothia parasitica*, the causative agent of chestnut blight is cytoplasmically transmitted (van Alfen *et al.*, 1975). Hypovirulent (H) strains are curative and can arrest the expansion of cankers, apparently by transfer of cytoplasmic determinants for hypovirulence to the former prevalent virulent (V) strains. Since their first discovery by Biraghi (1951) H strains have gradually spread in Italy where chestnut blight has lost its epidemic character and a host–pathogen balance appears to have been established (Mittempergher, 1979). Hypovirulent strains have also been isolated in France (Grente, 1965) and in the USA (Anagnostakis, 1978).

All H isolates of *Endothia parasitica* so far examined have been found to contain dsRNA (Day *et al.*, 1977), whereas most V strains, including ascospore isolates, lack dsRNA. On examination by polyacrylamide gel electrophoresis, several patterns of dsRNA components have been observed, but all H strains appear to contain at least one component in the mol. wt. range $5-7 \times 10^6$; American and European strains are distinct (J. A. Dodds, unpublished). In at least one strain the dsRNA is associated with club-shaped virus-like particles (Dodds, 1979); particles of similar morphology have been reported to be associated with a severe disease of *Agaricus bisporus* (Lesemann & Koenig, 1977). H strains do not have reduced vigour as saprophytes, but in axenic culture frequently revert to the V phenotype, presumably due to loss of the cytoplasmic H-determinant. Single conidial isolates derived from an American H-strain were found to be of three types: (*a*) identical to the parent in hypovirulence and dsRNA content; (*b*) more pathogenic than the parent, but still hypovirulent and curative, with a reduction in the quantity and number of dsRNA components; (*c*) indistinguishable from normal V strains and lacking dsRNA (Dodds & Elliston, 1978). J. A. Dodds & J. E. Elliston (unpublished) have shown that a distinctive dsRNA pattern can be moved by anastomosis from an American H-strain into a methionine-requiring V-strain and then back to a dsRNA-free single conidial isolate of the original H-strain. In all cases the presence of the distinctive dsRNA conferred a specific level of hypovirulence and its absence caused the strain to be virulent.

It is clear that very persuasive correlative evidence exists to indicate that dsRNA determines hypovirulence in *Endothia parasitica*. There are still problems; for example, in all cases it has not been possible to correlate levels of hypovirulence with patterns or quantity of dsRNA and one Italian wild-type V-strain containing dsRNA has been described (van Alfen, Bowman & Simmons, 1979). However analysis of dsRNA components so far has been only on the basis of molecular weight and it is likely that some *E. parasitica* strains may contain mutant dsRNAs, differing in their contributions to hypovirulence. Unequivocal proof of the role of dsRNA components as determinants of hypovirulence will require infection or transfection of dsRNA-free V strains with virus-like particles or dsRNA isolated from H strains.

H strains of *Endothia parasitica* have potential as a means of controlling chestnut blight, and their curative properties have been amply demonstrated. The main problems appear to be vegetative incompatibility (Anagnostakis, 1977) and induction of natural spread of H isolates. Ascospores lack dsRNA and are invariably virulent, and H isolates produce fewer conidia than their V counterparts. The natural spread of H strains in Italy may have resulted from an, as yet, unrecognised vector. Prospects for using H strains for re-establishment of the American chestnut in the USA have been discussed by Day & Dodds (1979) and by Anagnostakis (1978).

OUTLOOK

Many dsRNA virus infections of fungi are undoubtedly latent. However, it is becoming increasingly evident that dsRNA components in a number of instances may determine a fungus phenotype. Because of their intracellular mode of transmission the dsRNA molecules may be regarded as plasmids. dsRNA molecules have been proved to encode killer toxins and may be the causative agents of several fungus diseases. Fungus hypovirulence, without loss of saprophytic ability, may be encoded on dsRNA and under certain conditions hypovirulent strains may play an important role in establishing a host–parasite balance. Such dsRNA-determined alterations of fungus phenotype may be more widespread than has hitherto been realised. There are three main reasons for this view. (1) Much of the early work on dsRNA viruses utilised strains from culture collections; such strains may have been selected *inter alia* for their good growth characteristics and stability and hence only non-pathogenic virus infections would be likely to be found. As an example, diseased isolates of *Rhizoctonia solani* were noticed as early as 1965 (Castanho & Butler, 1978a). However such cultures either died or were discarded, and only healthy, pathogenic cultures were retained in the culture collection. (2) If determinants for a particular phenotype are encoded on satellite dsRNA molecules distinct from those essential for virus replication, latent infections may result from loss of such dsRNA molecules during intracellular transmission, including sexual or asexual spore formation. Such latent infections could persist indefinitively in a particular strain. Thus discovery of a latent virus infection in a particular fungus could indicate the existence (or past existence) of other strains of the fungus, carrying satellite dsRNA molecules, which confer a particular phenotype. (3) As more newly isolated strains are being studied it is becoming clear that many different patterns of dsRNA components may be found in different strains of a species. Complementation or exclusion interactions following hyphal anastomosis between different strains could result in the creation of new phenotypes. Significant further progress will require a greater understanding of the molecular biology of dsRNA virus replication and the role of host factors in virus replication and in its control, and of the dsRNA components and their gene products which determine particular fungus phenotypes, as well as those involved in latent infections.

I am very grateful to the following investigators, who generously supplied me with reprints, copies of manuscripts prior to publication and unpublished information: Drs J. Adler, S. L. Anagnostakis, R. F. Bozarth, J. Bruenn, H. Bussey, R. W. Detroy, J. A. Dodds, S. A. Ghabrial, M. Hollings, J. F. Hopper, Y. Koltin, P. A. Lemke. C. S. McLaughlin, N. K. van Alfen, R. B. Wickner, and T. W, Young.

REFERENCES

ADLER, J. (1979). Screening for viruses in human pathogenic fungi. In *Fungal Viruses*, ed. H. P. Molitoris, H. A. Wood & M. Hollings. Berlin, Heidelberg & New York: Springer Verlag, in press.

AIGLE, M. & LACROUTE, F. (1975). Genetical aspects of [URE 3], a non-mitochondrial, cytoplasmically inherited mutation in yeast. *Molecular and General Genetics*, **136**, 327–35.

AL-AIDROOS, K. & BUSSEY, H. (1978). Chromosomal mutants of *Saccharomyces cerevisiae* affecting the cell wall binding site for killer factor. *Canadian Journal of Microbiology*, **24**, 228–37.

AL-AIDROOS, K., SOMERS, J. M. & BUSSEY, H. (1973). Retention of cytoplasmic killer determinants in yeast cells after removal of mitochondrial DNA by ethidium bromide. *Molecular and General Genetics*, **122**, 323–30.

ALMOND, M. R. (1979). Biochemical characterisation of viruses isolated from the phytopathogenic fungus, *Gaeumannomyces graminis* var. *tritici*. PhD Thesis, University of London.

ALMOND, M. R., BUCK, K. W. & RAWLINSON, C. J. (1978). The virus complex of *Gaeumannomyces graminis* var. *tritici*. *Bulletin of the British Mycological Society*, **12**, 115.

ANAGNOSTAKIS, S. L. (1977). Vegetative incompatibility in *Endothia parasitica*. *Experimental Mycology*, **1**, 306–16.

ANAGNOSTAKIS, S. L. (1978). *The American chestnut: new hope for a fallen giant*, Bulletin 777, 9 pp. New Haven: The Connecticut Agricultural Experimental Station.

BALTIMORE, D. (1969). Replication of picornaviruses. In *The Biochemistry of Viruses*, ed. H. B. Levy, pp. 101–76. New York: Marcel Dekker.

BALTIMORE, D., AMBROS, V., FLANEGAN, J., PETTERSON, R. & ROSE, J. (1978). Poliovirus 5' protein and viral replication. *Abstracts of the Fourth International Congress for Virology*, pp. 62–3. Wageningen: Centre for Agricultural Publishing and Documentation.

BANERJEE, A. K., ABRAHAM, G. & COLONNO, R. J. (1977). Vesicular stomatitis virus: mode of transcription. *Journal of General Virology*, **34**, 1–8.

BANKS, G. T., BUCK, K. W., CHAIN, E. B., DARBYSHIRE, J. E. & HIMMELWEIT, F. (1969a). Virus-like particles in penicillin producing strains of *Penicillium chrysogenum*. *Nature, London*, **222**, 89–90.

BANKS, G. T., BUCK, K. W., CHAIN, E. B., DARBYSHIRE, J. E. & HIMMELWEIT, F. (1969b). *Penicillium cyaneo-fulvum* virus and interferon stimulation. *Nature, London*, **223**, 155–8.

BANKS, G. T., BUCK, K. W., CHAIN, E. B., DARBYSHIRE, J. E., HIMMELWEIT, F., RATTI, G., SHARPE, T. J. & PLANTEROSE, D. N. (1970). Antiviral activity of double-stranded RNA from a virus isolated from *Aspergillus foetidus*. *Nature, London*, **227**, 505–7.

BANKS, G. T., BUCK, K. W., CHAIN, E. B., HIMMELWEIT, F., MARKS, J. E., TYLER, J. M., HOLLINGS, M., LAST, F. T. & STONE, O. M. (1968). Viruses in fungi and interferon stimulation. *Nature, London*, **218**, 542–5.

BARNARD, J. P. & PETITJEAN, A. M. (1978). In-vitro synthesis of double-stranded RNA by *Drosophila* X virus purified virions. *Biochemical and Biophysical Research Communications*, **83**, 763–70.

BARTON, R. J. (1978). *Mycogone perniciosa* virus. *Report of the Glasshouse Crops Research Institute*, **1977**, 133.

BARTON, R. J. & HOLLINGS, M. (1979). Purification and some properties of two viruses infecting the cultivated mushroom *Agaricus bisporus*. *Journal of General Virology*, **42**, 231–40.

BENIGNI, R., IGNAZZITTO, G. & VOLTERRA, L. (1977). Double-stranded ribonucleic acid viruses in *Penicillium citrinum*. *Applied and Environmental Microbiology*, **34**, 811–14.

BERK, A. J. & SHARP, P. A. (1978a). Structure of the adenovirus 2 early mRNAs. *Cell*, **14**, 695–711.

BERK, A. J. & SHARP, P. A. (1978b). Spliced early mRNAs of Simian virus 40. *Proceedings of the National Academy of Sciences, USA*, **75**, 1274–8.

BEVAN, E. A. & HERRING, A. J. (1976). The killer character in yeast: preliminary studies of virus-like particle replication. In *Genetics, Biogenetics and Bioenergetics of Mitochondria*, ed. W. Bandelow, R. J. Schweyen, D. Y. Thomas, K. Wolf & F. Kaudewitz, pp. 153–62. Berlin: Walter de Gruyter.

BEVAN, E. A., HERRING, A. J. & MITCHELL, D. J. (1973). Preliminary characterisation of two species of dsRNA in yeast and their relationship to the 'killer' character. *Nature, London*, **245**, 81–6.

BEVAN, E. A. & MITCHELL, D. J. (1979). Killer systems of yeast. In *Viruses and Plasmids of Fungi*, ed. P. A. Lemke, pp. 161–99. New York: Marcel Dekker.

BEVAN, E. A. & SOMERS, J. M. (1969). Somatic segregation of the killer (k) and neutral (n) cytoplasmic genetic determinants in yeast. *Genetical Research*, **14**, 71–7.

BIRAGHI, A. (1951). *Caraterri de resistanze in* Castanea sativa *nei confronti di* Endothia parasitica, 5 pp. Firenze: Inst. Patologica Forestale Agraria.

BORRE, E., MORGANTINI, L. E., ORTALI, V. & TONOLO, A. (1971). Production of lytic plaques of viral origin in *Penicillium*. *Nature, London*, **229**, 568–9.

BOSTIAN, K. A., HOPPER, J. E., ROGERS, D. T. & TIPPER, D. J. (1979). Translational analysis of the killer-associated virus-like particle dsRNA genome of *Saccharomyces cerevisiae*: M dsRNA encodes toxin. *Cell*, in press.

BOZARTH, R. F. (1972). Mycoviruses: a new dimension in microbiology. *Environmental Health Perspectives*, **2**, 23–9.

BOZARTH, R. F. (1977). Biophysical and biochemical characterisation of virus-like particles containing a high molecular weight dsRNA from *Helminthosporium maydis*. *Virology*, **80**, 149–57.

BOZARTH, R. F. (1979). The physio-chemical properties of dsRNA mycoviruses. In *Viruses and Plasmids of Fungi*, ed. P. A. Lemke. New York: Marcel Dekker, in press.

BOZARTH, R. F. & HARLEY, E. H. (1976). The electrophoretic mobility of double-stranded RNA in polyacrylamide gels as a function of molecular weight. *Biochimica et Biophysica Acta*, **432**, 329–35.

BOZARTH, R. F., WOOD, H. A. & GOENAGA, A. (1972). Virus-like particles from a culture of *Diplocarpon rosae*. *Phytopathology*, **62**, 493.

BOZARTH, R. F., WOOD, H. A. & MANDELBROT, A. (1971). The *Penicillium stoloniferum* virus complex: two similar double-stranded RNA virus-like particles in a single cell. *Virology*, **45**, 516–23.

BRITTEN, R. J., GRAHAM, D. E. & NEUFELD, B. R. (1974). Analysis of repeating DNA sequences by reassociation. In *Methods in Enzymology*, ed. L. Grossman & K. Moldave, vol. 29, pp. 363–466. New York: Academic Press.

BROOKS, D. H. (1965). Root infection by ascospores of *Ophiobolus graminis* as a factor in epidemiology of the take-all disease. *Transactions of the British Mycological Society*, **48**, 237–48.

BRUENN, J. & KANE, W. (1978). Relatedness of the double-stranded RNAs present in yeast virus-like particles. *Journal of Virology*, **26**, 762–72.

BRUENN, J. & KEITZ, B. (1976). The 5′ ends of yeast killer factor RNAs are pppGp. *Nucleic Acids Research*, **3**, 2427–36.

BUCK, K. W. (1975). Replication of double-stranded RNA in particles of *Penicillium stoloniferum* virus S. *Nucleic Acids Research*, **2**, 1889–1902.

BUCK, K. W. (1977). Biochemical and biological implications of double-stranded RNA mycoviruses. In *Biologically Active Substances: Exploration and Exploitation*, ed. D. A. Hems, pp. 121–48. Chichester: John Wiley & Sons.

BUCK, K. W. (1978). Semi-conservative replication of double-stranded RNA by a virion-associated RNA polymerase. *Biochemical and Biophysical Research Communications*, **84**, 639–45.

BUCK, K. W. (1979a). Fungal viruses, double-stranded RNA and interferon induction. In *Viruses and Plasmids of Fungi*, ed. P. A. Lemke, pp. 1–43. New York: Marcel Dekker.

BUCK, K. W. (1979b). Replication of double-stranded RNA mycoviruses. In *Viruses and Plasmids of Fungi*, ed. P. A. Lemke, pp. 93–160. New York: Marcel Dekker.

BUCK, K. W. (1979c). Virion-associated RNA polymerases of double-stranded RNA mycoviruses. In *Fungal Viruses*, ed. H. P. Molitoris, H. A. Wood & M. Hollings, pp. 62, 77. Berlin, Heidelberg & New York: Springer Verlag.

BUCK, K. W. & GIRVAN, R. F. (1977). Comparison of the biophysical and bio-chemical properties of *Penicillium cyaneo-fulvum* virus and *Penicillium chryso-genum* virus. *Journal of General Virology*, **34**, 145–54.

BUCK, K. W., GIRVAN, R. F. & RATTI, G. (1973). Two serologically distinct double-stranded ribonucleic acid viruses isolated from *Aspergillus niger*. *Biochemical Society Transactions*, **1**, 1138–40.

BUCK, K. W. & KEMPSON-JONES, G. F. (1974). Capsid polypeptides of two viruses isolated from *Penicillium stoloniferum*. *Journal of General Virology*, **22**, 441–5.

BUCK, K. W. & RATTI, G. (1975). Biophysical and biochemical properties of two viruses isolated from *Aspergillus foetidus*. *Journal of General Virology*, **27**, 211–24.

BUSSEY, H. & SHERMAN, D. (1973). Yeast killer factor: ATP leakage and coordinate inhibition of macromolecular synthesis in sensitive cells. *Biochimica et Biophysica Acta*, **298**, 868–75.

BUSSEY, H. & SKIPPER, N. (1975). Membrane-mediated killing of *Saccharomyces cerevisiae* by glycoproteins from *Torulopsis glabrata*. *Journal of Bacteriology*, **124**, 476–83.

CASEY, J. & DAVIDSON, N. (1977). Rates of formation and thermal stabilities of RNA:DNA and DNA:RNA duplexes at high concentrations of formamide. *Nucleic Acids Research*, **4**, 1539–52.

CASTANHO, B. & BUTLER, E. E. (1978a). Rhizoctonia decline: a degenerative disease of *Rhizoctonia solani*. *Phytopathology*, **68**, 1505–10.

CASTANHO, B. & BUTLER, E. E. (1978b). Rhizoctonia decline: studies on hypoviru-lence and potential use in biological control. *Phytopathology*, **68**, 1511–14.

CASTANHO, B., BUTLER, E. E. & SHEPHERD, R. J. (1978). The association of double-stranded RNA with *Rhizoctonia* decline. *Phytopathology*, **68**, 1515–18.

CATEN, C. E. (1972). Vegetative incompatibility and cytoplasmic inheritance in fungi. *Journal of General Microbiology*, **72**, 221–9.

CATEN, C. E. & HANDLEY, L. (1978). 'Vegetative death' syndrome in *Aspergillus glaucus*. *Bulletin of the British Mycological Society*, **12**, 114.

CHATER, K. F. & MORGAN, D. H. (1974). Ribonucleic acid synthesis by isolated viruses of *Penicillium stoloniferum*. *Journal of General Virology*, **24**, 307–17.

COHN, M. S., TABOR, C. W. & TABOR, H. (1978). Isolation and characterisation of *Saccharomyces cerevisiae* mutants deficient in *S*-adenosyl methionine decarboxyl-ase, spermidine and spermine. *Journal of Bacteriology*, **134**, 208–13.

COHN, M. S., TABOR, C. W., TABOR, H. & WICKNER, R. B. (1978). Spermidine or spermine requirement for killer double-stranded RNA plasmid replication in yeast. *Journal of Biological Chemistry*, **253**, 5225–7.

COOPER, P. D. (1977). Genetics of picornaviruses. In *Comprehensive Virology*, ed. H. Fraenkal-Conrat & R. R. Wagner, vol. 9, pp. 133–207. New York: Plenum Press.

COPLIN, D. L., VAN ETTEN, J. L., KOSKI, R. K. & VIDAVER, A. K. (1975). Inter-mediates in the biosynthesis of double-stranded ribonucleic acids of bacteriophage φ6. *Proceedings of the National Academy of Sciences, USA*, **72**, 849–53.

CROSS, R. K. & FIELDS, B. N. (1977). Genetics of reoviruses. In *Comprehensive Virology*, ed. H. Fraenkel-Conrat & R. R. Wagner, vol. 9, pp. 291–340. New York: Plenum Press.

DAVIDSON, E. H., HOUGH, B. R., KLEIN, W. H. & BRITTEN, R. J. (1975). Structural genes adjacent to interspersed repetitive DNA sequence. *Cell*, 4, 217–39.

DAWSON, W. O. & WHITE, J. L. (1978). Characterisation of a temperature-sensitive mutant of tobacco mosaic virus deficient in the synthesis of all RNA species. *Virology*, 90, 209–13.

DAWSON, W. O. & WHITE, J. L. (1979). A temperature-sensitive mutant of tobacco mosaic virus deficient in synthesis of single-stranded RNA. *Virology*, 93, 104–10.

DAY, P. R. & DODDS, J. A. (1979). Viruses of plant pathogenic fungi. In *Viruses and Plasmids of Fungi*, ed. P. A. Lemke, pp. 201–38. New York: Marcel Dekker.

DAY, P. R., DODDS, J. A., ELLISTON, J. E., JAYNES, R. A. & ANAGNOSTAKIS, S. L. (1977). Double-stranded RNA in *Endothia parasitica*. *Phytopathology*, 67, 1393–6.

DEL VECCHIO, V. G., DIXON, C. & LEMKE, P. A. (1978). Immune electron microscopy of virus-like particles of *Agaricus bisporus*. *Experimental Mycology*, 2, 138–44.

DEMARINI, D. M., KURTZMAN, C. P., FENNELL, D. I., WORDEN, K. A. & DETROY, R. W. (1977). Transmission of PsV-F and PsV-S mycoviruses during conidiogenesis of *Penicillium stoloniferum*. *Journal of General Microbiology*, 100, 59–64.

DETROY, R. W. & WORDEN, K. A. (1979). Interactions of fungal viruses and secondary metabolites. In *Fungal Viruses*, ed. H. P. Molitoris, H. A. Wood & M. Hollings, Berlin, Heidelberg & New York: Springer Verlag, in press.

DIELEMAN-VAN ZAAYEN, A. (1979). Mushroom viruses. In *Viruses and Plasmids of Fungi*, ed. P. A. Lemke, pp. 239–324. New York: Marcel Dekker.

DIELEMAN-VAN ZAAYEN, A. & TEMMINK, J. H. M. (1968). A virus disease of cultivated mushrooms in the Netherlands. *Netherlands Journal of Plant Pathology*, 74, 48–51.

DOBOS, P., HALLETT, R., KELLS, D. T. C., HILL, B. J., BECHT, H. & TENINGES, D. (1978). Biochemical and biophysical studies of five animal viruses with bisegmented dsRNA genomes. *Abstracts of the Fourth International Congress for Virology*, p. 335. Wageningen: Centre for Agricultural Publishing and Documentation.

DODDS, J. A. (1979). Double-stranded RNA and virus-like particles in *Endothia parasitica*. In *American Chestnut Symposium Proceedings*, ed. W. MacDonald. West Virginia University Agricultural Experiment Station and United States Department of Agriculture, in press.

DODDS, J. A. & ELLISTON, J. E. (1978). Association between double-stranded RNA and hypovirulence in an American strain of *Endothia parasitica*. *Abstracts of the Third International Congress for Plant Pathology*, p. 57. Hamburg: Paul Parey.

ELLIS, L. F. & KLEINSCHMIDT, W. J. (1967). Virus-like particles of a fraction of statolon, a mould product. *Nature, London*, 215, 649–50.

FARRELL, P. J., BALKOW, K., HUNT, T. & JACKSON, R. J. (1977). Phosphorylation of initiation factor eIF-2 and the control of reticulocyte protein synthesis. *Cell*, 187–200.

FIERS, W., CONTRERAS, R., HAEGEMAN, G., ROGIERS, R., VAN DE WOORDE, A., VAN HEUVERSWYN, H., VAN HERREWEGH, H., VOLCKAERT, G. & YSEBAERT, M. (1978). Complete nucleotide sequence of SV40 DNA. *Nature, London*, 273, 113–20.

FINK, G. R. & STYLES, C. A. (1972). Curing of a killer factor in *Saccharomyces cerevisiae*. *Proceedings of the National Academy of the Sciences, USA*, 69, 2846–9.

FLANEGAN, J. B. & BALTIMORE, D. (1977). Poliovirus-specific primer-dependent RNA polymerase able to copy poly (A). *Proceedings of the National Academy of Sciences, USA*, **74**, 3677–80.

FRIED, H. M. & FINK, G. R. (1978). Electron microscopic heteroduplex analysis of 'killer' double-stranded RNA species of yeast. *Proceedings of the National Academy of Sciences, USA*, **75**, 4224–8.

FURUICHI, Y., MORGAN, M., MUTHUKRISHNAN, S. & SHATKIN, A. J. (1975). Reovirus messenger RNA contains a methylated, blocked 5′-terminal structure, m⁷G(5′)-ppp(5′)GmpCp-. *Proceedings of the National Academy of Sciences, USA*, **72**, 362–6.

FURUICHI, Y., MUTHUKRISHNAN, S. & SHATKIN, A. J. (1975). 5′-terminal m⁷G(5′)-ppp(5′) Gmp *in vivo*: identification in reovirus genome RNA. *Proceedings of the National Academy of Sciences, USA*, **72**, 742–5.

GHABRIAL, S. A., SANDERLIN, R. S. & CALVERT, L. A. (1979). Morphology and virus-like particle content of *Helminthosporium victoriae*. *Phytopathology*, in press.

GLANVILLE, M., LACHMI, B., SMITH, A. E. & KAARIANEN, L. (1978). Tryptic peptide mapping of the nonstructural proteins of Semliki Forest virus and their precursors. *Biochimica et Biophysica Acta*, **518**, 497–506.

GRENTE, J. (1965). Les formes hypovirulentes d'*Endothia parasitica* et les espoirs de lutte contre le chancre du châtaignier. *Comptes Rendus Hebdomadaire des Séances de l'Académie d'Agriculture de France*, **51**, 1033–7.

GRIMALDI, G., GUARDIOLA, J. & MARTINI, G. (1978). Fungal extrachromosomal DNA and its maintenance and expression in *E. coli* K-12. *Trends in Biochemical Sciences*, **3**, 248–9.

HARDY, K. G. (1975). Colicinogeny and related phenomena. *Bacteriological Reviews*, **39**, 464–515.

HASEGAWA, T. (1975) (ed.). *Proceedings of the First Intersectional Congress of the International Association of Microbiological Societies*, vol. 3. Tokyo: Science Council of Japan.

HASTIE, N. D., BRENNAN, V. & BRUENN, J. A. (1978). No homology between double-stranded DNA and nuclear DNA of yeast. *Journal of Virology*, **28**, 1002–5.

HERRING, A. J. & BEVAN, E. A. (1975). Double-stranded RNA containing particles from the yeast *Saccharomyces cerevisiae* and their relationship to the killer character. In *Molecular Biology of Nucleocytoplasmic Relationships*, ed. S. Puiseux-Dao, pp. 149–54. Amsterdam: Elsevier Scientific Publishing Company.

HERRING, A. J. & BEVAN, E. A. (1977). Yeast virus-like particles possess a capsid-associated single-stranded RNA polymerase. *Nature, London*, **268**, 464–6.

HOLLAND, I. B. (1975). Physiology of colicin action. In *Advances in Microbial Physiology*, ed. A. H. Rose & D. W. Tempest, vol. 12, pp. 56–139. London: Academic Press.

HOLLINGS, M. (1962). Viruses associated with a die-back disease of cultivated mushroom. *Nature, London*, **196**, 962–5.

HOLLINGS, M. (1978). Mycoviruses – viruses that infect fungi. *Advances in Virus Research*, **22**, 3–53.

HOLLINGS, M. & STONE, O. M. (1969). Viruses in fungi. *Science Progress, Oxford*, **57**, 371–91.

HOLLINGS, M. & STONE, O. M. (1971). Viruses that infect fungi. *Annual Review of Phytopathology*, **9**, 93–118.

HOLM, C. A., OLIVER, S. G., NEMAN, A. N., HOLLAND, L. E., MCLAUGHLIN, C. S., WAGNER, E. K. & WARNER, R. C. (1978). The molecular weight of yeast P1 double-stranded RNA. *Journal of Biological Chemistry*, **253**, 8332–6.

HOPPER, J. E., BOSTIAN, K. A., ROWE, L. B. & TIPPER, D. J. (1977). Translation of the L-species dsRNA genome of the killer-associated virus-like particle of *Saccharomyces cerevisiae*. *Journal of Biological Chemistry*, **252**, 9010–7.

HUANG, A. S. & BALTIMORE, D. (1977). Defective interfering animal viruses. In *Comprehensive Virology*, ed. H. Fraenkel-Conrat & R. R. Wagner, vol. 10, pp. 73–116. New York: Plenum Press.

IKEGAMI, M. & FRAENKEL-CONRAT, H. (1978). RNA-dependent RNA polymerase of tobacco plants. *Proceedings of the National Academy of Sciences, USA*, **75**, 2122–4.

JAY, G., ABRAMS, W. R. & KAEMPFER, R. (1974). Resistance of bacterial protein synthesis to double-stranded RNA. *Biochemical and Biophysical Research Communications*, **60**, 1357–64.

JOKLIK, W. K. (1974). Reproduction of reoviridae. In *Comprehensive Virology*, ed. H. Fraenkel-Conrat & R. R. Wagner, vol. 2, pp. 231–334. New York: Plenum Press.

KAMEN, R. I. (1975). Structure and function of the Qβ RNA replicase. In *RNA Phages*, ed. N. D. Zinder, pp. 203–34. New York: Cold Spring Harbor Laboratory.

KANDEL, J. & KOLTIN, Y. (1978). Killer phenomenon in *Ustilago maydis*: comparison of the killer proteins. *Experimental Mycology*, **2**, 270–8.

KHANDJIAN, E. W., TURIAN, G. & EISEN, H. (1977). Characterisation of the RNA mycovirus infecting *Allomyces arbuscula*. *Journal of General Virology*, **35**, 415–24.

KOLTIN, Y. (1977). Virus-like particles in *Ustilago maydis*: mutants with partial genomes. *Genetics*, **86**, 527–34.

KOLTIN, Y. (1979). Fungal viruses and killer factors: *Ustilago maydis* killer proteins. In *Fungal Viruses*, ed. H. P. Molitoris, H. A. Wood & M. Hollings. Berlin, Heidelberg & New York: Springer Verlag, in press.

KOLTIN, Y. & DAY, P. R. (1975). Specificity of *Ustilago maydis* killer proteins. *Applied Microbiology*, **30**, 694–6.

KOLTIN, Y. & DAY, P. R. (1976a). Inheritance of killer phenotypes and double-stranded RNA in *Ustilago maydis*. *Proceedings of the National Academy of Sciences, USA*, **73**, 594–8.

KOLTIN, Y. & DAY, P. R. (1976b). Suppression of the killer phenotype in *Ustilago maydis*. *Genetics*, **82**, 629–37.

KOLTIN, Y. & KANDEL, J. S. (1978). Killer phenomenon in *Ustilago maydis*: the organisation of the viral genome. *Genetics*, **88**, 267–76.

KOLTIN, Y., MAYER, I. & STEINLAUF, R. (1978). Killer phenomenon in *Ustilago maydis*: mapping the viral functions. *Molecular and General Genetics*, **166**, 181–6.

KORNBERG, A. (1977). Multiple stages in the enzymatic replication of DNA. *Biochemical Society Transactions*, **5**, 359–74.

LAMPSON, G. P., TYTELL, A. A., FIELD, A. K., NEMES, M. M. & HILLEMAN, M. R. (1967). Inducers of interferon and host resistance. I. Double-stranded RNA from extracts of *Penicillium funiculosum*. *Proceedings of the National Academy of Sciences, USA*, **58**, 782–9.

LECHNER, R. L. & KELLY, T. J. (1977). The structure of replicating adenovirus 2 DNA molecules. *Cell*, **12**, 1007–20.

LEIBOWITZ, M. J. & WICKNER, R. B. (1976). A chromosomal gene required for killer plasmid expression, mating and spore maturation in *Saccharomyces cerevisiae*. *Proceedings of the National Academy of Sciences, USA*, **73**, 2061–5.

LEIBOWITZ, M. J. & WICKNER, R. B. (1978). *pet* 18: a chromosomal gene required for cell growth and for the maintenance of mitochondrial DNA and the killer plasmid of yeast. *Molecular and General Genetics*, **165**, 115–21.

LEMKE, P. A. (1975). Biochemical and biological aspects of fungal viruses. In *Proceedings of the First International Congress of the International Association of Microbiological Societies*, ed. T. Hasegawa, vol. 3, pp. 380–95. Tokyo: Science Council of Japan.

LEMKE, P. A. (1976). Viruses of eukaryotic microorganisms. *Annual Review of Microbiology*, **30**, 105–45.

LEMKE, P. A. (1977). Fungal viruses in agriculture. In *Virology in Agriculture*, ed. T. O. Diener & J. A. Romberger, pp. 159–75. Montclair, New Jersey: Allanheld Osman & Co.

LEMKE, P. A. (1979 (ed.)). *Viruses and Plasmids of Fungi*. New York: Marcel Dekker.

LEMKE, P. A. & NASH, C. H. (1974). Fungal viruses. *Bacteriological Reviews*, **38**, 29–56.

LEMKE, P. A., NASH, C. A. & PIEPER, S. W. (1973). Lytic plaque formation and variation in virus titre among strains of *Penicillium chrysogenum*. *Journal of General Microbiology*, **76**, 265–75.

LESEMANN, D. E. & KOENIG, R. (1977). Association of club-shaped virus-like particles with a severe disease of *Agaricus bisporus*. *Phytopathologische Zeitschrift*, **89**, 161–9.

LHOAS, P. (1971). Infection of protoplasts from *Penicillium stoloniferum* with double-stranded RNA viruses. *Journal of General Virology*, **13**, 365–7.

LINDBERG, G. D. (1959). A transmissible disease of *Helminthosporium victoriae*. *Phytopathology*, **49**, 29–32.

LIVINGSTON, D. M. (1977). Inheritance of the 2 micron DNA plasmid from *Saccharomyces*. *Genetics*, **86**, 73–84.

MARINO, R., SAKSENA, K. N., SCHULER, M., MAYFIELD, J. E. & LEMKE, P. A. (1976). Double-stranded RNA from *Agaricus bisporus*. *Applied and Environmental Microbiology*, **31**, 433–8.

MAULE, A. P. & THOMAS, P. D. (1973). Strains of yeast lethal to brewery yeast. *Journal of the Institute of Brewing*, **79**, 137–41.

MINDICH, L., SINCLAIR, J. F. & COHEN, J. (1976). The morphogenesis of bacteriophage ϕ6: particles formed by nonsense mutants. *Virology*, **75**, 224–31.

MITCHELL, D. J., HERRING, A. J. & BEVAN, E. A. (1976). Virus uptake and interaction in yeasts. In *Microbial and Plant Protoplasts*, ed. J. F. Peberdy, A. H. Rose, H. J. Rogers & E. C. Cocking, pp. 91–105. London: Academic Press.

MITTEMPERGHER, L. (1979). The *Endothia parasitica* epidemic in Italy. In *American Chestnut Symposium Proceedings*, ed. W. MacDonald. West Virginia University Agricultural Experiment Station and United States Department of Agriculture, in press.

MOFFITT, E. M. & LISTER, R. M. (1975). Application of a serological test for detecting double-stranded RNA mycoviruses. *Phytopathology*, **65**, 851–9.

MOLITORIS, H. P., WOOD, H. A. & HOLLINGS, M. (1979) (eds). *Fungal Viruses*. Berlin, Heidelberg & New York: Springer Verlag, in press.

MOSSOP, D. W. & FRANKI, R. I. B. (1978). Survival of a satellite RNA *in vivo* and its dependence on cucumber mosaic virus for its replication. *Virology*, **86**, 562–6.

NASH, C. H., DOUTHART, R. J., ELLIS, L. F., VAN FRANK, R. M., BURNETT, J. P. & LEMKE, P. A. (1973). On the mycophage of *Penicillium chrysogenum*. *Canadian Journal of Microbiology*, **19**, 97–103.

NAUMOV, T. I. & NAUMOVA, G. I. (1973). The comparative genetics of yeast. XII. Study of antagonistic interrelations in *Saccharomyces*. *Genetika*, **9**, 85–90.

OLIVER, S. G., MCCREADY, S. J., HOLM, C., SUTHERLAND, P. A., MCLAUGHLIN, C. S. & COX, B. S. (1977). Biochemical and physiological studies of the yeast virus-like particle. *Journal of Bacteriology*, **130**, 1303–9.

PALFREE, R. G. E. & BUSSEY, H. (1979). Yeast killer toxin: purification and characterisation. *Biochemistry*, **93**, 487–93.

PALLETT, I. H. (1976). Interactions between fungi and their viruses. In *Microbial and Plant Protoplasts*, ed. J. F. Peberdy, A. H. Rose, H. J. Rogers & E. C. Cocking, pp. 107–24. London: Academic Press.

PARTRIDGE, J. E., VIDAVER, A. K. & VAN ETTEN, J. L. (1978). In-vitro transcription of bacteriophage φ6 double-stranded RNA. *Abstracts of the Annual Meeting of the American Society for Microbiology*, S53, 221.

PASSMORE, E. L. & FROST, R. R. (1974). The detection of virus-like particles in mushrooms and mushroom spores. *Phytopathologische Zeitschrift*, **80**, 85–7.

PAUCHA, E., MELLOR, A., HARVEY, R., SMITH, A. E., HEWICK, R. M. & WATERFIELD, M. D. (1978). Large and small tumour antigens from simian virus 40 have identical amino termini mapping at 0.65 map units. *Proceedings of the National Academy of Sciences, USA*, **75**, 2165–9.

PELHAM, H. R. B. (1978). Leaky AUG termination codon in tobacco mosaic virus RNA. *Nature, London*, **272**, 469–71.

PHILLISKIRK, G. & YOUNG, T. W. (1975). The occurrence of killer character in yeasts of various genera. *Antonie van Leeuwenhoek, Journal of Microbiology and Serology*, **41**, 147–51.

POLLACK, T. J., TESSMAN, I. & TESSMAN, E. S. (1978). Potential for variability through multiple gene products of bacteriophage φX174. *Nature, London*, **274**, 34–7.

PUHALLA, J. E. (1968). Compatibility reactions on solid medium and interstrain inhibition in *Ustilago maydis*. *Genetics*, **60**, 461–74.

RATTI, G. & BUCK, K. W. (1972). Virus particles in *Aspergillus foetidus*: a multicomponent system. *Journal of General Virology*, **14**, 165–75.

RATTI, G. & BUCK, K. W. (1975). RNA polymerase activity in double-stranded ribonucleic acid virus particles from *Aspergillus foetidus*. *Biochemical and Biophysical Research Communications*, **66**, 706–11.

RATTI, G. & BUCK, K. W. (1978). Semi-conservative transcription in particles of a double-stranded RNA mycovirus. *Nucleic Acids Research*, **5**, 3843–54.

RATTI, G. & BUCK, K. W. (1979). Transcription of double-stranded RNA in virions of *Aspergillus foetidus* virus S. *Journal of General Virology*, **42**, 59–72.

RAWLINSON, C. J. & BUCK, K. W. (1980). Viruses in *Gaeumannomyces* and *Phialophora* spp. In *Biology and Control of Take-all*, ed. M. J. C. Asher & P. J. Shipton. New York & London: Academic Press, in press.

372   K. W. BUCK

RAWLINSON, C. J., HORNBY, D., PEARSON, V. & CARPENTER, J. M. (1973). Virus-like
    particles in the take-all fungus. *Gaeumannomyces graminis*. *Annals of Applied
    Biology*, **74**, 197–209.
RAWLINSON, C. J. & MUTHYALU, G. (1976). Virus-infected isolates of *G. graminis*
    var. *tritici* in Barnfield soil. *Report of the Rothamsted Experimental Station for
    1976*, pp. 255–6.
REIJNDERS, L. (1978). The origin of multicomponent small ribonucleoprotein
    viruses. *Advances in Virus Research*, **23**, 79–102.
RIMON, A. & HASELKORN, R. (1978). Temperature-sensitive mutants of bacterio-
    phage $\phi6$ defective in both transcription and replication. *Virology*, **89**, 218–28.
ROBERTSON, H. D. & MATTHEWS, M. B. (1973). Double-stranded RNA as an
    inhibitor of protein synthesis of Krebs II ascites cells. *Proceedings of the National
    Academy of Sciences. USA*, **70**, 225–9.
ROGERS, D. & BEVAN, E. A. (1978). Group classification of killer yeasts based on
    cross reactions between strains of different species and origin. *Journal of General
    Microbiology*, **105**, 199–202.
ROMANOS, M. A., RAWLINSON, C. J., ALMOND, M. R. & BUCK, K. W. (1979).
    Production of fungal growth inhibitors by isolates of *Gaeumannomyces graminis*
    var. *tritici*. *Transactions of the British Mycological Society*, in press.
SAKSENA, K. N. & LEMKE, P. A. (1978). Viruses in fungi. In *Comprehensive Virology*,
    ed. H. Fraenkel-Conrat & R. R. Wagner, vol. 12, pp. 103–43. New York:
    Plenum Publishing Corporation.
SANDERLIN, R. S. & GHABRIAL, S. A. (1978). Physicochemical properties of two
    distinct types of virus-like particles from *Helminthosporium victoriae*. *Virology*,
    87, 142–51.
SANGER, F., AIR, G. M., BARRELL, B. G., BROWN, N. L., COULSON, A. R., FIDDES,
    J. C., HUTCHINSON, C. A., SLOCOMBE, P. M. & SMITH, M. (1977). Nucleotide
    sequence of bacteriophage ϕX174 DNA. *Nature, London*, **265**, 687–95.
SCHALLER, K. & NOMURA, M. (1976). Colicin E2 is a DNA endonuclease. *Proceed-
    ings of the National Academy of Sciences, USA*, **73**, 3989–3.
SENIOR, B. W. & HOLLAND, I. B. (1971). Effect of colicin E3 upon the 30S ribosomal
    subunit of *Escherichia coli*. *Proceedings of the National Academy of Sciences
    USA*, **68**, 959–63.
SHIELDS, D. & BLOBEL, G. (1978). Efficient cleavage and segregation of nascent
    presecretory proteins in a reticulocyte lysate supplemented with microsomal
    membranes. *Journal of Biological Chemistry*, **253**, 3753–6.
SILVERSTEIN, S. C., CHRISTMAN, J. K. & ACS, G. (1976). The reovirus replicative
    cycle. *Annual Review of Biochemistry*, **45**, 375–408.
SINDEN, J. W. & HAUSER, E. (1950). Report on two new mushroom diseases. *Mush-
    room Science*, **1**, 96–100.
SKIPPER, N. & BUSSEY, H. (1977). Mode of action of yeast toxins: energy require-
    ment for *Saccharomyces cerevisiae* killer toxin. *Journal of Bacteriology*, **129**,
    668–77.
SMITH, K. M. (1967). *Insect Virology*, p. 146. New York & London: Academic
    Press.
SMITH, K. M. (1977). *Plant Viruses*, 6th edn., p. 30. London: Chapman & Hall.
SOMERS, J. M. (1973). Isolation of suppressive sensitive mutants from killer and
    neutral strains of *Saccharomyces cerevisiae*. *Genetics*, **74**, 571–9.

SOMERS, J. M. & BEVAN, E. A. (1968). The inheritance of the killer character in yeast. *Genetical Research*, **13**, 71–83.

STAHL, U., LEMKE, P. A., TUDZYNSKI, P., KUCK, V. & ESSER, K. (1978). Evidence for plasmid-like DNA in a filamentous fungus, the ascomycete *Podospora anserina*. *Molecular and General Genetics*, **162**, 341–3.

STILL, P. E., DETROY, R. W. & HESSELTINE, C. W. (1975). *Penicillium stoloniferum* virus: altered replication in ultraviolet derived mutants. *Journal of General Virology*, **27**, 275–81.

SUGIURA, M. & MIURA, K. (1977). Transcription of double-stranded RNA by *Escherichia coli* DNA-dependent RNA polymerase. *European Journal of Biochemistry*, **73**, 179–84.

SWEENEY, K. T., TATE, A. & FINK, G. R. (1976). A study of the transmission and structure of double-stranded RNAs associated with the killer phenomenon in *Saccharomyces cerevisiae*. *Genetics*, **84**, 27–42.

TAGG, J. R., DAJANI, A. S. & WANNAMAKER, L. W. (1976). Bacteriocins of Gram-Positive bacteria. *Bacteriological Reviews*, **40**, 722–56.

TIMMIS, K., CABELLO, F. & COHEN, S. N. (1974). Utilisation of two distinct modes of replication by a hybrid plasmid constructed in vivo from separate replicons. *Proceedings of the National Academy of Sciences, USA*, **71**, 4556–60.

TOH-E, A., GUERRY, P. & WICKNER, R. B. (1978). Chromosomal superkiller mutants of *Saccharomyces cerevisiae*. *Journal of Bacteriology*, **136**, 1002–7.

TOH-E, A. & WICKNER, R. B. (1979). A mutant killer plasmid whose replication depends on a chromosomal 'superkiller' mutation. *Genetics*, in press.

TRINCI, A. P. J. (1971). Influence of the width of the peripheral growth zone on the radial growth rate of fungal colonies on solid media. *Journal of General Microbiology*, **67**, 325–44.

VAN ALFEN, N. K., BOWMAN, J. T. & SIMMONS, J. R. (1979). The segregation of an Italian virulent isolate of *Endothia parasitica* into H and V Types. In *American Chestnut Symposium Proceedings*, ed. W. MacDonald. West Virginia University Agricultural Experiment Station and United States Department of Agriculture, in press.

VAN ALFEN, N. K., JAYNES, R. A., ANAGNOSTAKIS, S. L. & DAY, P. R. (1975). Chestnut blight: biological control by transmissible hypovirulence in *Endothia parasitica*. *Science*, **189**, 890–1.

VAN ETTEN, J., LANE, L., GONZALEZ, C., PARTRIDGE, J. & VIDAVER, A. (1976). Comparative properties of bacteriophage φ6 and φ6 nucleocapsid. *Journal of Virology*, **18**, 652–8.

VAN ETTEN, J. L., VIDAVER, A. K., KOSKI, R. K. & SEMANCIK, J. S. (1973). RNA polymerase activity associated with bacteriophage φ6. *Journal of Virology*, **12**, 464–71.

VILLA-KOMAROFF, L., GUTTMAN, N., BALTIMORE, D. & LODISH, H. F. (1975). Complete translation of poliovirus RNA in a eukaryotic cell-free system. *Proceedings of the National Academy of Sciences, USA*, **72**, 4157–61.

VODKIN, M. (1977). Homology between double-stranded RNA and nuclear DNA of yeast. *Journal of Virology*, **21**, 516–21.

VODKIN, M. (1979). Fungal viruses and killer factors (*Saccharomyces cerevisiae*). In *Fungal Viruses*, ed. H. P. Molitoris, H. A. Wood & M. Hollings. Berlin, Heidelberg & New York: Springer Verlag, in press.

VODKIN, M., KATTERMAN, F. & FINK, G. R. (1974). Yeast killer mutants with altered double-stranded ribonucleic acid. *Journal of Bacteriology*, 117, 681–6.

WEINER, A. M. & WEBER, K. (1971). Natural read-through at the UGA termination signal of Qβ coat protein cistron. *Nature, New Biology*, 234, 206–9.

WHITE, J. L. & MURAKISHI, H. H. (1977). In-vitro replication of tobacco mosaic virus RNA in tobacco callus cultures: solubilisation of membrane-bound replicase and partial purification. *Journal of Virology*, 21, 484–92.

WICKNER, R. B. (1974). Chromosomal and non-chromosomal mutations affecting the 'killer character' of *Saccharomyces cerevisiae*. *Journal of Bacteriology*, 117, 681–6.

WICKNER, R. B. (1976a). Killer of *Saccharomyces cerevisiae*: a double-stranded ribonucleic acid plasmid. *Bacteriological Reviews*, 40, 757–73.

WICKNER, R. B. (1976b). Mutants of the killer plasmid of *Saccharomyces cerevisiae* dependent on chromosomal diploidy for expression and maintenance. *Genetics*, 82, 273–85.

WICKNER, R. B. (1977). Deletion of mitochondrial DNA bypassing a chromosomal gene needed for maintenance of the killer plasmid of yeast. *Genetics*, 87, 441–52.

WICKNER, R. B. (1978). Twenty-six chromosomal genes needed to maintain the killer double-stranded RNA plasmid of *Saccharomyces cerevisiae*. *Genetics*, 88, 419–25.

WICKNER, R. B. & LEIBOWITZ, M. J. (1976a). Chromosomal genes essential for replication of a double-stranded RNA plasmid of *Saccharomyces cerevisiae*: the killer character of yeast. *Journal of Molecular Biology*, 105, 427–43.

WICKNER, R. B. & LEIBOWITZ, M. J. (1976b). Two chromosomal genes required for killing expression in killer strains of *Saccaromyces cerevisiae*. *Genetics*, 82, 429–42.

WICKNER, R. B. & LEIBOWITZ, M. J. (1977). Dominant chromosomal mutation bypassing chromosomal genes needed for killer RNA plasmic replication in yeast. *Genetics*, 87, 453–69.

WOOD, H. A. (1973). Viruses with double-stranded RNA genomes. *Journal of General Virology*, 20, 61–85.

WOODS, D. R. & BEVAN, E. A. (1968). Studies on the nature of the killer factor produced by *Saccharomyces cerevisiae*. *Journal of General Microbiology*, 51, 115–26.

WOOD, H. A. & BOZARTH, R. F. (1972). Properties of virus-like particles of *Penicillium chrysogenum*: one double-stranded RNA molecule per particle. *Virology*, 47, 604–9.

WOOD, H. A. & BOZARTH, R. F. (1973). Heterokaryon transfer of virus-like particles and a cytoplasmically inherited determinant in *Ustilago maydis*. *Phytopathology*, 63, 1019–21.

WOOD, H. A., BOZARTH, R. F., ADLER, J. & MACKENZIE, D. W. (1974). Proteinaceous virus-like particles from an isolate of *Aspergillus flavus*. *Journal of Virology*, 13, 532–4.

WOODS, D. R., ROSS, I. W. & HENDRY, D. A. (1974). A new killer factor produced by a killer/sensitive yeast strain. *Journal of General Microbiology*, 81, 285–9.

YAMASHITA, S., DOI, Y. & YORA, K. (1973). Intracellular appearance of *Penicillium chrysogenum* virus. *Virology*, 55, 445–52.

YAMASHITA, S., DOI, Y. & YORA, K. (1975). Electron microscopic study of several fungal viruses. In *Proceedings of the First Intersectional Congress of the Inter-*

national Association of Microbiological Sciences, ed. T. Hasegawa, vol. 3, pp. 340–50. Tokyo: Science Council of Japan.

YOUNG, C. S. H. & COX, B. S. (1971). Extrachromosomal elements in a super-suppressor system of yeast. I. A nuclear gene controlling the inheritance of the extrachromosomal elements. *Heredity*, **26**, 413–22.

YOUNG, T. W. & PHILLISKIRK, G. (1977). The production of a yeast killer factor in the chemostat and the effects of killer yeast in mixed continuous culture with a sensitive strain. *Journal of Applied Bacteriology*, **43**, 425–36.

YOUNG, T. W. & YAGIU, M. (1978). A comparison of the killer character in different yeasts and its classification. *Antonie van Leewenhoek, Journal of Microbiology and Serology*, **44**, 59–77.

ZIMMERN, D. & WILSON, T. M. A. (1976). Location of the origin of reassembly on tobacco mosaic virus RNA and its relation to stable fragment. *FEBS Letters*, **71**, 294–8.

PATTERN FORMATION AND ITS REGENERATION IN THE PROTOZOA

B. C. GOODWIN

Developmental Biology Group, School of Biological Sciences, University of Sussex, Brighton, UK

INTRODUCTION

Micro-organisms have played a very significant role in the development of concepts which shape much of contemporary biological research. This is a truism for molecular biology, but when one realises that an organism such as *Dictyostelium discoideum* is included among the protozoa (order Amoebina), then it becomes evident that ideas in developmental biology have also been profoundly influenced by concepts deriving from the study of members of this diverse group. In fact, a rather basic prejudice of many contemporary developmental biologists, that the subject is really restricted to the study of the metazoa and the metaphyta, is reinforced by the observation that *D. discoideum* shows characteristic developmental behaviour such as pattern formation and differentiation only in its multicellular form. What I shall be arguing in this paper is that this is an unfortunate view which has led to a number of misconceptions regarding the nature of the developmental process, and hence to a misunderstanding of the type of process that is involved in morphogenesis. Amongst the protozoa, *D. discoideum* is an exception to the rule, i.e. that pattern formation and its regeneration occur in single cells. What I shall try to show is that, at a particular level of analysis, the morphogenetic phenomena that occur in the strictly unicellular protozoa are to be understood in exactly the same terms as those occurring in the metazoa.

This has, of course, been said before, by people much more familiar with the details of protozoology than I. These include early workers such as Lillie (1896) and Morgan (1901), and more recent ones such as Tartar (1961), Sonneborn (1970) and Frankel (1974). I am deeply indebted to their work and insight, which I have found extremely illuminating. My only excuse for taking up a theme which has been more competently handled in detail by others is that I would like to describe pattern formation and its regeneration in different terms to those previously essayed; an approach which emphasises the universality of morphogenetic processes throughout the phyla. This accords with the classical view but has tended to be neglected in recent years.

REGENERATION: PATTERNS AND FIELDS

The discovery of the regenerative powers of hydroid polyps by Tremblay (1744), who demonstrated that fragments obtained by cuts in any plane could give rise to a new, miniature whole organism which then grew to normal size with completely normal powers of sexual and asexual reproduction, had a profound influence upon the biological ideas of that age. In the latter half of the nineteenth century, this remarkable capacity of living organisms to undergo regeneration after being subdivided in various ways was extended to the level of the single cell by studies on helioza, and then by Nussbaum (1884) on ciliates. There are limits to the size of a fragment that will regenerate, it being claimed that in *Hydra* the minimum size is about 1/20th of the original linear dimensions (about 1/400th of the volume); while Tartar (1961) found that the smallest fragment of *Stentor coeruleus* which can regenerate is 1/123rd of the volume of a normal organism. These represent marked regenerative powers whereby a part can give rise to a whole. There are other constraints besides size; fragments of ciliate protozoa require the presence of a macronucleus for regeneration, while some species such as *Euplotes patella* require also a micronucleus. The unicellular green alga *Acetabularia*, on the other hand, can regenerate a cap in the absence of a nucleus. However, the point which I wish to make is that small fragments of organisms can regenerate miniature wholes which have the

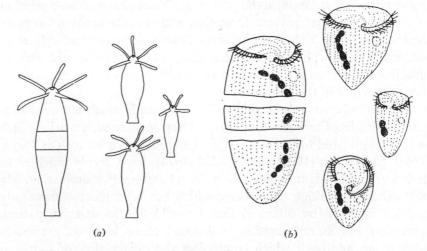

(a)　　　　　　　　　(b)

Fig. 1. Dissection of the whole organism into parts is followed by the regeneration of small entire organisms with typical form from the parts in (a) *Hydra*; (b) *Stentor* (from Westphal, 1976).

same basic morphology as the adult, whether these fragments consist of cells or parts of cells (Fig. 1). This property of regulation, i.e. pattern invariance irrespective of size, I take to be a basic feature of regeneration which any general model of pattern formation must be able to account for. Models based upon chemical reactions and diffusion, so-called reaction–diffusion processes (Turing, 1952; Gierer & Meinhardt, 1972), have difficulty with this phenomenon because the wavelengths of the standing waves of chemicals which underly all spatial patterning in such descriptions are fixed by the diffusion constants of the 'morphogens'. Regulation in such models depends upon the assumption that the diffusion constants which appear in the equations are not really constants, but are particular functions of other variables (Harrison & Lacalli, 1978), these functions themselves representing how the regenerating entity determines its spatial dimensions and so scales the diffusion 'constants'. This complicates the equations even further.

In seeking simplicity (and then distrusting it, as Whitehead counselled), some colleagues and I have been investigating the possibility that organismic patterns can be described by functions which are solutions of a very general field equation named after the great eighteenth century mathematician Laplace. The property which Laplace's equation describes is a very simple one: at every point within the boundaries of a field, the value of the field is the arithmetic mean of the values at equidistant neighbouring points. Mathematically, this is written in cartesian co-ordinates as

$$\nabla^2 U = \frac{\partial^2 U}{\partial x^2} + \frac{\partial^2 U}{\partial y^2} + \frac{\partial^2 U}{\partial z^2} = 0,$$

where $U(x, y, z)$ defines the field value at a point in space. This equation defines a particular type of smoothness in the variation of the field over a domain, and represents the property that if the field is disturbed in any way the values at all points will adjust themselves so as to restore the condition of smoothness described by the equation. The spatial pattern defined by the functions which are solutions of Laplace's equation, known as harmonic functions, are independent of the size of a domain, so that they naturally have the property of 'regulation'. Thus a purely local rule of spatial averaging results in a particular type of global order.

This extremely simple and powerful principle describes many physical fields. Can it be applied also to developmental fields? Experimental studies on regeneration in *Hydra*, insects, and amphibians (Lawrence, 1970; Webster, 1971; Bryant, 1975) suggest that a local averaging process occurs in developing tissues, and a model incorporating this rule has been applied with impressive success to regenerative processes in insect imaginal discs and limbs, and amphibian limbs (French, Bryant & Bryant, 1976). This chapter explores the consequences of using the rule in a particular form in the study of protozoan morphogenesis. The beauty of the principle is that it leads to a theory that is free of particular assumptions about the molecular or cellular basis of the developmental process: no specific microscopic mechanism is assumed, though there are many which conform to it, including diffusion and cell–cell interactions of particular types. Not knowing what molecular or cellular processes are involved in the spatial organisation of organisms, it is best to remain agnostic about detail and to use rules derived from phenomenology. This has the consequence that the resulting theory applies as well to the field properties of single cells as to those of multicellular organisms. No assumptions need be made about cellular interactions being the basis of pattern formation, as is done by Wolpert (1971), for example; and this allows one to escape from the restrictions imposed by a concept such as positional information, which in Wolpert's definition must be 'interpreted' by a nuclear genome in every part of the field, the interpretation resulting in cell differentiation. Frankel (1974) has tried to apply the notion of positional information to pattern formation in the ciliate protozoa, and discovered that in doing so he had to re-define the concept of 'interpretation', basically altering the logic of Wolpert's definition, which requires a genome to carry out this process (Wolpert & Lewis, 1975). Frankel's objective was to demonstrate in protozoa the existence of global fields which regulate and have all the other basic features of those that exist in the metazoa. In my opinion he succeeded admirably, by describing similarities between spatial organising principles in these groups and establishing a view of the relation between the global properties of fields and their local effects in a manner that re-emphasises classical views. In so doing, he provided a general description of field behaviour, equally applicable to unicellulars and multicellulars, which is in accordance with Driesch's (1908) conclusions deriving from comparative observations on the regenerative behaviour of *Stentor* and various metazoa, that the elements of his harmonious equipotential system must be subcellular. My concurrence with this view is expressed elsewhere (Goodwin, 1977; Goodwin & Pateromichelakis, 1979). The problem now is to attempt a detailed description of the properties of these postulated fields, and to illustrate how they may be applied to the study of pattern formation in the protozoa.

THE MORPHOGENETIC FIELD IN
TETRAHYMENA PYRIFORMIS

If the protozoa and the metazoa do obey essentially similar principles of pattern formation, then it should be possible to show that the field patterns which are observed in, say *Tetrahymena* (generally taken to be typical of the class Ciliata, of which *Stentor* is a more specialised member) can be described by the same class of function as those used to describe the early development of a vertebrate embryo such as the newt *Ambystoma maculata* (also taken to be typical of its class). This may appear to be a rather extreme form of ontogeny recapitulating phylogeny. However, it will emerge that this 'recapitulation' has nothing to do with descent from a common ancestor, a concept which was so dear to Haeckel's heart, enamoured as he was of Darwin's interpretation of all biological process in historical, contingent terms. Instead, our attention will be directed towards a much more fundamental principle of biological organisation, independent of historical contingency, and based upon a general law of the type sought by pre-Darwinian rational morphologists such as Cuvier and Owen.

Fig. 2. The typical form of *Tetrahymena pyriformis*, showing ciliary meridians and buccal cavity with its ciliature. (From Mackinnon & Hawes, 1961.)

The basic morphology of *Tetrahymena pyriformis* is shown in Fig. 2. According to Mackinnon & Hawes (1961) this ciliate 'embodies the most primitive type of organisation found in the Hymenostomida'. The essential features are the arrangement of the cilia on meridians, usually with a gentle spiralling which may be either to the left or to the right, and the buccal cavity or oral apparatus located near the anterior end. The overall shape varies greatly, from pear to egg to cucumber, but in general it is spheroidal with the anterior end more pointed than the posterior. As a result of the spiral arrangement of the cilia, the organism swims with a spiral or helical motion. The overall pattern which must be explained or described, then, is the polarity of the organism, the shape of the ciliary meridians, and the position of the oral apparatus. Whatever principles of explanation are used for this they must be able to provide an explanation of mutant forms, such as that described by Jerka-Dziadosz & Frankel (1979) which has two oral apparatuses with a specific relationship to one another, but the cilia are still arranged normally on meridians.

The descriptions of these morphological features will be based for simplicity upon a spherical shape for the organism. A great variety of experimental work has pointed to the cortex (the ectoplasm) as the major site of polarity and pattern formation, so that whatever it is that determines the morphology is largely confined to this thin surface layer of the organism. I shall therefore restrict, initially, the field description to the surface of a sphere, which requires only two variables to determine location, latitude (θ) and longitude (φ). However, when we come to a consideration of secondary oral primordium formation and cell division, which involves the whole of the cell, we shall require the use of a three-dimensional field description, using the third spatial variable, r (radius).

The restriction of the basic pattern to the cell surface, together with the fact that the organism has polarity so that the anterior (or north) pole is different from the posterior (or south) pole, is sufficient to select from among those functions satisfying Laplace's equation (more generally, Poisson's equation) on the sphere a unique one which describes the basic polar organisation of the cortical field. This is

$$U_0(\theta) = \alpha \ln \cot(\theta/2), \tag{1}$$

where α is a parameter whose value determines the strength of this polarising field, shown in Fig. 3. The function U_0 takes the value $-\infty$ at $\theta = \pi$, which defines the posterior pole, and increases continuously over the surface to $+\infty$ at $\theta = 0$, the anterior pole. It thus defines at every point of the surface, except at the poles, a sense of direction or a polarity given by the slope of the function (technically, its gradient: $du/d\theta = -\alpha/\sin\theta$). The minimum and maximum values of $\pm\infty$ at the poles are, of course, mathematical idealisations, to be interpreted more generally as extreme values of the field. At every point on the surface other than the poles, which define the field boundaries, the value of U_0 is the average of the values at equidistant neighbouring points, as is required for a solution of Laplace's equation. This polarising field is taken to determine the global spatial ordering of the cilia. The more general case in which the field lines or ciliary meridians have a spiral orientation is defined by the function

$$U(\theta, \varphi) = \alpha \ln \cot(\theta/2) + \beta\varphi, \tag{2}$$

a right-handed spiral being generated when β is negative and a left-handed spiral when β is positive. Thus the spiral is the general solution as observed, while a non-spiral polarity corresponds to the special case in which $\beta = 0$.

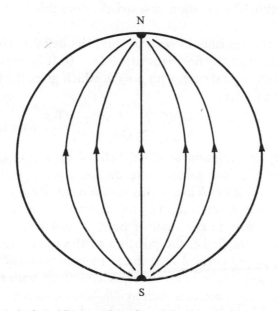

Fig. 3. Polarised meridians on the sphere defined by the basic polar field.

The next problem is to describe the existence and the position of the oral apparatus. This structure corresponds to a break or a hole in the surface, and we need to have some general and plausible description of why and how such a structure appears in a cortical field. A clue is obtained from the observation that prior to division, the first sign of the development of a new oral apparatus is the appearance of an 'anarchic field' of basal bodies that will later form the organised structure of the mouth ciliature. This initially disorganised domain, which invades ciliary meridians and interrupts their continuity, strongly suggests an initial loss of polarity in this region followed by the development of order as the field expands. Loss of polarity is what characterises a singularity known as a 'saddle point' in a Laplacean field. Since the presence of an oral apparatus on one side of *Tetrahymena pyriformis* shows that the organism has broken the symmetry described by the polarising function U_0, it is necessary to use a solution with the appropriate asymmetry to describe this structure. Again, there is only one solution of Laplace's equation on the surface of a sphere with this asymmetry, namely $\cot(\theta/2)\cos\varphi$, which we add to the general polar solution (2) to give the field

$$U(\theta, \varphi) = a \ln \cot(\theta/2) + \beta\varphi - b \cot(\theta/2) \cos\varphi, \tag{3}$$

where b is another parameter (the reason for the negative sign will be evident shortly). The question now arises: does this field have a saddle point?

Since polarity vanishes at such a point, its defining characteristic is that the gradient of the field is zero: $\nabla U = 0$. It is convenient first to consider the non-spiral case, with $\beta = 0$, which gives the following pair of equations to solve:

$$-a \tan(\theta/2) + b \cos\varphi = 0, \tag{4}$$

$$b \cot(\theta/2) \sin\varphi = 0. \tag{5}$$

There is only one solution to these, defined by $\varphi = 0$, $\tan(\theta/2) = b/a$. This is a point on the sphere lying on longitude $\varphi = 0$ (if we had a positive sign in front of b in (2), this would have given $\varphi = \pi$), and on a degree of latitude, θ, determined by b/a. Since the mouth in *Tetrahymena pyriformis* lies close to the anterior pole, θ must be small and in consequence b/a is small; i.e. the position of the mouth tells us that the contribution to the cortical field made by $\cot(\theta/2)\cos\varphi$ in (2), known as the first harmonic solution, is weak compared with the polar field given by U_0. Many protozoa such as *Didinium nasutum* and *Lacrimaria olor* (Fig. 4) do not have any asymmetry of the type found in *T. pyriformis* so that $b = 0$ in (3), the former showing radial symmetry about the polar axis while the latter has a distinct and highly elongated spiral form. The basic morphology of these forms is therefore described solely by the polar field U_0 in the form of equation (1) for *Didinium* and equation (2) for *Lacrymaria*. On the other hand, there are ciliates with a stronger asymmetry than *T. pyriformis*, such as *Blepharisma lateritium* and *Paramecium caudatum* (Fig. 5). The position of the buccal cavity mid-way between the poles in *P. caudatum* indicates that the contribution of the first harmonic solution to the field is as strong as that of the polar field in this species (i.e., θ has a value close to $\pi/2$, so that $\tan(\theta/2) \simeq 1$ and hence $b \simeq a$). Most ciliate protozoa thus appear to have basic body patterns which are describable by the function (3), a linear combination of the two most fundamental solutions of Laplace's equation on the sphere, different forms being given by different values of the

(a)   (b)

Fig. 4. Two ciliates lacking the asymmetry associated with a lateral buccal cavity; (a), (*Didinium nasutum*), with radial symmetry about the polar axis (from Westphal, 1976); (b), *Lacrymaria olor* with spiral symmetry (from Sleigh, 1973).

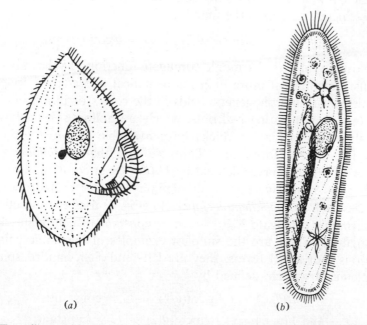

(a)   (b)

Fig. 5. Two ciliates with strong lateral asymmetry: (a) *Blepharisma lateritium*; (b), *Paramecium caudatum*. (From Westphal, 1976.)

parameters α, β and b. To get a little more insight into the structure of these fields and how much (or how little) they can describe, let us look more closely at the pattern described by the field in equation (3) and its relation to *T. pyriformis* morphology.

GLOBAL AND LOCAL INFLUENCES IN
TETRAHYMENA PYRIFORMIS

The field defined by equation (3) can be studied by looking at the pattern of flow lines which describe the polarity of the field at any point on the surface, as shown in Fig. 3 for the radially symmetric polar field. An interesting property of solutions of Laplace's equation in two dimensions, such as those restricted to the surface of a sphere (two variables, θ and φ, r being constant) is that they always occur in pairs which have the property that their gradients are orientated at right angles to one another. (When you walk along a level contour on a hill, you are always walking in a direction at right angles to the arrow which points in the direction of steepest ascent up the hill; these two directions correspond to the gradients of the two solutions.) Thus there is besides $U(\theta, \varphi)$ another solution, $V(\theta, \varphi)$, which has the property that its level contours, defined by $V(\theta, \varphi) = \text{constant}$, coincide with the curve along which $U(\theta, \varphi)$ increases. This is the function

$$V(\theta, \varphi) = +\beta \ln \cot(\theta/2) - \alpha\varphi + b \cot(\theta/2) \sin\varphi. \qquad (6)$$

The existence of these harmonic conjugate functions, as they are called, tells us something of more than mathematical interest; it implies that organisms with morphogenetic fields in the cortex or in a thin cellular shell (as in regenerating hydroids, which are hollow spheroids with a thin shell two cell-layers thick) have available to them orthogonal fields for pattern formation. Thus we find that *Didinium nasutum* locates its cilia on curves defined by $U_0(\theta) = \alpha \ln \cot(\theta/2) = \text{constant}$, which gives simply $\theta = \text{constant}$, defining degrees of latitude; while a species such as *Amphileptus claparedel* (Fig. 6) uses curves defined by the harmonic conjugate $V_0(\varphi) = \alpha \varphi = \text{constant}$, which define degrees of longitude. These are the simplest examples of harmonic conjugate functions. For spiral forms, they are left- and right-handed spirals of complementary pitch, defined by

$$U(\theta, \varphi) = \alpha \ln \cot(\theta/2) + \beta\varphi = \text{constant},$$
$$\text{and } V(\theta, \varphi) = +\beta \ln \cot(\theta/2) - \alpha\varphi = \text{constant}.$$

The fact that the detailed molecular architecture of the ciliature in the cortex of a ciliate protozoan shows both antero-posterior polarity and right–left asymmetry appears also to reflect the existence of these orthogonal conjugate fields. And we shall see later that the gastrulation field

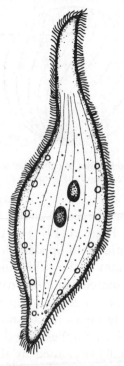

Fig. 6. *Amphileptus clarapadel*, a ciliate with non-spiral polarised meridians. (From Sleigh, 1973.)

of the newt requires the use of both families of functions defined by equations (3) and (6), with $\beta = 0$ (non-spiral patterns) to define the presumptive fate map of the early gastrula. Organisms with spiral cleavage and a spiral body pattern such as snails have $\beta \neq 0$, the handedness of the spiral being defined by the sign of this parameter.

Fig. 7 shows members of the family of curves defined by $V(\theta, \varphi) =$ constant, each curve corresponding to a different value of the constant. These curves correspond to the ciliary meridians or kineties in *Tetrahymena pyriformis*. The distance between these kineties, and hence their number for a given size of cell, is generally assumed to be determined by local rules of kinetid and ciliary territory formation (Sonneborn, 1970), which a global field theory of the type described here cannot treat without further specific assumptions.

In Fig. 7, the saddle point on the degree of longitude $\varphi = 0$ is taken to define the centre of the oral apparatus in *Tetrahymena pyriformis*.

Again, the detailed ciliature of the mouth (see Fig. 2) must arise from local interaction rules, but the handedness of the structures is dictated either

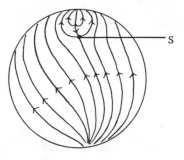

Fig. 7. Members of one of the two families of curves defined on the sphere by equations (3) and (6). The arrows define the direction of increasing U (θ, φ) S, which establishes posterio-anterior polarity saddle point.

by that of the spiral field or by local assembly constraints. The observations of Jerka-Dziadosz & Frankel (1979) on the mutant clone CU-127, in which two oral apparatuses with opposite handedness occur, indicate that the relevant determinant is the field, since here the same macro-molecular elements are assembled into mirror-symmetric structures in one cell. In the case of a normal cell, I assume that it is the handedness of the spiral field which determines the asymmetry of the oral ciliature.

The field geometry of Fig. 7 implies that the oral apparatus and the ciliary meridians are spatially organised by the same global field. The invariance of the positioning of the new oral apparatus relative to these meridians during cell division, to be discussed in the next section, certainly suggests some such type of underlying co-ordinate control system. However, Frankel (1979) has recently identified a mutant called 'disorganised' which when homozygous shows a severe disruption of the normal pattern of cilia, particularly at high temperatures (39.5 °C), although the oral primordium in a dividing cell appears in approximately the normal position. This may be interpreted to imply a dissociation of separate mechanisms responsible for the positioning of the oral apparatus and of the ciliary meridians, which would of course falsify the unitary field picture presented above. An alternative interpretation is that the disorganised mutant is one in which a constituent of the ciliary territories is altered in such a way that there is a failure of these elements to respond to the field influences which normally order them in rows, disorganisation being then a consequence of a disturbed assembly process in a cell whose primary field is normal. It might be possible to distinguish between these alternatives by reciprocal grafts between disorganised and wild-type cells. The direct lesson of this mutant, however, as pointed out by Frankel (personal communication), is that 'although the orientation of the ciliary meridians and the position of the new oral apparatus may be under a co-ordinated underlying control, they probably have no *direct* causal relationship to one another'. This view is compatible with the field description presented in this paper.

CELL DIVISION IN *TETRAHYMENA PYRIFORMIS*

The initial stages of the process resulting in the generation of two daughter cells in *Tetrahymena pyriformis* are shown in Fig. 8. The first evidence of preparation for division is the appearance of basal bodies at a particular point on the cell surface (Fig. 8a). These then multiply

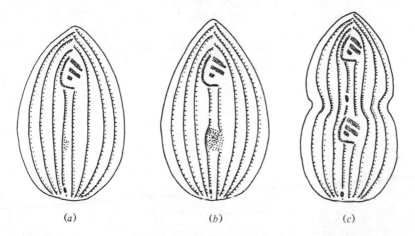

   (a)   (b)   (c)

Fig. 8. Stages in the process of cell division in *Tetrahymena*. (a), Basal bodies initiating the new oral ciliature have appeared; (b), the anarchic field has formed; (c) the oral membrenelles have assembled. (From Sleigh, 1973.)

to give an expanding anarchic field which disturbs the polar organisation of the local ciliary meridians, and then this oral field becomes organised into the typical ciliature of the mouth. At the same time, the cell begins to constrict centrally, leading to the generation of two cells. It was this observation of polar disturbance by the anarchic field of proliferating basal bodies, which led us to the conclusion that the oral apparatus arises at a saddle point in the cortical field, where polarity vanishes. We must now see if we can preserve and extend this description within the context of a process which involves not only the cell surface, but the depth of the organism as well.

 Solutions of Laplace's equations in three dimensions are known as solid harmonics, as distinguished from the surface harmonic solutions which we have been using so far to describe the cortical field. From the perspective of the present treatment of the organism, it is natural to attempt to describe cell division in terms of a field which extends throughout the cytoplasm. Just as the surface fields have been assumed to define polarities which direct the assembly of microtubules, kineto-somes, kinetodesmal fibres, and the other elements which make up the surface pattern of the ciliate protozoa, so a solid field is assumed to direct the assembly and orientation of the elements making up the mitotic spindle and other structures associated with cell division. These structures are bipolar, so a natural candidate for the organising field is one with a similar type of symmetry relative to the postero-anterior axis. This leads us to the solution

$$U(r, \theta) = \frac{r^2}{4}(3 \cos2\theta + 1), \tag{7}$$

which has the symmetry shown in Fig. 9. The function is positive in the regions at the poles and negative in the central band, with radial sym-

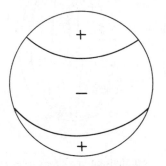

Fig. 9. Domains where the zonal harmonic defined by equation (7) takes positive and negative values.

metry about the axis which reflects the symmetry of the mitotic spindle which it is assumed to organise. The variable r in equation (7) is the radius. Another solution with the same symmetry as this is given by putting r^{-3} in place of r^2. However, this gives a singularity at the centre of the organism, for which there does not appear to be any evidence, so the alternative solution was chosen. This has the property that as the cell grows, corresponding to increasing r, the solid field increases in strength at the cell surface. This could be interpreted as a contributory factor to the initiation of cell division. Let us now see how the division field (7), together with the cortical field defined by equation (3), can give some insight into the process of cell duplication.

Adding the solid to the surface field, we get the following function:

$$U(r, \theta, \varphi) = a \ln \cot(\theta/2) + \beta\varphi - b \cot(\theta/2) \cos\varphi + \frac{dr^2}{4}(3 \cos 2\theta + 1), \quad (8)$$

where d is a fourth parameter. The question again arises whether on the surface of the sphere defined by $r = $ constant there are any saddle points which could define the site of the anarchic field of basal body proliferation where the mouth of the posterior daughter cell will form. We already know that in the absence of the solid harmonic ($d = 0$), there is a single saddle point which is taken to define the position of the parental mouth. However, as the cell prepares for division, we may assume that the solid harmonic grows in strength and we now need to see if there are further solutions of the equation $\nabla U = 0$, which defines saddle points where surface polarity vanishes. For convenience we take $\beta = 0$ initially (non-spiral kineties). The equations that must now be solved are

$$\frac{-\tan(\varphi/2)}{\sin^2(\theta/2)} + \frac{b\cos\varphi}{2\sin^2(\theta/2)} - k \sin 2\theta = 0 \quad (9)$$

and

$$\frac{b \cot(\theta/2) \sin\varphi}{\sin\theta} = 0, \quad (10)$$

where $k = 3dr^2/2$. Equation (10) gives $\varphi = 0$ or π, and again only the former gives a solution in the first equation when k is moderately small compared with a, which we assume always to be the case. We then find that there can be three solutions to equation (9), two of which define saddle points located near the anterior pole and below (posterior to) but not far from the equator. The exact positions depend upon parameter values. The saddle point near the anterior pole coincides with the general position of the parental mouth, while the second one, below the equator and on the same meridian, is taken to define the position of the secondary mouth primordium, the anarchic field. Solving equations (9) and (10) with $\beta \neq 0$ gives the same result except that now the saddle points lie on a spiral meridian. As the anarchic field of the second mouth expands beyond the saddle point, it extends into the asymmetric parental field and so assumes the same asymmetry. We must now see how the mirror-reflection observed by Jerka-Dziadosz & Frankel (1979) in their mutant can be explained.

MIRROR IMAGE DUPLICATION IN
TETRAHYMENA PYRIFORMIS

The typical form shown by clone CU127, carrying a mutation in a single genetic locus called janus, is essentially one of bilateral symmetry as shown in Fig. 10, which illustrates the positions of parental daughter oral apparatuses in a view looking down on the anterior pole of a cell preparing for division. This view may be regarded as a projection of the organism from the posterior pole onto a plane which is tangent to the anterior pole and perpendicular to the antero-posterior axis. It is convenient to carry out the projection mathematically in order to see how the pattern of the mutant can come about in terms of the fields used in this analysis. Let us first of all consider only the cortical or surface field in the inter-division mutant, which has two oral apparatuses located at an angle shown in Fig. 10. How can this be described?

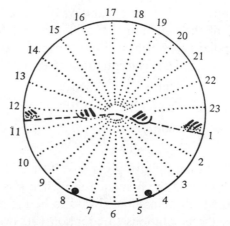

Fig. 10. Diagram of the bilaterally symmetric janus mutant of *Tetrahymena thermophila* showing the relative positions of the secondary buccal cavities in a cell preparing for division. The numbered dotted lines are the kineties. (From Jerka-Dziadosz & Frankel, 1979.)

A stereographic projection of a sphere actually projects the surface onto the whole of the tangent plane. This allows us to study the cortical fields in terms of functions of a complex variable, $z = re^{i\phi}$, where now r measures distance from the origin in the plane (the point of tangency with the anterior pole of the sphere), φ measures angular distance from a reference line on the plane, as in polar co-ordinates, and $i = \sqrt{-1}$. This procedure gives the two functions corresponding to U and V in equations (3) and (6) simultaneously, as complex conjugate (equivalently, harmonic conjugate) functions. In these terms, a normal *Tetrahymena pyriformis* cortical field is described by the function

$$W(z) = -a \ln z - \frac{b}{z}, \tag{11}$$

where a is a complex number ($\alpha + i\beta$) if the field has spiral meridians and is a real number if the meridians are straight. This function separates into real and imaginary parts as follows

$$W(z) = U(r, \varphi) + i\,V(r, \varphi)$$

$$= -\alpha \ln r + \beta \varphi - \frac{b}{r}\cos\varphi + i[\beta \ln r - \alpha\,\varphi + \frac{b}{r}\sin\varphi] \qquad (12)$$

The stereographic projection of the sphere is defined by $r = \tan(\theta/2)$, with φ unchanged, relating position on the sphere (θ, φ) to position on the plane (r, φ). When r is replaced by $\tan(\theta/2)$ in equation (12), we recover the functions $U(\theta, \varphi)$ and $V(\theta, \varphi)$ defined by equations (3) and (6). Use of the complex variable makes the search for saddle points much easier, since they are now solutions of the equation $dW/dz = 0$. Since the janus mutant has mirror-symmetry, we need to use a field which has this form. This is obtained simply by adding the next harmonic solution to the field which describes the normal organism. In terms of the original variables on the sphere, this second surface harmonic solution is $\cot^2\theta/2 \cos 2\varphi$. In the complex plane, it is just the next power of the variable $1/z$, so we use the function

$$W(z) = -\,a \ln z - \frac{b}{z} + \frac{c}{z^2} \qquad (13)$$

To get an idea of what this additional harmonic does to the field, let us first consider equation (13) with $b = 0$ (no first harmonic). The saddle points are then given by solutions of

$$\frac{dW}{dz} = \frac{-a}{z} - \frac{2c}{z^3} = 0$$

or $\qquad az^2 + 2c = 0.$

This gives $z = \pm\,i\sqrt{(2c/a)}$ which are points located $180°$ apart on the imaginary axis, at equal distance from the origin. On the sphere, this field would then give saddle points (interpreted as oral apparatuses) located on meridians at $180°$ to one another and at the same distance from the anterior pole. Jerka-Dziadosz & Frankel (1979) reported that such forms are included in the mutant population. Then we could say that in such individuals, the field is characterised by the presence of the second harmonic solution instead of the first. These two forms, then, are very closely related to one another, since they are described by adjacent members of the same harmonic series of solutions. The asymmetry of the mouth ciliature in those individuals with oral apparatuses $180°$ apart should have the same handedness or be indeterminate, depending upon whether the meridians are spiral or straight, respectively; i.e., they should not show mirror-inversion of these structures, as do those in which the oral apparatuses are less than $180°$ apart. Let us see how these may be explained.

We now return to equation (13) and look at the effect of the simultaneous presence of both harmonics on the position of the saddle points. These are defined by the solutions of the equations

$$\frac{\mathrm{d}W}{\mathrm{d}z} = \frac{-a}{z} + \frac{b}{z^2} - \frac{2c}{z^3} = 0,$$

or $\qquad az^2 - bz + 2c = 0$, giving the solutions

$$z = \frac{b \pm \sqrt{(b^2 - 8ac)}}{2a} = \frac{b \pm i\sqrt{8ac - b^2}}{2a}$$

Now what we notice is that for small values of b, these points are in just the relation described by the oral apparatuses in Fig. 10, the exact angle between them being determined by the value of a, b, and c. Furthermore, it can be shown that, because the saddle points are now both located on the same side of the imaginary axis, the fields in their neighbourhoods will be mirror-inversions of one another so that the ciliature of the oral apparatuses should reflect this, as they do. Thus we see that, in terms of the field description presented here, mutant forms are to be explained in terms of the presence of harmonic solutions other than those which normally occur. We may say that the complete set of surface and solid harmonics are potentially available to every organism for the specification of its form, since they satisfy the basic field constraint which defines organismic morphology (Laplace's equation). The pattern which is found in any individual is then the result of a selection and stabilisation process whereby particular solutions are specified by a procedure which in field theory is called the definition of boundary conditions. Exactly how this occurs in organisms remains an extremely interesting and important problem. Evidently, gene products can influence this specification of boundary conditions, but according to the theory described here, the genes in no sense *generate* the form of the field. They contribute to the selection of spatial fields belonging to the potential set defined as solutions of Laplace's equation, which set is generated by the rule of local spatial averaging. This rule describes a property of the living organism, not of the genes.

There is an immediate prediction that this view of pattern formation gives us. It is that there should be other ways of obtaining heritable abnormal forms apart from the selection of genetic mutants, since any procedure which alters boundary conditions in a manner compatible with the constraints operating in a given species should give a possible reproducible morphology. In particular, it is to be expected that microsurgical modification could produce stable mirror-symmetric forms of the general type characteristic of the *Tetrahymena* janus mutant. This is indeed the case, as described by Tchang & Pang (1965) for the hypotrich ciliate *Stylonchia mytilus*. Frankel & Jenkins (1979), in their discussion of the field implications of their genetic analysis, discuss this example and consider the 'extreme possibility . . . that the jan allele, rather than "creating" a morphogenetic field of reversed asymmetry, brings a precisely positioned pre-existing field above a threshold of phenotypic manifestation. However, as there is evidence that the janus condition can be expressed in cells of inbred strain A as well as B cytoplasmic ancestry, such a "silent" pre-existing field, if it exists, must be present in *T. thermophila* of diverse natural sources.' This perspicacious observation anticipates the view presented in this paper, which, however, takes the hypothesis two steps further in the same direction. Not only is it proposed that the mirror-symmetric field exists in silent or potential form in diverse *Tetrahymena* types, but in all the ciliates, along with all the other 'silent' fields defined by solutions of Laplace's equation. But furthermore, these fields are potentially available to all living organisms, and in fact define the set from which all organismic morphology is derived. This is, of course, a conjecture which needs extensive investigation, and the present paper does no more than make a start in exploring some of its consequences in relation to the protozoa.

To complete the field treatment of the mirror-symmetric *Tetrahymena* field, we need a description of cell division in the janus mutant, wherein two new saddle points are generated in the correct positions relative to the original ones defining the pair of parental oral apparatuses. Since the solid harmonic used to describe the division field is defined in three dimensions we cannot use a two-dimensional projection to study it and we must use a function like equation (8) but with the next harmonic added:

$$U(u, \ \theta, \ \varphi) = a \ \ln \cot(\theta/2) + \beta\varphi - b \ \cot(\theta/2) \ \cos\varphi + c \ \cot^2(\theta/2)$$
$$\cos 2\varphi + \frac{dr^2}{4} (3 \cos 2\theta + 1). \tag{14}$$

As might be expected, this gives the required two extra saddle points, located on the same meridians as the originals, so long as the parameters lie in a particular range. What this analysis does not explain is why one of these new oral apparatuses is normal, but the second shows various degrees of structural abnormality. These may be explicable in relation to a more detailed analysis of the contribution of the spiral element to field asymmetries in the neighbourhood of the saddle points. There are also morphological features of normal *Tetrahymena* which remain undescribed by this treatment, such as the position of the contractile vacuole pore. Indeed, it is clear that the cursory field treatment of protozoan morphology given here raises many unanswered questions. Before drawing some general conclusions, I shall pursue briefly two

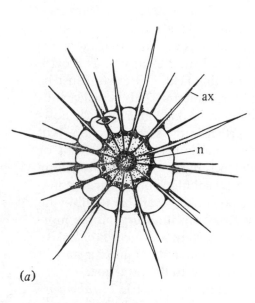

(a)

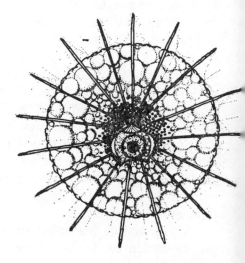

(b)

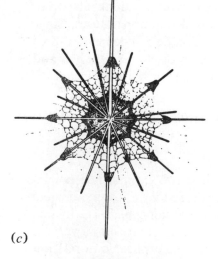

(c)

(d)

Fig. 11. Typical forms of the Heliozoa (*a. Actinophrys sol*) and Radiolarians (*b*), *Aulacantha scolymantha*; (*c*) *Acanthometron elasticum*; (*d*), *Hexacontium asteraconthion*). (From Sleigh, 1973.) n; Nucleus; ax, axopodium.

more issues. The first relates to the morphology of the orders Heliozoa and Radiolaria, and the second is a demonstration of how the same principles used to study protozoan morphology can be applied to early vertebrate embryogenesis, revealing some surprising homologies.

THE MORPHOLOGY OF HELIOZOANS AND RADIOLARIANS

The most distinctive feature of these beautiful forms is their radial symmetry, and the property that their structures extend to the centre of the cell, where the nucleus is situated. Characteristic morphologies are shown in Fig. 11. Such forms immediately invite description by solid spherical harmonics, which as we have seen describe fields extending throughout the depth of the sphere. An example of a field with the general type of symmetry found in these forms is shown in Fig. 12. If

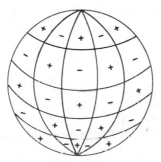

Fig. 12. The tesseral harmonic $P^4{}_8$, showing the domains where the function takes positive and negative values.

we were to locate skeletal elements in the centre of each domain labelled with a positive sign, which defines a local maximum of the field on a sphere of fixed radius, with these elements extending from a point near the centre of the sphere to a point some distance beyond the surface, then one would have the basic morphology required. Different harmonics give different numbers of such domains, with different symmetries. These can be added together with different parameters defining their relative strengths, so that skeletal elements of different size can be represented. However, there are basic symmetries in the solid harmonics which cannot be violated, so that there are definite limitations on the range of forms which can be described. This remains an area for future investigation.

THE FIELD OF THE EARLY AMPHIBIAN EMBRYO

My final area of concern in this chapter is the relationship between the fields that have been used to describe protozoan morphology, and those underlying embryogenesis in higher organisms, taking the newt as an example. There are essentially two dominant aspects of early amphibian morphogenesis, namely cleavage and the formation of the presumptive field in the gastrula. The first of these can be described by a sequence of solid harmonics that define in a precise manner the orientation of the cleavage planes. Since the principle of description is the same as that used to describe cell division in *Tetrahymena*, I shall do no more than list the first few fields in the order of their appearance. The convention used is that the positions of the centrioles defining the axis of the mitotic spindle are always near the centre of a domain marked with a positive sign. Thus first cleavage is described by the function

$$U(r, \theta, \varphi) = r^2 \sin^2\theta \cos 2\varphi, \tag{15}$$

which has the field structure in Fig. 13. This shows the regions where the field takes positive and negative values, the maximum positive values

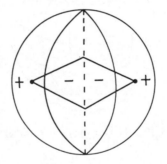

Fig. 13. The domains where the sectoral harmonic P^2_2 takes positive and negative values, showing how the mitotic spindle is assumed to be orientated in relation to the field.

being located in the centre of the appropriately marked domains, and similarly for minima. The cleavage plane is shown as a dotted line bisecting the spindle, drawn as a lozenge with the centrioles at the apices. Second cleavage is described by the same field as equation (15) with a negative sign in front of it so that positive and negative domains are interchanged, giving a cleavage plane at right angles to the first but still containing the polar axis. Third cleavage is given by the same field as that used to describe cell division in *Tetrahymena* (see Fig. 9), namely:

$$U(r, \theta) = \frac{r^2}{4}(3 \cos2\theta + 1).$$

The asymmetry of this cleavage in the amphibian is assumed to result from the addition of the cortical polar field of the egg, not considered so far because it has no effect on the first two cleavages, due to their orientation. This surface field is the same as that used to describe polarity in the ciliate protozoa without spiral kineties, so that the complete field at third cleavage is defined by:

$$a \ln \cot(\theta/2) + \frac{br^2}{4}(3 \cos2\theta + 1). \tag{16}$$

The first term in this field produces asymmetry, making the upper domain of Fig. 10 larger and the lower smaller, thus shifting the field towards the animal pole. Because these solid harmonics are transient, we do not consider the appearance of saddle points on the surface since their duration is not sufficient to bring about any morphological effect. Fourth cleavage is defined by $U(r, \theta, \varphi) = r^4 \sin^4 \theta \cos 4\varphi$, and so on. There are definite symmetries which must be obeyed. A detailed study of this and the gastrulation field is being made in collaboration with J. L. Rius, and will be presented elsewhere.

The general point I wish to make is that this description uses the same functions as those applied to protozoan morphology, but there is now a specific temporal order in the appearance of the harmonics. This implies particular constraints on the processes which are beyond the scope of the present analysis. Indeed, the whole question of the time-dependence of developmental processes, including cell division, is of fundamental significance and requires extensive investigation. This carries the field theory over into the fourth dimension, which is essential for an adequate treatment of pattern formation.

So far the description of the early amphibian field has involved only solid harmonics and the primitive polar solution, as in equation (16). However, the appearance of the grey crescent after fertilisation and the eventual emergence of the dorsal lip of the blastopore in the centre of this region tells us of a developing bilateral symmetry in the field. It is assumed that this grows in strength during blastula formation. Further-more, as cleavage proceeds and higher and higher solid harmonics are added, the solid field loses its initial organisation and becomes rather chaotic, as is observed by the progressive loss of organised cleavage planes and the loss of cell division synchrony. The solid field also weakens as the blastula becomes a hollow ball of cells. We are then left essentially with a surface or a cortical field defined by the primitive polar solution and the first harmonic which is assumed to be initiated by sperm entry, giving bilateral symmetry. This is defined by the field

$$U(\theta, \varphi) = a \ln \cot(\theta/2) - b \cot(\theta/2) \cos \varphi. \tag{17}$$

This is a familiar function, since it is the same as that used to describe the cortical field in *Tetrahymena*. It has a single saddle point located at tan $\theta/2 = b/a$. This is now taken to define the dorsal lip of the blasto- pore, which is the homologue in the developing amphibian of the oral apparatus in *Tetrahymena*. By the same reasoning, it is just where surface polarity disappears that we expect to find the appearance of a new structure developing along the radial axis, this time initiated by radially elongating bottle cells which, having lost their surface polarity, respond

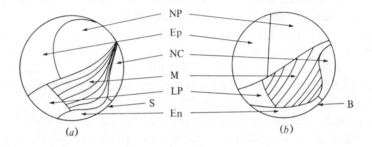

Fig. 14. (*a*) Members of the orthogonal families of curves on the sphere, showing the correspondence between them and (*b*) the presumptive areas of the newt gastrula. NP, neural plate; Ep, epidermis; NC, notochord; M, myotomes; LP, lateral plate mesoderm; En, endoderm; S, saddle point; B, blastopore lip.

to the residual field of the solid harmonics. The function (17) should define not only the position of the lip of the blastopore as a saddle point, but also the presumptive fate map of the early gastrula, that which defines how the different regions of the field will develop as embryogenesis proceeds. However, we (B. C. Goodwin & L. E. H. Trainor, unpublished) have found that in order to get a correspondence between the predicted and the observed field, it is necessary to move the anterior pole of the field into a position closer to the blastopore lip. The result is the field shown in Fig. 14, which includes an outline of the experimentally-determined fate map for comparison. What one sees is that some members of the family of curves which in *Tetrahymena* define ciliary meridians, now define inter-somitic boundaries while, members of the orthogonal family of functions describe other presumpt- tive tissue boundaries. The basic method of description, however, applies to both of these extremely different types of organism, one unicellular and the other multicellular.

CONCLUSIONS

The theory of pattern formation described in this chapter is based upon the idea that organismic morphology arises from the existence of a basic spatial organising constraint in living organisms, the rule of spatial averaging described above (p. 379). This rule generates Laplace's equation, so that biological form must be expressed by solutions of this equation known as harmonic functions. The particular solutions which describe the morphology of any species are determined by the boundary conditions that characterise the morphogenetic field of the species, and these are specified by genes, by other determinants such as cortical and cytoplasmic states, and by organelle structure. All organisms, then, are fields of the same generic type and it is this property which defines the living domain, not historical descendence from a hypothetical common ancestral type. Thus the biological realm is defined by a rational principle or a rule of organisation, not by historical continuity from a postulated chance event, the origin of life.

This is a view of organisms which the rational morphologists of the eighteenth and nineteenth centuries described, and which occasional and rather isolated figures of the present century such as Driesch, D'Arcy Thompson, and Waddington continued to believe in. It is a view which places the concept of the organism and biological organisation at the centre of the biological stage, emphasising the integrity of the whole which in the present treatment is defined as a field. The phenomena of generation and regeneration are then consequences of the basic properties of these organismic fields; they are not the result of molecular properties such as semi-conservative replication of DNA and the template mechanism of protein synthesis, remarkable as these processes are. These molecular activities constitute essential aspects of the living organism as we know it today, but they are inadequate to account for organismic morphology, as has been argued at length elsewhere (G. C. Webster & B. C. Goodwin, unpublished).

How does one deduce the properties of generation and regeneration from fields of the type described? We have seen that pattern formation or morphogenesis (generation) may be understood as the appearance within organisms of particular harmonic solutions which generate particular basic forms. The treatment of protozoan morphology presented has dealt exclusively with what might be called archetypal or germinal patterns, based upon essentially spherical form, and nothing has been said concerning the transformations of this form which characterise the morphogenetic process of different species. Such a description requires an interpretation of the harmonic field in terms of forces which organise the assembly of the constituents in the various species into their characteristic structures. It seems likely that this could be handled by concepts relating to curvature, but this is an area for future study. So also is the development of a time-dependent field theory wherein the temporal order of the harmonics arises from some kind of non-linear coupling between them, for example. As stated above, the field theory raises as many questions as it answers, and a complete description of generation of organismic form in field terms remains to be written, always assuming that this is possible. But at least the descriptive basis of such a theory may now be somewhat clearer.

Coming now to the question of regeneration, there is a very important property of harmonic functions which provides a kind of existence theorem for the regenerative process. If a harmonic function is defined over any part of a domain, such as a bit of the sphere in Fig. 7, then the function can be uniquely reconstructed over the whole of the sphere. Thus the part contains the whole in a specific mathematical sense, and one has here an analogue of the familiar holograph. We may then use this to deduce the regenerative properties of organisms, defined as harmonic fields: from a part, the whole can be regenerated. This does not mean that *any* part of any organism can regenerate the whole, since there are many constraints which operate in organisms besides the general field rules that we have considered. But the fundamental property of harmonic fields described above may be seen as that which confers on organisms the potential for regeneration.

A final remark concerns the nature of the fields that have been proposed as a basic defining characteristic of living organisms, and how they can be studied experimentally. Are they to be seen as essentially electrical or magnetic in nature, so that the harmonic functions are describing the same phenomena in biology as they do in physics? Or are we here dealing with a new type of field, characteristic of living organisms and defining their state in a manner which shows basic similarities to physical fields but which differs from these in some fundamental sense? For although the description of biological form presented here has emphasised formal similarities between physical and biological fields, in so far as both are assumed to satisfy the same time-independent equation (i.e. Laplace's), nevertheless there are many differences. These relate primarily to time-dependent behaviour, the generative and regenerative processes whereby a new individual with typical form is produced from a parent, as in cell division or embryo-genesis, or a whole from a part. It is here that the distinctive properties of biological fields are likely to emerge most clearly, and this has been left virtually untouched in the present treatment.

But the question remains whether the field properties of living organisms are reducible to those of physics, or whether they are irreducible and exhibit their own distinctive phenomenology. My belief is that one must remain agnostic on this until there is sufficient evidence to provide an informed conjecture. This will come from the application of a field analysis to the phenomenology of biological form, and attempts to deduce this from physical and chemical principles. I have no doubt that the phenomenology provided by the protozoa will continue to be an extremely rich testing-ground for the conjectures advanced here; and further, that developmental biologists will gain invaluable insights into the properties of metazoan fields as a result of studying those that occur in the protozoa. For I believe that the protozoa have explored and utilised virtually all of the fields available to them, and hence to higher organisms. The difference is that their surface harmonics are restricted to a single surface, the cell cortex, whereas the metazoa have discovered how to form several surfaces or sheets, the germ layers, with different fields in each. This allows for the appearance of a greater complexity through the combined consequence of interaction. However, the basic method of pattern formation, involving the interaction of both surface and solid harmonics as used above to describe protozoan morphogenesis is, in my opinion, an essential principle operating in the generation of biological form at all levels of organismic complexity.

I owe a debt of gratitude to many people for assistance in the development of the ideas which form the basis of the analysis presented, but particularly to G. C. Webster, F. W. Cummings and J. Frankel.

REFERENCES

BRYANT, P. J. (1975). Regeneration and duplication in imaginal discs. In *Cell Patterning*, ed. R. Porter & J. Rivers), *Ciba Foundation Symposium*, vol. 29, p. 71. Amsterdam: Elsevier.

DRIESCH, H. (1908). *Science and Philosophy of the Organism*. London: A. & C. Black.

FRANKEL, J. (1974). Positional information in unicellular organisms. *Journal of Theoretical Biology*, **47**, 439–81.

FRANKEL, J. (1979). An analysis of cell-surface patterning in *Tetrahymena*. In *Determinants of Spatial Organization*, pp. 215–46. New York & London: Academic Press.

FRANKEL, J. & JENKINS, L. M. (1979). A mutant of *Tetrahymena thermophila* with a partial mirror-image duplication of cell surface pattern. II. Nature of genic control. *Journal of Embryology and Experimental Morphology*, **49**, 203–27.

FRENCH, V., BRYANT, P. J. & BRYANT, S. V. (1976). Pattern regulation in epimorphic fields. *Science*, **193**, 969–81.

GIERER, A. & MEINHARDT, H. (1972). A theory of biological pattern formation. *Kybernetik*, **12**, 30–9.

GOODWIN, B. C. (1977). Mechanics, fields, and statistical mechanics in developmental biology. *Proceedings of the Royal Society of London, Series B*, **199**, 407–14.

GOODWIN, B. C. & PATEROMICHELAKIS, S. (1979). The role of electrical fields, ions, and the cortex in the morphogenesis of *Acetabularia*. *Planta, Berlin*, **145**, 427–35.

HARRISON, L. G. & LACALLI, T. C. (1978). Hyperchirality: a mathematically convenient and biochemically possible model for the kinetics of morphogenesis. *Proceedings of the Royal Society of London, Series B*, **202**, 361–97.

JERKA-DZIADOSZ, M. & FRANKEL, J. (1979). A mutant of *Tetrahymena thermophila* with a partial mirror-image duplication of cell surface pattern. I. Analysis of the phenotype. *Journal of Embryology and Experimental Morphology*, **49**, 167–202.

LAWRENCE, P. A. (1970). Polarity and patterns in the postembryonic development of insects. *Advances in Insect Physiology*, **7**, 197–265.

LILLIE, F. R. (1896). On the smallest parts of *Stentor* capable of regeneration: a contribution on the limits of divisibility of living matter. *Journal of Morphology*, **12**, 239–49.

MACKINNON, D. L. & HAWES, R. S. J. (1961). *An Introduction to the Study of Protozoa*. Oxford: Clarendon Press.

MORGAN, T. H. (1901). *Regeneration*. London: Macmillan.

NUSSBAUM, M. (1884). Über spontane und künstliche Zelltheilung. Sitzungberichte niederrhein. *Gesellschaft Nat.-u . Heilk. Bonn*, **41**, 259–63.

SLEIGH, M. A. (1973). *The Biology of Protozoa*. London: Edward Arnold.

SONNEBORN, T. M. (1970). Gene action in development. *Proceedings of the Royal Society of London, Series B*, **176**, 347–66.

TARTAR, V. (1961). *The Biology of* Stentor. Oxford: Pergamon Press.

TCHANG TSO-RUN, N. & PANG YAN-BIN (1965). Conditions for the artificial induction of monster jumbles of *Stylochia mytilus* which are capable of reproduction. *Scientia Sinica*, **14**, 1332–8.

TREMBLAY, A. (1744). *Philosophical Transactions of the Royal Society*, XLIII, No. 474.

TURING, A. M. (1952). The chemical basis of morphogenesis. *Philosophical Transactions of the Royal Society, Series B*, **237**, 37–72.

WEBSTER, G. C. (1971). Morphogenesis and pattern formation in hydroids. *Biological Reviews*, **46**, 1–46.

WESTPHAL, A. (1976). *Protozoa*. Glasgow: Blackie.

WOLPERT, L. (1971). Positional information and pattern formation. *Current Topics in Developmental Biology*, **6**, 183–224.

WOLPERT, L. & LEWIS, J. H. (1975). Towards a theory of development. *Federation Proceedings*, **34**, 14–20.

BIOCHEMICAL REGULATION OF CELL DEVELOPMENT AND AGGREGATION IN *DICTYOSTELIUM DISCOIDEUM*

CLAUDE ROSSIER*, EVELINE EITLE, ROEL VAN DRIEL† AND GÜNTHER GERISCH†

Biozentrum der Universität Basel, Klingelbergstrasse 70, 4056 Basel, Switzerland

INTRODUCTION

The eukaryotic micro-organism *Dictyostelium discoideum* can now be grown in chemically defined media (Franke & Kessin, 1977) and provides a convenient model system for the study of cell interactions in development. Several hours after the removal of nutrients the cells aggregate into a multicellular body and eventually form a fruiting body consisting of two types of cellulose-walled cells, spores and stalk cells. As discovered by Konijn *et al.* (1967), cell aggregation involves a chemotactic response to cyclic AMP which is released from cells acting as sources of attraction. It has long been known that during aggregation a dynamic pattern of waves can be observed which travel from a central region towards the periphery of an aggregation territory (Arndt, 1937). This pattern can now be interpreted to mean that cyclic AMP is periodically synthesised, and cyclic-AMP production in one cell is coupled to that in adjacent ones (for review see Gerisch & Malchow, 1976). Cells stimulated by cyclic AMP release cyclic AMP after a minimal delay of about 15 seconds. This delay limits the speed of wave propagation. The cyclic-AMP induced release of cyclic AMP is part of a relay system (Shaffer, 1962). Input signals are amplified through responding cells by a factor of about 100. Signal amplification involves the binding of cyclic AMP to cell surface receptors, the activation of adenylate cyclase, and the release of the newly synthesised cyclic AMP into the extracellular space.

Another mechanism important for the assembly of cells into a multicellular body is intercellular adhesion. A specific glycoprotein is expressed on the cell surface when the cells change their adhesive properties at the beginning of aggregation, as has been shown by

---

* Present address: Friedrich Miescher Institut, P.O. Box 273, 4002 Basel, Switzerland.

† Present address: Max-Planck-Institut für Biochemie, 8033 Martinsried bei München, Germany.

immunochemical work (Müller & Gerisch, 1978). This antigen has been previously termed 'contact site A' (Beug, Katz & Gerisch, 1973). Univalent antibody fragments (Fab) directed against this glycoprotein block the end-to-end adhesion which is characteristic of aggregating cells. At this stage the cells are typically elongated and assemble into streams when they move towards aggregation centres. The type of cell adhesion which is characteristic of aggregation-competent cells is largely resistant to EDTA, in contrast to adhesion of growth phase cells (for review see Gerisch, 1980; Gerisch *et al.*, 1980). Thus the assay for EDTA-stable cell contacts is an easy way to determine maturation of the cell adhesion system during development of the cells towards the aggregation-competent state.

Another class of cell surface constituents believed to be implicated in cell adhesion are lectins. The amounts of two lectins with overlapping sugar specificity, called discoidin I and II, increase after the end of growth. These lectins exist mainly as soluble cytoplasmic proteins, but a fraction of them is present on the cell surface (for review see Rosen and Barondes, 1978).

Here we will discuss (1) the involvement of the periodic cyclic-AMP signal system in the control of cell development, (2) rapid adaptation of certain responses to constant concentrations of an agonist for cyclic-AMP receptors, (3) the advantages of replacing cyclic AMP by a slowly hydrolysed analogue, adenosine $3',5'$-phosphorothioate (cAMP-S), (4) the control of extracellular cyclic-nucleotide phosphodiesterase and an inhibitor of this enzyme by almost constant concentrations of cyclic AMP or cAMP-S, (5) the inefficiency of cyclic AMP in regulating discoidin, (6) the convergence of signal processing pathways from cyclic-AMP and folic-acid receptors.

Folic acid is known as a chemotactic agent for cells which have just started development towards the aggregation stage (Pan, Hall & Bonner, 1972). Aggregation-competent cells are highly sensitive to cyclic AMP, but little to folic acid. The discussion will stress the independent control of various biochemical changes during cell development from the growth phase to the aggregation stage. Some of these changes are stimulated by repetitive pulses of cyclic AMP and are inhibited by a continuously elevated level of cyclic AMP. Other changes are stimulated in response to both ways of cyclic-AMP administration. A third class of changes appears to be unresponsive to cyclic AMP.

Responses to cyclic AMP can be classed as long-term or short-term. A typical long-term response is the regulation of contact sites A, which requires several hours. Typical short-term responses are the activation of adenylate or guanylate cyclases (Roos & Gerisch, 1976; Mato & Malchow, 1978), and increase of the calcium permeability after a cyclic-AMP pulse (Wick, Malchow & Gerisch, 1978) as shown in Fig. 1. The cellular cyclic-GMP concentration increases sharply within

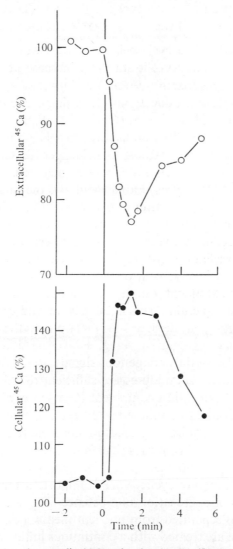

Fig. 1. Influx of calcium into cyclic-AMP stimulated cells. $^{45}Ca^{2+}$ was added to a cell suspension before the cells were stimulated by 2 μM cyclic AMP at 0-time. The calcium influx was calculated for a total cell volume in the suspension of 10%. The experiment was performed at 13 °C at which temperature the reaction was slowed down by a factor of about 2. Samples were taken off the stirred suspension and the cells quickly separated from the extracellular fluid. Counts of ^{45}Ca are given as per cent of pre-stimulation values. Data from Wick *et al.*, 1978.

the first few seconds after application of cyclic AMP, and returns to basal concentrations within one minute (Mato *et al.*, 1977; Wurster *et al.*, 1977).

CELL DEVELOPMENT IS STIMULATED BY PULSES OF CYCLIC AMP AND RETARDED BY NON-FLUCTUATING CONCENTRATIONS

The activity of adenylate cyclase is low in growth phase cells. In the axenically grown strain Ax-2 it starts to increase at about two hours after washing the cells free of nutrient medium (Klein, 1976). Thereafter the cells are capable of producing and releasing cyclic AMP periodically at intervals of six to nine minutes (Gerisch & Wick, 1975). The periodicity of cyclic-AMP production is due to the oscillatory control of adenylate cyclase activity (Roos, Scheidegger & Gerisch, 1977). In stirred suspensions, the cells synchronise their activities by means of their own cyclic-AMP pulses. Experimentally, the phase of the oscillations can be reset by application of cyclic AMP in pulses of 5 or 10 nM amplitude. Phase resetting by extracellular cyclic AMP indicates a functional connection between cyclic-AMP receptors and the system responsible for oscillations (for review, see Gerisch *et al.*, 1980).

In order to study involvement of the cyclic AMP signal system in the control of cell development, one has to choose conditions under which the system is 'out of action'. This is the case in wild-type cells harvested from nutrient medium at the late stationary phase: these cells are unable to develop to the aggregation stage. However, development can be restored by simulating the endogenous signals. Repetitive pulses of 5-nM amplitude given every six minutes are sufficient to make late stationary phase cells fully aggregation competent (Gerisch *et al.*, 1975a). Similar results are obtained in certain non-aggregating mutants which spontaneously do not acquire aggregation competence, but do so after stimulation by cyclic-AMP pulses (Darmon, Brachet & Pereira da Silva, 1975).

Pulsatile application of cyclic AMP is important for its positive effect on development of aggregation competence (Fig. 2). A negative effect is seen when cells separated from nutrient medium during the exponential growth phase are treated with a continuous influx of cyclic AMP but such cells spontaneously acquire aggregation competence after about six or seven hours (Plate 1). A continuous cyclic-AMP flux delays development. The continuous influx maintains a steady-state concentration resulting from the rate of influx of cyclic AMP and the rate of its hydrolysis by phosphodiesterases. Under the conditions of the experiment the calculated steady state concentrations of cyclic AMP are about 1 nM for an influx rate of 0.2 μmol 1^{-1} h^{-1}, and 3 nM for a rate of 1 μmol 1^{-1} h^{-1}. Only at the much higher level of 5×10^{-4}M has a non-fluctuating concentration of cyclic AMP a positive effect on cell development (Sampson, Town & Gross, 1978).

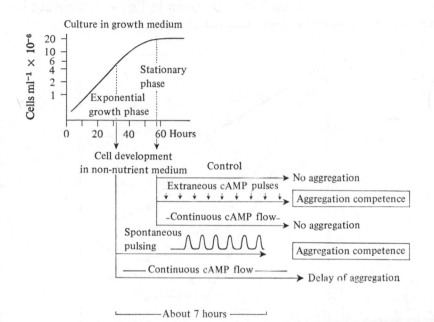

Fig. 2. Diagram of cyclic-AMP effects on cell development in *Dictyostelium discoideum* strain Ax-2. Cells harvested during the exponential growth phase and washed free of nutrient develop spontaneously into aggregation-competent ones. In a shaken suspension this takes about seven hours and is accompanied by the autonomous generation of periodic cyclic-AMP pulses. A continuous inflow of cyclic AMP delays development as shown in Plate 1. Cells harvested from the late stationary phase are unable to develop but can be stimulated to do so by exogenous pulses of cyclic AMP. From Gerisch (1980).

In order to relate advancement and delay of cell development to the regulation of a defined macromolecule, contact sites A have been measured. Their changes of activity reflect the positive and negative effects of cyclic AMP on development (Fig. 3).

Why are pulses of cyclic AMP required for the effective stimulation of cell development? This is answered by the finding that the activation of adenylate cyclase and certain other responses show rapid adaptation. In response to a step-wise increase of the stimulant concentration, the adenylate cyclase is transiently activated for two minutes, and declines to basal activity despite of the continued presence of high stimulant concentrations (Gerisch *et al.*, 1977). Since cyclic AMP is rapidly hydrolysed by phosphodiesterases produced by *Dictyostelium discoideum* cells, analogues of cyclic AMP are preferentially used as stimulants under conditions where their hydrolysis is negligible. When cyclic AMP is replaced by its slowly-hydrolysed thioanalogue, cAMP-S, one obtains a signal of virtually rectangular shape. The rise and fall of cyclic AMP produced in response to this signal is shown in Fig. 4. A transient increase of cyclic GMP is also observed and will be discussed below in

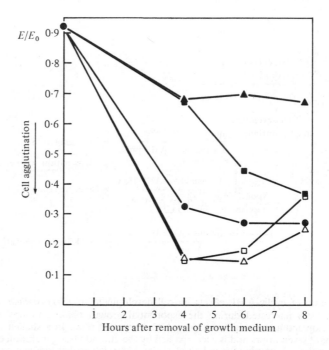

Fig. 3. The capacity of cells to form EDTA-stable contacts is developmentally regulated. In strain Ax-2 it is controlled by cyclic AMP. Cells were harvested from nutrient medium during exponential growth, washed and shaken at a density of 1×10^7 ml^{-1} in 0.017 M Soerensen phosphate buffer pH 6.0. After washing of the cells at the indicated times EDTA-stable adhesion was assayed by measuring light scattering values (E) under conditions of constant shear (Beug *et al.*, 1973). Values were normalised according to light scattering of completely dissociated but otherwise identical cells (E_0). Low values on the ordinate indicate strong intercellular adhesion in the presence of 10 mM EDTA. ●, control. ■, □, 0.2 μmol l^{-1} h^{-1} cyclic AMP administered either as a continuous influx (■) or in pulses every 6 min (□). ▲, △, 1 μmol l^{-1} h^{-1} cyclic AMP given as a continuous influx (▲) or in the form of pulses at intervals of 6 min (△).

connection with chemotaxis. An alternative to the use of slowly-hydrolysed analogues is a flow chamber in which cyclic-AMP concentrations can be easily modulated (Devreotes & Steck, 1979).

Why does an elevated steady concentration of cyclic AMP inhibit development? In cells where adenylate cyclase activity oscillates spon-

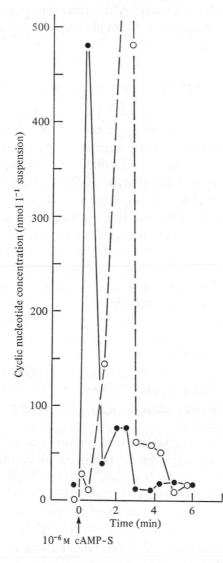

Fig. 4. A stepwise increase in the extracellular concentration of cAMP-S elicits transient increases in cyclic GMP (●) and cyclic AMP (○). A suspension of Ax-2 cells (10^8 cells ml^{-1}) that have been allowed to develop for 6 h was stimulated by adding cAMPS-S to a final concentration of 10^{-6}M. Medium: 0.017 M Soerensen phosphate buffer pH 6.0; temperature 23 °C. Cyclic-nucleotide concentrations were measured by using a radioimmunoassay.

taneously, oscillations are easily suppressed by the continuous presence of cyclic AMP. This explains the inhibitory action of a continuous influx of cyclic AMP on development. It is not absolutely clear, however, how a continuous influx of cyclic AMP interferes with spontaneous oscillations. An effect via cyclic AMP receptors is probable; it might result in the lack of feedback through oscillating extracellular cyclic AMP concentrations.

'FRUITY', A PHENOTYPE OF cAMP-S-RESISTANT MUTANTS

For the genetic analysis of cell aggregation it is a handicap that cyclic AMP is involved not only in chemotaxis of aggregating cells, but also in the developmental sequence leading to aggregation competence. Mutants defective in one component of the cyclic-AMP signal system are often pleiotropic since development ceases at an early stage of the interphase between growth and aggregation. As a consequence, such mutants lack a variety of markers characteristic of aggregation-competent cells. As a first step towards the isolation of mutants specifically defective in a certain component of the cyclic-AMP signal system we have screened for mutants that are capable of aggregating under conditions where the cyclic-AMP signal system is paralysed. The mutagenesis of such mutants is expected to yield secondary mutations which specifically affect single components of the cyclic AMP signal system, e.g. receptors or adenylate cyclase.

Developmental processes depending on cyclic-AMP pulses are blocked in the presence of cAMP-S (Rossier et al., 1979). Consequently wild-type cells do not aggregate on agar containing 5×10^{-7} M cAMP-S. A predominant class of mutants able to aggregate is characterised by numerous minute fruiting bodies and the absence of cell streams normally observed during aggregation (Plate 2). Since this phenotype has been previously described by Sussman (1955) as 'fruity', we have tested two authentic fruity mutants, fr-1 and fr-17, for cAMP-S resistance. Both mutants proved to aggregate and to form fruiting bodies in the presence of 5×10^{-7} M cAMP-S. Thus 'fruity' is a phenotype often associated with uncoupling of development from the inherent cyclic-AMP signal system.

An unexpected result was the finding of a cAMP-S-resistant wild-type strain, v-12. This strain produces and requires cyclic AMP as a chemotactic factor of aggregating cells. Aggregation and fruiting body formation is nevertheless possible in the presence of cAMP-S concentrations up to 1 mM, indicating that chemotaxis is not a requirement for aggregation. Obviously, cells can aggregate as a result of random movements and collisions (Rossier et al., 1979).

EXTRACELLULAR PHOSPHODIESTERASE IS REGULATED BY STEADY CONCENTRATIONS OF CYCLIC AMP OR cAMP-S

Cyclic AMP that is released into the extracellular space is rapidly hydrolysed by phosphodiesterases. This is important for the functioning of the periodic signal system since it prevents overloading of the receptors with cyclic AMP. A phosphodiesterase-negative mutant discovered by Darmon, Barra & Brachet (1978) does not show oscillations, and it does not develop to the aggregation-competent stage. When cyclic-AMP phosphodiesterase is added, oscillations are observed and the cells become able to aggregate.

Dictyostelium discoideum produces two or more phosphodiesterases which attack extracellular cyclic AMP: one on the cell surface (Malchow *et al.*, 1972), and one or two others in the extracellular space (Dicou & Brachet, 1979; Toorchen & Henderson, 1979). The activity of cell-surface phosphodiesterase increases during the interphase between growth and aggregation. Extracellular phosphodiesterase activity is high during the growth phase. After the end of growth, an inhibitor which blocks the extracellular phosphodiesterase is released into the extracellular space. The cell-surface phosphodiesterase is inhibitor-resistant under in-vivo conditions. After treatment of the membrane with detergent, it becomes sensitive, indicating the presence of an inhibitor-binding site, which is masked in the membrane-bound state (Malchow, Fuchila & Nanjundiah, 1975). The membrane-bound and extracellular phosphodiesterases as well as the inhibitor, are concanavalin A-binding glycoproteins (Eitle & Gerisch, 1977).

The increase of cell-surface phosphodiesterase during development is strongly enhanced by cyclic-AMP pulses and only weakly by a continuous influx (Roos, Malchow & Gerisch, 1977). Pulses of cAMP as well as a steady-state level of extracellular cyclic AMP in the nanomolar range are similarly potent in regulating extracellular phosphodiesterase and its inhibitor. Phosphodiesterase regulation behaves as a servo-control of the extracellular cyclic-AMP level. Raising the level results in an increase of phosphodiesterase activity and suppression of the inhibitor. The latter effect accounts only partially for the cyclic-AMP stimulated increase of phosphodiesterase activity (Yeh, Chan & Coukell, 1978; Gerisch, 1979). This increase probably involves the regulation of protein synthesis (Klein, 1975). It becomes detectable one to two hours after the beginning of cyclic-AMP application and continues for eight hours, provided that cyclic-AMP remains present

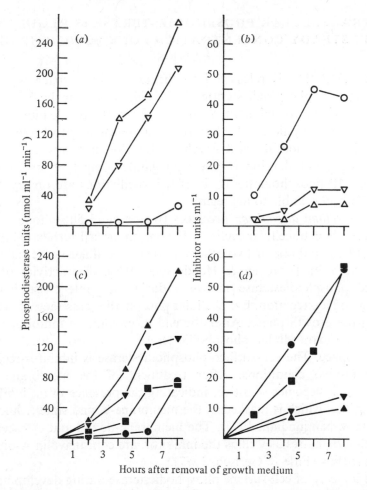

Fig. 5. Regulation of extracellular phosphodiesterase and its inhibitor by continuously applied cyclic AMP (*a*, *b*) or cAMP-S (*c*, *d*). Ordinate in *a*, *c*; activity of extracellular phosphodiesterase. Ordinate in *b*, *d*; units of phosphodiesterase inhibitor. ○, ●, controls. Flux rates of cyclic AMP were 5 μmol l^{-1} h^{-1} ($\triangledown$) or 25 μmol l^{-1} h^{-1} ($\triangle$). Flux rates of cAMP-S were 0.04 μmol l^{-1} h^{-1} (■); 0.2 μmol l^{-1} h^{-1} (▼); or 1 μmol l^{-1} h^{-1} (▲).

(Fig. 5*a*, *b*). After the cessation of cyclic-AMP inflow the secretion of inhibitor resumes indicating that the cells 'measure' cyclic AMP concentrations continuously over a period of hours. Steady state concentrations of cyclic AMP (Fig. 5*a*) calculated to stimulate extracellular phosphodiesterase are in the range of 5–100 nM.

If cyclic AMP is applied only once, a concentration of 10^{-4} M is required to obtain a substantial increase of extracellular phosphodiesterase activity (Klein, 1975). If one replaces cyclic AMP by its slowly-hydrolysed analogue, cAMP-S, a single application of 2 μmol l^{-1} induces a greater rise of extracellular phosphodiesterase activity than a 100-fold larger pulse of cyclic AMP (Fig. 6). When cAMP-S is continuously

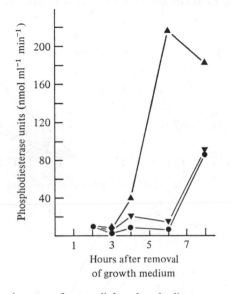

Fig. 6. Much stronger increase of extracellular phosphodiesterase activity in response to a single dose of cAMP-S than to one of cyclic AMP. The stimulants were added at 2 h. ●, control, ▼, 2 × 10^{-4}M cyclic AMP; ▲, 2 × 10^{-6}M cAMP-S.

injected into a cell suspension, a strong increase of the extracellular phosphodiesterase activity and suppression of the inhibitor is obtained with an influx rate of 2 × 10^{-7} mol l^{-1} h^{-1} (Fig. 5c, d).

FOLIC ACID FEEDS INTO THE SAME SIGNAL PROCESSING PATHWAYS AS CYCLIC AMP

Pan *et al.* (1972) have shown that shortly after the end of growth, folic acid causes a chemotactic response in *Dictyostelium discoideum*. The responsiveness declines in the course of cell development towards the aggregation-competent stage. As shown by Wurster and Schubiger (1977) folic acid pulses induce the cells to begin with periodic cyclic AMP production earlier than normal. Consequently, the interphase between the end of growth and the onset of aggregation competence is shortened. In contrast to pulses, continuous administration of folic acid does inhibit development similarly to cyclic AMP. The opposite effects of pulses and continuous fluxes on cell development can be quantified by the assay of EDTA-stable adhesion (Fig. 7c). Folic acid can replace cyclic AMP not only with respect to the regulation of cell development; it also effects the regulation of extracellular phosphodiesterase. The phosphodiesterase activity moderately increases in response to folic acid when it is applied in form of pulses or as a continuous flux (Fig. 7a).

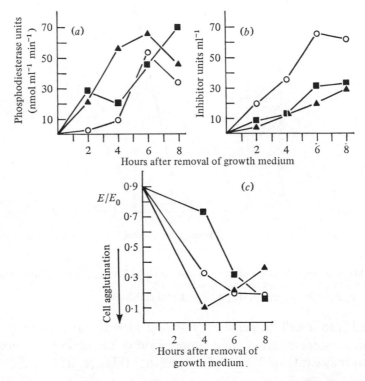

Fig. 7. Folic acid effects on the regulation of extracellular phosphodiesterase (a), of the inhibitor of this enzyme (b) and of contact sites A (c). ○, control; □, continuous inflow of 5 μ-mol l^{-1} h^{-1} folic acid; ▲, the same average amount of folic acid administered at intervals of 6 min in forms of pulses of 5×10^{-7}M amplitude.

The phosphodiesterase inhibitor is substantially suppressed by folic acid (Fig. 7b). The conclusion is that the signal processing pathways of folic acid and cyclic AMP converge. This holds for the pathways resulting in a rapid transient response as well as for those involved in the measurement of agonist concentrations over long periods of time.

CELLULAR LECTINS: THE INEFFICIENCY OF CYCLIC-AMP PULSES TO REGULATE THEM

In the wild-type strain NC-4 the activity of discoidin, a carbohydrate-binding protein, markedly increases when the cells acquire the ability to form EDTA-stable contacts. This has been tested by the agglutination of formalinised sheep erythrocytes (Rosen *et al.*, 1973). Later it was found that two different carbohydrate-binding proteins with similar sugar specificities do exist. Both agglutinate rabbit erythrocytes, whereas only discoidin I agglutinates sheep erythrocytes (Barondes *et al.*, 1978).

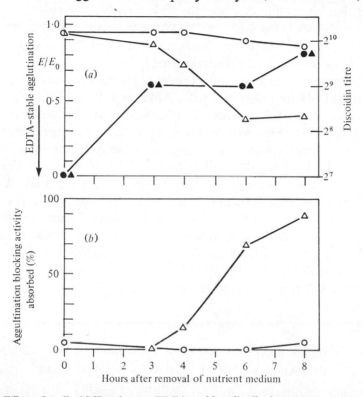

Fig. 8. Effect of cyclic-AMP pulses on EDTA-stable cell adhesion and discoidin I (*a*), and on contact sites A (*b*). The cells were harvested from nutrient medium during the late stationary phase (Gerisch *et al.*, 1975*a*). The cells were washed in 0.017 M Soerensen phosphate buffer pH 6.0 and shaken in the buffer with or without pulsatile cyclic-AMP application. Pulses of 1×10^{-8} M amplitude were administered every 5–7 min. (*a*) Discoidin I was determined by agglutination of formalinised sheep erythrocytes by dilution in steps of factor 2, according to Rosen *et al.* (1973). The undiluted extract was from 1.3×10^{7} cells ml^{-1} plus 0.4% erythrocytes (v/v). In controls the discoidin activity increased four fold during the first three hours and then levelled off (●). No significant effect of cyclic-AMP pulses on discoidin activity was observed (▼). (At 8 h the titres found in duplicates were 2^{9} and 2^{10}, respectively.) EDTA-stable contacts showed little increase within 8 h in controls (○), but a strong increase under cyclic-AMP stimulation (△). These contacts were assayed by measuring light scattering in a sample (*E*) as compared to a reference of completely dissociated cells (E_{0}). (*b*) Contact sites A were determined by Fab absorption according to Beug *et al.* (1973). The Fab was directed against membrane antigens of aggregation–competent cells. 0.2 mg Fab was incubated for 30 min with the membrane fraction from 2.5×10^{6} cells in a total volume of 50 μl. The contact site A inhibiting activity of the Fab was retitrated after absorption by measuring agglutination of aggregation-competent cells of strain v-12/M2 in the presence of Fab plus 10mM EDTA. ○, △ As for (*a*).

In a mutant specifically defective in discoidin I activity development is blocked at the stage of loose aggregates (Ray, Shinnick & Lerner, 1979).

The parallelism between discoidin regulation and maturation of the system responsible for cell adhesion has led Rosen *et al.* (1973) to suggest that interaction of discoidin with carbohydrate-containing cell-surface receptors is the basis of EDTA-stable contact formation. This argument does not hold for cyclic-AMP stimulated cells of strain Ax-2. As discussed above, Ax-2 cells harvested from nutrient medium during the late stationary phase do not become aggregation-competent, except when stimulated by pulses of cyclic AMP or folic acid. EDTA-stable contacts are regulated accordingly (Gerisch *et al.*, 1975a; Wurster & Schubiger, 1977) (Fig. 8a). Contact sites A, now identified with a specific glycoprotein, can be quantitated by their capacity to bind Fab which blocks EDTA-stable cell adhesion. The assay shows that late stationary phase cells spontaneously express contact sites A only weakly, but they express them strongly when stimulated by cyclic-AMP pulses (Fig. 8b). In contrast, discoidin I activity is not significantly different in cyclic-AMP stimulated and non-stimulated cells (Fig. 8a). The increase of discoidin I activity occurs earlier than EDTA-stable contact formation in the stimulated cells. Thus, regulation of discoidin I is not coupled to the cyclic-AMP controlled regulation of EDTA-stable contacts.

CYCLIC-GMP REGULATION: A SHORT-TERM RESPONSE PROBABLY INVOLVED IN CHEMOTAXIS

In principle, amoebae have two possibilities of sensing a spatial gradient of attractant: they can measure concentration differences over their surfaces, or they can extend pseudopods into different directions using them as sensors. During its extension a pseudopod can translate spatial concentration differences into changes of concentration with time (Fig. 9). It is an open question if, in the chemotactic response of amoeboid cells, such a temporal mechanism is involved. In chemotactic bacteria this is evidently the case (Berg & Brown, 1972; Macnab & Koshland, 1972). Their temporal response is based on adaptation due to the methylation of a set of membrane proteins (for review see Goy & Springer, 1978). One possibility of solving the question for amoebae is to search for chemical mediators of the chemotactic response, and to study their temporal behaviour after addition of attractant to living cells.

Extension of pseudopods towards a source of cyclic AMP can be observed within three to five seconds (Gerisch *et al.*, 1975*c*). Therefore, the search for intracellular mediators of the chemotactic response has been focussed on changes which occur within the first few seconds after the stimulation of cells by cyclic AMP or other attractants like folic acid. In *Dictyostelium discoideum* and related species, a transient increase of the cellular cyclic-GMP concentration has proven to be always associated with chemotactic responsiveness (Mato & Konijn, 1977; Wurster *et al.*, 1977; Wurster, Bozzaro & Gerisch, 1978).

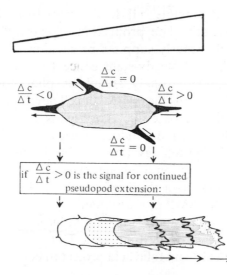

Fig. 9. Diagram to illustrate how an amoeboid cell might detect spatial concentration gradients by transforming them through extending pseudopods into changes of concentration in time. Modified from Gerisch *et al.*, (1975*b*). Block at top represents gradient.

A stepwise increase in concentration of the slowly-hydrolysed cAMP-S used as a stimulus results in a rapid, transient increase in intracellular cyclic-GMP concentration observed after stimulation by a pulse of cyclic AMP (Fig. 4). Thus cyclic-GMP regulation is induced by a temporal increase of the agonist concentration. This implies that a process undergoing adaptation within a minute is involved in the chemotactic response, provided that the cyclic-GMP change is a basis of this response.

The mechanism of cyclic-GMP action is still unclear. A specific cyclic-GMP binding protein in *Dictyostelium discoideum* cells is known, but not its function. (Rahmsdorf & Gerisch, 1978). One response to cyclic

AMP is immediately linked to the contractile system of the cells and may represent a later step in signal processing. This is a change in phosphorylation of myosin heavy chains (Rahmsdorf, Malchow & Gerisch, 1978; Plate 3). Phosphorylation of a myosin heavy chain is known to enhance the actin-activated ATPase activity of myosin in *Acanthamoeba* (Maruta & Korn, 1977). In mammalian cells, phosphorylation of a myosin light chain has the same effect (for review see Adelstein, 1978).

CONCLUSION

Pulses of cyclic AMP promote cell development towards the aggregation-competent stage, stimulate production of extracellular phosphodiesterase, and simultaneously suppress production of an inhibitor of this enzyme. An elevated steady-state concentration of cyclic AMP delays development, but strongly stimulates phosphodiesterase and suppresses its inhibitor. Thus the control of cell development on the one hand, and the regulation of cyclic AMP phosphodiesterase and its inhibitor on the other, are independent of each other. The activity of discoidin increases during cell development, but its control can also be uncoupled from the acquisition of aggregation competence. The regulation of discoidin represents a third type of control which is probably not linked to the cyclic-AMP signal system.

As a chemotactic agent, and also as an inducer of other responses, cyclic AMP can be replaced by folic acid. Thus the signal processing pathways for cyclic AMP and folic acid project to common intracellular targets.

Chemotactic responses are associated with an increase of the cyclic GMP concentration within the cells. When cells are stimulated by a cyclic-AMP pulse, cyclic GMP sharply increases and falls within one minute. This response is elicited by a sudden increase in cyclic AMP or folic acid concentration. Thus it is unimportant whether or not the extracellular stimulant is subsequently taken away i.e. hydrolysed, or remains continuously present. This result indicates adaptability of a step in signal processing from cell surface cyclic-AMP receptors to guanylate cyclase.

REFERENCES

ADELSTEIN, R. S. (1978). Myosin phosphorylation, cell motility and smooth muscle contraction. *Trends in Biochemical Sciences*, **3**, 27–30.
ARNDT, A. (1937). Rhizopodenstudien. III. Untersuchungen über *Dictyostelium mucoroides* Brefeld. *Wilhelm Roux Arch. Ent. W. Mech.*, **136**, 681–744.

BARONDES, S. H., ROSEN, S. D., FRAZIER, W. A., SIMPSON, D. L. & HAYWOOD, P. L. (1978). *Dictyostelium discoideum* agglutinins (Discoidins I and II). *Methods in Enzymology*, **50**, 306–12.

BERG, H. C. & BROWN, D. A. (1972). Chemotaxis in *Escherichia coli* analysed by three-dimensional tracking. *Nature, London*, **239**, 500–4.

BEUG, H., KATZ, F. E. & GERISCH, G. (1973). Dynamics of antigenic membrane sites relating to cell aggregation in *Dictyostelium discoideum*. *Journal of Cell Biology*, **56**, 647–58.

DARMON, M., BARRA, J. & BRACHET, P. (1978). The role of phosphodiesterase in aggregation of *Dictyostelium discoideum*. *Journal of Cell Science*, **31**, 233–43.

DARMON, M., BRACHET, P. & PEREIRA DA SILVA, L. H. (1975). Chemotactic signals induce cell differentiation in *Dictyostelium discoideum*. *Proceedings of the National Academy of Sciences, USA*, **72**, 3163–6.

DEVREOTES, P. N. & STECK, T. L. (1979). Cyclic 3′, 5′ AMP relay in *Dictyostelium*. II. Requirements for the initiation and termination of the response. *Journal of Cell Biology*, **80**, 300–9.

DICOU, E. L. & BRACHET, P. (1979). Multiple forms of an extracellular cyclic-AMP phosphodiesterase from *Dictyostelium discoideum*. *Biochimica et Biophysica Acta*, **578**, 232–42.

EITLE, E. & GERISCH, G. (1977). Implication of developmentally regulated concanavalin A binding proteins of *Dictyostelium* in cell adhesion and cyclic AMP regulation. *Cell Differentiation*, **6**, 339–46.

FRANKE, J. & KESSIN, R. (1977). A defined minimal medium for axenic strains of *Dictyostelium discoideum*. *Proceedings of the National Academy of Sciences, USA*, **74**, 2157–61.

GERISCH, G. (1979). Control circuits in cell aggregation and differentiation of *Dictyostelium discoideum*. In *Mechanisms of Cell Change*, ed. J. Ebert & T. Okada, pp. 225–39. New York: John Wiley & Sons.

GERISCH, G. (1980). Univalent antibody fragments as tools for the analysis of cell interactions in *Dictyostelium*. *Current Topics in Developmental Biology*, in press.

GERISCH, G., FROMM, H., HUESGEN, A. & WICK, U. (1975a). Control of cell–contact sites by cyclic AMP pulses in differentiating *Dictyostelium* cells. *Nature, London*, **255**, 547–9.

GERISCH, G., HÜLSER, D., MALCHOW, D. & WICK, U. (1975b). Cell communication by periodic cyclic-AMP pulses. *Philosophical Transactions of the Royal Society, Series B*, **272**, 181–92.

GERISCH, G., KRELLE, H., BOZZARO, S., EITLE, E. & GUGGENHEIM, R. (1980). Analysis of cell adhesion in *Dictyostelium* and *Polysphondylium* by the use of Fab. In *Cell Adhesion and Motility*, ed. A. Curtis & J. Pitts, Cambridge University Press.

GERISCH, G., MAEDA, Y., MALCHOW, D., ROOS, W., WICK, U. & WURSTER B. (1977). Cyclic AMP signals and the control of cell aggregation in *Dictyostelium discoideum*. In *Proceedings of the EMBO Workshop on Development and Differentiation in Cellular Slime Moulds*, ed. P. Cappuccinelli & J. M. Ashworth, pp. 105–24. Amsterdam: Elsevier/North-Holland Biomedical Press.

GERISCH, G. & MALCHOW, D. (1976). Cyclic AMP receptors and the control of cell aggregation in *Dictyostelium*. *Advances in Cyclic Nucleotide Research*, **7**, 49–68.

GERISCH, G., MALCHOW, D., HUESGEN, A., NANJUNDIAH, V., ROOS, W., WICK, U.

& HÜLSER, D. (1975c). Cyclic-AMP reception and cell recognition in *Dictyostelium discoideum*. In *Developmental Biology*, ed. D. McMahon & C. F. Fox, pp. 76–78. Menlo Park, California: W. A. Benjamin.

GERISCH, G., MALCHOW, D., ROOS, W. & WICK, U. (1979). Oscillations of cyclic nucleotide concentrations in relation to the excitability of *Dictyostelium* cells. *Journal of Experimental Biology*, **81**, 33–47.

GERISCH, G. & WICK, U. (1975). Intracellular oscillations and release of cyclic AMP from *Dictyostelium* cells. *Biochemical and Biophysical Research Communications*, **65**, 364–70.

GOY, M. F. & SPRINGER, M. S. (1978). In search of the linkage between receptor and response: The role of a protein methylation reaction in bacterial chemotaxis. In *Taxis and Behavior. Receptors and Recognition, Series B*, vol. 5, pp. 3–34. London: Chapman and Hall.

KLEIN, C. (1975). Induction of phosphodiesterase by cyclic adenosine 3′:5′-monophosphate in differentiating *Dictyostelium discoideum* amoebae. *Journal of Biological Chemistry*, **250**, 7134–8.

KLEIN, C. (1976). Adenylate cyclase activity in *Dictyostelium discoideum* amoebae and its changes during differentiation. *FEBS Letters*, **68**, 125–8.

KONIJN, T. M., VAN DEN MEENE, J. G. C., BONNER, J. T. & BARKLEY, D. S. (1967). The acrasin activity of adenosine-3′,5′-cyclic phosphate. *Proceedings of the National Academy of Sciences, USA*, **58**, 1152–4.

MACNAB, R. M. & KOSHLAND, D. E. JR. (1972). The gradient-sensing mechanism in bacterial chemotaxis. *Proceedings of the National Academy of Science, USA*, **69**, 2509–12.

MALCHOW, D., FUCHILA, J. & NANJUNDIAH, V. (1975). A plausible role for a membrane-bound cyclic AMP phosphodiesterase in cellular slime mold chemotaxis. *Biochimica et Biophysica Acta*, **385**, 421–8.

MALCHOW, D., NÄGELE, B., SCHWARZ, H. & GERISCH, G. (1972). Membrane-bound cyclic AMP phosphodiesterase in chemotactically responding cells of *Dictyostelium discoideum*. *European Journal of Biochemistry*, **28**, 136–42.

MARUTA, H. & KORN, E. D. (1977). Purification from *Acanthamoeba castellanii* of proteins that induce gelation and synthesis of F-actin. *Journal of Biological Chemistry*, **252**, 399–402.

MATO, J. M. & KONIJN, T. M. (1977). Chemotactic signals and cyclic GMP accumulation in *Dictyostelium*. In *Proceedings of the EMBO Workshop on Development and Differentiation in Cellular Slime Moulds*, ed. P. Cappuccinelli & J. M. Ashworth, pp. 93–103. Amsterdam: Elsevier/North-Holland Biomedical Press.

MATO, J. M., KRENS, F. A., VAN HAASTERT, P. J. & KONIJN, T. M. (1977). 3′:5′-cyclic AMP-dependent 3′:5′-cyclic GMP accumulation in *Dictyostelium discoideum*. *Proceedings of the National Academy of Sciences, USA*, **74**, 2348–51.

MATO, J. M. & MALCHOW, D. (1978). Guanylate cyclase activation in response to chemotactic stimulation in *Dictyostelium discoideum*. *FEBS Letters*, **90**, 119–22.

MÜLLER, K. & GERISCH, G. (1978). A specific glycoprotein as the target site of adhesion blocking Fab in aggregating *Dictyostelium* cells. *Nature, London*, **274**, 445–9.

PAN, P., HALL, E. M. & BONNER, J. T. (1972). Folic acid as second chemotactic substance in the cellular slime moulds. *Nature New Biology*, **237**, 181–2.

RAHMSDORF, H. J. & GERISCH, G. (1978). Specific binding proteins for cyclic AMP and cyclic GMP in *Dictyostelium discoideum*. *Cell Differentiation*, **7**, 249–57.

RAHMSDORF, H. J., MALCHOW, D. & GERISCH, G. (1978). Cyclic AMP-induced phosphorylation in *Dictyostelium* of a polypeptide comigrating with myosin heavy chains. *FEBS Letters*, **88**, 322–6.

RAY, J., SHINNICK, T. & LERNER, R. (1979). A mutation altering the function of a carbohydrate binding protein blocks cell–cell cohesion in developing *Dictyostelium discoideum*. *Nature, London*, **279**, 215–21.

ROOS, W. & GERISCH, G. (1976). Receptor-mediated adenylate cyclase activation in *Dictyostelium discoideum*. *FEBS Letters*, **68**, 170–2.

ROOS, W., MALCHOW, D. & GERISCH, G. (1977). Adenylyl cyclase and the control of cell differentiation in *Dictyostelium discoideum*. *Cell Differentiation*, **6**, 229–39.

ROOS, W., SCHEIDEGGER, C. & GERISCH, G. (1977). Adenylate cyclase activity oscillations as signals for cell aggregation in *Dictyostelium discoideum*. *Nature, London*, **266**, 259–61.

ROSEN, S. D. & BARONDES, S. H. (1978). Cell adhesion in the cellular slime molds. In *Specificity of Embryological Interactions. Receptors and Recognition Series B*, vol. 4, pp. 235–64.

ROSEN, S. D., KAFKA, J. A., SIMPSON, D. L. & BARONDES, S. H. (1973). Developmentally regulated, carbohydrate-binding protein in *Dictyostelium discoideum*. *Proceedings of the National Academy of Sciences, USA*, **70**, 2554–7.

ROSSIER, C., GERISCH, G., MALCHOW, D. & ECKSTEIN, F. (1979). Action of a slowly hydrolysable cyclic AMP analogue on developing cells of *Dictyostelium discoideum*. *Journal of Cell Science*, **35**, 321–38.

SAMPSON, J., TOWN, C. & GROSS, J. (1978). Cyclic AMP and the control of aggregative phase gene expression in *Dictyostelium discoideum*. *Developmental Biology*, **67**, 54–64.

SHAFFER, B. M. (1962). The Acrasina. *Advances in Morphogenesis*, **2**, 109–82.

SUSSMAN, M. (1955). 'Fruity' and other mutants of the cellular slime mould, *Dictyostelium discoideum*: a study of developmental aberrations. *Journal of General Microbiology*, **13**, 295–309.

TOORCHEN, D. & HENDERSON, E. J. (1979). Characterization of multiple extracellular cAMP-phosphodiesterase forms in *Dictyostelium discoideum*. *Biochemical and Biophysical Research Communications*, **87**, 1168–75.

WICK, U., MALCHOW, D. & GERISCH, G. (1978). Cyclic-AMP stimulated calcium influx into aggregating cells of *Dictyostelium discoideum*. *Cell Biology International Reports*, **2**, 71–9.

WURSTER, B., BOZZARO, S. & GERISCH, G. (1978). Cyclic GMP regulation and responses of *Polysphondylium violaceum* to chemoattractants. *Cell Biology International Reports*, **2**, 61–9.

WURSTER, B. & SCHUBIGER, K. (1977). Oscillations and cell development in *Dictyostelium discoideum* stimulated by folic acid pulses. *Journal of Cell Science*, **27**, 105–14.

WURSTER, B., SCHUBIGER, K., WICK, U. & GERISCH, G. (1977). Cyclic GMP in *Dictyostelium discoideum*. Oscillations and pulses in response to folic acid and cyclic AMP signals. *FEBS Letters*, **76**, 141–4.

YEH, R. P., CHAN, F. K. & COUKELL, M. B. (1978). Independent regulation of the extracellular cyclic AMP phosphodiesterase-inhibitor system and membrane differentiation by exogenous cyclic AMP in *Dictyostelium discoideum*. *Developmental Biology*, **66**, 361–74.

EXPLANATION OF PLATES

PLATE 1

Aggregation of Ax-2 cells pre-exposed to pulses or to a continuous influx of cyclic AMP. Cells were washed free of nutrient medium during the exponential growth phase, washed, and shaken in 0.017 M Soerensen phosphate buffer pH 6.0 at 23 °C at a density of 1×10^7 cells ml^{-1}. After 4 h $(a-e)$ or 6 h $(f-j)$ the cells were washed again in order to remove any added cyclic AMP, and transferred onto a glass surface. Half an hour thereafter photographs were taken. a, f; controls in which the cells developed without cyclic AMP addition. In the 4-h sample the cells were just at the beginning of aggregation, in the 6-h sample aggregation competence was fully developed. b, g; cells treated by a continuous influx of 0.2 μmol l^{-1} h^{-1} cyclic AMP. $c. h$; treated with a continuous influx of 1 μmol l^{-1} h^{-1} In both cases acquisition of aggregation competence was delayed. d, i; the same average amount of cyclic AMP as in b, g, but applied every 6 min in pulses of 2×10^{-8}M amplitude. e, j; the same as c, h, but cyclic AMP applied in pulses of 1×10^{-7}M amplitude at periods of 6 min. The advancement of cell development is indicated by aggregation in the 4-h samples.

PLATE 2

Colonies of *Dictyostelium discoideum* strain Ax-2 after mutagenesis. Cells were plated together with *Escherichia coli* on nutrient agar containing 5×10^{-7}M cAMP-S. After incubation for six days at 23 °C wild-type colonies appeared as phenocopies of non-aggregating mutants. A cAMP-S-resistant mutant had formed numerous small fruiting bodies.

PLATE 3

Changes of myosin phosphorylation after stimulation of intact cells with a pulse of 3 nM cyclic AMP. The cells were incubated at 11 °C in order to slow down the responses. Samples were withdrawn from the cell suspension and incubated in a detergent-containing solution of [^{32}P]ATP. Autoradiography after SDS-polyacrylamide gel electrophoresis shows increased incorporation of ^{32}P in myosin heavy chains at 45–75 s after stimulation. Thereafter incorporation returns to basal values. The lane on the left shows the Coomassie-blue stained proteins of the cell lysate with the predominant bands of myosin heavy chains (210 kd) and actin (42 kd). Data from Rahmsdorf, Malchow & Gerisch (1978).

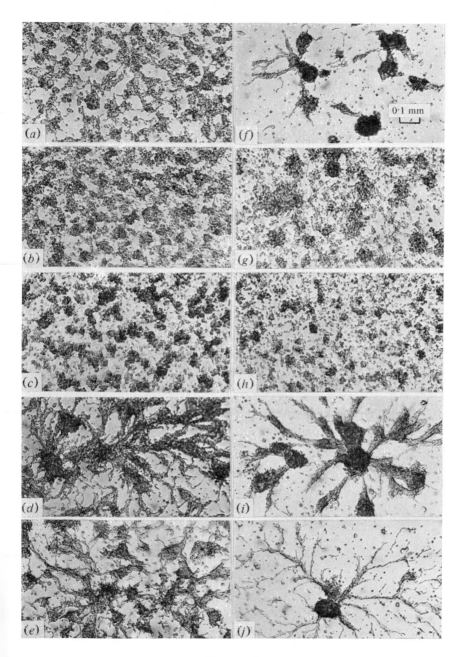

PLATE 1

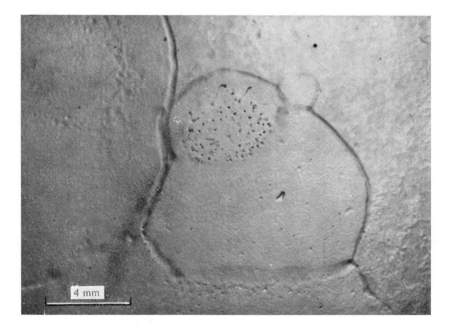

PLATE 2

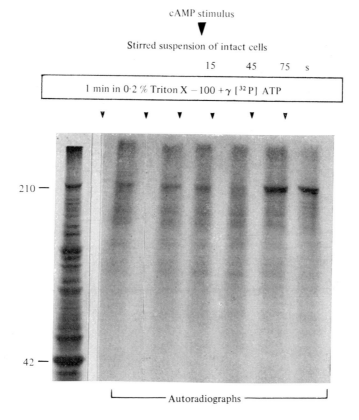

cAMP stimulus

▼

Stirred suspension of intact cells

15   45   75   s

1 min in 0·2 % Triton X − 100 + γ [^{32}P] ATP

210 —

42 —

Autoradiographs

PLATE 3

INDEX

Acanthamoeba castellanii: actin of, (amino-acid sequence compared with mammalian actins) 255, (copolymerisation with muscle myosins) 258–9, 259–60, (profilin and polymerisation of) 256–8; actomyosin in, 253, 254; gelactin of, and association of actin filaments, 258; mitochondrial DNA of, 148, 149, 157; myosins of, 261–3; peptide maps of myosins of, 266–7; phosphorylation in regulation of myosins, 264–5, 267, 268; respiratory activity of, in relation to cell cycle, 153

Acanthometron elasticum, 396

Acetabularia, regeneration in, 378

acetate, as end-product of hydrogenosome metabolism, 130, 131, 132, 135

N-acetylglucosamine: activates chitin synthase, 231; incorporated by microsomal preparations from *Coprinus*, 228; rate of incorporation of, as measure of rate of cell wall synthesis, 210; UDP compound of, as substrate for chitin synthesis, 231

N-acetylglucosaminidase, in cell walls of *Aspergillus nidulans*, 240

acetyl phosphate, formed from acetyl CoA in prokaryotes only, 130, 132

Achlya ambisexualis, enzymes in vesicles in, 226, 241

Achlya bisexualis: generation time of, 9; sex hormones of, 15

Acholeplasma laidlawii, pimaricin and membrane of, 115

acrasins, attractants produced by *Dictyostelium* spp., 16

actin, composing microfilaments of eukaryotes, found also in some prokaryotes, 4–5; in *Acanthamoeba, see under Acanthamoeba*; associated with chromosome fibres, 79, 93; in *Physarum*, 253

actinomycetes, colonies of, 18

actinomycin D, inhibits sexual sensitisation and co-stimulation in *Tetrahymena*, 309

Actinophrys sol, 396

actomyosin: in *Acanthamoeba*, 253, 254; in motile processes in eukaryote cells, 253; in movement of chromosomes to poles? 57; question of functioning of systems of, 268–9; ultrastructural and biochemical approaches to processes involving, 254

S-adenosyl methionine decarboxylase,

mutants of *Saccharomyces* deficient in, lose M-dsRNA, 348

adenylate cyclase of *Dictyostelium*: activated in aggregation, 405, 407; low in growth-phase cells, increases on removal of nutrients, 408; oscillatory control of activity of, 408; transiently activated by cAMP pulse, 409, 410–11, suppressed by continuous presence of cAMP, 411–12

Agaricus bisporus (mushroom): chitosomes in, 227; dieback disease of, caused by group of viruses, 358; dsRNA virus in, 333, 334; uridine nucleotides in, 232

agglutinins, of plus and minus types, on tips of flagella of *Chlamydomonas* gametes, 303; interaction of opposite types of, 305–6

algae: chloroplasts of, 181–3; DNA C-values for, 25; primary production by, 32; substances produced by, inducing agglutination and cell surface interactions, 15; as symbionts in lichens, 20, 181, and in protozoa, corals, flatworms, molluscs, 20

Allomyces arbusculus, dsRNA virus in, 334

Allomyces macrogynus: chitin synthase of, 229; chitosomes in, 227; microfibrils in cell wall of, 217

Allomyces spp.: avoiding reaction in zoospores of, 276; female sex attractant of (sirenin), 15

Alternaria kikuchiana (pear black-spot fungus): controlled by polyoxins, 108–9; dipeptides competitively inhibit uptake of polyoxins by, 109

Amoeba proteus: bacteria infecting a strain of, converted into essential endo-symbionts, 7

amoeboid movement, 12

cAMP, stimulant for aggregation in *Dictyostelium*, 16, 405; application of pulses of, sets phase of oscillation of adenylate cyclase activity, induces aggregation competence, 408, 409, and stimulates production of phosphodi-esterase and suppresses inhibitor, 413–14, 420; cell surface receptors for, 405; continuous flux of, inhibits development, 408, 409, 410–11, 420; long-term and short-term responses to, 406–7; temporary increase in cell content of, after pulse of cAMP-S, 410, 411; temporary increase in cell content of cGMP, after pulse of, 418–20